Stafford Library
Columbia College
1001 Rogers Street
Columbia, MO 65216

ARITHMETIC, ELEMENTARY ALGEBRA, GEOMETRY:
A Guided Approach

Other books in the Drooyan/Rosen series:

ARITHMETIC: A GUIDED APPROACH

ELEMENTARY ALGEBRA: A GUIDED APPROACH

ARITHMETIC, ELEMENTARY ALGEBRA, GEOMETRY: A Guided Approach

IRVING DROOYAN
Los Angeles Pierce College, emeritus

BILL ROSEN
Los Angeles Pierce College

JOHN WILEY & SONS
NEW YORK   CHICHESTER   BRISBANE   TORONTO   SINGAPORE

Acquisitions editor: Carolyn Moore
Cover designer: Karin Batten
Text designer: Karin Gerdes Kinchloe
Editorial supervisor: Martha Cooley
Copy editor: Virginia Dunn
Production supervisor: Philip McCaffrey

Copyright © 1986, by John Wiley & Sons, Inc.

All rights reserved. Published simultaneously in Canada.

Reproduction or translation of any part of
this work beyond that permitted by Sections
107 and 108 of the 1976 United States Copyright
Act without the permission of the copyright
owner is unlawful. Requests for permission
or further information should be addressed to
the Permissions Department, John Wiley & Sons.

Library of Congress Cataloging-in-Publication Data

Drooyan, Irving.
  Arithmetic, elementary algebra, geometry.

  Includes index.
  1. Arithmetic—1961–   . 2. Algebra.
3. Geometry. I. Rosen, Bill, 1949–   . II. Title.
QA107.D763   1986   513′.142   85-22735
ISBN 0-471-82129-2 (pbk.)

Printed in the United States of America

10 9 8 7 6 5 4 3 2 1

PREFACE

This worktext provides a *guided approach* to developmental mathematics. It includes an integrated treatment of arithmetic, elementary algebra, and basic topics of geometry. The mathematical properties that are introduced in Part I (Arithmetic) are integrated into Part II (Elementary Algebra), where algebra is developed as a generalized arithmetic, and in Part III (Geometry). The unifying elements in the three parts are the mathematical properties common to these subjects and the approach to solving applied problems.

PART I: ARITHMETIC

Problem-Solving Skills

The primary emphasis in Part I addresses problem-solving skills. In attending to such skills we have focused on learning when to apply one or another of the basic operations as well as on how to perform these operations. We have emphasized how to express simple word sentences as simple equations, and how to solve these equations by a few elementary rules. We have also stressed writing proportions as models for word sentences, and solving such proportions. In addition, a general method is introduced by which students can approach the variety of applied problems that are included in the text. Many of the problems are consumer oriented; they deal with comparison shopping, interest, discounts, depreciation, appreciation, and the like.

Computations with Whole Numbers and Decimals

Students who enroll in a developmental mathematics course at the college level are likely to have been exposed to whole numbers and decimals during their elementary or secondary schooling. Many of these students will know some of the algorithms for basic operations with these numbers; others may have to relearn all of these procedures. Because of students' diverse backgrounds and because their interest can easily be lost in an extensive repetition of this material, we have treated whole numbers and decimals together in Chapters 1 and 2. This organization will also help to build students' confidence early in the course before fractions are introduced in Chapter 3.

The integration of whole numbers and decimals in Chapters 1 and 2 will be particularly helpful to instructors who allow their students to use calculators for some or for all of the computational work in this text.

Considerable attention is given in these chapters to informal mental arithmetic and estimating answers—essential skills for students whether they use the algorithms for the basic operations, or a calculator.

Review Material

The important properties introduced are listed in a summary at the end of Part I. This summary is followed by cumulative review exercises.

PART II: ELEMENTARY ALGEBRA
Problem-Solving Skills

Part II includes a comprehensive treatment of elementary algebra, and continues the step-by-step approach to the solution of word problems using mathematical models that was introduced in Part I. Particular attention is given first to writing word phrases for the unknown quantities, and then to designating these quantities by variables and writing mathematical models in the form of equations.

Review Material

The important properties introduced are listed in a summary at the end of Part II. This summary is followed by cumulative review exercises.

PART III: GEOMETRY

In Part III, some basic topics of Euclidean geometry are introduced. These topics, which include commonly used mensuration formulas, have a general educational interest for students who have not completed a formal course in geometry at the high school level. The treatment does not include formal proofs.

FEATURES COMMON TO ALL THREE PARTS
Partially Solved Examples

The unique format of partially solved "Complete Examples" in the exercise sets makes this text ideal for students in independent study courses as well as for students in traditional courses. The "Complete Examples" actively engage students in learning and provide immediate reinforcement. Confidence and competence build as students progress with easy-to-understand explanations and a detailed, guided approach at every step.

Exercises, Answers, and Solutions

The topics introduced in the text are supported by carefully graded exercise sets. The odd- and even-numbered exercises in each set are paired and keyed to examples in the text. In addition, all of the topics are continuously reviewed in reviews at the end of each chapter. Answers are provided for the odd-numbered exercises in each section, and detailed solutions are provided for all exercises in the chapter reviews.

ANCILLARY MATERIALS

An Instructor's Manual is available that includes a complete testing program that can be readily duplicated for classroom use. A computerized testing program is also available.

Parts II and III of this worktext are modified versions of *Elementary Algebra with Geometry* by Irving Drooyan and William Wooton. We thank William Wooton for giving us permission to use parts of that text to prepare this edition.

Irving Drooyan
Bill Rosen

CONTENTS

Preface   page v

PART ONE   ARITHMETIC

Chapter 1 Whole Numbers and Decimals; Addition and Subtraction   3

- 1.1 Whole Numbers and Decimals   3
- 1.2 Rounding Off Numbers   8
- 1.3 Addition   11
- 1.4 Subtraction   17
- 1.5 Solving Equations   25
- 1.6 Problem Solving   32
- *Chapter Review*   37

Chapter 2 Whole Numbers and Decimals; Multiplication and Division   39

- 2.1 Multiplication   39
- 2.2 Special Products   46
- 2.3 Division   50
- 2.4 Special Quotients   60
- 2.5 Unit-Product Rule; Unit Quotient Rule   64
- 2.6 Solving Equations   71
- 2.7 Ratio and Proportion   77
- *Chapter Review*   84

CONTENTS

Chapter 3 Fractions 87

3.1 Factoring 87
3.2 Transforming Fractions 91
3.3 Least Common Denominator 96
3.4 Addition 101
3.5 Subtraction 110
3.6 Multiplication and Division 119
3.7 Decimal Equivalents of Fractions 126
Chapter Review 130

Chapter 4 Percent 133

4.1 Percent Equivalents 133
4.2 Types of Percent Problems 136
4.3 Applications of Percent 141
4.4 Simple Interest 148
Chapter Review 153

Chapter 5 Measurement 155

5.1 The United States System 155
5.2 Arithmetic of Denominate Numbers 161
5.3 The Metric System 165
5.4 United States-Metric Conversions 171
5.5 Applications 174
Chapter Review 186

Summary for Part I 188
Cumulative Review for Part I 190

PART TWO ALGEBRA

Chapter 6 Integers 195

6.1 Integers and their Graphs 195
6.2 Sums 201
6.3 Differences 204
6.4 Products and Quotients 208
6.5 Order of Operations 214
Chapter Review 217

CONTENTS

Chapter 7 Polynomials 219

7.1 Algebraic Expressions 219
7.2 Numerical Evaluation 225
7.3 Sums Involving Variables 228
7.4 Differences Involving Variables 234
7.5 Products Involving Variables 238
7.6 Quotients Involving Variables 241
Chapter Review 245

Chapter 8 First-Degree Equations and Inequalities 247

8.1 Solving Equations 247
8.2 Further Solutions of Equations 257
8.3 Solving Formulas 261
8.4 Applications 264
8.5 Inequalities 270
Chapter Review 277

Chapter 9 Products and Factors 279

9.1 The Distributive Law 279
9.2 Factoring Monomials from Polynomials 283
9.3 Binomial Products I 285
9.4 Factoring Trinomials I 289
9.5 Binomial Products II 295
9.6 Factoring Trinomials II 298
9.7 Factoring the Difference of Two Squares 303
9.8 Equations Involving Parentheses 305
9.9 Word Problems Involving Numbers 307
9.10 Applications 311
Chapter Review 318

Chapter 10 Properties of Fractions 319

10.1 Forms of Fractions; Graphical Representation 319
10.2 Reducing Fractions to Lowest Terms 324
10.3 Quotients of Polynomials 330
10.4 Building Fractions 335
10.5 Integer Exponents; Scientific Notation 339
Chapter Review 343

xi

CONTENTS

Chapter 11 Operations with Fractions   345

- 11.1 Products of Fractions   345
- 11.2 Quotients of Fractions   349
- 11.3 Sums and Differences of Fractions with Like Denominators   352
- 11.4 Sums of Fractions with Unlike Denominators   356
- 11.5 Difference of Fractions with Unlike Denominators   361
- 11.6 Complex Fractions   366
- 11.7 Fractional Equations   370
- 11.8 Applications   376
- *Chapter Review*   380

Chapter 12 First-Degree Equations and Inequalities in Two Variables   383

- 12.1 Solving Equations in Two Variables   383
- 12.2 Graphs of Ordered Pairs   390
- 12.3 Graphing First-Degree Equations   393
- 12.4 Intercept Method of Graphing   399
- 12.5 Slope of a Line   401
- 12.6 Equations of Straight Lines   405
- 12.7 Direct Variation   410
- 12.8 Inequalities in Two Variables   412
- *Chapter Review*   416

Chapter 13 Systems of Linear Equations   417

- 13.1 Graphical Solutions   417
- 13.2 Solving Systems by Addition I   420
- 13.3 Solving Systems by Addition II   424
- 13.4 Solving Systems by Substitution   427
- 13.5 Applications Using Two Variables   429
- *Chapter Review*   432
- *Review of Factoring*   433

CONTENTS

Chapter 14 Quadratic Equations 435

- 14.1 Solving Equations in Factored Form 435
- 14.2 Solving Quadratic Equations by Factoring I 438
- 14.3 Solving Quadratic Equations by Factoring II 442
- 14.4 Applications 446
- *Chapter Review* 450

Chapter 15 Radical Expressions 451

- 15.1 Radicals 451
- 15.2 Irrational Numbers 454
- 15.3 Simplifying Radical Expressions I 459
- 15.4 Simplifying Radical Expressions II 464
- 15.5 Fractions Involving Radical Expressions 467
- 15.6 Products of Radical Expressions 470
- 15.7 Quotients of Radical Expressions 473
- *Chapter Review* 478

Chapter 16 Solving Quadratic Equations by Other Methods 481

- 16.1 Extraction of Roots 481
- 16.2 Completing the Square 486
- 16.3 Quadratic Formula 490
- 16.4 Graphing Quadratic Equations in Two Variables 495
- *Chapter Review* 499
- *Summary for Part II* 500
- *Cumulative Review for Part II* 502

PART THREE GEOMETRY

Chapter 17 Elements of Geometry 507

- 17.1 Introductory Concepts 507
- 17.2 Intersecting Lines and Parallel Lines 512
- 17.3 Triangles 516
- 17.4 Congruent Triangles; Similar Triangles 523
- 17.5 Quadrilaterals 528
- 17.6 Circles 532

17.7 Solids   535
17.8 Trigonometric Ratios   538
17.9 Solving Triangles   542
Chapter Review   545

Odd-Numbered Answers   547
Solutions to All Chapter Review Exercises   582
Index   615
Appendix: Tables of Measurements   626
Table of Squares, Square Roots, and Prime Factors   inside back cover

PART I
ARITHMETIC

1 WHOLE NUMBERS AND DECIMALS; ADDITION AND SUBTRACTION

Some of the topics in this chapter may be familiar to you. However, your careful attention to these topics, in addition to ones that are new, will provide you with a foundation for success in the following chapters.

1.1 WHOLE NUMBERS AND DECIMALS

A number without nonzero digits to the right of the decimal point is a **whole number**; a number with nonzero digits to the right of the decimal point is not a whole number. For example, 18. and 236.00 are whole numbers; 18.75 and 236.02 are not whole numbers.

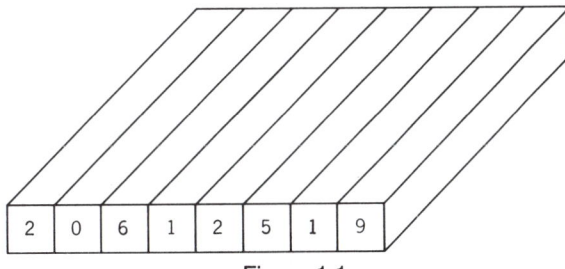

Figure 1.1

Each digit of a whole number has a place value, as shown in Figure 1.1, where

the "1" in the *tens* position equals 1 ten,

the "5" in the *hundreds* position equals 5 hundreds,

the "2" in the *thousands* position equals 2 thousands,

and so forth.

WHOLE NUMBERS AND DECIMALS; ADDITION AND SUBTRACTION

The number 20,612,519 shown in Figure 1.1 is read as

"twenty million, six hundred twelve thousand, five hundred nineteen,"

where a comma that appears three digits from the right is read by the word "thousand," and a comma that appears six digits from the right is read by the word "million."

Example 1 Read each number.

    a. 4,803   b. 4,803,154   c. 6,123,000

Solutions
a. Four thousand, eight hundred three.
b. Four million, eight hundred three thousand, one hundred fifty-four.
c. Six million, one hundred twenty-three thousand.

Example 2 Use digits to write each number.

    a. Four thousand, three hundred twenty-one.
    b. One million, two hundred fifty-eight thousand, three.
    c. Twelve million, sixty-four thousand, twenty five.

Solutions a. 4,321   b. 1,258,003   c. 12,064,025

It is customary to omit the comma in four-digit numbers, a practice that we shall follow in the remainder of this book. For example, the number 4,321 of Example 2a. is most often written as 4321.

Positions to the right of the decimal point are called *decimal places*. Although names are given to each decimal place, it is usually sufficient to be familiar with the names of the first four decimal-place values shown in Figure 1.2., where

the "9" in the *tenths* position equals 9 *tenths* or 90 *hundredths*,

the "8" in the *hundredths* position equals 8 hundredths or 80 *thousandths*, and

the "1" in the *thousandths* position equals 1 thousandth or 10 *ten-thousandths*.

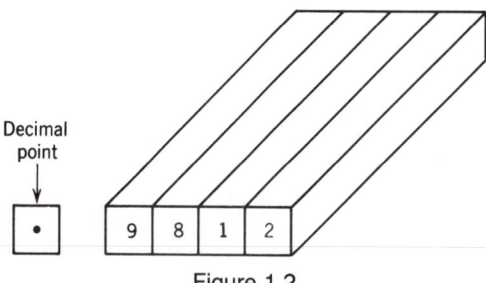

Figure 1.2

Numbers that involve a decimal point are called **decimal numbers,** or simply **decimals.** The number 7.9 is used to name the sum of 7 and 0.9. Similarly,

$$27.42 = 27 + 0.42 \quad \text{and} \quad 58.015 = 58 + 0.015$$

1.1 WHOLE NUMBERS AND DECIMALS

READING AND WRITING DECIMALS

> To read a decimal number:
> 1. Read the whole number part (if any).
> 2. Read "and" for the decimal point.
> 3. Read the number to the right of the decimal point and then read the name of the decimal place in which the right-hand digit appears.

Example 3 Read each decimal.

    a. 981.2   b. 98.12   c. 9.812   d. 0.9812

Solutions
a. Nine hundred eighty-one and two tenths (*one* decimal place corresponds to *tenths*).
b. Ninety-eight and twelve hundredths (*two* decimal places correspond to *hundredths*).
c. Nine and eight hundred twelve thousandths (*three* decimal places correspond to *thousandths*).
d. Nine thousand eight hundred twelve ten-thousandths (*four* decimal places correspond to *ten-thousandths*). See Figure 1.2.

Example 4 Use digits to write each decimal number.

    a. Five and six hundred two thousandths.
    b. Five thousand six hundred two ten-thousandths.
    c. Fifty-six and two hundredths.
    d. Fifty-six and two tenths.

Solutions
a. 5.602 (*thousandths* corresponds to *three* decimal places).
b. 0.5602 (*ten thousandths* corresponds to *four* decimal places).
c. 56.02 (*hundredths* corresponds to *two* decimal places).
d. 56.2 (*tenths* corresponds to *one* decimal place).

COMPARING DECIMALS

To decide which of two decimals is the greater, we compare the digits in the tenths place, in the hundredths place, and so on, until we reach a place where the digit in one number is greater than the corresponding digit in the other number.

Example 5 Which decimal is greater:

    a. 0.05 or 0.0499?   b. 0.013201 or 0.013199?

Solutions a. The first digits are identical:

Because 5 is greater than 4, the number 0.05 is greater than 0.0499.

5

WHOLE NUMBERS AND DECIMALS; ADDITION AND SUBTRACTION

b. The first three digits are identical:

$$0.013201$$
$$0.013199$$

Because 2 is greater than 1, the number 0.013201 is greater than 0.013199.

EXERCISES 1.1
Read each number. See Example 1.

Complete Examples
a. 25,308

   Twenty five _____(1)_____ three hundred _____(2)_____ .

b. 2,406,004

   Two million four hundred _____(3)_____ thousand _____(4)_____ .

| (1) thousand | (2) eight | (3) six | (4) four |

1. 483   2. 695   3. 2341   4. 5768   5. 9780   6. 1048
7. 69,179   8. 14,194   9. 62,590   10. 20,008   11. 779,210   12. 209,699
13. 277,565   14. 912,919   15. 5,349,818   16. 1,860,270   17. 12,106,910   18. 35,002,962

Use digits to write each number. See Example 2.

Complete Examples
a. Four thousand thirty two. _____(1)_____

b. One million twenty thousand three hundred fourteen. _____(2)_____

| (1) 4032 | (2) 1,020,314 |

19. Seven thousand, seven hundred seven.
20. Five thousand, four hundred ninety-six.
21. Twenty-five thousand, thirty two.
22. Twenty-five thousand, three hundred thirty-two.

1.1 WHOLE NUMBERS AND DECIMALS

23. One hundred thousand, six hundred seventy.

24. Six hundred twelve thousand, twelve.

25. Six hundred twelve thousand, one hundred twenty.

26. Eight million, two hundred thousand, four hundred eight.

27. Ninety-eight million, seven hundred sixty-five thousand, four hundred thirty-two.

28. Forty million, four hundred thousand, four.

Read each decimal. See Example 3.

Complete Examples

a. 0.018

   Eighteen _____

   (1)

b. 22.05

   Twenty-two _____ five _____

   (2)   (3)

(1) thousandths   (2) and   (3) hundredths

29. 0.17
30. 0.36
31. 2.01
32. 3.42
33. 23.112
34. 31.075
35. 82.1006
36. 67.3442

Use digits to write each decimal number. See Example 4.

Complete Examples

a. Thirty seven ten thousandths.

   Ans. _____

   (1)

b. Three hundred fifteen and sixty seven hundredths.

   Ans. _____

   (2)

(1) 0.0037   (2) 315.67

37. Thirty-three thousandths.

38. One hundred fifteen thousandths.

39. Nine and two ten-thousandths.

40. Eighty-two and seventy ten-thousandths.

41. One and two hundred three ten-thousandths.

42. Six hundred six and six hundred six thousandths.

Specify which decimal number of each pair is greater. See Example 5.

Complete Examples

a. 0.034; 0.34

   0.34
   ⋮   Because 3 is greater than 0, _____ is greater than 0.034.
   0.034   (1)

(Continued)

7

WHOLE NUMBERS AND DECIMALS; ADDITION AND SUBTRACTION

b. 0.2977; 0.2978

0.2978
↑↑↑ :
↓↓↓
0.2977

Because 8 is greater than 7, _____ is greater than 0.2977.
(2)

---

(1) 0.34   (2) 0.2978

---

43. 0.25; 0.27   44. 0.88; 0.89   45. 0.056; 0.5499   46. 0.619; 0.62

47. 0.047; 0.00987   48. 0.011; 0.00681   49. 0.0012345; 0.0012354   50. 0.6999999; 0.7

51. 0.123405; 0.123408   52. 0.0016326; 0.0016236

Arrange each group of decimals in order of size, with the smallest first.

53. 0.55536, 0.55528, 0.55541   54. 0.61607, 0.61606, 0.61599

55. 0.802878, 0.802778, 0.802876   56. 0.961894, 0.961984, 0.962894

1.2 ROUNDING OFF NUMBERS

The result of a calculation that involves the cost of an item might be a number such as $23.418. The cost to the nearest cent is $23.42. We say that 23.418 has been "rounded off to the nearest hundredth" or "to the nearest cent."

The following procedure describes a way to round off numbers.

To round off numbers:

1. Underline the last digit to be kept; call it the *round-off digit*.
2. Consider the digit immediately to the right of the round-off digit; call it the *test digit*.

If the round-off digit is not in a decimal place, continue with steps 3 and 4:

3. If the test digit is *less than* 5 (0, 1, 2, 3 or 4), replace the test digit and all digits to the right of the test digit with zeros, but do not change the round-off digit.
4. If the test digit is 5 *or more* (5, 6, 7, 8 or 9), add 1 to the round-off digit, then replace the test digit and all digits to the right of it with zeros.

If the round-off digit is in a decimal place, continue with steps 3a and 4a:

3a. If the test digit is less than 5 (0, 1, 2, 3, or 4), discard it together with any digits to the right of it.
4a. If the test digit is 5 or more (5, 6, 7, 8 or 9), add 1 to the round-off digit, discard the test digit and all digits to the right of it.

1.2 ROUNDING OFF NUMBERS

For example, to round off 749.02586 to the nearest hundredth, the round-off digit 2 in the hundredths place is underlined first:

$$749.0\underline{2}586$$

with the 2 marked as the round-off digit and the 5 to its right marked as the test digit.

The test digit is 5, hence 1 is added to the round-off digit:

$$749.0\overset{3}{\cancel{2}}586$$

The round-off digit is in a decimal place, so all digits to the right of the round-off digit are dropped:

$$749.0\overset{3}{\cancel{2}}\cancel{5}\cancel{8}\cancel{6} \longrightarrow 749.03$$

Thus, 249.02586 is written as 249.03, to the nearest hundredth.

Example 1 In each case the round-off digit is underlined.

a. Round off to the nearest ten:

$$2\underline{4}9.02586 \longrightarrow 2\overset{5}{\cancel{4}}\overset{0}{\cancel{9}}.\cancel{0}\cancel{2}\cancel{5}\cancel{8}\cancel{6} \longrightarrow 250$$

b. Round off to the nearest tenth:

$$249.\underline{0}2586 \longrightarrow 249.0\cancel{2}\cancel{5}\cancel{8}\cancel{6} \longrightarrow 249.0$$

c. Round off to the nearest thousandth:

$$249.02\underline{5}86 \longrightarrow 249.02\overset{6}{\cancel{5}}\cancel{8}\cancel{6} \longrightarrow 249.026$$

Round-off instructions sometimes specify a given number of decimal places.

Example 2 a. Round off to two decimal places:

$$13.4\underline{9}63 \longrightarrow 13.\overset{5}{\cancel{4}}\overset{0}{\cancel{9}}\cancel{6}\cancel{3} \longrightarrow 13.50$$

b. Round off to three decimal places:

$$13.49\underline{6}3 \longrightarrow 13.496\cancel{3} \longrightarrow 13.496$$

WHOLE NUMBERS AND DECIMALS; ADDITION AND SUBTRACTION

EXERCISES 1.2

Round off each number to the nearest: a. Ten, b. Tenth, c. Hundredth, d. Thousandth.
See Example 1.

Complete Examples

57.3456

a. 5̲7.3456 — round-off digit; test digit is 5 or more

 5̸7̸.̸3̸4̸5̸6̸ ⟶ _____ (1)   [6 above]

b. 57.3̲456 — round-off digit; test digit less than 5

 57.3̸4̸5̸6̸ ⟶ _____ (2)

c. 57.34̲56 — round-off digit; test digit is 5 or more

 57.3̸4̸5̸6̸ ⟶ _____ (3)   [5 above]

d. 57.345̲6 — round-off digit; test digit is 5 or more

 57.34̸5̸6̸ ⟶ _____ (4)   [6 above]

| (1) 60 | (2) 57.3 | (3) 57.35 | (4) 57.346 |

---

1. 14.7742
2. 21.6344
3. 76.2825
4. 54.6079
5. 69.8991
6. 32.8196
7. 45.9098
8. 89.8982
9. 70.9597
10. 58.9595
11. 19.9505
12. 97.9396

Round off each number to a. one, b. two, and c. three decimal places. See Example 2.

Complete Examples

3.05362

a. 3.0̲5362 ⟶ 3.0̸5̸3̸6̸2̸ ⟶ _____ (1)   [1 above]

b. 3.05̲362 ⟶ 3.05̸3̸6̸2̸ ⟶ _____ (2)

10

1.3 ADDITION

c. $3.05\underline{3}62 \longrightarrow 3.05\overset{4}{\cancel{3}}\cancel{6}\cancel{2} \longrightarrow \underline{}$
(3)

(1) 3.1	(2) 3.05	(3) 3.054

13. 1.9069
14. 2.2591
15. 0.91994
16. 0.65232

17. 0.09857
18. 0.07579
19. 6.1695
20. 4.2945

Round off each amount to the nearest cent (two decimal places).

21. $1.285
22. $17.074
23. $8.395
24. $25.008

25. $47.291
26. $252.297
27. $57.999
28. $59.999

Tax payments to the Internal Revenue Service may be rounded off to the nearest dollar. For example, $22.49 is rounded to $22, and $22.51 is rounded to $23. Round off each tax amount to the nearest dollar (to the nearest whole number).

29. $38.37
30. $38.51
31. $509.62
32. $509.42

33. $600.89
34. $600.49
35. $1234.19
36. $1234.79

37. In the fall of a recent year there were approximately 2,146,000 students enrolled in private colleges in the United States. Round this off a. to the nearest hundred thousand, and b. to the nearest ten thousand.

38. The liftoff weight of the manned spacecraft Apollo 15 was 107,142 pounds. Round this off a. to the nearest ten thousand, and b. to the nearest thousand.

39. From 1950 to 1970 the number of new passenger cars imported into the United States each year increased by 1,992,000. Round this off a. to the nearest hundred thousand, and b. to the nearest ten thousand.

40. During the recent year the railroads in the United States earned approximately $1,124,600,000 carrying transportation equipment. Round this off a. to the nearest million, and b. to the nearest ten million.

1.3 ADDITION

The result of an operation of addition on two numbers is called a **sum** (or **total**).

BASIC ADDITION FACTS

The sum of two numbers between 1 and 9 and including 1 and 9 can be obtained from Table 1.1 on page 12.

WHOLE NUMBERS AND DECIMALS; ADDITION AND SUBTRACTION

Table 1.1
Basic Addition Facts

+	1	2	3	4	5	6	7	8	9
1	2	3	4	5	6	7	8	9	10
2	3	4	5	6	7	8	9	10	11
3	4	5	6	7	8	9	10	11	12
4	5	6	7	8	9	10	11	12	13
5	6	7	8	9	10	11	12	13	14
6	7	8	9	10	11	12	13	14	15
7	8	9	10	11	12	13	14	15	16
8	9	10	11	12	13	14	15	16	17
9	10	11	12	13	14	15	16	17	18

For example,

$$\begin{array}{c} 5 \\ +2 \\ \hline 7 \end{array} \quad \text{and} \quad \begin{array}{c} 7 \\ +6 \\ \hline 13 \end{array}$$

$$\text{or} \quad 5 + 2 = 7 \qquad \text{or} \quad 7 + 6 = 13$$

As shown in these examples, numbers to be added can be written in a vertical or a horizontal form.

You are probably familiar with the basic addition facts in Table 1.1. If not, you should try to memorize them as soon as possible.

PROPERTIES OF ADDITION

There are two properties of addition that are sometimes useful in doing computations. The first is known as the **commutative property.**

> **The order in which numbers are added does not change the sum.**

For example,

$$5 + 7 = 12 \quad \text{and} \quad 7 + 5 = 12$$

This example also shows that the commutative property enables us to *check* an addition by *adding the numbers in reverse order*.

The second property concerns the addition of 0 to a number.

> **The sum of any number and zero is the number.**

For example,

$$8 + 0 = 8 \quad \text{and} \quad 0 + 0.23 = 0.23$$

1.3 ADDITION

COMPUTING SUMS

Sums that involve numbers not shown in Table 1.1 can be computed by using special arithmetic procedures. In such cases the numbers should be written so that units digits are above units digits, ten digits are above tens digits, etc.

Example 1 a. Add $17 + 32$.   b. Add $1.035 + 4.62$.

Solutions a. First arrange the numbers vertically, with the 1 above the 3 and the 7 above the 2. Then add the units digits, and next add the tens digits.

$$\begin{array}{r} 17 \\ +32 \\ \hline 49 \end{array}$$

b. The corresponding units are properly oriented by aligning the decimal points. Then first add the digits farthest to the right (4.62 may be written as 4.620), followed by those to the left, and continuing in this fashion.

$$\begin{array}{r} 1.035 \\ +4.620 \\ \hline 5.655 \end{array}$$

In the above examples the sum of the digits in any column was less than 10. If this sum is ten or more, an extra step is required as shown in the following example.

Example 2 a. Add $67 + 58$.   b. Add $16.24 + 7.99$.

Solutions

a.
$$\begin{array}{r} 1 \\ 67 \\ +58 \\ \hline 125 \end{array}$$

$7 + 8 = 15 =$ 1 ten and 5 units

$1 + 6 + 5 = 12$ tens $= 1$ hundred and 2 tens

b. First write the numbers with the decimal points aligned. Then start by adding the hundredths digits.

$$\begin{array}{r} 11\ 1 \\ 16.24 \\ +7.99 \\ \hline 24.23 \end{array}$$

$4 + 9 = 13$ hundredths $= 1$ tenth $+ 3$ hundredths
$1 + 2 + 9 = 12$ tenths $= 1$ unit $+ 2$ tenths
$1 + 6 + 7 = 14$ units $= 1$ ten $+ 4$ units
$1 + 1 = 2$ tens

The extra steps required in the above examples are referred to as *carrying*. The next example illustrates this process when three numbers are added.

WHOLE NUMBERS AND DECIMALS; ADDITION AND SUBTRACTION

Example 3 Add 1089 + 526 + 498.

Solution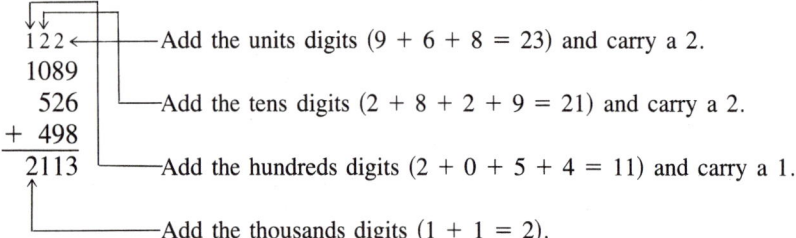

ESTIMATING SUMS

An important skill that you can develop as you study arithmetic is that of *estimating* answers to calculations. One method that can be used to estimate an answer involves the use of rounded-off numbers. It is usually convenient to use rounded-off numbers that consist of only *one* nonzero digit.

Example 4 *Estimate* each sum.

    a. 83 + 35 + 27   b. 867 + 147 + 558   c. $74.23 + $29.50 + $19.63

Solutions a. Round off each number to the nearest ten and mentally add the rounded-off numbers:

$$80 + 40 + 30 = 150$$

The estimated sum is 150.

b. Round off each number to the nearest hundred and mentally add the rounded-off numbers:

$$900 + 100 + 600 = 1600$$

The estimated sum is 1600.

c. Round off each number to the nearest ten dollars and mentally add the rounded-off numbers:

$$\$70 + \$30 + \$20 = \$120$$

The estimated sum is $120.

EXERCISES 1.3

Find each sum mentally. (If you have not learned the basic addition facts, refer to Table 1.1.)

	a.	b.	c.	d.	e.	f.	g.
1.	9 6	8 3	5 5	7 9	5 8	0 7	8 8
2.	9 4	3 9	8 9	7 5	7 0	6 5	4 6

14

1.3 ADDITION

	a.	b.	c.
3.	3 + 4 + 5	8 + 1 + 7	6 + 2 + 5
4.	5 + 4 + 5	7 + 1 + 7	1 + 8 + 4
5.	2 + 5 + 6 + 0	3 + 3 + 0 + 7	8 + 0 + 1 + 5
6.	2 + 5 + 0 + 9	8 + 0 + 5 + 4	3 + 2 + 9 + 0

Find each sum. See Example 1.

Complete Examples

a.   156
    +332

    ┌─ 6 + 2 units
    ├─ 5 + 3 tens
    └─ 1 + 3 _____   Ans. _____
              (1)   (2)

b.   32.05
    + 7.632

    ┌─ 0 + 2 thousandths
    ├─ 5 + 3 hundredths
    ├─ 0 + 6 tenths
    ├─ 2 + 7 units
    └─ 3 + 0 _____   Ans. _____
              (3)   (4)

(1) hundreds   (2) 488   (3) tens   (4) 39.682

7. 46	8. 31	9. 132	10. 605	11. 2.3	12. 6.8
+23	+52	+437	+293	+5.5	+3.1

13. 23.4 + 56.1   14. 16.5 + 71.3   15. 11.076 + 3.712   16. 5.213 + 14.651

Find each sum. See Examples 2 and 3.

Complete Example

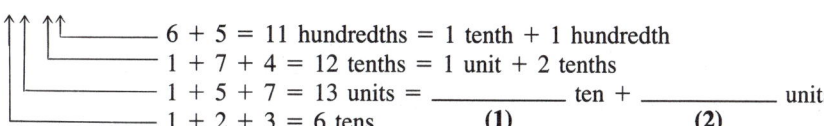

          ┌─ carry 1 ten
          ├─ carry 1 unit
    1 1 1 ├─ carry 1 tenth
   25.76
   37.45

    ┌─ 6 + 5 = 11 hundredths = 1 tenth + 1 hundredth
    ├─ 1 + 7 + 4 = 12 tenths = 1 unit + 2 tenths
    ├─ 1 + 5 + 7 = 13 units = _____ ten + _____ units
    └─ 1 + 2 + 3 = 6 tens   (1)   (2)

Ans. _____
      (3)

(1) 1   (2) 3   (3) 63.21

15

WHOLE NUMBERS AND DECIMALS; ADDITION AND SUBTRACTION

17.	46	18.	35	19.	393	20.	330
	28		48		33		95
21.	7049	22.	6319	23.	6162	24.	9908
	7742		5173		8258		5607
25.	9.12	26.	3.32	27.	5739	28.	4673
	8.46		0.55		6487		7857
	0.88		7.03		3598		2762
29.	97.30	30.	5.573	31.	92,739	32.	8976
	53.84		3.859		42,736		478
	6.82		4.966		96,833		7328
	7.45		8.426		47,485		98
							9663

33. 98 + 427 + 75
34. 23 + 333 + 62
35. 6.76 + 3.2 + 432
36. 1.8 + 500 + 7.62
37. 924 + 7766 + 96 + 9984
38. 455 + 8546 + 56 + 5041
39. 9.7 + 161.1 + 7.86 + 94.53
40. 45.34 + 96.17 + 9.66 + 63.4
41. 58.10 + 734.9 + 106 + 798
42. 30.986 + 0.812 + 234 + 420

43. The following chart shows the number of sales made each day at each of four different branches of a department store.
    a. Find the weekly total for each branch.
    b. Find the daily total for all branches.
    c. Find the total sales at all branches for the week.

44. The following table shows the scores made by four golf players in five consecutive games.
    a. Which player had the lowest total score? The highest total?
    b. In which round did the four players have the lowest total score?

Branch	Mon	Tues	Wed	Thurs	Fri	Weekly Total
A	2846	3079	4200	4483	5037	___
B	1381	1575	2103	2874	3697	___
C	987	994	1007	1246	2548	___
D	3037	3421	3863	3804	4251	___
Daily Total	___	___	___	___	___	

Player	Bob	Beatrice	Connie	Darryle	Round Totals
Round 1	80	81	84	76	___
Round 2	75	73	81	80	___
Round 3	74	72	74	79	___
Round 4	71	70	71	77	___
Round 5	74	75	78	74	___
Totals	___	___	___	___	

Find the missing dimension in each of the following drawings:

45.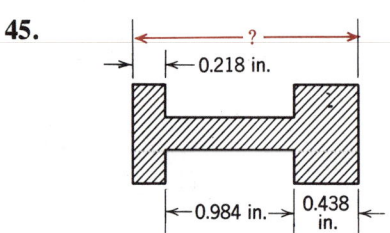

46.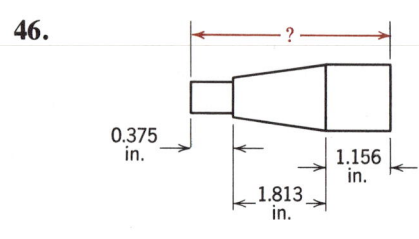

Use rounded-off numbers to estimate each sum. See Example 4.

Complete Example

    923 + 465 + 378

Round off each number to the nearest hundred and mentally add the rounded-off numbers.

    900 + _____ + _____
            (1)   (2)

Ans. _____
     (3)

(1) 500	(2) 400	(3) 1800

47. 28 + 34 + 45   48. 64 + 79 + 83   49. 1.1 + 2.2 + 3.3
50. 4.4 + 5.5 + 6.6   51. 54.1 + 20.3 + 65.2   52. 38.9 + 43.1 + 10.8
53. 962 + 843 + 734   54. 577 + 688 + 799   55. 1025 + 2025 + 3025
56. 2500 + 3400 + 4300

1.4 SUBTRACTION

The operation of **subtraction** is related to the addition operation and is called the *inverse* of addition. For example, when we subtract 2 from 7, we seek a number that when added to 2 gives a sum of 7. From our knowledge of the addition facts in Table 1.1 on page 12, we obtain 5, called the **difference** of 2 subtracted from 7. The operation of subtraction is designated by the symbol "−"; the numbers that are to be subtracted can be written in vertical or horizontal form. For example,

$$\begin{array}{r}7\\-2\\\hline 5\end{array} \quad \text{or} \quad 7 - 2 = 5$$

In Section 1.3 we noted that the operation of addition is commutative. Thus, 7 + 2 equals 2 + 7. Note that this property does *not* apply to subtraction. For example, 7 − 2 equals 5; however, 2 − 7 does not represent a whole number or a decimal.

COMPUTING DIFFERENCES

If the numbers involved in a subtraction problem contain more than a single digit, we proceed much as in addition. We write the numbers so that units digits are over units digits, tens digits are over tens digits, etc.

WHOLE NUMBERS AND DECIMALS; ADDITION AND SUBTRACTION

Example 1   a. Subtract: $756 - 32$.   b. Subtract: $8.96 - 1.7$.

Solutions   a. Writing the numbers in vertical form so that corresponding units are aligned, we first subtract the units digits, then the tens digits, and last the hundreds digits.

$$\begin{array}{r} 756 \\ -\;32 \\ \hline 724 \end{array}$$

b. The corresponding units are properly oriented by aligning the decimal points. We then start the process by subtracting the hundredths digits, where we first have written 1.7 as 1.70.

$$\begin{array}{r} 8.96 \\ -\;1.70 \\ \hline 7.26 \end{array}$$

In addition we sometimes have the extra step of carrying. In subtraction we sometimes have an extra step of *borrowing*.

Example 2   Subtract: $932 - 247$.

Solution   First we write the numbers in the vertical form.

$$\begin{array}{r} 932 \\ -\;247 \end{array}$$

Since 7 is greater than 2 in the units position, we must first consider 932 as

$$9 \text{ hundreds} + 3 \text{ tens} + 2 \text{ units}.$$

Because 3 tens equals 2 tens and 10 units, this sum can also be written as

$$9 \text{ hundreds} + 2 \text{ tens} + 12 \text{ units},$$

where we have *borrowed* 1 ten and changed it to 10 units. This can be shown as follows:

$$\begin{array}{r} 2\;\;12\;\; \\ 9\;\cancel{3}\;\cancel{2} \\ -\;2\;4\;7 \end{array} \quad \text{Now subtracting the units digits gives} \quad \begin{array}{r} 2\;\;12\;\; \\ 9\;\cancel{3}\;\cancel{2} \\ -\;2\;4\;7 \\ \hline 5 \end{array} \quad (12 - 7 = 5)$$

Since 4 is greater than 2 in the tens position, we change 9 hundreds to 8 hundreds and 10 tens and "borrow" the 10 tens. This can be shown as follows:

$$\begin{array}{r} 12\;\;\;\;\; \\ 8\;\cancel{2}\;12\;\; \\ \cancel{9}\;\cancel{3}\;\cancel{2} \\ -\;2\;4\;7 \\ \hline 6\;8\;5 \end{array} \quad (12 - 4 = 8 \text{ and } 8 - 2 = 6)$$

Hence $932 - 247 = 685$.

1.4 **SUBTRACTION**

Example 3 Subtract: $12.031 - 3.125$.

Solution First align the decimal points.

$$\begin{array}{r} \overset{2\ 11}{1\,2\,.\,0\,\cancel{3}\,\cancel{1}} \\ -\ \ 3\,.\,1\,2\,5 \\ \hline 6 \end{array}$$ Since 5 is greater than 1, we borrow 1 hundredth = 10 thousandths.

$$\begin{array}{r} \overset{2\ 11}{1\,2\,.\,0\,\cancel{3}\,\cancel{1}} \\ -\ \ 3\,.\,1\,2\,5 \\ \hline 0\ 6 \end{array}$$

$$\begin{array}{r} \overset{1\ 10\ 2\ 11}{1\,\cancel{2}\,.\,\cancel{0}\,\cancel{3}\,\cancel{1}} \\ -\ \ 3\,.\,1\,2\,5 \\ \hline .\,9\,0\,6 \end{array}$$ Since 1 is greater than 0, we borrow 1 unit = 10 tenths.

$$\begin{array}{r} \overset{11}{} \\ \overset{0\ \cancel{1}\ 10\ 2\ 11}{\cancel{1}\,\cancel{2}\,.\,\cancel{0}\,\cancel{3}\,\cancel{1}} \\ -\ \ \ 3\,.\,1\,2\,5 \\ \hline 8\,.\,9\,0\,6 \end{array}$$ Since 3 is greater than 1, we borrow 1 ten = 10 units.

Hence $12.031 - 3.125 = 8.906$.

One way to obtain the result of computations involving more than two numbers is to perform the operations on two numbers at a time in the order that they appear from left to right.

Example 4 Compute: $11.34 - 1.78 - 2.53 + 4.62$.

Solution
$$\begin{array}{l} \underbrace{11.34 - 1.78}\ - 2.53 + 4.62 \\ \underbrace{9.56- 2.53}\ + 4.62 \\ 7.03+ 4.62 = 11.65 \end{array}$$

CHECKING SUBTRACTION

Because the operation of subtraction is the inverse operation of addition, a subtraction operation can be checked by addition. For example,

Computation: $25{,}483 - 10{,}008 = 15{,}475$

Check: $10{,}008 + 15{,}475 = 25{,}483$

Or, in vertical form,

$$\begin{array}{r} 25{,}483 10{,}008 \\ -10{,}008 +15{,}475 \\ \hline 15{,}475 25{,}483 \end{array}$$

WHOLE NUMBERS AND DECIMALS; ADDITION AND SUBTRACTION

THE ROLE OF ZERO

In Section 1.3 we observed that adding zero to any number results in the same number. Similarly, subtracting zero from any number results in the same number. For example,

$$892 - 0 = 892$$

Furthermore, *the difference of any number and itself is zero*. For example,

$$28 - 28 = 0 \quad \text{and} \quad 892 - 892 = 0$$

Now, consider an expression such as

$$55 + 28 - 28 \tag{1}$$

Because $28 - 28$ is equal to zero, we have

$$55 + 28 - 28 = 55 + 0$$
$$= 55$$

Expression (1) may sometimes appear in alternate forms, such as

$$28 + 55 - 28 \quad \text{or} \quad 28 - 28 + 55 \tag{2}$$

Note that the result for each expression is 55 because the inverse operations of addition and subtraction "undo" each other. The fact that addition and subtraction are inverse operations suggests the following rule:

> **If a number is added to and subtracted from a given number, the result is the given number.**

Thus, we do not need to use arithmetic procedures to do computations of the types shown above. We could have obtained the result for each expression in (2) by simply following this rule.

Example 5 Do each computation *mentally*.
   a. $234 + 98 - 98$ b. $234 - 98 + 98$ c. $98 + 234 - 98$

Solution In each case the result is 234.

ESTIMATING DIFFERENCES

In Section 1.3 we estimated sums by adding rounded-off numbers. Differences can be estimated in a similar manner.

Example 6 *Estimate* each difference.
   a. $94 - 39$ b. $406 - 297$ c. $\$337.11 - \$142.60$

1.4 SUBTRACTION

Solutions a. Round off each number to the nearest ten and mentally subtract the rounded-off numbers:

$$90 - 40 = 50$$

The estimated difference is 50.

b. Round off each number to the nearest hundred and mentally subtract the rounded-off numbers:

$$400 - 300 = 100$$

The estimated difference is 100.

c. Round off each number to the nearest hundred dollars and mentally subtract the rounded-off numbers:

$$\$300 - \$100 = \$200$$

The estimated difference is $200.

EXERCISES 1.4
Find each difference mentally. (Refer to Table 1.1 on page 12 as necessary.)

	a.	b.	c.	d.	e.	f.	g.
1.	9 −1	7 −2	7 −7	16 −8	3 −0	8 −8	16 −7
2.	6 −2	10 −3	11 −2	12 −7	17 −9	14 −7	10 −4

Find each difference. See Example 1. Check by addition.

Complete Example

8.56
−3.24

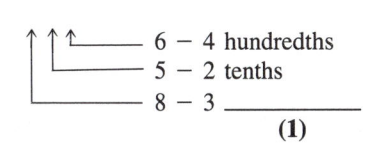

Check:
8.56 → 3.24
−3.24 ← +5.32
5.32 ←
      (3)

6 − 4 hundredths
5 − 2 tenths
8 − 3 _____
    (1)

Ans. _____
    (2)

(1) units   (2) 5.32   (3) 8.56

WHOLE NUMBERS AND DECIMALS; ADDITION AND SUBTRACTION

| 3. | 78 −61 | 4. | 94 −72 | 5. | 873 −612 | 6. | 648 −403 |

| 7. | 8.4 −3.2 | 8. | 9.7 −6.2 | 9. | 17.83 −2.41 | 10. | 14.49 −2.06 |

Find each difference. See Examples 2 and 3. Check by addition.

Complete Example

```
  73.2
 −26.5
```

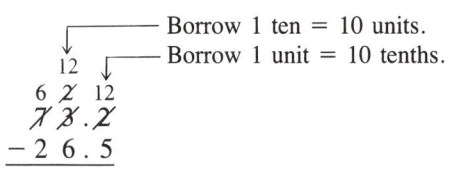

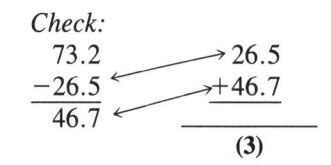

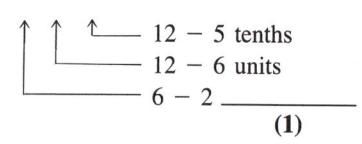

        ┌── Borrow 1 ten = 10 units.
        │ ┌── Borrow 1 unit = 10 tenths.
       12 │
     6 ⁄2 12
     7 ⁄3 . ⁄2
    − 2 6 . 5

    ↑ ↑ ↑ ── 12 − 5 tenths
        └── 12 − 6 units
            └── 6 − 2 ─────────
                              (1)

Check:
  73.2 ──→ 26.5
 −26.5 ←── +46.7
  46.7 ←── ─────
                (3)

Ans. _____
         (2)

(1) tens   (2) 46.7   (3) 73.2

| 11. | 73 −26 | 12. | 81 −34 | 13. | 438 −219 | 14. | 694 −357 | 15. | 8.34 −2.17 | 16. | 6.81 −4.18 |

| 17. | 12.86 −3.94 | 18. | 15.46 −7.82 | 19. | 742 −583 | 20. | 615 −247 | 21. | 62.34 −58.86 | 22. | 48.03 −17.26 |

| 23. | 48.842 −5.053 | 24. | 21.069 −8.182 | 25. | 44,394 −10,637 | 26. | 98,322 −63,188 | 27. | 4.002 −1.434 | 28. | 6.004 2.587 |

29. $10 − $3.47   30. $10 − $8.07   31. $20 − $6.08   32. $20 − $17.83

1.4 SUBTRACTION

Compute. See Example 4.

Complete Examples

a. $\underbrace{4721 + 242}\ - 3107$
   $\underbrace{4963 - 3107}_{(1)} = $ _____

b. $\underbrace{5.73 - 2.64}\ + 3.2 - 2.05$
   $\underbrace{3.09\ \ \ + 3.2}\ - 2.05$
   $\underbrace{}_{(2)} - 2.05 = \underline{}_{(3)}$

---

(1) 1856   (2) 6.29   (3) 4.24

---

33. $1150 - 87 - 225$
34. $5111 - 383 - 51$
35. $96.77 - 32.02 - 6.42$
36. $9.85 - 5.3 - 2.216$
37. $19.444 - 6.649 + 5.465$
38. $78.675 - 8.401 + 0.819$
39. $63,798 + 6499 - 5465$
40. $44,160 + 7812 - 8839$

41. The following chart shows the meter readings in kilowatt-hours (kwh) at several buildings at the beginning and end of a billing period. Find the number of kilowatt-hours of electrical energy used by each building during the period.

Building	a.	b.	c.	d.
Start	27,083	33,114	10,003	34,777
End	60,711	85,562	49,001	51,233
kwh	___	___	___	___

42. The following table shows the number of gallons of oil in four separate storage tanks before and after additional oil was pumped into each tank. Find the number of gallons of oil that were pumped into each tank.

Tank	a.	b.	c.	d.
Before	1723	10,071	9070	8334
After	90,046	81,043	20,431	32,256
Gallons	___	___	___	___

For each checking account statement, compute the closing balance. Add deposits to opening balance; add checks, and find the difference. If the sum of the checks is greater, the account is overdrawn.

---

43. Opening balance: $865.49

Date	Deposits	Checks
3/2	927.55	216.23
		49.82
3/7		31.20
3/12		153.38
3/15	211.22	
3/23		326.83
		171.50
3/30		446.25

44. Opening balance: $327.67

Date	Deposits	Checks
4/1	859.14	176.37
		20.40
		17.55
4/15		213.81
4/29		32.66
		59.88
4/30		89.85
		236.21

23

WHOLE NUMBERS AND DECIMALS; ADDITION AND SUBTRACTION

45. Opening balance: $542.29

Date	Deposits	Checks
5/3		99.84
		27.13
5/5		68.81
5/15	825.50	432.29
		505.40
5/25		17.87
		186.50
5/30		17.80

46. Opening balance: $3571.63

Date	Deposits	Checks
6/2		1,486.50
		615.75
6/10		903.40
6/12	350.00	615.75
6/15	125.00	
6/22		215.75
6/25	300.00	
6/29		215.50

Find the missing dimension in each of the following drawings, rounded off to the nearest hundredth of an inch.

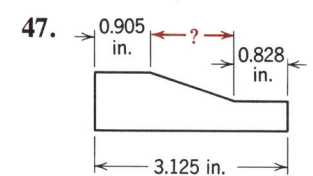

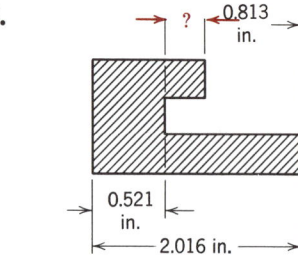

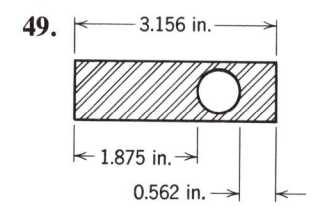

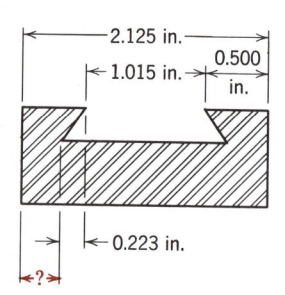

47. 0.905 in., ?, 0.828 in., 3.125 in.

48. ?, 0.813 in., 0.521 in., 2.016 in.

49. 3.156 in., 1.875 in., 0.562 in.

50. 2.125 in., 1.015 in., 0.500 in., 0.223 in., ?

Compute mentally. See Example 5.

Complete Examples

a. $4.39 + \underbrace{6.01 - 6.01}_{\text{Result is zero.}}$

Ans. _____
   (1)

b. $\underbrace{86.1 + 125 - 86.1}_{\text{Result is zero.}}$

Ans. _____
   (2)

---

(1) 4.39   (2) 125

51. $63{,}856 - 41 + 41$
52. $12{,}358 + 58 - 58$
53. $484 + 2375 - 484$
54. $682 + 23{,}407 - 682$
55. $4.03 - 4.03 + 57.239$
56. $5.9 - 5.9 + 23.508$

Use rounded-off numbers to estimate each of the following. See Example 6.

Complete Example

$$567.1 - 231.4$$

Round off each number to the nearest hundred and mentally subtract the rounded-off numbers.

$$600 - \underline{}_{(1)}$$

Ans. $\underline{}_{(2)}$

---

(1) 200   (2) 400

---

57. $81 - 42$
58. $91 - 33$
59. $981 - 783$
60. $622 - 428$
61. $1126 - 713$
62. $3917 - 1009$
63. $555.55 - 271.04$
64. $608.81 - 321.24$
65. $356.20 + 238.19 - 109.37$
66. $66.66 + 44.44 - 22.22$

1.5 SOLVING EQUATIONS

Statements of equality such as

$$28 + 6 = 34 \quad \text{and} \quad 28 - 6 = 22$$

are called **equations.** Some equations, such as

$$S = 28 + 34 \quad \text{and} \quad D = 54 - 17 \tag{1}$$

include letters that are used to represent numbers; such letters are sometimes called *unknowns* or *variables*.

SOLVING EQUATIONS

To solve an equation involving an unknown means to find a replacement value (number) for the unknown so that the equation will be a true statement. Such a

WHOLE NUMBERS AND DECIMALS; ADDITION AND SUBTRACTION

number is called a **solution** of the equation. Thus, if S is replaced by 62 in the equation

$$S = 28 + 34$$

we have

$$62 = 28 + 34$$

which is a true statement. Therefore, 62 is a solution of the equation.

In Equations (1) above, the values of S and D can be obtained by a direct computation of a sum and a difference, respectively. Thus,

$$S = 28 + 34 = 62 \quad \text{and} \quad D = 54 - 17 = 37$$

Sometimes, as in the equations

$$N + 3 = 8 \quad \text{and} \quad N - 28 = 356$$

the solutions cannot be obtained by a direct computation. The equation $N + 3 = 8$ is readily solved by use of our basic addition facts. If we replace N by 5 to obtain $5 + 3 = 8$, we see that 5 is the solution of the equation. However, basic addition facts will not immediately solve the equation $N - 28 = 356$. Hence, we need to develop methods for solving such equations.

EQUIVALENT EQUATIONS

Equations that have identical solutions are called **equivalent equations.** Thus, because 25 is the solution of each of the equations

$$N + 5 = 30 \quad \text{and} \quad N = 30 - 5$$

these equations are equivalent.

If the solution of a given equation is not evident by inspection, then we change the equation to an equivalent but simpler equation. This procedure can be repeated, if necessary, until the solution is obvious. The following rule can be used to change an equation into an equivalent equation.

> **If a number is added to, or subtracted from, each side of an equation, the resulting equation is equivalent to the original equation.**

We shall refer to this statement as the **addition-subtraction rule** for writing equivalent equations.

The goal to keep in mind when applying the addition-subtraction rule to solve an equation is to *obtain an equivalent equation in which the unknown appears by itself on only one side of the equation.* We accomplish this by using the fact that addition and subtraction are inverse operations that "undo" each other.

1.5 SOLVING EQUATIONS

Example 1 Solve.

    a. $N - 28 = 356$ b. $N + 249 = 312$

Solutions a.
$$N - 28 = 356 \qquad (2)$$

First, add 28 to each side of the equation to "undo" the subtraction of 28 in the left side. Thus,

$$N - 28 + 28 = 356 + 28$$

Since $N - 28 + 28$ equals N,

$$N = 356 + 28$$

from which

$$N = 384 \qquad (3)$$

Because Equations (2) and (3) are equivalent (have the same solution) and the solution of Equation (3) is obviously 384, the solution of Equation (2) is also 384. The solution can be *checked* by substituting 384 for N in the given Equation (2) and verifying that

$$384 - 28 \stackrel{?}{=} 356$$
$$356 = 356$$

b.
$$N + 249 = 312 \qquad (4)$$

First, subtract 249 from each side of the equation to "undo" the addition of 249 in the left side. Thus,

$$N + 249 - 249 = 312 - 249$$

Because $N + 249 - 249$ is equal to N,

$$N = 312 - 249$$

from which

$$N = 63 \qquad (5)$$

Because Equations (4) and (5) are equivalent and the solution of Equation (5) is obviously 63, the solution of Equation (4) is also 63.

    The solution can be *checked* by substituting 63 for N in the given Equation (4) and verifying that

$$63 + 249 \stackrel{?}{=} 312$$
$$312 = 312$$

    Sometimes we can simplify an equation by finding sums and differences before using the addition-subtraction rule.

WHOLE NUMBERS AND DECIMALS; ADDITION AND SUBTRACTION

Example 2   a.
$$N + 463 = \underbrace{970 - 119}$$ (6)
$$N + 463 = 851$$

Subtracting 463 from each side, we have

$$N + 463 - 463 = 851 - 463$$

Because $N + 463 - 463$ is equal to N,

$$N = \underbrace{851 - 463}$$
$$N = 388$$

The solution to Equation (6) is 388. We can check by substituting 388 for N in the given Equation (6) and verifying that

$$388 + 463 \stackrel{?}{=} 970 - 119$$
$$851 = 851$$

b.
$$89 = 62 + N - 57$$ (7)

By the commutative property of addition (Section 1.3) we may write

$$89 = N + \underbrace{62 - 57}$$
$$89 = N + 5$$

Next, subtracting 5 from each side of the equation, we have

$$89 - 5 = N + 5 - 5$$

Because $N + 5 - 5$ is equal to N,

$$\underbrace{89 - 5} = N$$
$$84 = N$$

The solution to Equation (7) is 84. We can check by substituting 84 for N in the given Equation (7) and verifying that

$$89 \stackrel{?}{=} 62 + 84 - 57$$
$$89 = 89$$

In the preceding examples we added numbers to, or subtracted numbers from, each side of the equations. The addition-subtraction rule can also be applied to add (or subtract) the unknown to each side of an equation. *This procedure is particularly appropriate when the unknown is subtracted from a number on one side of the equation.*

Example 3 $48 - N = 20$

Solution   Adding N to each side of the equation, we obtain

$$48 - N + N = 20 + N$$

1.5 SOLVING EQUATIONS

Because $48 - N + N$ equals 48,

$$48 = 20 + N$$

Next, subtracting 20 from each side,

$$48 - 20 = 20 + N - 20$$

The left side of the equation equals 28. Because $20 + N - 20$ equals N,

$$28 = N$$

We can check by substituting 28 for N in the given equation and verifying that

$$48 - 28 \stackrel{?}{=} 20$$
$$20 = 20$$

The addition-subtraction rule can also be used to solve equations that involve additions and subtractions of decimals.

Example 4   Solve $427.5 + N = 760.21$.

Solution   First subtract 427.5 from each side of the equation.

$$427.5 + N - 427.5 = 760.21 - 427.5$$

Because $427.5 + N - 427.5$ equals N,

$$N = 760.21 - 427.5$$
$$N = 332.71$$

Check by substituting 332.71 for N in the given equation and verifying that

$$427.5 + 332.71 \stackrel{?}{=} 760.21$$
$$760.21 = 760.21$$

Example 5   Solve $63.028 = N - 193.94$.

Solution   First add 193.94 to each side of the equation.

$$63.028 + 193.94 = N - 193.94 + 193.94$$

Because $N - 193.94 + 193.94$ equals N, we have

$$63.028 + 193.94 = N$$
$$256.968 = N$$

We can check by substituting 256.968 for N in the given equation and verifying that

$$63.028 \stackrel{?}{=} 256.968 - 193.94$$
$$63.028 = 63.028$$

WHOLE NUMBERS AND DECIMALS; ADDITION AND SUBTRACTION

EXERCISES 1.5
Solve each equation by inspection.

1. $D - 5 + 5 = 8$
2. $N - 11 + 11 = 25$
3. $36 - 36 + S = 18$
4. $209 - 209 + S = 117$
5. $19 + N - 19 = 32$
6. $153 + N - 153 = 47$
7. $68 = N - 51 + 51$
8. $94 = N - 342 + 342$

Solve each equation; check your answers. See Examples 1 and 2.

Complete Examples

a. $N - 36 = 143$

   Add 36 to each side of the equation.

   $N - 36 + 36 = 143 + \underline{}$
   (1)

   $N = \underline{}$
   (2)

   Check: $179 - 36 \stackrel{?}{=} 143$
   $143 = 143$

b. $N + 182 = 641$

   Subtract 182 from each side of the equation.

   $N + 182 - 182 = 641 - \underline{}$
   (3)

   $N = \underline{}$
   (4)

   Check: $459 + 182 \stackrel{?}{=} 641$
   $641 = 641$

(1) 36	(2) 179	(3) 182	(4) 459

9. $S - 47 = 219$
10. $S - 93 = 57$
11. $58 + N = 97$
12. $82 + N = 103$
13. $S - 210 = 704$
14. $S - 1328 = 631$
15. $888 = 730 + N$
16. $528 = 127 + N$
17. $N - 32 = 67 - 41$
18. $58 + N = 203 - 46$
19. $65 = 43 + N - 24$
20. $117 = 57 + N + 28$
21. $S + 315 = 58 + 342$
22. $S - 73 = 161 - 37$
23. $45 + 98 = D - 17$
24. $123 - 74 = D - 28$
25. $S - 521 = 632 + 145$
26. $S - 521 = 632 - 145$
27. $58 + D - 58 = 12$
28. $397 + D - 397 = 1234$
29. $763 = 114 + S - 114$
30. $511 = S + 268 - 268$

Solve each equation; check your answers. See Example 3.

Complete Examples

a. $146 - N = 78$

b. $123 - R - 64 = 20$

30

1.5 SOLVING EQUATIONS

Add N to each side of the equation.

$146 - N + N = 78 + N$

$\underline{\qquad}_{(1)} = 78 + N$

Subtract 78 from each side of the equation.

$146 - 78 = 78 + N - \underline{\qquad}_{(2)}$

$\underline{\qquad}_{(3)} = N$

Check: $146 - 68 \stackrel{?}{=} 78$

$78 = 78$

Subtract $(123 - 64)$ and add R to each side of the equation.

$59 - R + R = 20 + R$

$\underline{\qquad}_{(4)} = 20 + R$

Subtract 20 from each side of the equation.

$59 - 20 = 20 + R - 20$

$\underline{\qquad}_{(5)} = R$

Check: $123 - 39 - 64 \stackrel{?}{=} 20$

$20 = 20$

(1) 146 (2) 78 (3) 68 (4) 59 (5) 39

31. $86 - D = 44$
32. $321 = 419 - D$
33. $2804 = 3927 - R$
34. $1763 - R = 982$
35. $49 - N + 16 = 27$
36. $153 - N - 48 = 16$
37. $12{,}405 - R - 9321 = 2808$
38. $13{,}067 - R + 654 = 11{,}971$

Solve each equation; check your answers. See Examples 4 and 5.

Complete Examples

a. $N + 63.25 = 124.01$

$N + 63.25 - 63.25 = 124.01 - \underline{\qquad}_{(1)}$

$N = \underline{\qquad}_{(2)}$

Check: $60.76 + 63.25 \stackrel{?}{=} 124.01$
$124.01 = 124.01$

b. $N - 0.063 = 1.248$

$N - 0.063 + 0.063 = 1.248 + \underline{\qquad}_{(3)}$

$N = \underline{\qquad}_{(4)}$

Check: $1.311 - 0.063 \stackrel{?}{=} 1.248$
$1.248 = 1.248$

(1) 63.25 (2) 60.76 (3) 0.063 (4) 1.311

39. $3.327 + N = 7.099$
40. $5.685 + N = 8.599$
41. $N - 0.815 = 0.722$
42. $N - 0.964 = 0.248$
43. $1.478 = N - 6.656$
44. $8.707 = N - 1.33$
45. $8.7917 = 7.7341 + R$
46. $4.2206 = 3.5126 + R$
47. $18.06 + N = 12.9 + 25.31$
48. $10.73 + N = 48.1 + 2.635$

WHOLE NUMBERS AND DECIMALS; ADDITION AND SUBTRACTION

Solve each equation by inspection.

49. $N + 4 = 9$
50. $8 + D = 11$
51. $10 = R - 9$
52. $19 = R + 9$
53. $16 = 7 + S$
54. $T - 8 = 7$
55. $N - 9 = 8$
56. $D - 12 = 8$
57. $N + 5 + 9 = 20$
58. $N + 9 - 5 = 8$
59. $16 = S + 2 + 8$
60. $5 = D + 8 - 3$

1.6 PROBLEM SOLVING

Equations that include unknowns can be used as mathematical "models" for word sentences. Thus, the ability to write and solve equations is an invaluable aid for solving problems.

Suggestions for solving word problems:

1a. Read the problem and note particularly what is asked for. *Write* a short word phrase to describe this quantity.

b. Choose a letter (the unknown) to represent the number that you are trying to find. (Any letter can be used.)

2. Write an equation that involves the unknown and is a "model" for the word sentence in the problem.

3. Solve the equation.

4. Interpret the solution in terms of what is asked for in the problem.

Example 1   The sum of 1673 and a number is 5284. Find the number.

Solution   1. Express in a word phrase what you want to find; represent the unknown number by a letter, say N.

$$\text{A number: } N$$

2. The sum of 1673 and N is 5284

$$1673 + N = 5284 \tag{1}$$

3. Solve the equation. First, subtract 1673 from each side to obtain

$$1673 + N - 1673 = 5284 - 1673$$

Because $1673 + N - 1673$ equals N,

$$N = 5284 - 1673 \tag{2}$$
$$N = 3611$$

4. The number is 3611.

1.6 PROBLEM SOLVING

Because subtraction is not commutative, special care must be taken to write equations for word sentences that involve this operation.

Example 2 When 7.63 is subtracted from a certain number, the difference is 5.04. Find the number.

Solution

1. Express in a word phrase what you want to find; represent the unknown number by N.

$$\text{A number: } N$$

2. 7.63 subtracted from N is 5.04

$$N - 7.63 = 5.04 \qquad (1)$$

3. Solve the equation. First, add 7.63 to each side to obtain

$$N - 7.63 + 7.63 = 5.04 + 7.63$$

Because $N - 7.63 + 7.63$ equals N,

$$N = 7.63 + 5.04 \qquad (2)$$
$$N = 12.67$$

4. The number is 12.67.

The procedure illustrated by Examples 1 and 2 can also be used when the problem is concerned with a measure of a quantity instead of just a number.

Example 3 A college registrar's office needs to mail a "permit to enroll" to each of 8507 students. If the office staff mails 5378 permits the first week, how many more permits remain to be mailed?

Solution

1. Express in a word phrase what you want to find; represent it by a letter.

$$\text{The number of permits remaining: } P.$$

2. Express the sentence "The sum of 5378 and P equals 8507" as an equation:

$$5378 + P = 8507 \qquad (1)$$

3. Solve the equation. Subtracting 5378 from each side, we obtain

$$5378 + P - 5378 = 8507 - 5378$$

Because $5378 + P - 5378$ equals P,

$$P = 8507 - 5378 \qquad (2)$$
$$P = 3129$$

4. 3129 permits remain to be mailed.

It may sometimes be possible to use different equations for solving a given problem. In each of the above examples, you can obtain the same result if you start with Equation (2) rather than with Equation (1).

WHOLE NUMBERS AND DECIMALS; ADDITION AND SUBTRACTION

EXERCISES 1.6

Write an equation to represent each sentence, using the letter N for the unknown.

Complete Examples

a. The sum of a number and 17 is 85

   N + ___(1)___ = ___(2)___

   Ans. N + 17 = 85

b. If a number is subtracted from 12.1 the result is 6.3

   ___(3)___ − N = ___(4)___

   Ans. 12.1 − N = 6.3

(1) 17   (2) 85   (3) 12.1   (4) 6.3

1. The sum of a number and 107 is 118.
2. The sum of a number and 86 is 811.
3. The sum of 3.2 and a number is 6.1.
4. The sum of 4.3 and a number is 9.6.
5. A number is the sum of 248 and 76.
6. A number is the sum of 324 and 288.
7. The difference of 82 subtracted from a number is 49.
8. The difference of 209 subtracted from a number is 61.
9. If a number is subtracted from 9.87, the result is 7.42.
10. If a number is subtracted from 53.3, the result is 11.2.
11. The difference of 489 subtracted from 12,900 is a number.
12. The difference of 625 subtracted from 5087 is a number.

For each of Problems 13–40: a. Write an equation. b. Solve the equation.
For Problems 13–20, see Examples 1 and 2.

Complete Example

If 25.6 is subtracted from a certain number, the result is 13.3. Find the number.

a. A number: N
   25.6 subtracted from N is 13.3

   Ans. N − ___(1)___ = 13.3

b. N − 25.6 = 13.3
   N − 25.6 + 25.6 = 13.3 + ___(2)___

   N = ___(3)___

Ans. The number is 38.9.

(1) 25.6   (2) 25.6   (3) 38.9

1.6 PROBLEM SOLVING

13. The sum of 43 and a number is 126. Find the number.

14. The sum of what number and 147 is 168?

15. When 20.3 is subtracted from a certain number the difference is 37.6. Find the number.

16. The number 4.68 is obtained when a certain number is subtracted from 17.12. Find the number.

17. The sum 30.84 is obtained when 15.76 is added to a certain number. Find the number.

18. If 42.8 is added to a number, the sum is 97.4. Find the number.

19. The number 1712 is obtained when 468 is subtracted from a certain number. Find the number.

20. If a certain number is subtracted from 2154 the difference is 1976. Find the number.

For Problems 21–28, see Example 3.

Complete Example

A tire store sets a goal of 17,840 tires to be sold in a month. After three weeks, 13,554 tires are sold. How many more tires must be sold to meet the goal?

a.   The number of tires to be sold: N

The sum of 13,554 and N equals 17,840

Ans.   13,554 + _____(1)_____ = 17,840

b.   13,554 + N = 17,840

13,554 + N − 13,554 = 17,840 − _____(2)_____

N = _____(3)_____

Ans. The store must sell 4286 more tires.

(1) N (2) 13,554   (3) 4286

21. One football player has a record of 11,834 yards gained, and another a record of 9275 yards. How many more yards must the second player gain to equal the record of the first player?

22. A stadium holds 89,425 people. If 67,584 tickets have been sold to a ballgame, how many more tickets to the game can be sold?

23. A storage tank can hold 49,320 gallons of gasoline. If the tank contains 26,942 gallons, how many more gallons of gasoline can be stored in the tank?

24. A brick manufacturer has 53,800 bricks in a storage yard and 79,520 bricks in a second storage yard. If an order comes in for 220,000 bricks, how many more bricks must be manufactured to complete the order?

25. At the start of a nationwide tour, the mileage reading in a car was 49,307 miles. At the end of the tour the reading was 62,802 miles. How many miles were driven during the tour?

26. An athlete signed a one-year contract for $150,000 and received a first payment of $29,650. How much more will he receive?

27. The sticker price of a new car is $6382. This price does not include such extra charges as taxes, transportation, dealer preparation and the like. If the car sells for $6966, how much are the extra charges?

28. A football stadium in Brazil seats 155,000 people; the Superdome in New Orleans seats 71,827 people at a football game. Find the difference of their seating capacities.

WHOLE NUMBERS AND DECIMALS; ADDITION AND SUBTRACTION

Complete Examples

How much money must be deposited into a checking account so that checks for $209 and $126 can be issued and $200 will remain in the account at the end of the month?

a. The amount of money to be deposited: A

Ans.
$$A \text{ less } \$209 \text{ and less } \$126 \text{ equals } 200$$
$$A - 209 - \underline{}_{(1)} = 200$$

b.
$$A - 209 - 126 = 200$$
$$A - 209 - 126 + 126 + 209 = 200 + 126 + \underline{}_{(2)}$$
$$A = 200 + 126 + 209$$
$$A = \underline{}_{(3)}$$

Ans. $535 must be deposited into the account.

---

(1) 126   (2) 209   (3) 535

---

29. How much money must be deposited into a savings account so that $65, $120, and $250 can be withdrawn and $560 will remain in the account?

30. Find the opening amount in a checking account if $1426 remains in the account after checks for $870, $433, and $398 are issued.

31. Mr. Allen has $812 in a checking account. During the month he wrote checks for $75, $236, $49, and another check that he forgot to enter in his records. If $417 remained in his account at the end of the month, find the amount of the forgotten check.

32. A storage tank in a gasoline station contained 30,520 gallons one Monday morning and 11,824 gallons at closing time the following Friday. Find the number of gallons sold on Friday if 3528 gallons were sold on Monday, 3690 gallons sold on Tuesday, 4423 gallons sold on Wednesday, and 5295 gallons sold on Thursday.

33. If the sales figures for the Handy Hardware Shop are $1983.42, $2604.67 and $3258.95, how much sales income is needed in the fourth week to meet a goal of $10,000?

34. The Ace Sporting Goods Store sales figure for a four-week period is $10,834.29. Find the sales figure for the fourth week if the sales for the first three weeks are $2123.68, $3201.72, and $1991.35.

35. How much money must be deposited into a payroll account so that paychecks in the amounts $627.55, $849.42, $763.31, and $935.77 can be issued, and $3000 will remain in the account at the end of the month?

36. From a payroll account containing $6000, paychecks in the amounts $593.14, $728.95, $894.19 and a fourth check that was not entered in the records, were issued. If $3000 remained in the account at the end of the month, find the amount of the missing paycheck.

37. Three pieces with lengths 1.06, 2.12, and 1.88 inches, respectively, are cut from a rod that was 10.00 inches long. If .06 inch is wasted on each cut, how much is left of the original rod?

38. A tank contains 500 liters of solvent at the start of the week. During the week, solvent was pumped from the tank in the amounts of 24.6 liters, 87.3 liters, and 107.5 liters. How much remained in the tank at the end of the week?

39. A dealer in scrap iron had 250 tons of iron in stock at the start of the week. During the week, he made three purchases of 115 tons, 87 tons, and 106 tons, respectively. He also made two sales of 95 tons and 180 tons. How many tons were left at the end of the week?

40. On the first day of June, a woman has $987.15 in a bank account. During the month, she made deposits of $108.40, $216.32, and $50.60 and withdrawals of $89.96 and $500.17. How much remains in the account at the end of the month?

CHAPTER ONE REVIEW EXERCISES*

Read each number.

1. 2597
2. 791,209

For Exercises 3–6, write the numeral for each number.

3. Seven thousand, four hundred thirty-two.
4. Twenty-one thousand, two hundred six.
5. Thirty-four and five tenths.
6. Two hundred fifty two ten-thousandths.
7. Specify which is greater, 0.015452 or 0.015462.
8. Round off 5.8964 to the nearest:
   a. Tenth.   b. Hundredth.   c. Thousandth.
9. Round off $18.006 to the nearest cent.
10. Round off $157.49 to the nearest dollar.

Find each sum. Check by adding in reverse order.

11. 437
    581
    109

12. 54.98
    66.57
    78.62

13. 942 + 7667 + 8984

14. 1.4 + 12.2 + 17.4

Find each of the following: a. *To the nearest tenth.* b. *To the nearest hundredth.* c. *To the nearest thousandth.*

15. 19.7911 + 46.217 + 9.77612

16. 2.9507 + 68.57353 + 3.4142

Compute.

17. 5439
    −1103

18. 9.409 − 2.358

19. 5479 − 87 − 3324

20. 916.07 − 72.892 − 5.117 + 4.39

---

*Solutions are given to *all* chapter Review Exercises.

WHOLE NUMBERS AND DECIMALS; ADDITION AND SUBTRACTION

Solve each equation by inspection.

21. $D - 12 + 12 = 7$
22. $3.49 = 217 + D - 217$

Solve each equation.

23. $S - 15.5 = 40.7$
24. $1.82 = S - 9.05$
25. $518 + N - 9.7 = 672$

For Exercises 26 and 27 write an equation to represent the sentence, using the letter N for the unknown number.

26. The sum of a number and 701 is 811.
27. The difference of 2.8 subtracted from a number is 94.2.

For Exercises 28–30: a. Write an equation. b. Solve the equation.

28. What number added to 34 gives the sum 261?
29. If a certain number is subtracted from 1524, the difference is 976. Find the number.
30. A photographer wants to buy a camera at $347.69, a telephoto lens at $84.50, a zoom lens at $129.50, and a gadget bag at $15.75. What will be the total cost?

2 WHOLE NUMBERS AND DECIMALS; MULTIPLICATION AND DIVISION

In Chapter 1 we discussed some properties of addition and subtraction of whole numbers and decimals, and solved certain kinds of problems that involved these operations. In this section we first consider two more basic operations on these numbers. We then will solve other kinds of problems that involve these operations.

2.1 MULTIPLICATION

The operation of **multiplication** is used to find the **product** of two numbers; each of the numbers is called a **factor** of the product. The symbol "×" is used to indicate this operation.

BASIC MULTIPLICATION FACTS

The product of two numbers between 1 and 9 and including 1 and 9 can be obtained from Table 2.1.

Table 2.1
Basic Multiplication Facts

×	1	2	3	4	5	6	7	8	9
1	1	2	3	4	5	6	7	8	9
2	2	4	6	8	10	12	14	16	18
3	3	6	9	12	15	18	21	24	27
4	4	8	12	16	20	24	28	32	36
5	5	10	15	20	25	30	35	40	45
6	6	12	18	24	30	36	42	48	54
7	7	14	21	28	35	42	49	56	63
8	8	16	24	32	40	48	56	64	72
9	9	18	27	36	45	54	63	72	81

WHOLE NUMBERS AND DECIMALS; MULTIPLICATION AND DIVISION

For example,

$$\begin{array}{r}5\\ \times\ 3\\ \hline 15\end{array} \quad \text{and} \quad \begin{array}{r}7\\ \times\ 8\\ \hline 56\end{array}$$

or $3 \times 5 = 15$ or $7 \times 8 = 56$

As shown in these examples, numbers to be multiplied can be written in a vertical or a horizontal form.

You are probably familiar with the basic multiplication facts in Table 2.1 as well as with the addition facts in Table 1.1. If not, you should try to memorize these facts as soon as possible.

PROPERTIES OF MULTIPLICATION

Similar to addition, the operation of multiplication is commutative.

> **The order in which numbers are multiplied does not change the product.**

For example, from Table 2.1 we note that

$$8 \times 7 = 56 \quad \text{and} \quad 7 \times 8 = 56$$

The commutative property enables us to *check* a multiplication by multiplying the same factors in the reverse order.

With respect to products, both the numbers 1 and 0 play special roles.

> **The product of any number and 1 is the number.**

Thus,

$$7 \times 1 = 7 \quad \text{and} \quad 1 \times 86 = 86$$

> **The product of any number and 0 is 0.**

For example,

$$7 \times 0 = 0 \quad \text{and} \quad 0 \times 86 = 0$$

COMPUTING PRODUCTS

The product of two whole numbers, in which at least one of them contains more than a single digit, can be found as shown in the following examples.

2.1 MULTIPLICATION

Example 1 Multiply: 2×342.

Solution Recall that

$$342 = 3 \text{ hundreds} + 4 \text{ tens} + 2 \text{ units}$$

To multiply by 2, we must multiply the hundreds, tens, and units by 2. We begin with the units.

$$\begin{array}{r} 3\,4\,2 \\ \times \quad 2 \\ \hline 6\,8\,4 \end{array}$$

- 2×2 units = 4 units
- 2×4 tens = 8 tens
- 2×3 hundreds = 6 hundreds

Hence $2 \times 342 = 684$.

Example 2 Multiply: 3×684.

Solution As in the previous example, we begin with the units digits.

$$\begin{array}{r} 6\,8\,4 \\ \times \quad 3 \\ \hline 12 \end{array}$$

3×4 units = 12 units = 1 ten + 2 units

Because 3×4 units is more than 10 units, we must *carry* the 1 ten to the ten's place.

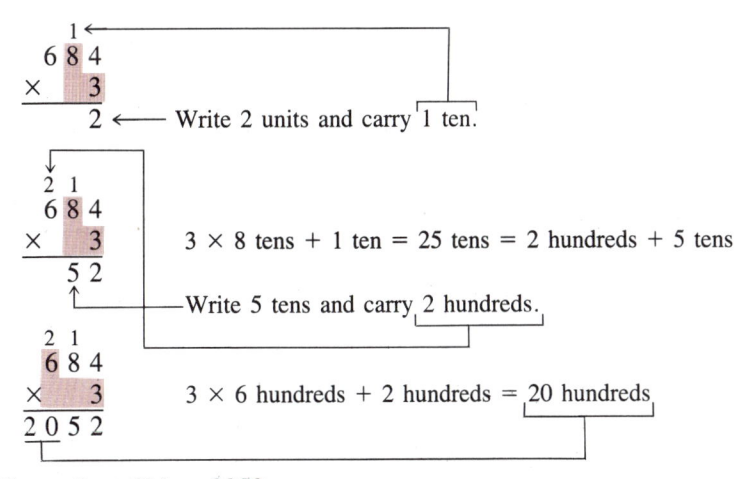

$$\begin{array}{r} 1 \\ 6\,8\,4 \\ \times \quad 3 \\ \hline 2 \end{array}$$

Write 2 units and carry 1 ten.

$$\begin{array}{r} 2\,1 \\ 6\,8\,4 \\ \times \quad 3 \\ \hline 5\,2 \end{array}$$

3×8 tens + 1 ten = 25 tens = 2 hundreds + 5 tens

Write 5 tens and carry 2 hundreds.

$$\begin{array}{r} 2\,1 \\ 6\,8\,4 \\ \times \quad 3 \\ \hline 2\,0\,5\,2 \end{array}$$

3×6 hundreds + 2 hundreds = 20 hundreds

Hence $3 \times 684 = 2052$.

Example 3 Multiply: 65×37.

Solution Since $65 = 60 + 5$, we multiply 37 by 5 and by 60 and then *add* the results.

$$\begin{array}{r} 3 \\ 3\,7 \\ \times \quad 5 \\ \hline 1\,8\,5 \text{ units} \end{array} \qquad \begin{array}{r} 4 \\ 3\,7 \\ \times \; 6\,0 \\ \hline 2\,2\,2\,0 \text{ units} \end{array} \qquad \begin{array}{r} 1 \\ 1\,8\,5 \\ +\,2\,2\,2\,0 \\ \hline 2\,4\,0\,5 \end{array}$$

In the above example, we have written 60×37 as 2220 units. Since 60 is 6 tens,

41

WHOLE NUMBERS AND DECIMALS; MULTIPLICATION AND DIVISION

we could write 60 × 37 as 6 × 37 tens = 222 tens and omit the final zero. In fact, this is usually done, and the product looks like this:

$$\begin{array}{r} 37 \\ \times\,65 \\ \hline 185 \\ 222\square \\ \hline 2405 \end{array}$$ ⟵ Zero omitted

The product of two numbers, at least one of which is a decimal, can be found by first proceeding as in the above example and then placing the decimal point in the proper place. *The number of decimal places in the product is the sum of the number of decimal places in the numbers being multiplied.*

Example 4 Multiply: 0.6 × 7.19.

Solution First find 6 × 719.

$$\begin{array}{r} 15 \\ 719 \\ \times\ \ \ 6 \\ \hline 4314 \end{array}$$

Now, insert the decimal point as shown below. Usually the work appears like this:

$$\begin{array}{r} 7.19 \\ \times 0.6 \\ \hline 4.314 \end{array}$$ ⟵ Two digits right of decimal point
⟵ One digit right of decimal point
⟵ 2 + 1 = 3 digits right of decimal point

Hence 0.6 × 7.19 = 4.314.

Example 5 Multiply: 0.013 × 0.708.

Solution First multiply 13 × 708; then, since 0.013 and 0.708 each have three decimal places, the product will have *six* decimal places. Note that the zero to the left of the decimal point in each factor is not used here.

$$\begin{array}{r} 0.708 \\ \times 0.013 \\ \hline 2\ \ 124 \\ 7\ \ 08\ \ \ \\ \hline .009\ 204 \end{array}$$

—Insert two zeros, since there must be *six* decimal places in the product.

Hence 0.013 × 0.708 = 0.009204.

ROUNDING OFF PRODUCTS

The methods that we used to round off numbers in Chapter 1 can be used to round off products.

2.1 MULTIPLICATION

Example 6 Compute 2.034×21.42: a. To the nearest tenth; b. To the nearest hundredth; c. To the nearest thousandth.

Solution $2.034 \times 21.42 = 43.56828$

    a. 43.6   b. 43.57   c. 43.568

EXERCISES 2.1

Compute each product mentally. If you have not learned the basic multiplication facts, refer to Table 2.1.

	a.	b.	c.	d.	e.	f.
1.	2 ×5	3 ×6	4 ×7	5 ×8	6 ×9	8 ×4
2.	4 ×9	7 ×4	5 ×9	7 ×7	6 ×6	5 ×6
3.	7 ×5	6 ×7	8 ×7	9 ×8	5 ×0	1 ×5
4.	8 ×1	8 ×0	0 ×9	1 ×9	6 ×1	6 ×0

Find each product in Exercises 5–32. See Examples 1 and 2.

Complete Examples

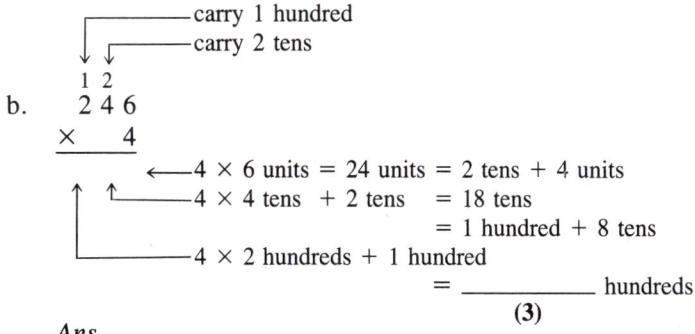

a.   1 2 2
     ×   4

        ← 4 × 2 units = 8 units
        ── 4 × 2 tens = 8 tens
        ── 4 × 1 hundred = 4 _____ (1)

Ans. _____
    (2)

b.   carry 1 hundred
     carry 2 tens
     1 2
     2 4 6
   ×   4

    ← 4 × 6 units = 24 units = 2 tens + 4 units
    ── 4 × 4 tens + 2 tens = 18 tens
                        = 1 hundred + 8 tens
    ── 4 × 2 hundreds + 1 hundred
                        = _____ hundreds
                            (3)

Ans. _____
    (4)

(1) hundreds   (2) 488   (3) 9   (4) 984

WHOLE NUMBERS AND DECIMALS; MULTIPLICATION AND DIVISION

5. 21
     ×4

6. 33
     ×2

7. 413
     ×2

8. 312
     ×3

9. 37
     ×3

10. 28
     ×4

11. 135
     ×5

12. 214
     ×7

See Example 3.

Complete Examples

a. 26
   ×32
   ―――
   52 ←―― 2 × 26
   78☐ ←―― 30 × 26
         or 3 tens × ―――――
                      (1)

   Ans. ―――――
          (2)

b. 316
   ×28
   ―――
   2528 ←―― 8 × 316
   632☐ ←―― 20 × 316
          or 2 tens × ―――――
                       (3)

   Ans. ―――――
          (4)

(1) 26   (2) 832   (3) 316   (4) 8848

13. 14
     ×12

14. 21
     ×15

15. 341
     ×14

16. 412
     ×21

17. 132 × 42

18. 268 × 56

19. 4320 × 16

20. 5043 × 24

21. 124 × 204

22. 256 × 402

23. 102 × 108

24. 504 × 203

See Example 4.

Complete Example

4.86 ←―― Two digits right of decimal point
×0.04 ←―― Two digits right of decimal point

        Product has ――――― digits right of decimal point.
                       (1)

Ans. ―――――
       (2)

(1) 4 or four   (2) 0.1944

25. 6.8
     ×0.6

26. 4.7
     ×0.5

27. 0.32
     ×0.4

28. 0.47
     ×0.3

44

2.1 MULTIPLICATION

29.	2.04	30.	3.07	31.	4.012	*32.	5.007
	×0.02		×0.04		×0.03		×0.05

See Example 5.

Complete Example

2.06 ⟵ Two digits right of decimal point
×0.35 ⟵ Two digits right of decimal point
10 30 ⟵ 5 × 206
61 8☐ ⟵ 30 × 2.06

Product has _____ digits right of decimal point.
(1)

Ans. _____
(2)

---

(1) 4 or four   (2) 0.7210 or 0.721

33.	2.5	34.	4.7	35.	4.32	36.	5.07
	×1.3		×3.2		×2.4		×8.4
37.	4.12	38.	6.28	39.	0.032	40.	0.041
	×0.42		×0.31		×0.12		×0.32

Find each product: a. to the nearest tenth. b. To the nearest hundredth. c. To the nearest thousandth. See Example 6.

Complete Example

16.3×0.036

```
   16.3
 × 0.036
   978
   489
  .5868
```

Ans. a. 0.5̲868 ⟶ _____
                        (1)

b. 0.58̲68 ⟶ _____
                        (2)

c. 0.586̲8 ⟶ _____
                        (3)

---

(1) 0.6   (2) 0.59   (3) 0.587

45

WHOLE NUMBERS AND DECIMALS; MULTIPLICATION AND DIVISION

41. 14.9×0.042 **42.** 16.21×3.071 **43.** 0.063×4.9

44. 0.0011×39.8 **45.** $4.2 \times 0.04 \times 2.1$ **46.** $3.07 \times 0.05 \times 3.4$

2.2 SPECIAL PRODUCTS

A simple procedure can be used to find, by inspection, products that involve factors such as 0.1, 0.01, 0.001, and so on. For example, consider the products

$67.5 \times 0.1 = 6.75$, $67.5 \times 0.01 = 0.675$, and $67.5 \times 0.001 = 0.0675$,

and note that in each case the only change between the factor 67.5 and the resulting product is the placement of the decimal point. Results such as these suggest the following rules.

To multiply by	Move the decimal point
0.1	one place to the left
0.01	two places to the left
0.001	three places to the left
etc.	etc.
	Add or drop zeros as needed

Example 1 a. 143.27×0.1 b. 0.975×0.01 c. 702×0.001
    $= 14.327$ $= 0.00975$ $= 0.702$

A procedure similar to the above can be used to find products that involve factors such as 10, 100, 1000, and so on. Consider the products

$5.38 \times 10 = 53.8$, $5.38 \times 100 = 538$, and $5.38 \times 1000 = 5380$,

and note that in each case the only change between the factor 5.38 and the resulting product is the placement of the decimal point. Results such as those suggest the following rules.

To multiply by	Move the decimal point
10	one place to the right
100	two places to the right
1000	three places to the right
etc.	etc.
	Add or drop zeros as needed

Example 2 a. 473×10 b. 58.6×100 c. $0.0274 \times 1000 = 027.4$
    $= 4730$ $= 5860$ $= 27.4$

2.2 SPECIAL PRODUCTS

ESTIMATING PRODUCTS

Consider the product 70 × 600 = 42,000. This product can be obtained mentally. Observe that 70 has *one zero to the right* of the digit 7, whereas 600 has *two zeros to the right* of the digit 6 and 42,000 has *three zeros to the right* of the digits 42. Also, note that 42 is the product of the nonzero digits 7 and 6. Hence,

$$\begin{pmatrix} \text{one} \\ \text{zero} \end{pmatrix} + \begin{pmatrix} \text{two} \\ \text{zeros} \end{pmatrix} = \begin{pmatrix} \text{three} \\ \text{zeros} \end{pmatrix}$$

$$70 \times 600 = 42{,}000$$

This rapid procedure can be used to find products in which the nonzero digits can be obtained mentally.

Example 3 a. 40 × 80   b. 200 × 30 × 40

Solutions a. 4 × 8 = 32. Hence,   b. 2 × 3 × 4 = 24. Hence,

$$\begin{pmatrix} \text{one} \\ \text{zero} \end{pmatrix} + \begin{pmatrix} \text{one} \\ \text{zero} \end{pmatrix} = \begin{pmatrix} \text{two} \\ \text{zeros} \end{pmatrix}$$

$$40 \times 80 = 3200$$

$$\begin{pmatrix} \text{two} \\ \text{zeros} \end{pmatrix} + \begin{pmatrix} \text{one} \\ \text{zero} \end{pmatrix} + \begin{pmatrix} \text{one} \\ \text{zero} \end{pmatrix} = \begin{pmatrix} \text{four} \\ \text{zeros} \end{pmatrix}$$

$$200 \times 30 \times 40 = 240{,}000$$

In Sections 1.3 and 1.4 we used rounded-off numbers to estimate sums and differences. We can estimate products by using rounded-off numbers together with the above technique for finding products mentally.

Example 4 *Estimate* each product.

a. 37 × 81   b. 217 × 26 × 39

Solutions a. First round off each factor to the nearest ten. Then mentally find the product.

$$\begin{pmatrix} \text{one} \\ \text{zero} \end{pmatrix} + \begin{pmatrix} \text{one} \\ \text{zero} \end{pmatrix} = \begin{pmatrix} \text{two} \\ \text{zeros} \end{pmatrix}$$

$$40 \times 80 = 3200$$

The estimated product is 3200.

b. First round off 217 to the nearest hundred and 26 and 39 to the nearest ten. Then mentally find the product.

$$\begin{pmatrix} \text{two} \\ \text{zeros} \end{pmatrix} + \begin{pmatrix} \text{one} \\ \text{zero} \end{pmatrix} + \begin{pmatrix} \text{one} \\ \text{zero} \end{pmatrix} = \begin{pmatrix} \text{four} \\ \text{zeros} \end{pmatrix}$$

$$200 \times 30 \times 40 = 240{,}000$$

The estimated product is 240,000.

Rounded-off numbers may also be used to estimate products involving decimals.

WHOLE NUMBERS AND DECIMALS; MULTIPLICATION AND DIVISION

Example 5   *Estimate* each product.

　　a.　58.79×7　　　　　　　　b.　42.1×76

Solutions　a.　First round off 58.79 to a number with only one nonzero digit: 60. Then mentally find the product.

$$\begin{pmatrix} \text{one} \\ \text{zero} \end{pmatrix} + \begin{pmatrix} \text{no} \\ \text{zeros} \end{pmatrix} = \begin{pmatrix} \text{one} \\ \text{zero} \end{pmatrix}$$

$$60 \;\times\; 7 \;=\; 420$$

The estimated product is 420.

　　b.　First, round off 42.1 and 76 to numbers that involve only one nonzero digit: 42.1 to 40 and 76 to 80. Then mentally find the product.

$$\begin{pmatrix} \text{one} \\ \text{zero} \end{pmatrix} + \begin{pmatrix} \text{one} \\ \text{zero} \end{pmatrix} = \begin{pmatrix} \text{two} \\ \text{zeros} \end{pmatrix}$$

$$40 \;\times\; 80 \;=\; 3200$$

The estimated product is 3200.

EXPONENTIAL NOTATION

Finding the product of two or more *identical* factors, such as

$$6 \times 6 \times 6 \times 6 \times 6$$

is called *raising a number to a power*. The above product can be more concisely represented by the symbol

$$6^5,$$

read "six to the fifth power," where the number 5 indicates how many times the number 6 is used as a factor. Here 6 is called the **base**, and 5 is called the **exponent**. The notation itself is called **exponential notation**.

There are special names, **squared** and **cubed**, for the second and third powers of a number.

Example 6　a.　$6 \times 6 = 6^2$ (read "six to the second power," or "six squared").
　　　　　　b.　$6 \times 6 \times 6 = 6^3$ (read "six to the third power," or "six cubed").
　　　　　　c.　$6 \times 6 \times 6 \times 6 = 6^4$ (read "six to the fourth power").

Example 7　Compute each power.

　　a.　3^2　　　　　　　　　　　　b.　2^5

Solutions　a.　$3^2 = 3 \times 3$　　　　　　b.　$2^5 = 2 \times 2 \times 2 \times 2 \times 2$
　　　　　　　　　　$= 9$　　　　　　　　　　　　　　　$= 32$

2.2 SPECIAL PRODUCTS

Of particular interest are powers with base 10. For example,

$$10^2 = 10 \times 10 \qquad 10^3 = 10 \times 10 \times 10 \qquad 10^4 = 10 \times 10 \times 10 \times 10$$
$$= 100 \qquad \qquad = 1000 \qquad \qquad\quad = 10{,}000$$

Such numbers are called *powers of ten*.

EXERCISES 2.2

Find each product mentally. See Examples 1 and 2.

Complete Examples

a. $25.62 \times 0.01 \longrightarrow 25.62 =$ _____
    (1)

b. $25.62 \times 1000 \longrightarrow 25.620 =$ _____
    (2)

(1) 0.2562   (2) 25,620

1. 89.5×0.01
2. 895×0.1
3. 0.895×0.01
4. 0.895×0.1
5. 0.895×0.001
6. 0.0895×0.001
7. 0.0895×0.1
8. 0.0895×0.01
9. 704×100
10. 7.04×1000
11. 7.04×10
12. 0.704×100
13. 0.704×1000
14. 0.0704×10

Compute each product mentally. See Example 3.

Complete Examples

a. $60 \times 80 =$ ___ ___
    6×8
    1 zero + 1 zero

b. $300 \times 20 \times 70 =$ ___ ___
    $3 \times 2 \times 7$
    2 zeros + 1 zero + 1 zero

Ans. _____
     (1)

Ans. _____
     (2)

(1) 4800   (2) 420,000

15. 50×90
16. 500×400
17. $700 \times 5{,}000$
18. $800 \times 2{,}000$
19. $30 \times 20 \times 50$
20. $300 \times 200 \times 500$

49

WHOLE NUMBERS AND DECIMALS; MULTIPLICATION AND DIVISION

Use rounded-off numbers to estimate each product. See Examples 4 and 5.

Complete Examples

a. 363×28

Round off 363 to the nearest hundred; round off 28 to the nearest ten.

$400 \times \underline{\qquad}$
$\quad\quad\quad\quad$ (1)

This estimated product equals $\underline{\qquad}$.
$\quad\quad\quad\quad\quad\quad\quad\quad\quad\quad\quad\quad$ (2)

b. 48.31×83

Round off 48.31 and 83 to the nearest ten.

$50 \times \underline{\qquad}$
$\quad\quad\quad\quad$ (3)

This estimated product equals $\underline{\qquad}$.
$\quad\quad\quad\quad\quad\quad\quad\quad\quad\quad\quad\quad$ (4)

---

(1) 30   (2) 12,000   (3) 80   (4) 4000

---

21. 11×28
22. 48×19
23. 59×31
24. 215×28
25. 48.33×6
26. 78.6×7
27. 692.07×8
28. 971.85×5
29. 56.03×22
30. 29.98×59
31. 528.73×18
32. 1995×19

Find each power. See Examples 6 and 7.

Complete Examples

a. 4^3

$4^3 = \underbrace{4 \times 4 \times 4}_{\text{Three factors of 4.}}$

$= \underline{\qquad}$
$\quad\quad$ (1)

b. $(3.2)^2$

$(3.2)^2 = \underbrace{3.2 \times 3.2}_{\text{Two factors of 3.2.}}$

$= \underline{\qquad}$
$\quad\quad$ (2)

---

(1) 64   (2) 10.24

---

33. 2^3
34. 3^2
35. 3^3
36. 2^4
37. 4^2
38. 5^2
39. 10^2
40. 10^3
41. 3^5
42. 4^3
43. 7^3
44. 5^3
45. $(1.2)^3$
46. $(2.1)^3$
47. $(12.3)^2$
48. $(21.2)^2$

2.3 DIVISION

The operation of **division** is related to the multiplication operation and is called the *inverse* of multiplication. For example, when we divide 56 by 8, we seek a number whose product with 8 is 56. From our knowledge of the multiplication facts in Table 2.1 on page 39 we obtain 7, called the **quotient** of 56 divided by 8. The number

2.3 DIVISION

56 is called the **dividend,** and 8 is called the **divisor.** The operation of division is designated by the symbol "÷" or by a horizontal bar. For example,

$$\underset{\text{dividend}}{56} \div \underset{\text{divisor}}{8} = \overset{\text{quotient}}{7} \quad \text{or} \quad \frac{\overset{\text{dividend}}{56}}{\underset{\text{divisor}}{8}} = \overset{\text{quotient}}{7}$$

In Section 2.1 we noted that the operation of multiplication is commutative. This property does *not* apply to division. For example, 56 ÷ 8 does not equal 8 ÷ 56.

QUOTIENTS INVOLVING ZERO

It is not possible to divide by zero. To see why this is so, consider an example, say 5 ÷ 0. If there were a quotient Q such that

$$\frac{5}{0} = Q$$

then a check of the result tells us that $0 \times Q$ must be equal to 5. But, for any number Q, $0 \times Q$ equals 0, not 5. Hence, we conclude that $\frac{5}{0}$ does not represent a number. *Division by zero is not possible.* Note, however, that 0 can be divided by any nonzero number to obtain the quotient 0. For example,

$$\frac{0}{5} = 0 \quad \text{because} \quad 5 \times 0 = 0.$$

and

$$\frac{0}{25} = 0 \quad \text{because} \quad 25 \times 0 = 0$$

COMPUTING QUOTIENTS

Quotients of whole numbers and decimals in which one or both of the numbers to be divided is *not* shown in Table 2.1 can be computed by using the method illustrated in the following examples.

Example 1 Divide: $69 \div 3 \quad \left(\frac{69}{3}\right)$.

Solution Since 69 = 6 tens + 9 units, we must divide 6 tens and 9 units each by 3. We begin by dividing first the tens and then the units. The following format is commonly used.

$$3 \overline{\smash{\big)}\, 6\,9}^{\,2} \quad \longleftarrow \begin{array}{l} 6 \div 3 = 2 \text{ tens} \\ \text{tens} \end{array}$$

$$3 \overline{\smash{\big)}\, 6\,9}^{\,2\,3} \quad \longleftarrow \begin{array}{l} 9 \div 3 = 3 \text{ units} \\ \text{units} \end{array}$$

Hence 69 ÷ 3 = 23.

WHOLE NUMBERS AND DECIMALS; MULTIPLICATION AND DIVISION

Example 2 Divide: $273 \div 7$ $\left(\dfrac{273}{7}\right)$.

Solution Here $273 = 2$ hundreds $+ 7$ tens $+ 3$ units, and we are to divide each by 7. However, since the number of hundreds is less than 7, we must consider

$$273 = 27 \text{ tens} + 3 \text{ units}$$

Now the number of tens is greater than 7, and we can proceed.

$$\begin{array}{r} 3 \\ 7\,)\overline{273} \\ 21 \\ \hline 6 \end{array}$$

⟵ Use 3, since 3×7 is less than 27 whereas 4×7 is greater than 27.
⟵ $3 \times 7 = 21$ tens
⟵ $27 - 21 = 6$ tens remaining

$$\begin{array}{r} 3 \\ 7\,)\overline{273} \\ 21\downarrow \\ \hline 63 \end{array}$$

⟵ Bring down the 3, which gives 63 units.

Now divide 63 units by 7.

$$\begin{array}{r} 39 \\ 7\,)\overline{273} \\ 21 \\ \hline 63 \\ 63 \\ \hline 0 \end{array}$$

⟵ $63 \div 7 = 9$
⟵ $9 \times 7 = 63$ units
⟵ $63 - 63 = 0$ units

Hence $273 \div 7 = 39$.

Example 3 Divide: $3302 \div 26$ $\left(\dfrac{3302}{26}\right)$.

Solution Here,

$$3302 = 3 \text{ thousands} + 3 \text{ hundreds} + 0 \text{ tens} + 2 \text{ units}$$

but because we are dividing by 26, we must have a starting number equal to or greater than 26. So we use

$$3302 = 33 \text{ hundreds} + 0 \text{ tens} + 2 \text{ units}$$

$$\begin{array}{r} 1 \\ 26\,)\overline{3302} \\ 26 \\ \hline 7 \end{array}$$

⟵ Use 1, since 1×26 is less than 33 and 2×26 is greater than 33.
⟵ $1 \times 26 = 26$ hundreds
⟵ $33 - 26 = 7$ hundreds

Now bring down the next digit.

$$\begin{array}{r} 1 \\ 26\,)\overline{3302} \\ 26\downarrow \\ \hline 70 \end{array}$$

⟵ 70 tens

2.3 DIVISION

Next, divide 26 into 70.

$$\begin{array}{r} 1\ 2 \\ 2\ 6\ )\overline{3\ 3\ 0\ 2} \\ 2\ 6 \\ \hline 7\ 0 \\ 5\ 2 \\ \hline 1\ 8 \end{array}$$

← Use 2, since 2 × 26 is less than 70 whereas 3 × 26 is greater than 70.

← 2 × 26 = 52 tens

← 70 − 52 = 18 tens

Now bring down the next digit.

$$\begin{array}{r} 1\ 2 \\ 2\ 6\ )\overline{3\ 3\ 0\ 2} \\ 2\ 6 \\ \hline 7\ 0 \\ 5\ 2 \\ \hline 1\ 8\ 2 \end{array}$$

Next, divide 26 into 182.

$$\begin{array}{r} 1\ 2\ 7 \\ 2\ 6\ )\overline{3\ 3\ 0\ 2} \\ 2\ 6 \\ \hline 7\ 0 \\ 5\ 2 \\ \hline 1\ 8\ 2 \\ 1\ 8\ 2 \\ \hline 0 \end{array}$$

← 182 ÷ 26 = 7

← 7 × 26 = 182

Hence 3302 ÷ 26 = 127.

The next example involves a little complication.

Example 4   Divide: $29{,}162 \div 14 \quad \left(\dfrac{29{,}162}{14}\right).$

Solution   Since the first two digits, 29, form a number greater than 14, we begin there and proceed as above. However, when we bring down the next digit, there is a complication

$$\begin{array}{r} 2 \\ 1\ 4\ )\overline{2\ 9\ 1\ 6\ 2} \\ 2\ 8 \\ \hline 1\ 1 \end{array}$$

This is 11 thousands. However, since 11 is less than 14, we cannot divide by 14. This simply means that there are zero thousands in the quotient. Hence, *we place a zero in the quotient and bring down the next digit and divide* 116 *by* 14.

$$\begin{array}{r} 2\ 0\ 8 \\ 1\ 4\ )\overline{2\ 9\ 1\ 6\ 2} \\ 2\ 8 \\ \hline 1\ 1\ 6 \\ 1\ 1\ 2 \\ \hline 4 \end{array}$$

Use 8, since 8 × 14 is less than 116, while 9 × 14 is greater than 116.

← 8 × 14 = 112

← 116 − 112 = 4 tens

(Continued)

WHOLE NUMBERS AND DECIMALS; MULTIPLICATION AND DIVISION

$$
\begin{array}{r}
2\,0\,8\,3 \\
14\overline{)2\,9\,1\,6\,2} \\
2\,8 \\
\overline{1\,1\,6} \\
1\,1\,2 \\
\overline{4\,2} \\
4\,2 \\
\overline{0}
\end{array}
$$
⟵ Divide 42 by 14 (42 ÷ 14 = 3).

⟵ 3 × 14 = 42

Hence 29,162 ÷ 14 = 2,083.

A decimal can be divided by a whole number, as shown in the next example. Simply place a decimal point in the quotient directly above the decimal point as shown, and then divide as if both were whole numbers.

Example 5   Divide: $6.9 \div 3$ $\left(\dfrac{6.9}{3}\right)$.

Solution $3\overline{)6.9}$ ⟵ Place decimal point directly above the decimal point in 6.9.

2.3
$3\overline{)6.9}$ Divide as in the examples above.

Hence 6.9 ÷ 3 = 2.3.

Example 6   Divide: $38 \div 5$ $\left(\dfrac{38}{5}\right)$.

Solution   Since 38 is not exactly divisible by 5, we write 38 as 38.0 and proceed to divide as in the example above.

$$
\begin{array}{r}
7. \\
5\overline{)38.0} \\
35 \\
\overline{3}
\end{array}
$$
⟵ Use 7, since 7 × 5 is less than 3 whereas 8 × 5 is greater than 38.

⟵ (38 − 35 = 3 units).

$$
\begin{array}{r}
7. \\
5\overline{)38.0} \\
35\downarrow \\
\overline{3\,0}
\end{array}
$$
⟵ Bring down 0.

$$
\begin{array}{r}
7.6 \\
5\overline{)38.0} \\
35 \\
\overline{3\,0} \\
3\,0 \\
\overline{0}
\end{array}
$$
⟵ Divide 30 by 5 (30 ÷ 5 = 6).

⟵ 6 × 5 = 30

Hence 38 ÷ 5 = 7.6.

Example 7   Divide: $63.75 \div 15$ $\left(\dfrac{63.75}{15}\right)$.

2.3 DIVISION

Solution

$$\begin{array}{r} 4.25 \\ 15 \overline{\smash{\big)}\ 63.75} \\ \underline{60} \\ 37 \\ \underline{30} \\ 75 \\ \underline{75} \\ 0 \end{array}$$

Hence $63.75 \div 15 = 4.25$.

The above procedure is applicable whenever the divisor is a whole number (not zero). If a decimal is divided by a *decimal*, both numbers must be changed as shown in the next example, so that we divide by a whole number.

Example 8 Divide: $0.0756 \div 0.012$ $\left(\dfrac{0.0756}{0.012}\right)$.

Solution Move *each* decimal point three places to the right so that the divisor 0.012 becomes a whole number. This is equivalent to multiplying each number by 1000 and will result in the same quotient.

$0.012 \overline{\smash{\big)}\ 0.075\,6}$ ——— Move decimal points three places.

Now divide.

$$\begin{array}{r} 6.3 \\ 12 \overline{\smash{\big)}\ 75.6} \\ \underline{72} \\ 36 \\ \underline{36} \\ 0 \end{array}$$

Hence $0.0756 \div 0.012 = 6.3$.

Example 9 Divide: $22.5 \div 6.25$ $\left(\dfrac{22.5}{6.25}\right)$.

Solution First move the decimal points, as in the above example.

$6.25 \overline{\smash{\big)}\ 22.50}$ ⟵——— Insert as many zeros as required to move decimal point two places so that the divisor is a whole number.

The process usually looks like this:

$$\begin{array}{r} 3. \\ 6.25 \overline{\smash{\big)}\ 22.50} \\ \underline{18\ 75} \\ 3\ 75 \end{array}$$

(Continued)

WHOLE NUMBERS AND DECIMALS; MULTIPLICATION AND DIVISION

At this point, the answer is 3 with a remainder of 375. However, when dividing decimals, the quotient is usually written in decimal form, without a remainder. To do this, we must change the form.

$$
\begin{array}{r}
3.6 \\
6.25\,\overline{)22.50\,0} \\
\underline{18\ 75} \\
3\ 75\ 0 \\
\underline{3\ 75\ 0} \\
0
\end{array}
$$

← Add zeros to the right of the decimal point and bring down zero.

Hence $22.5 \div 6.25 = 3.6$.

Not all division problems will have a remainder of zero. Therefore, if a quotient in decimal form is required, it is sometimes necessary to stop the division at a certain point.

Example 10 Divide $23 \div 15.7 \left(\dfrac{23}{15.7}\right)$ and round off the quotient to the nearest tenth.

Solution Since the quotient is to be to the nearest tenth, we will divide to hundredths and then round off.

$$
\begin{array}{r}
1.46 \\
15.7\,\overline{)23.0\,00} \\
\underline{15\ 7} \\
7\ 3\ 0 \\
\underline{6\ 2\ 8} \\
1\ 0\ 20 \\
\underline{9\ 42} \\
78
\end{array}
$$

Hence, to the nearest tenth, $23 \div 15.7$ equals 1.5.

CHECKING DIVISION

Because division is the inverse operation of multiplication, a division operation can be checked by multiplication. For example,

Computation: $4171 \div 97 = 43$
Check: $97 \times 43 = 4171$

Or, in vertical form,

Computation: $\dfrac{4171}{97} = 43$ Check: $97 \times 43 = 4171$

Of course, a check should be made using a quotient before it has been rounded off.

2.3 DIVISION

EXERCISES 2.3
Compute mentally. Refer to Table 2.1 if necessary.

	a.	b.	c.	d.	e.	f.	g.
1.	$\dfrac{8}{2}$	$\dfrac{9}{3}$	$\dfrac{4}{2}$	$\dfrac{7}{7}$	$\dfrac{8}{4}$	$\dfrac{16}{0}$	$\dfrac{0}{16}$
2.	$\dfrac{0}{36}$	$\dfrac{10}{5}$	$\dfrac{15}{15}$	$\dfrac{27}{9}$	$\dfrac{36}{6}$	$\dfrac{24}{0}$	$\dfrac{9}{9}$
3.	$\dfrac{40}{10}$	$\dfrac{60}{10}$	$\dfrac{80}{10}$	$\dfrac{80}{20}$	$\dfrac{0}{20}$	$\dfrac{20}{0}$	$\dfrac{7}{7}$
4.	$\dfrac{42}{6}$	$\dfrac{42}{0}$	$\dfrac{42}{7}$	$\dfrac{0}{7}$	$\dfrac{48}{8}$	$\dfrac{45}{9}$	$\dfrac{45}{5}$

Find each quotient in Exercises 5–36. See Examples 1 and 2.

Complete Examples

a. $64 \div 4$

$$4\overline{)64}$$

Use 1, since 1×4 is less than 6 whereas 2×4 is greater than 6.

$4 \leftarrow 1 \times 4 = 4$ tens
$2 \leftarrow 6 - 4 = 2$ tens

$24 \div 4 = \underline{}$ units **(1)**

$$4\overline{)64}$$
$$\underline{4}\downarrow$$
$$24$$
$$\underline{24} \leftarrow 6 \times 4 = 24 \text{ units}$$
$$0 \leftarrow 24 - 24 = 0 \text{ units}$$

Ans. _____
   (2)

b. $370 \div 5$

$$5\overline{)370}$$

Use 7, since 7×5 is less than 37 whereas 8×5 is greater than 37.

$35 \leftarrow 7 \times 5 = 35$ tens
$2 \leftarrow 37 - 35 = 2$ tens

$20 \div 5 = \underline{}$ units **(3)**

$$5\overline{)370}$$
$$\underline{35}\downarrow$$
$$20$$
$$\underline{20} \leftarrow 4 \times 5 = 20 \text{ units}$$
$$0 \leftarrow 20 - 20 = 0 \text{ units}$$

Ans. _____
   (4)

---

(1) 6   (2) 16   (3) 4   (4) 74

---

5. $48 \div 4$ 6. $96 \div 3$ 7. $682 \div 2$ 8. $848 \div 4$
9. $468 \div 3$ 10. $864 \div 4$ 11. $575 \div 5$ 12. $496 \div 4$

WHOLE NUMBERS AND DECIMALS; MULTIPLICATION AND DIVISION

See Examples 3 and 4.

Complete Example

$$\frac{4030}{26}$$

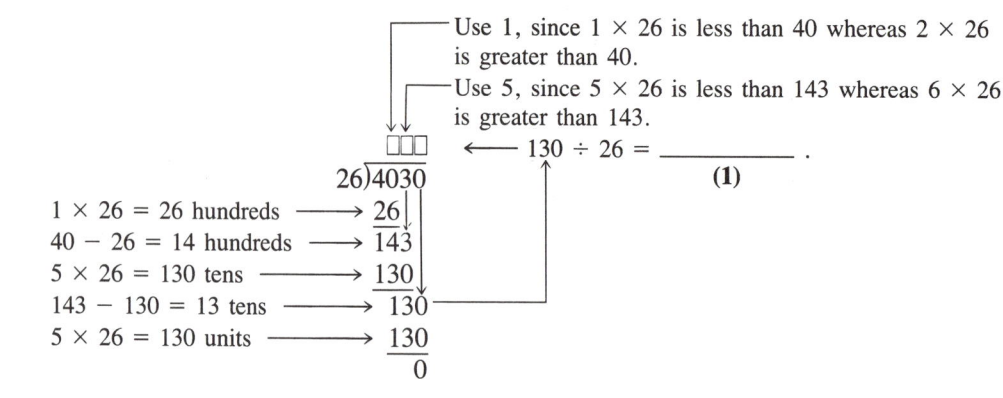

Use 1, since 1 × 26 is less than 40 whereas 2 × 26 is greater than 40.
Use 5, since 5 × 26 is less than 143 whereas 6 × 26 is greater than 143.

```
         □□□
    26)4030         ← 130 ÷ 26 = _____ .
1 × 26 = 26 hundreds  →  26                    (1)
40 − 26 = 14 hundreds →  143
5 × 26 = 130 tens     →  130
143 − 130 = 13 tens   →  130
5 × 26 = 130 units    →  130
                          0
```

Ans. _____
(2)

---

(1) 5   (2) 155

---

13. $\dfrac{377}{29}$ 14. $\dfrac{294}{21}$ 15. $\dfrac{2520}{15}$ 16. $\dfrac{2916}{12}$

17. $\dfrac{10{,}164}{84}$ 18. $\dfrac{12{,}482}{79}$ 19. $\dfrac{14{,}040}{312}$ 20. $\dfrac{27{,}495}{423}$

See Examples 5, 6, and 7.

Complete Example

$$\frac{42}{8}$$

Since 42 is not divisible by 8, write 42 as 42.00.

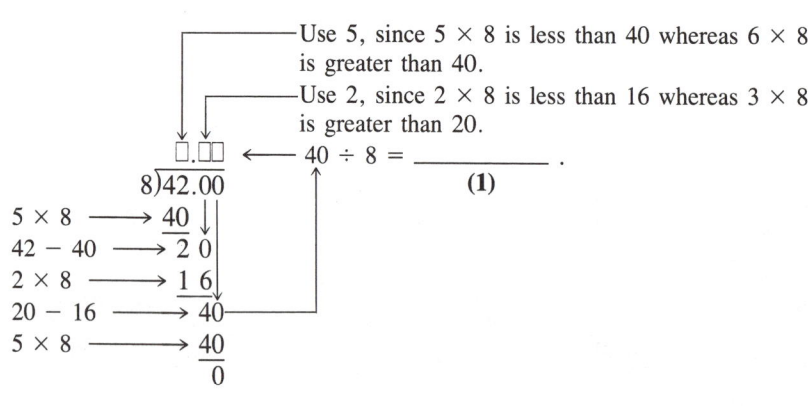

Use 5, since 5 × 8 is less than 40 whereas 6 × 8 is greater than 40.
Use 2, since 2 × 8 is less than 16 whereas 3 × 8 is greater than 20.

```
           □.□□       ← 40 ÷ 8 = _____ .
       8)42.00                    (1)
5 × 8    →  40
42 − 40  →   2 0
2 × 8    →   1 6
20 − 16  →    40
5 × 8    →    40
               0
```

Ans. _____
(2)

---

(1) 5   (2) 5.25

58

2.3 DIVISION

21. $\dfrac{9.2}{4}$ 22. $\dfrac{5.4}{3}$ 23. $\dfrac{31.57}{7}$ 24. $\dfrac{50.58}{9}$

25. $\dfrac{67.5}{15}$ 26. $\dfrac{97.2}{18}$ 27. $\dfrac{394.88}{32}$ 28. $\dfrac{1790.95}{43}$

See Examples 8 and 9.

Complete Example

$55.728 \div 3.24$ or $\dfrac{55.728}{3.24}$

Write in the form $3.24\overline{)55.728}$; move each decimal point two places and divide.

```
                                    ┌─ Use 1, since 1 × 324 is less than 557
                                    │  whereas 2 × 324 is greater than 557.
                                    │  ┌─ Use 7, since 7 × 324 is less than
                                    │  │  2332, whereas 8 × 324 is greater than 2332.
                        □□.□ ←────── 648 ÷ 324 = _____ .
                   3.24)55.72 8                    (1)
    1 × 324  ───→     32 4
    557 − 324 ───→    23 32
    7 × 324  ───→     22 68
    2332 − 2268 ──→      64 8
    2 × 324  ───→        64 8
                            0
```

Ans. _____
     (2)

(1) 2   (2) 17.2

29. $6.85 \div 0.5$ 30. $8.64 \div 0.4$ 31. $71.91 \div 4.7$ 32. $593.14 \div 9.4$

33. $229.9 \div 0.38$ 34. $374.4 \div 0.78$ 35. $4.3776 \div 0.019$ 36. $10.9752 \div 0.024$

Find each quotient: a. To the nearest tenth. b. To the nearest hundredth. See Example 10.

Complete Example

$39.6 \div 1.7$ or $\dfrac{39.6}{1.7}$

(Continued)

59

WHOLE NUMBERS AND DECIMALS; MULTIPLICATION AND DIVISION

Divide to thousandths and then round off.

```
        2 3.294
1.7)3 9.6 000
    3 4
    ─────
      5 6
      5 1
      ─────
        50
        34
        ─────
        160
        153
        ─────
          70
          68
          ─────
           2
```

Ans. a. 23.294 to the nearest tenth is _____ .
  (1)

b. 23.294 to the nearest hundredth is _____ .
  (2)

(1) 23.3   (2) 23.29

37. $23.1 \div 6.1$

38. $14.9 \div 4.2$

39. $0.23 \div 0.7$

40. $0.59 \div 0.9$

41. $\dfrac{14.25}{0.61}$

42. $\dfrac{23.46}{0.72}$

2.4 SPECIAL QUOTIENTS

A simple procedure can be used to divide a number by 10 or by powers of 10. Consider the quotients

$814.6 \div 10 = 81.46$, $814.6 \div 100 = 8.146$,   and $814.6 \div 1000 = 0.8146$

and note that in each case, the only change between the dividend 814.6 and the quotient is the placement of the decimal point. Results such as these suggest the following rules.

To divide by	Move the decimal point
10	one place to the left
100	two places to the left
1000	three places to the left
etc.	etc.
	Add or drop zeros as needed.

Example 1 a. $41.7 \div 10 = 4.17$ or $\dfrac{41.7}{10} = 4.17$

b. $0.79 \div 100 = 0.0079$ or $\dfrac{0.79}{100} = 0.0079$

60

2.4 SPECIAL QUOTIENTS

c. $2815 \div 1000 = 2815. \div 1000$
$= 2.815$, or $\frac{2815}{1000} = 2.815$

QUOTIENTS EQUAL TO ONE

If any nonzero number is divided by itself, the quotient is always 1. For example,

$$\frac{5}{5} = 1 \quad \text{and} \quad \frac{138}{138} = 1$$

Now, recall from Section 2.1 that the product of any number and 1 is that same number, and consider the expression

$$\frac{5}{5} \times N \qquad (1)$$

where N represents *any* number. Because $\frac{5}{5}$ is equal to one and because $1 \times N$ is equal to N, we have that

$$\frac{5}{5} \times N = 1 \times N = N$$

Expression (1) may sometimes appear in such alternate forms as

$$\frac{5 \times N}{5} \quad \text{or} \quad 5 \times \frac{N}{5} \quad \text{or} \quad \frac{N \times 5}{5} \qquad (2)$$

Each of these expressions equals N. These results also follow from the fact that multiplication and division are inverse operations and therefore "undo" each other. In general we have the following rule.

> **If a given number is multiplied and divided by a nonzero number, the result is the given number.**

Example 2 Do each computation *mentally*.

a. $\frac{18}{18} \times 123$ b. $\frac{18 \times 123}{18}$ c. $18 \times \frac{123}{18}$

Solutions In each case the result is 123.

ESTIMATING QUOTIENTS

In section 2.1 we computed certain products mentally by a technique involving "adding" a number of zeros. A similar technique involving "subtracting" a number of zeros can be used to compute whole number quotients.

WHOLE NUMBERS AND DECIMALS; MULTIPLICATION AND DIVISION

Example 3 Do each computation *mentally*.

a. $400 \div 20$ b. $16,000 \div 40$

Solutions a. $4 \div 2 = 2$. Hence, b. $16 \div 4 = 4$. Hence,

$$\binom{\text{two}}{\text{zeros}} - \binom{\text{one}}{\text{zero}} = \binom{\text{one}}{\text{zero}}$$

$$400 \div 20 = 20$$

$$\binom{\text{three}}{\text{zeros}} - \binom{\text{one}}{\text{zero}} = \binom{\text{two}}{\text{zeros}}$$

$$16,000 \div 40 = 400$$

We can estimate quotients by using rounded-off numbers together with the above technique of mentally finding quotients.

Example 4 *Estimate* each quotient.

a. $437 \div 23$ b. $17,600 \div 28$

Solutions a. We first round off 437 to 400 and 23 to 20. Then,

$$400 \div 20 = 20$$

The estimated quotient is 20.

b. We first round off 17,600 to 18,000 and 28 to 30. Then,

$$18,000 \div 30 = 600$$

The estimated quotient is 600.

EXERCISES 2.4

Find each quotient by inspection. See Example 1.

Complete Examples

a. $24.2 \div 10$
   $2\,4.2 \div 10$
   Ans. _____
   (1)

b. $24.2 \div 100$
   $24.2 \div 100$
   Ans. _____
   (2)

c. $\dfrac{24.2}{1000}$

   Ans. _____
   (3)

(1) 2.42 (2) 0.242 (3) 0.0242

1. $407 \div 10$
2. $407 \div 100$
3. $407 \div 1000$
4. $38.6 \div 10$
5. $38.6 \div 100$
6. $38.6 \div 1000$
7. $\dfrac{0.53}{10}$
8. $\dfrac{0.53}{100}$

62

2.4 SPECIAL QUOTIENTS

9. $\dfrac{0.53}{1000}$
10. $\dfrac{7099}{10}$
11. $\dfrac{7099}{100}$
12. $\dfrac{7099}{1000}$

Find each quotient by inspection. See Example 2.

13. $\dfrac{6 \times 18}{6}$
14. $8 \times \dfrac{95}{8}$
15. $\dfrac{143 \times 11}{11}$
16. $68 \times \dfrac{12}{12}$

17. $\dfrac{57 \times 33}{33}$
18. $235 \times \dfrac{94}{94}$
19. $26 \times \dfrac{409}{26}$
20. $\dfrac{47 \times 963}{47}$

Find each quotient by inspection. See Example 3.

Complete Examples

a. $15{,}000 \div 30$

$15{,}000 \div 30 =$ ___
$15 \div 3$
3 zeros − 1 zero = ___ zeros
(1)

Ans. ___
(2)

b. $2400 \div 60$

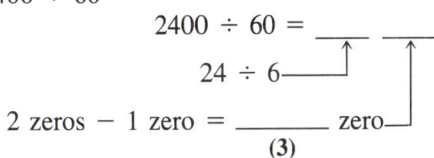

$2400 \div 60 =$ ___
$24 \div 6$
2 zeros − 1 zero = ___ zero
(3)

Ans. ___
(4)

(1) 2 or two   (2) 500   (3) 1 or one   (4) 40

21. $600 \div 30$
22. $800 \div 40$
23. $5000 \div 50$
24. $1800 \div 90$
25. $18{,}000 \div 90$
26. $18{,}000 \div 90$

Estimate each quotient. See Example 4.

Complete Example

a. $12{,}398 \div 372$

Round off 12,398 to 12,000. Round off 372 to 400.

$12{,}000 \div 400 =$ ___
$12 \div 4$
3 zeros − 2 zeros = ___ zero
(1)

Ans. ___
(2)

b. $\$8035.19 \div 23$

Round off 8035.19 to 8000. Round off 23 to 20.

$8000 \div 20 =$ ___
$8 \div 2$
3 zeros − 1 zero = ___ zeros
(3)

Ans. ___
(4)

(1) 1 or one   (2) 30   (3) 2 or two   (4) $400

63

WHOLE NUMBERS AND DECIMALS; MULTIPLICATION AND DIVISION

27. 798 ÷ 38
28. 567 ÷ 27
29. 936 ÷ 32
30. 5264 ÷ 52
31. 3896 ÷ 16
32. 15,694 ÷ 826
33. $56.12 ÷ 6
34. $77.94 ÷ 8
35. $42.20 ÷ 5
36. $855.31 ÷ 9
37. $757.19 ÷ 4
38. $757.19 ÷ 2
39. $604.38 ÷ 10
40. $604.28 ÷ 20
41. $604.90 ÷ 30
42. $3912.90 ÷ 18
43. $3912.90 ÷ 12
44. $3912.90 ÷ 79

2.5 UNIT-PRODUCT RULE; UNIT-QUOTIENT RULE

In this section we introduce two rules that are useful for solving certain types of problems.

UNIT-PRODUCT RULE

Consider the problem

"If one pencil costs 8¢, how much do 12 pencils cost?"

To solve the problem, we may correctly reason:

"If one pencil costs 8¢, then 12 pencils cost 12 times as much."

We then compute the product

(number of pencils) × (cost per pencil) = (cost of 12 pencils)
12 × 8 = 96

Hence, 12 pencils cost 96¢. Note that in this problem we are given the cost of *one pencil* and asked for the cost of *more than one pencil*.

As another example, consider the problem

"If a recipe calls for four cups of flour for one loaf of bread, how many cups of flour are needed for 15 loaves?"

To solve this problem, we have

(number of loaves) × (cups per loaf) = (total cups of flour)
15 × 4 = 60

Hence, 15 loaves require 60 cups of flour. In this problem we are given the flour needed for *one loaf* and asked for the amount of flour needed for *more than one loaf*.

While the above problems are not difficult, their solutions show a pattern that suggests the following rule, which can be used to solve more complicated problems.

64

2.5 UNIT-PRODUCT RULE; UNIT-QUOTIENT RULE

> **UNIT-PRODUCT RULE**
> If a given number is associated with one unit, the number associated with more than one such unit is equal to the product
>
> (number of units) × (number associated with one unit)

The unit referred to in the unit-product rule can be one pencil, one cup, one mile, one gallon, one dollar, and so on.

Example 1 If an automobile can travel 18 miles per gallon of gasoline, how far can it travel on 29 gallons?

Solution
1. Express in words what you want to find, and represent it by a letter.

   Miles the car can travel on 29 gallons: M

2. Write an equation that represents the condition on M.

   The phrase "18 miles per gallon" means "18 miles on one gallon." Thus, 18 is the number associated with *one unit* (1 gallon). Because the number M is associated with *more than one unit* (29 gallons), by the unit-product rule:

   (number of miles) = (number of gallons) × (miles per gallon)
   M = 29 × 18

3. Solve the equation by a direct computation.

   $$M = 29 \times 18 = 522$$

4. The car can travel 522 miles on 29 gallons of gasoline.

UNIT-QUOTIENT RULE

We use the unit-product rule to solve problems in which we are given a number associated with *one unit* and asked to find the number associated with *more than one unit*. We now consider a problem in which we are given a number associated with *more than one unit* and asked to find the number associated with *one unit*.

"If 8 oranges cost 96¢, find the cost of one orange."

To solve this problem, we may correctly reason:

"If 96¢ is the cost of 8 oranges, then we must separate 96¢ into eight equal parts to find the cost of one orange."

Then,

$$96 \div 8 = \frac{96}{8} = 12$$

WHOLE NUMBERS AND DECIMALS; MULTIPLICATION AND DIVISION

Thus, the cost of one orange is 12¢. Note that in this problem we are given the cost of *more than one orange* and asked to find the cost of *one orange*. Note also that the answer is obtained by computing the quotient:

(total cost of 12 oranges) ÷ (number of oranges)
96 ÷ 8

As another example, consider the problem

"If an automatic typewriter types 720 lines in nine minutes, how many lines can it type in one minute?"

To solve this problem, we may reason:

"If 720 lines can be typed in nine minutes, then we must separate 720 into nine equal parts in order to find how many lines can be typed in one minute."

Then,

$$720 \div 9 = \frac{720}{9} = 80$$

Thus, 80 lines can be typed in one minute. In this problem we are given the number of lines typed in *more than one minute* and asked to find the number of lines typed in *one minute*. Note that the answer is obtained by computing the quotient:

(total number of lines typed) ÷ (number of minutes)
720 ÷ 9

These examples suggest the following rule.

UNIT-QUOTIENT RULE
If a given number is associated with more than one unit, the number associated with one such unit is equal to the quotient

$$\frac{\text{number associated with more than one unit}}{\text{number of units}}$$

When using the unit-quotient rule we must be careful to divide by the correct divisor. Many word problems that can be solved by applying the above rule contain phrases that use the word "per." As a guide, you may find it helpful to remember that the word that follows "per" indicates the divisor. For example,

"miles per hour"   indicates   "miles divided by hours"
"miles per gallon" indicates   "miles divided by gallons"
"dollars per pound" indicates   "dollars divided by pounds"

Example 2 If an airplane flies 2425 miles in five hours, find the speed of the airplane in miles per hour.

2.5 UNIT-PRODUCT RULE; UNIT-QUOTIENT RULE

Solution

1. Express in words what you want to find, and represent it by a letter.

   Speed of the airplane in miles per hour: S

2. Write an equation that represents the condition on S.

   The number 2425 is associated with *more than one unit* (five hours). Because the number S is associated with *one unit* (one hour), the unit-quotient rule is used. The phrase "miles per hour" suggests "miles divided by hours."

   (miles per hour) = (miles in five hours) ÷ (number of hours)
   $$S = 2425 \div 5$$

3. Solve the equation by a direct computation.

   $$S = 2425 \div 5 = 485$$

4. The airplane flies 485 miles per hour.

   The above examples were solved by using either the unit-product rule or the unit-quotient rule. Some problems are solved by using both rules.

Example 3 A chemical supply company sells 6.4 pounds of a floor cleaning powder for $8.80. Find the cost of 18.75 pounds of powder.

Solution First we find the unit cost—the cost of one pound of powder—and then use this result to find the cost of 18.75 pounds of powder.

1. Let C represent the cost per pound. The number 8.80 is assocaited with 6.4 pounds; the number C is associated with 1 pound. Hence, the unit-quotient rule is used. The phrase "cost per pound" suggests "dollars divided by pounds." Thus,

   $$C = \frac{8.80}{6.4} = 1.375$$

2. Let T represent the cost of 18.75 pounds. The number 1.375 (from Step 1) is associated with 1 pound; the number T is associated with 18.75 pounds. By the unit-product rule,

   $$T = 18.75 \times 1.375 = 25.78125$$

   The cost of 18.75 pounds is $25.78 to the nearest cent.

EXERCISES 2.5

For Problems 1-8: a. Write an equation. b. Solve the equation. See Example 1.

Complete Example

If a person buying an automobile pays $127 per month, how much will she pay over a period of 12 months?

Amount paid in 12 months: A

(Continued)

WHOLE NUMBERS AND DECIMALS; MULTIPLICATION AND DIVISION

a. Use the unit product rule.

$$(\text{amount}) = \begin{pmatrix}\text{number} \\ \text{of months}\end{pmatrix} \times \begin{pmatrix}\text{dollars} \\ \text{per month}\end{pmatrix}$$

Ans. A = _____ × _____
         (1)   (2)

b. Solve the equation by direct computation.

$A = 12 \times 127 =$ _____
                              (3)

Ans. She pays $ _____ in twelve months.
                    (3)

---

(1) 12   (2) 127   (3) $1524

---

1. If an automobile can travel 19 miles per gallon of gasoline, how far can it travel on 23 gallons?

2. If a motorcycle can travel 67 miles per gallon of gasoline, how far can it travel on 4 gallons?

3. If a typist can type 58 words per minute, how many words can he type in 45 minutes?

4. If a machine can address 385 letters per hour, how many letters can it address in 8 hours?

5. If a computer's high-speed printer can type 125 lines per minute, how many lines can it type in 35 minutes?

6. A fast-food restaurant is able to serve 135 customers per hour. How many customers can it serve in 18 hours?

7. If a man pays $175 per month for an apartment, how much does he pay over a period of 12 months?

8. If a woman spends $18 per week for transportation to and from her job, how much does she spend in 26 weeks?

In Exercises 9–12 round off to the nearest cent.

9. Find the total cost for each item.

Description	Quantity	Cost (each)	Total cost
Chair	24	$ 19.95	_____
Table	16	$ 39.75	_____
Couch	12	$326.30	_____
Television	36	$349.27	_____

10. Find the total pay of each employee.

Employee	Hourly rate	Hours worked	Total pay
Eaton, A.	$7.52	38.75	_____
Frye, B.	$6.39	52.5	_____
Gold, C.	$8.73	61.6	_____
Hardy, G.	$5.97	19.25	_____

11. The following table lists the results of a study of average costs per month at four different colleges. Find the total cost at each college for a *nine*-month school year.

College	Tuition and fees	Room and board	Books and supplies	Personal expenses	Cost (9 mos.)
A	$653.33	$347.22	$ 86.67	$300.60	_____
B	829.44	450.00	107.44	345.56	_____
C	374.44	362.22	80.00	277.78	_____
D	218.89	333.33	76.67	272.22	_____

12. The following table lists average utility costs per month for each of four families. Find the total cost for each family for *four* months.

Family	Water	Electricity	Telephone	Gas	Cost (4 mos.)
A	$19.75	$34.53	$29.42	$61.27	_____
B	19.42	31.19	26.38	70.68	_____
C	16.37	28.44	38.73	94.76	_____
D	17.04	40.39	34.25	82.18	_____

2.5 UNIT-PRODUCT RULE; UNIT-QUOTIENT RULE

Each of the following items sells for the price given. To the nearest cent, find either the cost per ounce or the cost per pound. See Example 2.

Complete Example

An 8-ounce bottle of ketchup sells for 73¢.

Cost per ounce: C

Use the unit-quotient rule.

$$\begin{pmatrix} \text{cost per} \\ \text{ounce} \end{pmatrix} = \begin{pmatrix} \text{total cost of} \\ \text{8 ounces} \end{pmatrix} \div \begin{pmatrix} \text{number of} \\ \text{ounces} \end{pmatrix}$$

$$C = \underline{}_{(1)} \div \underline{}_{(2)}$$

Solve the equation by direct computation.

$$C = 73 \div 8 = \underline{}_{(3)}$$

Ans. To the nearest cent, the cost per ounce is $\underline{}_{(4)}$ ¢.

---

(1) 73   (2) 8   (3) 9.125   (4) 9¢

---

13. 10 ounces of frozen spinach for 35¢.

14. 24-ounce bottle of syrup for 99¢.

15. 25 pounds of dog food for $4.67.

16. 20 pounds of plant food for $1.77.

17. 50 pounds of salt for $1.17.

18. 20 pounds of charcoal for $2.19.

19. The chart gives the number of months for which various automobile batteries are guaranteed to operate, and the cost of each battery. Find the cost per month of each battery.

	A	B	C	D
Months	24	36	48	60
Cost	$17.88	$19.88	$23.88	$31.88
Cost per mo.				

20. The results of mileage tests on four different cars are listed. Find the results in miles per gallon for each car.

	A	B	C	D
Distance (miles)	9613	11,806	14,275	18,431
Gasoline (gal)	490.5	524.7	1089.6	675.2
Miles per gal				

69

WHOLE NUMBERS AND DECIMALS; MULTIPLICATION AND DIVISION

For Exercises 21–26, See Example 3.

Complete Example
A soft drink company sells 24 ounces of its soda for $1.02. Find the cost of 50.5 ounces of soda.

Step 1

Cost of one ounce of soda: C.

Use the unit-quotient rule.

$$C = 1.02 \div 24 = \underline{\qquad}$$
(1)

Step 2

Cost of 50.5 ounces of soda: T.

Use the unit-product rule.

$$T = 50.5 \times 0.0425 = \underline{\qquad}$$
(2)
(From Step 1)

Ans. The cost of 50.5 ounces, to the nearest cent, is $\underline{\qquad}$.
(3)

---

(1) 0.0425   (2) 2.14625   (3) $2.15

---

21. A 12.5-pound package of laundry detergent sells for $2.33. To the nearest cent, find the cost of a. 27.8 pounds. b. 142.5 pounds.

22. To the nearest tenth of a pound, find how much of the detergent of Exercise 21 can be bought for a. $29.95. b. $85.50.

23. If an automobile runs for 521.6 miles on 27.5 gallons of gasoline, how far (to the nearest tenth of a mile) will it run on a. 35.4 gallons b. 136.8 gallons?

24. If 45 floor tiles cost $16.65, find the cost of a. 260 floor tiles. b. 580 floor tiles.

25. If 3 yards of shelf lining costs $1, find the cost of a. 5 yards b. 31 yards.

26. If 4 ounces of yarn costs 59¢, find the cost of a. 65 ounces. b. 38 ounces.

Exercises 27–34 involve either the unit-product rule or the unit-quotient rule.

27. A compositor can set 175 pieces of type per minute. At this rate, how many pieces can he set in 25 minutes?

28. A gas station services 37 cars per hour. How many cars can it service in 18 hours?

29. A computer's high-speed printer can print 3300 lines in 25 minutes. How many lines can it print per minute?

30. One of the fastest printers made can print 30,000 words in 60 seconds. How many words per second is that?

31. A woman pays $87.50 per month for a computer. How much does she pay over a period of 24 months?

32. A man pays $9 per week for parking privileges. How much is his yearly cost (52 weeks)?

33. If 704 yards of wire weigh 16 pounds, how many yards weigh 1 pound?

34. One of the fastest rates of selling gasoline was 3116 gallons sold from a single pump in 24 hours. To the nearest tenth, how many gallons per hour was this?

In a large Western city, residential users pay $0.032 for each kilowatt-hour (kwh) of electricity for the first 150 kwh; $0.0214 per kwh for the next 250 kwh; $0.0189 for the next 600 kwh; $0.0169 for each additional kwh. Compute the cost for each of the following amounts of electricity.

Complete Example

$$1050 \text{ kwh}$$

Total cost: C

Use the unit-product rule.

Cost of first 150 kwh: $150 \times \$0.032 = \$4.80$

Cost of next 250 kwh: $250 \times \$0.0214 = \$5.35$

Cost of next 600 kwh: $600 \times \$0.0189 = \$11.34$

Find number of kwh that remain.

$$1050 - 150 - 250 - 600 = \underline{}_{(1)}$$

Cost of last 50 kwh: $50 \times \$0.0169 = 0.845$

Find the total cost.

$$C = 4.80 + 5.35 + 11.34 + 0.845 = \underline{}_{(2)}$$

Ans. To the nearest cent, the cost is $\underline{}_{(3)}$.

(1) 50   (2) 22.335   (3) $22.34

35. 1126 kwh
36. 1295 kwh
37. 1652 kwh
38. 1800 kwh
39. 2000 kwh
40. 2140 kwh

2.6 SOLVING EQUATIONS

Recall from Section 1.6 that the addition-subtraction rule can be used to change an equation to an equivalent equation (the solutions are the same). The following rule also enables us to change an equation into an equivalent equation.

If each side of an equation is multiplied by, or divided by, the same nonzero number, the resulting equation is equivalent to the original equation.

71

WHOLE NUMBERS AND DECIMALS; MULTIPLICATION AND DIVISION

Example 1 Solve: $\dfrac{N}{5} = 15$.

Solution First, applying the multiplication-division rule and multiplying each side of the equation by 5 to "undo" the division in the left side, we obtain

$$5 \times \dfrac{N}{5} = 15 \times 5$$

Because $5 \times \dfrac{N}{5}$ equals N,

$$N = 15 \times 5$$
$$N = 75$$

The solution is 75. We can check by substituting 75 for N in the given equation and verifying that $\dfrac{75}{5} = 15$.

Example 2 Solve: $27 \times N = 918$.

Solution First, applying the multiplication-division rule and dividing each side of the equation by 27 to "undo" the multiplication in the left side, we obtain

$$\dfrac{27 \times N}{27} = \dfrac{918}{27}$$

Because $\dfrac{27 \times N}{27}$ equals N, we have

$$N = \dfrac{918}{27}$$
$$N = 34$$

The solution is 34. We can check by substituting 34 for N in the given equation and verifying that $27 \times 34 = 918$.

Example 3 Solve: $600 = 24 \times R \times 5$.

Solution First, we simplify the right side of the equation. We write

$$600 = R \times \underbrace{24 \times 5}$$

$$600 = R \times 120$$

Next, dividing each side of the equation by 120, we have

$$\frac{600}{120} = \frac{R \times 120}{120}$$

Because $\frac{R \times 120}{120}$ equals R,

$$\frac{600}{120} = R$$

$$5 = R$$

The solution is 5. We can check by substituting 5 for R in the given equation and verifying that $600 = 24 \times 5 \times 5$.

Sometimes we have to multiply and then divide both sides of an equation in order to obtain an equivalent equation in which the unknown appears by itself on only one side of the equation.

Example 4 Solve: $\frac{3 \times P}{4} = 75$.

Solution Multiplying each side of the equation by 4, we have

$$4 \times \frac{3 \times P}{4} = 75 \times 4$$

Because $4 \times \frac{3 \times P}{4}$ is equal to $3 \times P$,

Next, dividing each side of the equation by 3, we obtain

$$\frac{3 \times P}{3} = \frac{75 \times 4}{3}$$

Because $\frac{3 \times P}{3}$ is equal to P,

$$P = \frac{75 \times 4}{3}$$

$$P = 100$$

The solution is 100. We can check by substituting 100 for P in the original equation and verifying that $\frac{3 \times 100}{4} = 75$.

APPLICATIONS

Equations such as those illustrated in the above examples can be used to solve a variety of problems. You may want to review the list of steps suggested in Section 1.6 as a guide for solving the word problems in the exercises.

WHOLE NUMBERS AND DECIMALS; MULTIPLICATION AND DIVISION

EXERCISES 2.6
Solve each equation; check your answers. See Example 1.

Complete Example

$$\frac{N}{2.7} = 0.23$$

Multiply each side of the equation by _____(1)_____ .

$$2.7 \times \frac{N}{2.7} = 0.23 \times 2.7$$

$$N = \underline{}_{(2)}$$

Check: $\frac{0.621}{2.7} \stackrel{?}{=} 0.23$

$0.23 = 0.23$

(1) 2.7   (2) 0.621

1. $\frac{N}{8} = 13$
2. $\frac{N}{7} = 15$
3. $\frac{N}{1.6} = 4$
4. $\frac{N}{2.3} = 7$
5. $\frac{N}{3} = 21.2$
6. $\frac{N}{4} = 17.5$
7. $2.4 = \frac{N}{1.3}$
8. $1.2 = \frac{N}{3.5}$
9. $0.14 = \frac{N}{3.2}$
10. $0.21 = \frac{N}{5.3}$
11. $12.5 = \frac{N}{3.7}$
12. $15.2 = \frac{N}{4.8}$

Solve each equation; check your answers. See Example 2.

Complete Example

$$45 \times N = 153$$

Divide both sides of the equation by _____(1)_____ .

$$\frac{45 \times N}{45} = \frac{153}{45}$$

$$N = \underline{}_{(2)}$$

Check: $45 \times 3.4 \stackrel{?}{=} 153$

$153 = 153$

(1) 45   (2) 3.4

13. $5 \times N = 175$
14. $4 \times N = 644$
15. $N \times 33 = 1683$
16. $N \times 56 = 3080$
17. $71 \times N = 198.8$
18. $53 \times N = 206.7$
19. $6.11 = 1.3 \times N$
20. $14.31 = 2.7 \times N$

2.6 SOLVING EQUATIONS

21. $1.371 = 0.03 \times N$ **22.** $40.88 = 0.7 \times N$

Solve each equation; check your answers. See Example 3.

Complete Example

$$7 \times N \times 2.4 = 114.24$$

Simplify the left side of the equation.

$N \times \underbrace{7 \times 2.4} = 114.24$

$N \times \underline{\quad\quad} = 114.24$
      (1)

Divide each side of the equation by $\underline{\quad\quad}$.
                          (2)

$\dfrac{N \times 16.8}{16.8} = \dfrac{114.24}{16.8}$

$N = \underline{\quad\quad}$
     (3)

Check: $7 \times 6.8 \times 2.4 \stackrel{?}{=} 114.24$
                $114.24 = 114.24$

(1) 16.8   (2) 16.8   (3) 6.8

23. $9 \times R \times 3 = 972$ **24.** $5 \times R \times 9 = 765$ **25.** $4 \times T \times 1.2 = 30.24$

26. $6 \times T \times 5.3 = 130.38$ **27.** $59.878 = 1.3 \times N \times 4.7$ **28.** $22.464 = 2.4 \times N \times 5.2$

Solve each equation; check your answers. See Example 4.

Complete Example

$$\dfrac{18 \times P}{30} = 30.6$$

Multiply each side of the equation by $\underline{\quad\quad}$.
                             (1)

$\underline{\quad} \times \dfrac{18 \times P}{30} = 30.6 \times \underline{\quad}$

$18 \times P = \underline{\quad\quad}$
           (2)

Divide each side of the equation by $\underline{\quad\quad}$.
                           (3)

$\dfrac{18 \times P}{\quad} = \dfrac{918}{\quad}$

$P = \underline{\quad\quad}$
     (4)

Check: $\dfrac{18 \times \quad}{30} \stackrel{?}{=} 30.6$
            $30.6 = 30.6$

(1) 30   (2) 918   (3) 18   (4) 51

75

WHOLE NUMBERS AND DECIMALS; MULTIPLICATION AND DIVISION

29. $\dfrac{3 \times P}{23} = 18$

30. $\dfrac{7 \times P}{34} = 56$

31. $\dfrac{5 \times P}{8.1} = 700$

32. $\dfrac{9 \times P}{6.1} = 1080$

33. $\dfrac{1.9 \times P}{720} = 53.2$

34. $\dfrac{6.4 \times P}{571} = 198.4$

Solve each equation; check your answers. See Examples 1, 2, 3, or 4.

35. $\dfrac{D}{241} = 4216$

36. $\dfrac{N}{3252} = 419$

37. $5737 \times R = 1{,}939{,}106$

38. $1436 \times R = 960{,}684$

39. $10{,}413{,}726 = 87 \times R \times 97$

40. $3{,}881{,}556 = 422 \times R \times 63$

41. $\dfrac{1328 \times N}{683} = 9296$

42. $\dfrac{P \times 278}{7395} = 1390$

43. $\dfrac{N}{1944} = 8332$

44. $\dfrac{N}{6764} = 2088$

Solve each equation by inspection.

45. $\dfrac{N}{2} = 6$

46. $2 \times N = 6$

47. $3 \times R = 12$

48. $\dfrac{R}{3} = 8$

49. $15 = 3 \times D$

50. $9 = \dfrac{D}{3}$

51. $8 = \dfrac{T}{4}$

52. $8 = T \times 4$

53. $25 = 5 \times N$

54. $\dfrac{S}{6} = 3$

55. $2 \times 4 \times D = 24$

56. $\dfrac{5 \times D}{2} = 20$

57. $\dfrac{6 \times D}{3} = 20$

58. $20 = 5 \times D \times 2$

For Exercises 59–62, find the number N of items sold. In each problem: a. Write an equation. b. Solve.

Complete Example

a. Price per item, $17; total receipts, $13,209.

   The number of items sold: N

Use the unit-product rule.

   Ans. $17 \times N = $ _____
   (1)

76

b. Divide each side of the equation by ___ .

$$\frac{17 \times N}{} = \frac{13{,}209}{}$$

$$N = \frac{13{,}209}{17} \qquad (1)$$

$$N = \underline{}\\ (2)$$

Ans. 777 items were sold.   *Note:* The same result is obtained by starting directly with Equation (1) and using the unit-quotient rule.

(1) 13,209   (2) 777

---

59. Price per item, $104; total receipts, $66,456.

60. Price per item, $456; total receipts, $450,072.

61. Price per item, $37,570; total receipts, $30,769,830.

62. Price per item, $77,921; total receipts, $26,882,745.

63. If a certain number N is divided by 276, the result is 4017. Find the number N.

64. If a certain number N is multiplied by 611, the result is 3,490,643. Find the number N.

65. A store is having a sale on paint at $9 per gallon. At the end of one week the total receipts are $9036. How many gallons of paint have been sold?

66. At the end of a month-long sale of refrigerators, a store has total receipts of $237,015. If the refrigerators sell for $345 each, how many were sold during the month?

2.7 RATIO AND PROPORTION

In this section we introduce a certain kind of equation that is very useful for solving many different problems.

RATIOS

A **ratio** is a method for comparing two numbers. For example, if a person can walk two miles in the same amount of time that she can jog three miles, we say "the ratio of the distances traveled in the same time is 2 to 3," and write either the fraction $\frac{2}{3}$ or the colon notation 2:3 (read "two to three"). The numbers 2 and 3 are the **terms** of the ratio. In this text we will use only the fraction form of a ratio.

Example 1   a. The ratio of 7 to 9 can be expressed as $\frac{7}{9}$. The terms are 7 and 9.

b. The ratio of 1.5 to 1.75 can be expressed as $\frac{1.5}{1.75}$. The terms are 1.5 and 1.75.

WHOLE NUMBERS AND DECIMALS; MULTIPLICATION AND DIVISION

PROPORTIONS

A statement of equality between two ratios is called a **proportion.** Thus, the equation

$$\frac{3}{6} = \frac{4}{8}$$

is a proportion. The proportion is read "three is to six as four is to eight." More generally, if a, b, c, and d are nonzero numbers, then, in the proportion

$$\text{means} \overbrace{\frac{a}{b} = \frac{c}{d}} \text{extremes}$$

the terms a and d are called the **extremes** and the terms b and c are called the **means.**

Example 2 In the proportion $\frac{5}{3} = \frac{20}{12}$, the extremes are 5 and 12, the means are 3 and 20.

An important property of proportions is stated as follows:

> **CROSS-MULTIPLICATION RULE**
>
> **In a proportion, the product of the extremes is equal to the product of the means.**

The cross-multiplication rule asserts that

$$\text{if} \quad \frac{a}{b} = \frac{c}{d}, \quad \text{then} \quad a \times d = b \times c$$

Example 3 a. If $\frac{5}{7} = \frac{25}{35}$, then $5 \times 35 = 7 \times 25$; (175 = 175).

b. If $\frac{1.5}{2.25} = \frac{2}{3}$, then $1.5 \times 3 = 2.25 \times 2$; (4.5 = 4.5).

SOLVING PROPORTIONS

Given a proportion in which one of the terms is unknown, the cross-multiplication rule provides us with an equation that we can solve to find the unknown term. We refer to this process as *solving a proportion*.

Example 4 Solve: $\frac{20}{12} = \frac{N}{6}$.

Solution By the cross-multiplication rule,

$$20 \times 6 = 12 \times N$$

2.7 RATIO AND PROPORTION

Dividing each side by 12, we have

$$\frac{20 \times 6}{12} = \frac{\cancel{12} \times N}{\cancel{12}}$$

$$10 = N$$

USING PROPORTIONS

Consider the problem

"If eight oranges cost 56 cents, how much do six oranges cost?"

One way to solve this problem involves a two-step procedure. For the first step we apply the unit-quotient rule to find the cost of one orange:

$$56 \div 8 = 7$$

Thus, one orange costs 7 cents. For the second step we apply the unit-product rule to find the cost C of six oranges:

$$C = 6 \times 7 = 42$$

Thus, six oranges cost 42 cents.

Another method for solving problems similar to the above example combines both steps into one procedure that involves solving a proportion. To set up such a proportion, we first set up a table with a *cents* column and an *oranges* column, as in Figure 2.1*a*. Then, because C cents is paired with six oranges, we enter C as the numerator in the *cents* column and 6 as the numerator in the *oranges* column, as in Figure 2.1*b*. Next, because 56 cents is paired with eight oranges, we enter 56 as the denominator in the *cents* column and 8 as the denominator in the *oranges* column, as in Figure 2.1*c*

Cents	**Oranges**		**Cents**	**Oranges**		**Cents**	**Oranges**
—	=	—	$\underline{C}$	=	$\underline{6}$	$\dfrac{C}{56}$	=	$\dfrac{6}{8}$
(a)			(b)			(c)	

Figure 2.1

To complete the problem, we solve the proportion obtained in *c*. By the cross-multiplication rule,

$$C \times 8 = 56 \times 6$$

Dividing each side by 8, we have

$$\frac{C \times \cancel{8}}{\cancel{8}} = \frac{56 \times 6}{8}$$

$$C = 42$$

and, as before, six oranges cost 42 cents.

WHOLE NUMBERS AND DECIMALS; MULTIPLICATION AND DIVISION

Different choices can be made when setting up a proportion for solving a problem. For example, each of the proportions

$$\frac{56}{C} = \frac{8}{6} \quad \text{or} \quad \frac{6}{8} = \frac{C}{56} \quad \text{or} \quad \frac{8}{6} = \frac{56}{C} \quad \text{or} \quad \frac{8}{56} = \frac{6}{C}$$

can be used to solve the above problem correctly. However, the computations are usually simpler if *the unknown is the numerator on the left side* as in Figure 2.1c.

Example 5   A car can travel 136 miles on 8 gallons of gasoline. If the gasoline tank of the car holds 19 gallons, how far can the car travel on one tankful?

Solution   Letting M represent the number of miles the car can travel on 19 gallons (one tankful), we prepare a *miles* column and a *gallons* column. Next, we enter M as the numerator in the *miles* column and 19 as the numerator in the *gallons* column.

Miles	Gallons		Miles	Gallons		Miles	Gallons
—	=	—	M	=	19	$\dfrac{M}{136}$	= $\dfrac{19}{8}$

(a) (b) (c)

Then, because the car travels 136 miles on 8 gallons, we enter 136 as the denominator in the *miles* column and 8 as the denominator in the *gallons* column, and proceed to solve the proportion obtained in c. By the cross-multiplication rule,

$$M \times 8 = 136 \times 19$$

Dividing each side by 8, we have

$$\frac{M \times \cancel{8}}{\cancel{8}} = \frac{136 \times 19}{8}$$

$$M = 323$$

The car can travel 323 miles on one tankful of gasoline.

Example 6   If a machine assembler earns $25 for 7.5 hours of work, how much can he earn in 42 hours?

Solution   Letting D represent the amount earned in 42 hours, we first prepare a *dollars* column and an *hours* column. Because D is paired with 42, we enter D as the numerator in the *dollars* column and 42 as the numerator in the *hours* column. Next, because $25 is paired with 7.5 hours, we enter 25 as the denominator in the *dollars* column, 7.5 as the denominator in the *hours* column, and proceed to solve the resulting proportion in c.

Dollars	Hours		Dollars	Hours		Dollars	Hours
—	=	—	D	=	42	$\dfrac{D}{25}$	= $\dfrac{42}{7.5}$

(a) (b) (c)

2.7 RATIO AND PROPORTION

By the cross-multiplication rule,

$$D \times 7.5 = 25 \times 42,$$

Divide each side by 7.5

$$\frac{D \times 7.5}{7.5} = \frac{25 \times 42}{7.5},$$

$$D = 140.$$

The employee can earn $140 in 42 hours.

EXERCISES 2.7

Write each ratio as a fraction. See Example 1.

a. 4 to 7

Ans. _____
   (1)

b. 1.63 to 2.49

Ans. _____
   (2)

(1) $\frac{4}{7}$ (2) $\frac{1.63}{2.49}$

1. 5 to 18
2. 7 to 45
3. 17. to 8
4. 12 to 5
5. 8.87 to 4.31
6. 5.68 to 3.47

Verify that the product of the extremes is equal to the product of the means. See Examples 2 and 3.

Complete Example

a. $\frac{5}{6} = \frac{15}{18}$

The extremes are 5 and 18.
The means are 6 and _____ .
                        (1)

$$5 \times 18 \stackrel{?}{=} 6 \times 15$$

$$90 = \underline{}_{(2)}$$

b. $\frac{3}{8} = \frac{18.6}{49.6}$

The extremes are 3 and 49.6.
The means are 8 and _____ .
                        (3)

$$3 \times 49.6 \stackrel{?}{=} 8 \times 18.6$$

$$148.8 = \underline{}_{(4)}$$

(1) 15   (2) 90   (3) 18.6   (4) 148.8

81

WHOLE NUMBERS AND DECIMALS; MULTIPLICATION AND DIVISION

7. $\dfrac{7}{8} = \dfrac{14}{16}$
8. $\dfrac{12}{17} = \dfrac{36}{51}$
9. $\dfrac{3.5}{9.5} = \dfrac{7}{19}$

10. $\dfrac{21}{31} = \dfrac{10.5}{15.5}$
11. $\dfrac{12}{17} = \dfrac{15.6}{22.1}$
12. $\dfrac{6.3}{4.6} = \dfrac{17.01}{12.42}$

Solve each proportion. Give each answer to the neartest tenth unless the answer is a whole number. See Example 4.

Complete Example

a. $\dfrac{3}{7} = \dfrac{21}{N}$

Use the cross-multiplication rule.

$3 \times N = 7 \times \underline{\quad\quad}$ (1)

a. Divide each side of the equation by 3.

$\dfrac{\cancel{3} \times N}{\cancel{3}} = \dfrac{7 \times 21}{3}$

$N = \dfrac{7 \times 21}{3}$

$N = \underline{\quad\quad}$ (2)

b. $\dfrac{7}{9} = \dfrac{14.7}{N}$

Use the cross-multiplication rule.

$7 \times N = 9 \times \underline{\quad\quad}$ (3)

Divide each side of the equation by 7.

$\dfrac{\cancel{7} \times N}{\cancel{7}} = \dfrac{9 \times 14.7}{7}$

$N = \dfrac{9 \times 14.7}{7}$

$N = \underline{\quad\quad}$ (4)

---

(1) 21   (2) 49   (3) 14.7   (4) 18.9

---

13. $\dfrac{27}{36} = \dfrac{N}{4}$
14. $\dfrac{33}{21} = \dfrac{N}{7}$
15. $\dfrac{11}{N} = \dfrac{35}{105}$
16. $\dfrac{14}{N} = \dfrac{154}{55}$

17. $\dfrac{33}{N} = \dfrac{77}{231}$
18. $\dfrac{91}{N} = \dfrac{26}{182}$
19. $\dfrac{27.3}{294} = \dfrac{N}{210}$
20. $\dfrac{135}{18.9} = \dfrac{N}{35}$

21. $\dfrac{261.8}{22.1} = \dfrac{154}{N}$
22. $\dfrac{700.7}{165} = \dfrac{6.37}{N}$
23. $\dfrac{N}{6} = \dfrac{77}{42}$
24. $\dfrac{N}{18} = \dfrac{45}{27}$

25. $\dfrac{N}{12.5} = \dfrac{27}{40}$
26. $\dfrac{N}{6.25} = \dfrac{18}{20}$

2.7 RATIO AND PROPORTION

For Exercises 27–40: a. Write a proportion. b. Solve the proportion. See Examples 5 and 6.

Complete Example
A tire stores sells three tires for $107.70. Find the cost of nine tires.

$$\text{Cost of nine tires: } C$$

a. Prepare a "cost" column and a "tires" column. Enter the correct numbers for the proportion; C is paired with 9 tires, and $107.70 is paired with 3 tires.

$$\begin{array}{cc} \text{Cost} & \text{Tires} \\ \dfrac{C}{107.70} = & \dfrac{9}{\underline{\qquad}} \\ & (1) \end{array}$$

Ans. $\dfrac{C}{107.70} = \dfrac{9}{3}$

b. Solve the proportion.

$$C \times 3 = 9 \times \underline{\qquad}_{(2)}$$

$$\dfrac{C \times \cancel{3}}{\cancel{3}} = \dfrac{9 \times 107.70}{3}$$

$$C = \underline{\qquad}_{(3)}$$

Ans. The nine tires will cost $323.10

---

(1) 3   (2) 107.70   (3) 323.10 or 323.1

---

27. If seven apples cost 42¢, find the cost of four apples.

28. If 9 bananas cost 72¢, find the cost of 15 bananas.

29. If 8 ounces of cheese cost 96¢, find the cost of 20 ounces.

30. If 6 loaves of bread cost $1.74, find the cost of 14 loaves.

31. A store offers men's shirts on sale at three shirts for $14.97. Find the cost of five shirts.

32. A plant nursery sells some plants at 5 for $4.65. Find the cost of 12 such plants.

33. An automobile parts dealer sells tires at three for $76.90. Find the cost of eight tires.

34. If 4 calculator batteries sell for $2.99, find the cost of 10 batteries.

35. Three packages of camera flash bulbs sell for $6.24. Find the cost of 10 packages.

36. A piece of aluminum building panel 6 feet long is priced at $7.95. Find the price of 26 feet of paneling.

37. A 54-inch by 36-inch photograph is to be reduced in size so that the 36-inch side becomes 12 inches long. How long will the 54-inch side become?

38. A linotype operator can set type for 11 pages of a book in seven hours. How many hours are needed to set type for a book with 484 pages?

83

WHOLE NUMBERS AND DECIMALS; MULTIPLICATION AND DIVISION

39. The "speed enforcement arrest index" is used by a certain state highway patrol to discourage drivers from speeding. The index is given by the ratio 30 to 1, which means that officers are expected to issue 30 speeding citations for each injury or fatal accident on a given part of the highway. Solve for the missing data using this index.

	a.	b.	c.	d.
Number of citations	330	510	___	___
Injuries or accidents	___	___	14	8

40. In a recent survey of four cities, it was found that eight out of nine families owned television sets. Solve for the missing data, using the results of the survey.

	a.	b.	c.	d.
Number of families	23,400	127,647	___	___
Number of TV sets	___	___	11,080	32,032

41. If a car uses 12 gallons of gasoline to travel 216 miles, how many gallons are needed to travel 2970 miles?

42. If the car of Exercise 41 has a gasoline tank that holds 23 gallons, how far can the car travel on a tankful of gasoline?

43. A 6-inch by 8-inch photograph is to be enlarged so that the 6-inch side becomes 24 inches long. How long will the 8-inch side become?

44. A faulty automobile speedometer reads 36 miles per hour when the true speed is 32 miles per hour. Solve for the missing data:

	a.	b.	c.	d.
Speedometer	54	63	___	___
True speed	___	___	64	40

45. A length of one inch on a road map corresponds to 32 actual highway miles. Determine how many highway miles correspond to each map length indicated in the table.

	a.	b.	c.	d.	e.
Map (inches)	4	5	8	___	___
Highway miles	___	___	___	224	352

46. On a floor plan of a house and lot, 1 inch of drawing corresponds to four actual feet. Find the house and lot dimensions for each of the following lengths on the floor plan.

a. Front of the house: 23 inches.
b. Side of the house: 18 inches.
c. Front of the lot: 37 inches.
d. Side of the lot: 56 inches.

CHAPTER TWO REVIEW EXERCISES

1. Compute: 469 × 38.
2. Compute: 3.71 × 46.73. Find the product a. To the nearest hundredth b. To the nearest thousandth.
3. Find each product mentally.
   a. 9.85 × 0.1   b. 40.7 × 10   c. 0.987 × 0.01

CHAPTER TWO REVIEW EXERCISES

4. Use rounded-off numbers to estimate each product.

    a. 403 × 58   b. 39.2 × 29

5. Compute each power.

    a. 9^2 b. $(2.4)^2$ c. $(4.1)^3$

6. Compute:

    a. $\dfrac{1022}{14}$ b. $\dfrac{16.017}{5.7}$

7. Find the quotient 976.6 ÷ 7.35:

    a. To the nearest tenth   b. To the nearest hundredth.

8. Compute mentally:

    a. 68.3 ÷ 100   b. $\dfrac{3.1}{1000}$

9. Find each quotient by inspection.

    a. $1.5 \times \dfrac{62}{1.5}$ b. $\dfrac{4.9 \times 3.1}{3.1}$

10. Use rounded-off numbers to estimate each quotient.

    a. 609 ÷ 22   b. 598.2 ÷ 19

For Exercises 11–14: a. Write an equation. b. Solve the equation. Use any letter for the unknown value.

11. If an automobile can travel 27 miles per gallon of gasoline, how far can it travel on 16 gallons?
12. If a man pays $108 a month for an automobile, how much does he pay over a period of 36 months?
13. A dealer paid $6072 for a shipment of 24 television sets, each the same model. Find the cost of each set.
14. An automatic machine produces 540 items in 15 minutes. How many items are produced per minute?

Solve each equation.

15. $\dfrac{N}{32.1} = 26.7$ 16. $285 \times D = 118{,}845$

17. $8 \times R \times 6 = 28.32$ 18. $\dfrac{28 \times P}{27} = 420$

For each word problem: a. Write a proportion. b. Solve the proportion.

19. If 12 pears cost 72¢, find the cost of 8 pears.
20. A plant nursery sells plants at 6 for $5.22. Find the cost of 10 such plants.
21. If 4 small batteries sell for $3.16, find the cost of 14 batteries.
22. A photograph that is 4 inches wide and 6 inches long is enlarged so that the 4-inch side becomes 10 inches. How long will the 6-inch side become?

3 FRACTIONS

In Chapter 2 we saw how a quotient $n \div d$ of two numbers could be written in the form $\frac{n}{d}$. Quotients that are written in this form are called **fractions.** The number n is called the **numerator** and the number d is called the **denominator.** These numbers are referred to as the **terms** of the fraction. If the numerator is less than the denominator, the fraction is called a **proper fraction;** if the numerator is equal to or greater than the denominator, the fraction is called an **improper fraction.** Thus,

$$\frac{3}{8} \quad \text{and} \quad \frac{15}{16}$$

are proper fractions and

$$\frac{8}{8} \quad \text{and} \quad \frac{16}{15}$$

are improper fractions. Any whole number can be written as an improper fraction simply by writing that number over 1 as a denominator. For example, 8 can be written as the fraction $\frac{8}{1}$.

In order to do computations with fractions, it is important to know how to write some whole numbers in a special form introduced in Section 3.1.

3.1 FACTORING

When a number such as 75 is written as the product of 3×25, we say that 3×25 is a *factored form* of 75. The process of changing a number to factored form is called **factoring.**

If one whole number is divided by a second whole number so that the quotient is a whole number, then the second whole number is an *exact divisor* of the first. Thus, because

$$75 \div 3 = 25$$

and 25 is a whole number, 3 is an exact divisor of 75.

FRACTIONS

An exact divisor of a whole number is also called a *factor* of the number. For example, 25 and 3 are two of the factors of 75.

Example 1   a.   15 is a factor of 75 because 75 ÷ 15 = 5, and 5 is a whole number.

b.   4 is not a factor of 75 because 75 ÷ 4 = 18.75, and 18.75 is not a whole number.

COMPLETELY FACTORED FORM

A whole number greater than 1 that has no exact divisor except for itself and 1 is called a **prime number.** The first 10 prime numbers are

$$2, 3, 5, 7, 11, 13, 17, 19, 23, 29$$

Whole numbers greater than 1 that are not prime numbers are called **composite numbers.** For example, 15 is a composite number because it is exactly divisible by 3 and by 5. When a composite number is written as a product of prime numbers, it is said to be *completely factored,* and this product is called the **completely factored form** of the number. Such factored forms can sometimes be found by inspection, together with a knowledge of multiplication facts given in Table 2.1. Each of the factors in the completely factored form is called a **prime factor**.

Example 2   a.   21 = 3 × 7   b.   25 = 5 × 5   c.   30 = 2 × 3 × 5

d.   4 × 5 is not the completely factored form of 20 because 4 is not a prime number.

The following rules can be used to decide whether or not a given number can be exactly divided by 2 or 3 or 5; they can be helpful in writing the completely factored form of some numbers.

Divisibility Rules

1. A whole number is divisible by 2 if the last digit to the right is an even digit (0, 2, 4, 6, 8).
2. A whole number is divisible by 3 if the sum of its digits is divisible by 3.
3. A whole number is divisible by 5 if the last digit to the right is 0 or 5.

Example 3   a.   36 is divisible by 2 because the last digit to the right is an even digit, 6.
b.   36 is divisible by 3 because 3 + 6 = 9, and 9 is divisible by 3.
c.   35 is divisible by 5 because the last digit to the right is 5.

The completely factored form of a number may include the same factor more than once.

Example 4   a.   25 = 5 × 5   b.   63 = 3 × 3 × 7   c.   16 = 2 × 2 × 2 × 2

If a number N cannot be readily factored by inspection or by using the divisibility rules for 2, 3, and 5, we can try each prime factor in turn (more than once, if necessary) to see if it is an exact divisor. The process ends when N is completely factored or when we find that N is prime.

3.1 FACTORING

Example 5 Write 308 in completely factored form.

Solution First, note that 308 is divisible by 2 because the last digit, 8, is an even digit. Thus,

$$308 \div 2 = 154 \quad \text{or} \quad 308 = 2 \times 154$$

Next, note that 154 is divisible by 2. Thus,

$$154 \div 2 = 77 \quad \text{or} \quad 154 = 2 \times 77$$

Hence,

$$308 = 2 \times 154$$
$$= 2 \times 2 \times 77$$

Now, observe that 77 is not divisible by 2, or 3, or 5. Divide 77 by the next greatest prime number, 7:

$$77 \div 7 = 11 \quad \text{or} \quad 77 = 7 \times 11$$

Hence, 308 can be factored as follows:

$$308 = 2 \times 154$$
$$= 2 \times 2 \times 77$$
$$= 2 \times 2 \times 7 \times 11$$

Because 2, 7, and 11 are prime numbers, the completely factored form of 308 is $2 \times 2 \times 7 \times 11$ (or $2^2 \times 7 \times 11$), and because multiplication is commutative, any arrangement of the factors 2, 2, 7, and 11 is correct.

EXERCISES 3.1

Determine whether the first number is a factor of the second number. See Example 1.

Complete Example

a. 3; 246

$$246 \div 3 = \underline{\qquad}$$
$$\quad\quad\quad\quad (1)$$

82 is a whole number so
3 <u>is/is not</u> a factor of 246.
$\quad\;\,$(2)

b. 8; 202

$$202 \div 8 = \underline{\qquad}$$
$$\quad\quad\quad\quad (3)$$

25.25 is not a whole number so
8 <u>is/is not</u> a factor of 202.
$\quad\;\,$(4)

(1) 82	(2) is	(3) 25.25	(4) is not

1. 4; 244
2. 9; 342
3. 3; 142
4. 8; 196
5. 21; 455
6. 19; 493
7. 29; 696
8. 41; 615

FRACTIONS

By the rules used in Example 3, determine if each number is divisible by 2, by 3, by 5.

Complete Examples

a. 816 is/is not divisible by 2 because the last digit to the right is an even digit, 6.
    (1)

b. 816 is/is not divisible by 3 because 8 + 1 + 6 = 15, and 15 is divisible by 3.
    (2)

c. 816 is/is not divisible by 5 because the last digit to the right is neither 0 nor 5.
    (3)

(1) is   (2) is   (3) is not

9. 120	10. 130	11. 72	12. 82
13. 225	14. 126	15. 20,712	16. 17,235

Write each number in completely factored form. See Example 4.

Complete Examples

a. $12 = 3 \times 2 \times$ ___(1)___   b. $24 = 2 \times 2 \times 2 \times$ ___(2)___   c. $15 =$ ___(3)___

(1) 2   (2) 3   (3) 5×3 or 3×5

17. 4	18. 8	19. 27	20. 9
21. 54	22. 72	23. 81	24. 48

Write each number in completely factored form. See Example 5.

Complete Examples

a. $100 = 2 \times 50$
     $= 2 \times 2 \times$ ___(1)___
     $= 2 \times 2 \times$ ___(2)___

b. $693 = 3 \times 231$
     $= 3 \times 3 \times 77$
     $= 3 \times 3 \times$ ___×___
                        (3)

(1) 25   (2) 5×5 or 5^2 (3) 7×11 or 11×7

3.2 TRANSFORMING FRACTIONS

25. 272	**26.** 315	**27.** 176	**28.** 405
29. 528	**30.** 972	**31.** 8250	**32.** 7560

In each of the following exercises, only one number is prime. Which number is it?

33. a. 2187 b. 1409 c. 9305 d. 2114   **34.** a. 2043 b. 5245 c. 1663 d. 3002

3.2 TRANSFORMING FRACTIONS

EQUAL FRACTIONS

Figure 3.1a shows a line segment of length one. The same segment is divided into two equal parts in Figure 3.1b, into four equal parts in Figure 3.1c, and into eight equal parts in Figure 3.1d. Observe that, in Figure 3.1c, two of the four equal parts (or $\frac{2}{4}$ of the segment) and in Figure 3.1d, four of the eight equal parts (or $\frac{4}{8}$ of the segment) are also the same as $\frac{1}{2}$ of the segment.

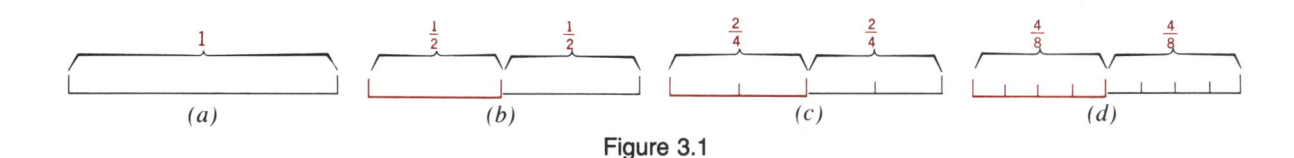

Figure 3.1

We say that $\frac{4}{8}$, $\frac{2}{4}$, and $\frac{1}{2}$ are *equal fractions* and write

$$\frac{4}{8} = \frac{2}{4} = \frac{1}{2}$$

which indicates that fractions can be written in different forms. Note that

$$\frac{1 \times 2}{2 \times 2} = \frac{2}{4} \quad \text{and} \quad \frac{1 \times 4}{2 \times 4} = \frac{4}{8}$$

Hence

$$\frac{1}{2} = \frac{1 \times 2}{2 \times 2} = \frac{1 \times 4}{2 \times 4}$$

or

$$\frac{1}{2} = \frac{2}{4} = \frac{4}{8}$$

FRACTIONS

Results such as this suggest the following rule, which enables us to change the form of a fraction.

> **FUNDAMENTAL PRINCIPLE OF FRACTIONS**
> If both the numerator and the denominator of a given fraction are multiplied or divided by the same nonzero number, the resulting fraction is equal to the given fraction.

This principle is expressed in symbols as

$$\frac{n}{d} = \frac{n \times k}{d \times k} \quad \text{or} \quad \frac{n}{d} = \frac{n \div k}{d \div k}$$

where k does not equal 0.

Example 1 a. $\frac{1}{4} = \frac{1 \times 9}{4 \times 9} = \frac{9}{36}$; $\frac{1}{4}$ and $\frac{9}{36}$ are equal fractions.

b. $\frac{12}{36} = \frac{12 \div 4}{36 \div 4} = \frac{3}{9}$; $\frac{12}{36}$ and $\frac{3}{9}$ are equal fractions.

REDUCING FRACTIONS

If each of two whole numbers has the same whole number as a factor, that factor is called a *common factor* (or *common divisor*) of the two numbers. Thus, in Example 1b. above, 4 is a common factor of 12 and 36. A fraction in which the numerator and denominator do not have any common factors is said to be in **lowest terms.**

> To reduce a fraction to lowest terms:
> 1. Factor the numerator and the denominator.
> 2. "Divide out" common factors.

Slash bars (/) are frequently used on common factors to indicate which factors are being divided out.

Example 2 a. $\frac{6}{10} = \frac{\cancel{2} \times 3}{\cancel{2} \times 5} = \frac{3}{5}$ b. $\frac{9}{30} = \frac{\cancel{3} \times 3}{2 \times \cancel{3} \times 5} = \frac{3}{10}$

Note that in the above examples, $\frac{\cancel{2}}{\cancel{2}}$ and $\frac{\cancel{3}}{\cancel{3}}$ show that a factor of 1 is "removed" from the entire quotient, and the value of the quotient is not changed. You may find it helpful to show the factor "1" above and below each factor being divided out. Examples 2a. and 2b. would appear as

$$\frac{6}{10} = \frac{\overset{1}{\cancel{2}} \times 3}{\underset{1}{\cancel{2}} \times 5} = \frac{3}{5} \quad \text{and} \quad \frac{9}{30} = \frac{\overset{1}{\cancel{3}} \times 3}{2 \times \underset{1}{\cancel{3}} \times 5} = \frac{3}{10}$$

3.2 TRANSFORMING FRACTIONS

Example 3 We sometimes need to divide out more than one common factor when reducing a fraction to lowest terms.

a. $\dfrac{12}{32} = \dfrac{\cancel{2} \times \cancel{2} \times 3}{\cancel{2} \times \cancel{2} \times 2 \times 2 \times 2} = \dfrac{3}{8}$

b. $\dfrac{72}{360} = \dfrac{\cancel{2} \times \cancel{2} \times \cancel{2} \times \cancel{3} \times \cancel{3}}{\cancel{2} \times \cancel{2} \times \cancel{2} \times \cancel{3} \times \cancel{3} \times 5} = \dfrac{1}{5}$

Note that if all the factors in the numerator of a fraction are divided out, as in Example 3b., the factor 1 must be written in the numerator of the reduced fraction.

BUILDING FRACTIONS

In Example 1a. we saw that the fundamental principle of fractions can be used to write

$$\frac{1}{4} = \frac{1 \times 9}{4 \times 9} = \frac{9}{36}$$

Note that each term of $\dfrac{9}{36}$ is greater than the corresponding term of $\dfrac{1}{4}$. We say that the fraction $\dfrac{1}{4}$ has been *built to higher terms* and we refer to the factor 9 as the *building factor*. Most often, we need to build a fraction to higher terms with the requirement that the resulting fraction will have a specified denominator. For example, to build the fraction $\dfrac{3}{4}$ to an equal fraction with the denominator 64, we need to find a building factor k such that

$$\frac{3 \times k}{4 \times k} = \frac{?}{64}$$

A comparison of the denominators indicates that we want a number k such that $4 \times k$ will be equal to 64. By inspection, or by dividing 64 by 4, we obtain the building factor $k = 16$. Thus,

$$\frac{3}{4} = \frac{3 \times 16}{4 \times 16} = \frac{48}{64}$$

To build a fraction to higher terms:

1. Obtain the building factor by inspection or by dividing the new denominator by the given denominator.
2. Multiply the numerator and denominator of the given fraction by the building factor.

Example 4 Build the fraction $\dfrac{5}{8}$ to a fraction with denominator 32.

Solution
1. The new denominator is 32, the original denominator is 8. Thus the building factor, $32 \div 8$, is 4.
2. $\dfrac{5}{8} = \dfrac{5 \times 4}{8 \times 4} = \dfrac{20}{32}$

FRACTIONS

Example 5 $\frac{11}{12} = \frac{?}{180}$

Solution 1. The new denominator is 180, the original denominator is 12. Thus the building factor, 180 ÷ 12, is 15.

2. $\frac{11}{12} = \frac{11 \times 15}{12 \times 15} = \frac{165}{180}$

The fundamental principle of fractions can also be used to change a whole number to a fraction with a specified denominator.

Example 6 a. $1 = \frac{?}{8}$ b. $7 = \frac{?}{8}$

Solutions a. $1 = \frac{1}{1} = \frac{1 \times 8}{1 \times 8} = -$ b. $7 = \frac{7}{1} = \frac{7 \times 8}{1 \times 8} = \frac{56}{8}$

EXERCISES 3.2
Reduce each fraction to lowest terms. See Examples 1, 2, and 3.

Complete Examples

a. $\frac{15}{25} = \frac{3 \times \cancel{5}}{5 \times \cancel{5}}$
 = _____
    (1)

b. $\frac{63}{84} = \frac{3 \times \cancel{3} \times \cancel{7}}{2 \times 2 \times \cancel{3} \times \cancel{7}}$
 = _____
    (2)

(1) $\frac{3}{5}$ (2) $\frac{3}{4}$

1. $\frac{8}{14}$ 2. $\frac{22}{33}$ 3. $\frac{30}{34}$ 4. $\frac{12}{15}$ 5. $\frac{25}{35}$ 6. $\frac{15}{25}$

7. $\frac{35}{85}$ 8. $\frac{48}{72}$ 9. $\frac{15}{48}$ 10. $\frac{40}{65}$ 11. $\frac{12}{42}$ 12. $\frac{18}{54}$

13. $\frac{24}{32}$ 14. $\frac{16}{64}$ 15. $\frac{27}{81}$ 16. $\frac{18}{27}$ 17. $\frac{16}{44}$ 18. $\frac{24}{40}$

19. $\frac{36}{60}$ 20. $\frac{42}{72}$ 21. $\frac{36}{64}$ 22. $\frac{32}{56}$ 23. $\frac{38}{52}$ 24. $\frac{56}{88}$

3.2 TRANSFORMING FRACTIONS

Build each fraction or whole number to an equal fraction with the given denominator. See Examples 4, 5, and 6.

Complete Examples

a. $\dfrac{3}{8} = \dfrac{}{56}$

   Building factor: $56 \div 8 = \underline{}_{(1)}$.

   $\dfrac{3 \times 7}{8 \times 7} = \underline{}_{(2)}$

b. $6 = \dfrac{}{12}$

   Write 6 as $\dfrac{6}{1}$.

   $\dfrac{6 \times 12}{1 \times 12} = \underline{}_{(3)}$

(1) 7   (2) $\dfrac{21}{56}$ (3) $\dfrac{72}{12}$

25. $\dfrac{1}{2} = \dfrac{}{24}$
26. $\dfrac{2}{3} = \dfrac{}{21}$
27. $\dfrac{3}{5} = \dfrac{}{35}$
28. $\dfrac{4}{7} = \dfrac{}{56}$
29. $\dfrac{5}{8} = \dfrac{}{40}$

30. $\dfrac{5}{9} = \dfrac{}{54}$
31. $\dfrac{7}{12} = \dfrac{}{36}$
32. $\dfrac{5}{11} = \dfrac{}{77}$
33. $\dfrac{11}{18} = \dfrac{}{90}$
34. $\dfrac{3}{14} = \dfrac{}{70}$

35. $\dfrac{9}{16} = \dfrac{}{144}$
36. $\dfrac{7}{24} = \dfrac{}{96}$
37. $\dfrac{3}{8} = \dfrac{}{128}$
38. $\dfrac{5}{6} = \dfrac{}{120}$
39. $\dfrac{2}{7} = \dfrac{}{105}$

40. $\dfrac{2}{15} = \dfrac{}{180}$
41. $\dfrac{5}{18} = \dfrac{}{108}$
42. $\dfrac{5}{12} = \dfrac{}{240}$
43. $\dfrac{1}{9} = \dfrac{}{135}$
44. $\dfrac{1}{13} = \dfrac{}{390}$

45. $1 = \dfrac{}{12}$
46. $4 = \dfrac{}{16}$
47. $5 = \dfrac{}{16}$
48. $6 = \dfrac{}{20}$

Solve each problem. Express answers in fraction form.

Complete Example

If 10 students in a class of 25 received A's, what "fraction" of the class:
a. Received A's?   b. Did not receive A's?

a. The ratio of A students to the total number of students is the fraction

$$\dfrac{10}{25} = \dfrac{2 \times 5}{5 \times 5} = \underline{}_{(1)}$$

Ans. $\dfrac{2}{5}$ of the class received A's.

95

FRACTIONS

b. If ten students received A's, then 25 − 10 = _____ did not receive A's.
$${\scriptstyle(2)}$$

$$\frac{15}{25} = \frac{3 \times 5}{5 \times 5} = \underline{}$$
$${\scriptstyle(3)}$$

Ans. $\frac{3}{5}$ of the class did not receive A's.

---

(1) $\frac{2}{5}$ (2) 15 (3) $\frac{3}{5}$

---

49. If 4 of the 18 players of a basketball squad "fouled out" during a game, what "fraction" of the squad fouled out?

50. If 30 out of 150 light bulbs in a sign are burned out, what "fraction" of the light bulbs are still able to burn?

51. In an arithmetic class, 14 students are male and 22 are female. a. What "fraction" of the class are male? b. Female?

52. The number of students in a history class that received each grade are as shown in the table below. What fraction of the class earned each of the grades?

Grade	A	B	C	D	F	Total Number of Grades
Number of students	8	11	27	6	4	___
Fraction of class	___	___	___	___	___	

3.3 LEAST COMMON DENOMINATOR

Fractions with identical denominators are called *like fractions*. The number that is the denominator of each of two or more like fractions is called the **common denominator** of the fractions. Thus,

$\frac{3}{6}$ and $\frac{4}{6}$ are like fractions with common denominator 6;

$\frac{6}{12}$ and $\frac{8}{12}$ are like fractions with common denominator 12;

$\frac{12}{24}$ and $\frac{16}{24}$ are like fractions with common denominator 24.

Now note that the pair of fractions $\frac{1}{2}$ and $\frac{2}{3}$ with different denominators can be expressed as

$\frac{3}{6}$ and $\frac{4}{6}$, or $\frac{6}{12}$ and $\frac{8}{12}$, or $\frac{12}{24}$ and $\frac{16}{24}$

where each pair of fractions has a common denominator. These results suggest that unlike fractions may be expressed as like fractions with various common denomi-

3.3 LEAST COMMON DENOMINATOR

nators. The least of these common denominators is called the **least common denominator** of the fractions (abbreviated **LCD**). It is the least number that is exactly divisible by each of the original denominators. Thus, the LCD for $\frac{1}{2}$ and $\frac{2}{3}$ is 6 because 6 is the least number that is exactly divisible by 2 and by 3.

If the least common denominator cannot be found by inspection the following procedure can be used.

To find the LCD of a set of unlike fractions:

1. Express each denominator in completely factored form, aligning common factors where possible.
2. Write a product whose factors are each of the different prime factors that occur in any of the denominators, and include each of these factors the greatest number of times that it appears in any one of the given denominators.

Example 1   a. Find the LCD for $\frac{2}{3}$ and $\frac{3}{5}$.   b. Write $\frac{2}{3}$ and $\frac{3}{5}$ as like fractions, with their LCD as the common denominator.

Solutions   a. Each denominator is prime. The LCD must contain each of the factors 3 and 5 once. Hence the LCD is

$$3 \times 5 = 15$$

b. $\frac{2}{3} = \frac{?}{15}$ and $\frac{3}{5} = \frac{?}{15}$. Because

$$15 \div 3 = 5 \quad \text{and} \quad 15 \div 5 = 3$$

the respective building factors are 5 and 3. Use the fundamental principle of fractions.

$$\frac{2}{3} = \frac{2 \times 5}{3 \times 5} = \frac{10}{15} \quad \text{and} \quad \frac{3}{5} = \frac{3 \times 3}{5 \times 3} = \frac{9}{15}$$

Example 2   a. Find the LCD for $\frac{5}{6}$ and $\frac{7}{15}$.   b. Write $\frac{5}{6}$ and $\frac{7}{15}$ as like fractions, with their LCD as the common denominator.

Solutions   a. Factor each denominator; align common factors. The LCD must contain each of the factors 2, 3, and 5 once.

$$6 = 2 \times 3$$
$$15 = 3 \times 5$$
$$\text{LCD:} \quad 2 \times 3 \times 5 = 30$$

b. $\frac{5}{6} = \frac{?}{30}$ and $\frac{7}{15} = \frac{?}{30}$. Because

$$30 \div 6 = 5 \quad \text{and} \quad 30 \div 15 = 2$$

the respective building factors are 5 and 2. Use the fundamental principle of fractions.

$$\frac{5}{6} = \frac{5 \times 5}{6 \times 5} = \frac{25}{30} \quad \text{and} \quad \frac{7}{15} = \frac{7 \times 2}{15 \times 2} = \frac{14}{30}$$

FRACTIONS

Example 3   a. Find the LCD for $\frac{3}{4}$, $\frac{5}{8}$, and $\frac{7}{10}$.   b. Write $\frac{3}{4}$, $\frac{5}{8}$, and $\frac{7}{10}$ as like fractions, with their LCD as the common denominator.

Solutions   a. Factor each denominator; align common factors. The LCD must contain the factor 2 three times (because 8 contains 2 as a factor three times), and the factor 5 once.

$$\begin{cases} 4 = 2 \times 2 \\ 8 = 2 \times 2 \times 2 \\ 10 = 2 \times 5 \\ \text{LCD:} = 2 \times 2 \times 2 \times 5 = 40 \end{cases}$$

b. $\frac{3}{4} = \frac{?}{40}$, $\frac{5}{8} = \frac{?}{40}$, and $\frac{7}{10} = \frac{?}{40}$. Because

$$40 \div 4 = 10, \quad 40 \div 8 = 5, \quad \text{and} \quad 40 \div 10 = 4$$

the respective building factors are 10, 5, and 4. Use the fundamental principle of fractions.

$$\frac{3}{4} = \frac{3 \times 10}{4 \times 10} = \frac{30}{40}, \quad \frac{5}{8} = \frac{5 \times 5}{8 \times 5} = \frac{25}{40}, \quad \text{and} \quad \frac{7}{10} = \frac{7 \times 4}{10 \times 4} = \frac{28}{40}$$

COMPARING FRACTIONS

Figure 3.2 shows a line segment of length one, divided into five parts. Observing that two of the five parts, taken together, is less than four of the five parts, we say that $\frac{2}{5}$ is less than $\frac{4}{5}$ (or $\frac{4}{5}$ is greater than $\frac{2}{5}$).

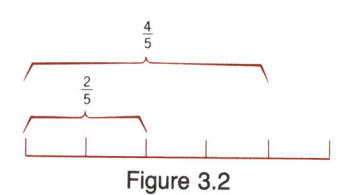

Figure 3.2

In general, if m is less than n, the fraction $\frac{m}{d}$ is less than the fraction $\frac{n}{d}$.

We sometimes need to arrange two or more fractions in order, from least to greatest (or greatest to least). If the fractions are like fractions, we arrange them in order by starting with the fraction with the least numerator. If the fractions are unlike, we first change them to like fractions.

Example 4   Write each set of fractions in order, from least to greatest.

   a. $\frac{7}{9}$, $\frac{1}{9}$, and $\frac{4}{9}$ b. $\frac{1}{9}$, $\frac{5}{12}$, and $\frac{2}{15}$

Solutions   a. Because 1 is less than 4 and 4 is less than 7, we write

$$\frac{1}{9}, \frac{4}{9}, \frac{7}{9}$$

3.3 LEAST COMMON DENOMINATOR

b. We first find the LCD of the three fractions. The building factors are: $180 \div 9 = 20$, $180 \div 12 = 15$, and $180 \div 15 = 12$.

$$\begin{cases} 9 = 3 \times 3 \\ 12 = 2 \times 2 \times 3 \\ 15 = 3 \times 5 \\ \text{LCD: } 2 \times 2 \times 3 \times 3 \times 5 = 180 \end{cases}$$

Hence,

$$\frac{1}{9} = \frac{1 \times 20}{9 \times 20} = \frac{20}{180}, \quad \frac{5}{12} = \frac{5 \times 15}{12 \times 15} = \frac{75}{180}, \text{ and}$$

$$\frac{2}{15} = \frac{2 \times}{15 \times} = \frac{24}{180}$$

Because 20 is less than 24 and 24 is less than 75, we write

$$\frac{20}{180}, \frac{24}{180}, \frac{75}{180} \text{ or } \frac{1}{9}, \frac{2}{15}, \frac{5}{12}$$

EXERCISES 3.3

Write each group of fractions as a group of like fractions with the LCD as the common denominator. See Examples 1, 2 and 3.

Complete Example

$$\frac{2}{7}, \frac{3}{5}$$

Each denominator is prime. The LCD is $7 \times 5 = 35$. Build each fraction.

$$35 \div 7 = 5; \quad 35 \div 5 = 7$$

The respective building factors are 5 and _____ .
 (1)

$$\frac{2}{7} = \frac{2 \times 5}{7 \times 5} = \underline{} \quad ; \quad \frac{3}{5} = \frac{3 \times 7}{5 \times 7} = \underline{}$$
 (2) (3)

Ans. $\frac{10}{35}, \frac{21}{35}$

(1) 7   (2) $\frac{10}{35}$ (3) $\frac{21}{35}$

1. $\frac{2}{3}, \frac{3}{4}$ 2. $\frac{1}{3}, \frac{2}{5}$ 3. $\frac{1}{4}, \frac{2}{7}$ 4. $\frac{3}{5}, \frac{5}{6}$ 5. $\frac{2}{3}, \frac{1}{6}$ 6. $\frac{1}{2}, \frac{5}{16}$

FRACTIONS

7. $\frac{7}{1}, \frac{1}{6}$
8. $\frac{5}{1}, \frac{2}{9}$
9. $\frac{5}{6}, \frac{3}{10}$
10. $\frac{3}{8}, \frac{1}{20}$
11. $\frac{5}{12}, \frac{3}{30}$
12. $\frac{3}{16}, \frac{5}{24}$

Complete Example

$\frac{1}{6}, \frac{5}{12}, \frac{4}{9}$

Find the LCD. The LCD must contain the factor 2 two times and the factor 3 two times. The LCD is _____(1)_____ .

$$\begin{cases} 6 = 2 \times 3 \\ 12 = 2 \times 3 \times 2 \\ 9 = 3 \times 3 \\ \text{LCD:} \quad 2 \times 3 \times 2 \times 3 \end{cases}$$

Build each fraction:

$$36 \div 6 = 6; \quad 36 \div 12 = 3; \quad 36 \div 9 = 4$$

The respective building factors are 6, 3, and _____(2)_____ .

$$\frac{1}{6} = \frac{1 \times 6}{6 \times 6} = \frac{6}{36}; \quad \frac{5}{12} = \frac{5 \times 3}{12 \times 3} = \underline{}_{(3)}; \quad \frac{4}{9} = \frac{4 \times 4}{9 \times 4} = \underline{}_{(4)}$$

Ans. $\frac{6}{36}, \frac{15}{36}, \frac{16}{36}$

---

(1) 36   (2) 4   (3) $\frac{15}{36}$ (4) $\frac{16}{36}$

---

13. $\frac{1}{4}, \frac{2}{3}, \frac{1}{6}$
14. $\frac{1}{6}, \frac{5}{18}, \frac{5}{9}$
15. $\frac{7}{15}, \frac{1}{30}, \frac{2}{5}$
16. $\frac{1}{20}, \frac{3}{20}, \frac{7}{15}$

17. $\frac{5}{18}, \frac{1}{9}, \frac{1}{12}$
18. $\frac{5}{6}, \frac{1}{18}, \frac{5}{12}$
19. $\frac{5}{1}, \frac{2}{15}, \frac{3}{20}$
20. $\frac{3}{1}, \frac{11}{12}, \frac{5}{16}$

21. $\frac{6}{1}, \frac{1}{6}, \frac{5}{8}$
22. $\frac{1}{16}, \frac{1}{18}, \frac{4}{1}$
23. $\frac{5}{24}, \frac{3}{1}, \frac{1}{18}$
24. $\frac{1}{36}, \frac{1}{24}, \frac{2}{1}$

Write each set of fractions in order, from the least to the greatest. See Example 4.

Complete Examples

a. $\frac{6}{7}, \frac{2}{7}, \frac{5}{7}$

2 is less than 5 and 5 is less than 6.

Ans. $\frac{2}{7}, \underline{}_{(1)}, \frac{6}{7}$

100

3.4 ADDITION

b. $\frac{5}{12}, \frac{1}{6}, \frac{4}{9}$

As shown on page 100, $\frac{5}{12} = \frac{15}{36}$, $\frac{1}{6} = \frac{6}{36}$, and $\frac{4}{9} =$ _____
(2)

6 is less than 15 and 15 is less than _____ .
(3)

Ans. $\frac{6}{36}, \frac{15}{36}, \frac{16}{36}$ or $\frac{1}{6}, \frac{5}{12},$ _____
(4)

---

(1) $\frac{5}{7}$ (2) $\frac{16}{36}$ (3) 16   (4) $\frac{4}{9}$

---

25. $\frac{5}{11}, \frac{2}{11}, \frac{3}{11}$ 26. $\frac{1}{15}, \frac{7}{15}, \frac{2}{15}$ 27. $\frac{5}{12}, \frac{7}{12}, \frac{1}{12}$ 28. $\frac{3}{8}, \frac{1}{8}, \frac{5}{8}$

29. $\frac{1}{6}, \frac{1}{12}, \frac{5}{24}$ 30. $\frac{1}{5}, \frac{3}{10}, \frac{4}{15}$ 31. $\frac{3}{7}, \frac{5}{14}, \frac{10}{21}$ ✓ 32. $\frac{3}{4}, \frac{5}{8}, \frac{15}{16}$

33. $\frac{2}{3}, \frac{3}{4}, \frac{7}{12}$ 34. $\frac{5}{6}, \frac{11}{12}, \frac{3}{4}$ 35. $\frac{9}{16}, \frac{15}{32}, \frac{7}{8}$ 36. $\frac{7}{12}, \frac{19}{36}, \frac{37}{72}$

Solve each problem. Express answers in fraction form.

37. Three packages of sugar weigh $\frac{13}{16}$, $\frac{3}{4}$, and $\frac{7}{8}$ of a pound, respectively. Which package weighs the most?

38. A slot in a machine is $\frac{1}{2}$-inch wide. Which of the bars with the following widths will not fit into the slot?

$\frac{31}{64}$ inch $\frac{15}{32}$ inch $\frac{5}{8}$ inch $\frac{9}{16}$ inch

39. If the head of a bolt is $\frac{1}{2}$ inch wide, which of the following size wrenches are too small to turn the bolt?

$\frac{8}{16}$ inch $\frac{15}{32}$ inch $\frac{31}{64}$ inch $\frac{5}{8}$ inch

40. Auger bits (used to drill holes in wood) are numbered to indicate sixteenths of an inch. Thus, a #7 auger bit drills a $\frac{7}{16}$-inch hole. What number auger bit drills:

   a. A $\frac{1}{4}$-inch hole?   b. A 1-inch hole?

3.4 ADDITION

The sum of two or more like fractions can be written as a single fraction by the following rule.

FRACTIONS

> **The sum of two or more like fractions is a fraction with the same denominator and a numerator equal to the sum of the numerators of the original fractions.**

In symbols, the sum of two like fractions is represented by

$$\frac{a}{d} + \frac{b}{d} = \frac{a+b}{d}$$

Example 1 a. $\dfrac{1}{8} + \dfrac{3}{8} = \dfrac{1+3}{8}$
$= \dfrac{4}{8} = \dfrac{1}{2}$

b. $\dfrac{4}{15} + \dfrac{2}{15} + \dfrac{7}{15} = \dfrac{4+2+7}{15}$
$= \dfrac{13}{15}$

Note that, as in Example 1(a), we follow the customary practice of writing the answer in lowest terms.

> To find the sum of two or more unlike fractions:
> 1. Find the lowest common denominator LCD of the given fractions.
> 2. Change the given fractions into like fractions with the LCD as the common denominator.
> 3. Find the sum of the resulting like fractions.
> 4. Reduce to lowest terms.

Example 2 $\dfrac{5}{12} + \dfrac{3}{10}$

Solution Because $12 = 2 \times 2 \times 3$ and $10 = 2 \times 5$, the LCD of the fractions is

$$2 \times 2 \times 3 \times 5 = 60$$

Since

$$\frac{5}{12} = \frac{5 \times 5}{2 \times 5} = \frac{25}{60} \quad \text{and} \quad \frac{3}{10} = \frac{3 \times 6}{10 \times 6} = \frac{18}{60},$$

we have

$$\frac{5}{12} + \frac{3}{10} = \frac{25}{60} + \frac{18}{60}$$

$$= \frac{25+18}{60} = \frac{43}{60}$$

Example 3 $\dfrac{2}{15} + \dfrac{1}{2} + \dfrac{4}{9}$

Solution Because $15 = 3 \times 5$, $2 = 2$, and $9 = 3 \times 3$, the LCD of the fractions is

$$3 \times 3 \times 2 \times 5 = 90.$$

3.4 ADDITION

Since

$$\frac{2}{15} = \frac{2 \times 6}{15 \times 6} = \frac{12}{90}, \qquad \frac{1}{2} = \frac{1 \times 45}{2 \times 45} = \frac{45}{90}, \quad \text{and} \quad \frac{4}{9} = \frac{4 \times 10}{9 \times 10} = \frac{40}{90},$$

we have

$$\frac{2}{15} + \frac{1}{2} + \frac{4}{9} = \frac{12}{90} + \frac{45}{90} + \frac{40}{90}$$

$$= \frac{12 + 45 + 40}{90} = \frac{97}{90}$$

MIXED NUMBERS

It is customary to write a sum such as $3 + \frac{4}{7}$ as $3\frac{4}{7}$, omitting the + symbol. Numbers such as $3\frac{4}{7}$ that consist of a whole number part and a fractional part are called **mixed numbers.** When necessary, such numbers can be written as improper fractions. For example,

$$3\frac{4}{7} = 3 + \frac{4}{7} = \frac{3}{1} + \frac{4}{7}$$

The LCD for $\frac{3}{1}$ and $\frac{4}{7}$ is 7. Hence,

$$3\frac{4}{7} = \frac{3}{1} + \frac{4}{7} = \frac{3 \times 7}{1 \times 7} + \frac{4}{7}$$

$$= \frac{21}{7} + \frac{4}{7} = \frac{25}{7}$$

Example 4 a. $5\frac{3}{4} = 5 + \frac{3}{4}$ b. $11\frac{5}{16} = 11 + \frac{5}{16}$

$$= \frac{5 \times 4}{1 \times 4} + \frac{3}{4} \qquad\qquad = \frac{11 \times 16}{1 \times 16} + \frac{5}{16}$$

$$= \frac{20}{4} + \frac{3}{4} = \frac{23}{4} \qquad\qquad = \frac{176}{16} + \frac{5}{16} = \frac{181}{16}$$

The procedure shown in the above examples is sometimes shortened as follows:

$$5\frac{3}{4} = \frac{(5 \times 4) + 3}{4} = \frac{23}{4} \quad \text{and} \quad 11\frac{5}{16} = \frac{(11 \times 16) + 5}{16} = \frac{181}{16}$$

We can also change an improper fraction to a mixed number. For example, consider the fraction $\frac{23}{4}$. We first divide 23 by 4 to obtain the whole number 5 and a remainder. Because $4 \times 5 = 20$, there is a remainder of 3 ($23 - 20 = 3$); hence $\frac{23}{4} = 5\frac{3}{4}$. This result is sometimes obtained as follows:

$$\frac{23}{4} = \frac{20 + 3}{4} = \frac{20}{4} + \frac{3}{4}$$

$$= 5 + \frac{3}{4} = 5\frac{3}{4}$$

FRACTIONS

Example 5 Change each improper fraction to a mixed number.

a. $\dfrac{15}{4}$ 	b. $\dfrac{87}{16}$

Solutions a. 15 ÷ 4 equals 3 and a remainder. Hence, the whole number part of the mixed number is 3. Because 3 × 4 = 12 and 15 − 12 = 3, we have

$$\frac{15}{4} = \frac{12+3}{4} = \frac{12}{4} + \frac{3}{4}$$

$$= 3 + \frac{3}{4} = 3\frac{3}{4}$$

b. 87 ÷ 16 equals 5 and a remainder. Hence, the whole number part of the mixed number is 5. Because 5 × 16 = 80 and 87 − 80 = 7, we have

$$\frac{87}{16} = \frac{80+7}{16} = \frac{80}{16} + \frac{7}{16}$$

$$= 5 + \frac{7}{16} = 5\frac{7}{16}$$

ADDITION OF MIXED NUMBERS

An efficient procedure for adding mixed numbers involves separating the whole number and fractional parts.

Example 6
$$2\frac{7}{8} + 3\frac{3}{8} + 11\frac{5}{8} = (2 + 3 + 11) + \left(\frac{7}{8} + \frac{3}{8} + \frac{5}{8}\right)$$
$$= 16 + \frac{15}{8} = 16 + \left(\frac{8}{8} + \frac{7}{8}\right) = 16 + 1 + \frac{7}{8} = 17\frac{7}{8}.$$

If the fractional parts of the mixed numbers are unlike fractions, they are first changed into like fractions.

Example 7 $5 + 12\dfrac{5}{8} + 9\dfrac{7}{16}$

Solution The LCD of the fractional parts is 16. Hence,

$$5 + 12\frac{5}{8} + 9\frac{7}{16} = 5 + 12\frac{5 \times 2}{8 \times 2} + 9\frac{7}{16}$$

$$= 5 + 12\frac{10}{16} + 9\frac{7}{16}$$

$$= \left(5 + 12 + 9\right) + \left(\frac{10}{16} + \frac{7}{16}\right)$$

$$= 26 + \frac{17}{16} = 26 + 1 + \frac{1}{16} = 27\frac{1}{16}.$$

Computations involving mixed numbers are sometimes written in a vertical arrangement.

3.4 ADDITION

Example 8 a. $\quad 2\frac{7}{8}$

$\qquad 3\frac{3}{8}$

$\qquad 11\frac{5}{8}$

$\qquad 16\frac{15}{8} = 16 + 1 + \frac{7}{8} = 17\frac{7}{8}$

b. $\quad 5 \quad = \quad 5$

$\qquad 12\frac{5}{8} \;= 12\frac{10}{16}$

$\qquad 9\frac{7}{16} = \;\;9\frac{7}{16}$

$\qquad\qquad 26\frac{17}{16} = 26 + 1 + \frac{1}{16} = 27\frac{1}{16}$

EXERCISES 3.4

Express fractional answers in lowest terms and improper fractions as mixed numbers. Find each sum. See Examples 1 and 5.

Complete Examples

a. $\dfrac{7}{12} + \dfrac{3}{12} = \dfrac{7+3}{12}$

$\qquad\quad = \dfrac{10}{12} = \underline{\qquad}$
$\qquad\qquad\qquad\qquad\;$ (1)

b. $\dfrac{5}{11} + \dfrac{4}{11} + \dfrac{7}{11} = \dfrac{5+4+7}{11}$

$\qquad\qquad\qquad = \dfrac{16}{11} = \underline{\qquad}$
$\qquad\qquad\qquad\qquad\qquad\quad$ (2)

(1) $\dfrac{5}{6}$ (2) $1\dfrac{5}{11}$

1. $\dfrac{2}{7} + \dfrac{3}{7}$ 2. $\dfrac{1}{9} + \dfrac{4}{9}$ 3. $\dfrac{1}{12} + \dfrac{5}{12}$ 4. $\dfrac{3}{10} + \dfrac{5}{10}$

5. $\dfrac{1}{16} + \dfrac{5}{16} + \dfrac{13}{16}$ 6. $\dfrac{5}{32} + \dfrac{7}{32} + \dfrac{29}{32}$ 7. $\dfrac{1}{15} + \dfrac{7}{15} + \dfrac{4}{15}$ 8. $\dfrac{5}{18} + \dfrac{17}{18} + \dfrac{11}{18}$

105

FRACTIONS

Find each sum. See Examples 2, 3, and 5.

Complete Examples

a. $\dfrac{3}{5} + \dfrac{5}{6}$

The LCD is $5 \times 6 =$ _____ .
$$(1)$$

$\dfrac{3 \times 6}{5 \times 6} + \dfrac{5 \times 5}{6 \times 5}$

$= \dfrac{18 + 25}{30}$

$= \dfrac{43}{30} = \dfrac{30 + 13}{30}$

$= \dfrac{30}{30} + \dfrac{13}{30} =$ _____
$$(2)$$

b. $\dfrac{1}{3} + \dfrac{1}{6} + \dfrac{4}{15}$

The LCD is $3 \times 2 \times 5 =$ _____
$$(3)$$

$\dfrac{1 \times 10}{3 \times 10} + \dfrac{1 \times 5}{6 \times 5} + \dfrac{4 \times 2}{15 \times 2}$

$= \dfrac{10 + 5 + 8}{30}$

$=$ _____
(4)

---

(1) 30 (2) $1\dfrac{13}{30}$ (3) 30 (4) $\dfrac{23}{30}$

---

9. $\dfrac{2}{3} + \dfrac{3}{4}$

10. $\dfrac{1}{3} + \dfrac{2}{5}$

11. $\dfrac{5}{8} + \dfrac{3}{10}$

12. $\dfrac{3}{8} + \dfrac{1}{20}$

13. $\dfrac{1}{12} + \dfrac{3}{20} + \dfrac{7}{15}$

14. $\dfrac{5}{18} + \dfrac{1}{27} + \dfrac{1}{12}$

15. $3 + \dfrac{11}{12} + \dfrac{5}{16}$

16. $6 + \dfrac{3}{20} + \dfrac{5}{32}$

Change each mixed number to an improper fraction. See Example 4.

Complete Examples

a. $4\dfrac{3}{7} = \dfrac{(4 \times 7) + 3}{7}$

$= \dfrac{28 + 3}{7}$

$=$ _____
(1)

b. $13\dfrac{3}{5} = \dfrac{(13 \times 5) + 3}{5}$

$= \dfrac{65 + 3}{5}$

$=$ _____
(2)

---

(1) $\dfrac{31}{7}$ (2) $\dfrac{68}{5}$

---

17. $5\dfrac{2}{3}$

18. $6\dfrac{2}{5}$

19. $12\dfrac{4}{5}$

20. $10\dfrac{1}{4}$

106

3.4 ADDITION

21. $21\frac{5}{16}$ **22.** $30\frac{3}{16}$ **23.** $29\frac{11}{32}$ **24.** $42\frac{27}{32}$

Change each improper fraction to a mixed number. See Example 5.

Complete Examples

a. $\frac{25}{7}$ equals 3 and a remainder of $25 - (3 \times 7)$.

$\frac{25}{7} = \frac{21 + 4}{7} = \frac{21}{7} + \underline{}$
(1)

$= 3 + \frac{4}{7} = \underline{}$
(2)

b. $\frac{26}{11}$ equals 2 and a remainder of $26 - (2 \times 11)$.

$\frac{26}{11} = \frac{22 + 4}{11} = \frac{22}{11} + \underline{}$
(3)

$= 2 + \frac{4}{11} = \underline{}$
(4)

(1) $\frac{4}{7}$ (2) $3\frac{4}{7}$ (3) $\frac{4}{11}$ (4) $2\frac{4}{11}$

25. $\frac{11}{3}$ **26.** $\frac{15}{4}$ **27.** $\frac{43}{8}$ **28.** $\frac{37}{8}$

29. $\frac{107}{16}$ **30.** $\frac{123}{16}$ **31.** $\frac{317}{32}$ **32.** $\frac{361}{32}$

Find each sum. See Example 6.

Complete Examples

a. $3\frac{1}{4} + 6\frac{1}{4}$

$= (3 + 6) + \left(\frac{1}{4} + \frac{1}{4}\right)$

$= 9 + \underline{}$
(1)

$= 9\frac{2}{4} = \underline{}$
(2)

b. $3\frac{2}{5} + 4\frac{3}{5} + 2\frac{1}{5}$

$= (3 + 4 + 2) + \left(\frac{2}{5} + \frac{3}{5} + \frac{1}{5}\right)$

$= 9 + \underline{}$
(3)

$= 9 + 1 + \frac{1}{5} = \underline{}$
(4)

(1) $\frac{2}{4}$ (2) $9\frac{1}{2}$ (3) $\frac{6}{5}$ (4) $10\frac{1}{5}$

107

FRACTIONS

33. $2\frac{2}{5} + 5\frac{1}{5}$

34. $1\frac{9}{32} + 2\frac{25}{32} + 3\frac{9}{32}$

35. $6\frac{3}{8} + 5\frac{1}{8}$

36. $2\frac{1}{8} + 3\frac{3}{8} + 6\frac{5}{8}$

37. $8 + 7\frac{11}{16} + 2\frac{9}{16}$

38. $2 + 3\frac{5}{8} + 6\frac{7}{8}$

Find each sum. See Examples 7 and 8.

Complete Example

$$2\frac{3}{4} + 5\frac{1}{12} + 4\frac{5}{6}$$

The LCD of the fractional parts is 12.

$$2\frac{3}{4} + 5\frac{1}{12} + 4\frac{5}{6} = 2\frac{3 \times 3}{4 \times 3} + 5\frac{1}{12} + 4\frac{5 \times 2}{6 \times 2}$$

$$= 2\frac{9}{12} + 5\frac{1}{12} + 4\frac{10}{12}$$

$$= \left(2 + 5 + \underline{}_{(1)}\right) + \left(\frac{9}{12} + \frac{1}{12} + \frac{10}{12}\right)$$

$$= 11 + \frac{20}{12} = 11 + \underline{}_{(2)} + \frac{8}{12}$$

$$= 12\frac{8}{12} = \underline{}_{(3)}$$

(1) 4   (2) 1   (3) $12\frac{2}{3}$

39. $3\frac{1}{4} + 1\frac{5}{8}$

40. $5\frac{3}{4} + 2\frac{7}{8}$

41. $1\frac{1}{15} + 2\frac{2}{3} + 1\frac{1}{12}$

42. $3\frac{1}{6} + 2\frac{5}{27} + 2\frac{5}{18}$

43. $3 + 5\frac{2}{3} + 6\frac{3}{4}$

44. $2 + 4\frac{1}{3} + 2\frac{1}{4}$

45. $\begin{array}{r} 4 \\ 5\frac{1}{3} \\ 2\frac{2}{9} \\ \hline \end{array}$

46. $\begin{array}{r} 2 \\ 3\frac{3}{4} \\ 4\frac{5}{8} \\ \hline \end{array}$

47. $\begin{array}{r} 2\frac{5}{16} \\ 3\frac{7}{8} \\ 1\frac{1}{4} \\ \hline \end{array}$

48. $\begin{array}{r} 5\frac{3}{4} \\ 2\frac{3}{8} \\ 1\frac{1}{2} \\ \hline \end{array}$

3.4 ADDITION

49. $12\frac{1}{4}$
    $24\frac{7}{8}$
    $2\frac{9}{16}$

50. $33\frac{11}{16}$
    $13\frac{3}{4}$
    $5\frac{5}{8}$

51. $12\frac{13}{16}$
    $16\frac{1}{4}$
    $5\frac{7}{8}$

52. $81\frac{1}{3}$
    $73\frac{1}{4}$
    $42\frac{7}{12}$

Find the missing dimension in each of the following drawings.

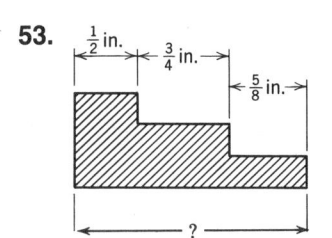

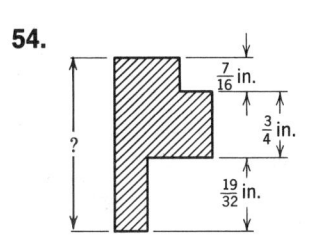

53. [Drawing with dimensions $\frac{1}{2}$ in., $\frac{3}{4}$ in., $\frac{5}{8}$ in., and ? for total length]

54. [Drawing with dimensions $\frac{7}{16}$ in., $\frac{3}{4}$ in., $\frac{19}{32}$ in., and ? for total height]

A building supply company sells four grades (A, B, C, and D) of crushed rock. The table shows the number of tons sold each day during one week.

Grade	Mon.	Tues.	Wed.	Thurs.	Fri.	Sat.
A	$18\frac{1}{2}$	$15\frac{3}{8}$	$20\frac{3}{4}$	25	$14\frac{1}{2}$	$10\frac{1}{8}$
B	$12\frac{1}{4}$	$16\frac{1}{2}$	$18\frac{3}{8}$	$17\frac{3}{4}$	$9\frac{7}{8}$	8
C	$6\frac{3}{8}$	$5\frac{3}{4}$	$12\frac{1}{4}$	$11\frac{3}{8}$	8	$6\frac{1}{2}$
D	$9\frac{3}{4}$	$10\frac{5}{8}$	22	$16\frac{1}{2}$	$17\frac{1}{4}$	$5\frac{3}{4}$

Find the total tonnage sold on the indicated day.

55. Monday
56. Tuesday
57. Wednesday
58. Thursday
59. Friday
60. Saturday

Find the total tonnage of each grade sold during the week.

61. Grade A
62. Grade B
63. Grade C
64. Grade D

FRACTIONS

3.5 SUBTRACTION

The difference of two fractions can be written as a single fraction by the following rule.

> **The difference of two like fractions is a fraction with the same denominator and a numerator equal to the difference of the numerators of the original fractions.**

In symbols, the difference of two like fractions is represented by

$$\frac{a}{d} - \frac{b}{d} = \frac{a - b}{d}$$

The procedure for subtracting two fractions can be extended to problems involving more than two fractions, as illustrated by Example 1(b) below.

Example 1   a. $\dfrac{3}{5} - \dfrac{2}{5} = \dfrac{3 - 2}{5}$ b. $\dfrac{19}{32} - \dfrac{11}{32} - \dfrac{2}{32} = \dfrac{19 - 11 - 2}{32}$

$= \dfrac{1}{5}$ $\phantom{b.\ \dfrac{19}{32} - \dfrac{11}{32} - \dfrac{2}{32}\ }= \dfrac{6}{32} = \dfrac{3}{16}$

To Subtract Unlike Fractions:

1. Find the least common denominator (LCD) of the given fractions.
2. Change the given fractions into like fractions with the LCD as the common denominator.
3. Subtract the resulting like fractions.
4. Reduce to lowest terms.

Example 2   a. $\dfrac{5}{12} - \dfrac{3}{10}$ b. $\dfrac{1}{2} - \dfrac{2}{9} - \dfrac{2}{15}$

Solutions   a. Because $12 = 2 \times 2 \times 3$ and $10 = 2 \times 5$, the LCD of the two fractions is

$$2 \times 2 \times 3 \times 5 = 60$$

Hence,

$$\frac{5}{12} - \frac{3}{10} = \frac{5 \times 5}{12 \times 5} - \frac{3 \times 6}{10 \times 6} = \frac{25}{60} - \frac{18}{60}$$

$$= \frac{25 - 18}{60} = \frac{7}{60}$$

b. Because $2 = 2$, $9 = 3 \times 3$, and $15 = 3 \times 5$, the LCD of the three fractions is

$$2 \times 3 \times 3 \times 5 = 90.$$

Hence,

$$\frac{1}{2} - \frac{2}{9} - \frac{2}{15} = \frac{1 \times 45}{2 \times 45} - \frac{2 \times 10}{9 \times 10} - \frac{2 \times 6}{15 \times 6}$$

3.5 SUBTRACTION

$$\frac{1}{2} - \frac{2}{9} - \frac{2}{15} = \frac{45}{90} - \frac{20}{90} - \frac{12}{90}$$
$$= \frac{45 - 20 - 12}{90} = \frac{13}{90}$$

SUBTRACTION OF MIXED NUMBERS

Some differences of mixed numbers can be found by finding the difference of the fractional parts; finding the difference of the whole number; and then writing the *sum* of these two results.

Example 3 a. $12\frac{7}{8} - 8\frac{3}{8} = (12 - 8) + \left(\frac{7}{8} - \frac{3}{8}\right)$

$$= 4 + \frac{4}{8}$$

$$= 4 + \frac{1}{2} = 4\frac{1}{2}$$

b. $6\frac{3}{4} - 2\frac{1}{4} - 1\frac{1}{4} = (6 - 2 - 1) + \left(\frac{3}{4} - \frac{1}{4} - \frac{1}{4}\right)$

$$= 3 + \frac{1}{4} = 3\frac{1}{4}$$

If the fractional parts of the mixed numbers are unlike fractions, they are first changed into like fractions.

Example 4 $29\frac{13}{16} - 11\frac{1}{4} - 8\frac{1}{8}$

Solution The LCD is 16. Hence,

$$29\frac{13}{16} - 11\frac{1}{4} - 8\frac{1}{8} = 29\frac{13}{16} - 11\frac{1 \times 4}{4 \times 4} - 8\frac{1 \times 2}{8 \times 2}$$

$$= 29\frac{13}{16} - 11\frac{4}{16} - 8\frac{2}{16}$$

$$= (29 - 11 - 8) + \left(\frac{13}{16} - \frac{4}{16} - \frac{2}{16}\right)$$

$$= 10 + \frac{7}{16} = 10\frac{7}{16}$$

"BORROWING" IN SUBTRACTION

In problems such as $8\frac{1}{5} - 3\frac{4}{5}$, where the fractional part of the second number is greater than the fractional part of the first number, we can *borrow* 1 from the whole number part of the first number, change it into a fraction with an appropriate denominator, such as $\frac{2}{2}, \frac{3}{3}$, etc., and then add it to the fractional part. Thus, to compute

$$8\frac{1}{5} - 3\frac{4}{5}$$

FRACTIONS

we first write

$$8\frac{1}{5} = 8 + \frac{1}{5} = (7 + 1) + \frac{1}{5}$$
$$= 7 + \left(\frac{5}{5} + \frac{1}{5}\right) = 7\frac{6}{5}$$

Now, using $7\frac{6}{5}$ in place of $8\frac{1}{5}$, we have

$$8\frac{1}{5} - 3\frac{4}{5} = 7\frac{6}{5} - 3\frac{4}{5} = (7 - 3) + \left(\frac{6}{5} - \frac{4}{5}\right)$$
$$= 4 + \frac{2}{5} = 4\frac{2}{5}$$

Example 5 $31\frac{5}{12} - 22\frac{3}{4}$.

Solution The LCD is 12. Hence,

$$31\frac{5}{12} - 22\frac{3}{4} = 31\frac{5}{12} - 22\frac{3 \times 3}{4 \times 3}$$
$$= 31\frac{5}{12} - 22\frac{9}{12}$$

Because $\frac{9}{12}$ is greater than $\frac{5}{12}$, we rewrite $31\frac{5}{12}$ as follows:

$$31\frac{5}{12} = 30 + 1 + \frac{5}{12} = 30 + \frac{12}{12} + \frac{5}{12}$$
$$= 30 + \frac{17}{12} = 30\frac{17}{12}$$

Now, using $30\frac{17}{12}$ in place of $31\frac{5}{12}$, we have

$$31\frac{5}{12} - 22\frac{3}{4} = 30\frac{17}{12} - 22\frac{9}{12} = (30 - 22) + \left(\frac{17}{12} - \frac{9}{12}\right)$$
$$= 8 + \frac{8}{12} = 8 + \frac{2}{3} = 8\frac{2}{3}$$

Another type of problem that requires "borrowing" is illustrated in the next example.

Example 6 $11 - 5\frac{1}{8}$.

Solution First, we write

$$11 = 10 + 1$$
$$= 10 + \frac{8}{8} = 10\frac{8}{8}.$$

Then

$$11 - 5\frac{1}{8} = 10\frac{8}{8} - 5\frac{1}{8} = (10 - 5) + \left(\frac{8}{8} - \frac{1}{8}\right)$$
$$= 5 + \frac{7}{8} = 5\frac{7}{8}$$

3.5 SUBTRACTION

A vertical arrangement is sometimes used to find differences.

Example 7 a. $31\frac{5}{12} = 31\frac{5}{12} = 30\frac{17}{12}$ b. $11 = 10\frac{8}{8}$
$-22\frac{3}{4} = -22\frac{9}{12} = -22\frac{9}{12}$ $-5\frac{1}{8} = -5\frac{1}{8}$
$8\frac{8}{12} = 8\frac{2}{3}$ $5\frac{7}{8}$

MIXED ADDITIONS AND SUBTRACTIONS

The methods of Section 3.4 and this section can be combined to compute mixed sums and differences.

Example 8 $16\frac{2}{3} - 4\frac{1}{12} + 10\frac{3}{4}$.

Solution The LCD is 12. Hence,

$$16\frac{2}{3} - 4\frac{1}{12} + 10\frac{3}{4} = 16\frac{2 \times 4}{3 \times 4} - 4\frac{1}{12} + 10\frac{3 \times 3}{4 \times 3}$$

$$= 16\frac{8}{12} - 4\frac{1}{12} + 10\frac{9}{12}$$

$$= (16 - 4 + 10) + \left(\frac{8}{12} - \frac{1}{12} + \frac{9}{12}\right)$$

$$= 22 + \frac{16}{12}$$

$$= 22 + 1 + \frac{4}{12} = 23\frac{1}{3}$$

EXERCISES 3.5

Express fractional answers in lowest terms and improper fractions as mixed numbers. Compute each difference. See Example 1.

Complete Examples

a. $\frac{6}{7} - \frac{2}{7}$

$= \frac{6-2}{7} = \underline{}$
(1)

b. $\frac{11}{16} - \frac{5}{16} - \frac{2}{16}$

$= \frac{11-5-2}{16} = \frac{4}{16} = \underline{}$
(2)

(1) $\frac{4}{7}$ (2) $\frac{1}{4}$

113

FRACTIONS

1. $\dfrac{6}{7} - \dfrac{2}{7}$
2. $\dfrac{3}{4} - \dfrac{1}{4}$
3. $\dfrac{9}{16} - \dfrac{5}{16}$
4. $\dfrac{17}{32} - \dfrac{3}{32}$
5. $\dfrac{27}{32} - \dfrac{11}{32} - \dfrac{13}{32}$
6. $\dfrac{23}{25} - \dfrac{7}{25} - \dfrac{6}{25}$
7. $\dfrac{15}{16} - \dfrac{9}{16} - \dfrac{2}{16}$
8. $\dfrac{10}{20} - \dfrac{2}{20} - \dfrac{4}{20}$

See Example 2.

Complete Example

$$\dfrac{11}{12} - \dfrac{1}{3} - \dfrac{3}{10}$$

The LCD is $2 \times 2 \times 3 \times 5 = \underline{}$.
(1)

$$\dfrac{11 \times 5}{12 \times 5} - \dfrac{1 \times 20}{3 \times 20} - \dfrac{3 \times 6}{10 \times 6}$$

Change each fraction into like fractions with denominator 60.

$$= \dfrac{55}{60} - \dfrac{20}{60} - \underline{}$$
(2)

Subtract numerators.

$$= \dfrac{55 - 20 - 18}{60} = \underline{}$$
(3)

(1) 60   (2) $\dfrac{18}{60}$ (3) $\dfrac{17}{60}$

9. $\dfrac{3}{4} - \dfrac{2}{3}$
10. $\dfrac{5}{6} - \dfrac{3}{5}$
11. $\dfrac{3}{8} - \dfrac{1}{20}$
12. $\dfrac{5}{24} - \dfrac{3}{16}$
13. $\dfrac{15}{16} - \dfrac{3}{8} - \dfrac{1}{4}$
14. $\dfrac{31}{32} - \dfrac{5}{16} - \dfrac{1}{2}$
15. $2 - \dfrac{5}{8} - \dfrac{1}{2}$
16. $3 - \dfrac{5}{16} - \dfrac{3}{20}$

See Example 3.

Complete Example

$$7\dfrac{11}{15} - 4\dfrac{2}{15} - 1\dfrac{4}{15}$$

$$= (7 - 4 - 1) + \left(\dfrac{11}{15} - \dfrac{2}{15} - \dfrac{4}{15}\right)$$

Find the difference of the whole number parts and the fractional parts.

$$= \underline{} + \dfrac{11 - 2 - 4}{15}$$
(1)

114

3.5 SUBTRACTION

$$= 2 + \underline{\qquad}_{(2)}$$

Add the two results.

$$= 2\frac{5}{15} = \underline{\qquad}_{(3)}$$

Reduce the fractional part.

---

(1) 2   (2) $\frac{5}{15}$ (3) $2\frac{1}{3}$

---

17. $5\frac{13}{16} - 2\frac{7}{16}$

18. $6\frac{2}{3} - 3\frac{1}{3}$

19. $8\frac{19}{32} - 4\frac{5}{32}$

20. $10\frac{5}{6} - 3\frac{1}{6}$

21. $5\frac{6}{7} - 2\frac{3}{7} - 1\frac{2}{7}$

22. $8\frac{8}{9} - 3\frac{2}{9} - 2\frac{5}{9}$

See Example 4.

Complete Example

$$10\frac{11}{12} - 3\frac{1}{4} - 1\frac{1}{2}$$

The LCD is $2 \times 2 \times 3 = \underline{\qquad}_{(1)}$.

$$10\frac{11}{12} - 3\frac{1 \times 3}{4 \times 3} - 1\frac{1 \times 6}{2 \times 6}$$

Change each fractional part into like fractions with denominator 12.

$$= 10\frac{11}{12} - 3\frac{3}{12} - 1\frac{6}{12}$$

$$= (10 - 3 - 1) + \left(\frac{11}{12} - \frac{3}{12} - \frac{6}{12}\right)$$

Find the difference of the whole number parts and the fractional parts.

$$= 6 + \underline{\qquad}_{(2)} = \underline{\qquad}_{(3)}$$

Add the two results and reduce the fractional part.

---

(1) 12   (2) $\frac{2}{12}$ (3) $6\frac{1}{6}$

---

23. $12\frac{25}{32} - 5\frac{1}{4} - 2\frac{3}{8}$

24. $10\frac{11}{12} - 3\frac{1}{4} - 1\frac{1}{3}$

25. $18\frac{51}{64} - 6\frac{5}{16} - 3\frac{3}{8}$

26. $21\frac{17}{18} - 8\frac{2}{3} - 2\frac{1}{6}$

27. $30\frac{23}{24} - 5\frac{1}{8} - 7\frac{2}{3}$

28. $25\frac{7}{8} - 4\frac{3}{16} - 2\frac{1}{4}$

FRACTIONS

See Example 5.

Complete Examples

a. $6\frac{3}{8} - 2\frac{1}{2}$

The LCD is $2 \times 2 \times 2 = $ _____ .
(1)

$6\frac{3}{8} - 2\frac{1 \times}{2 \times} = 6\frac{3}{8} - 2\frac{4}{8}$

Rewrite $6\frac{3}{8}$ as

$6\frac{3}{8} = 5 + 1 + \frac{3}{8} = 5 + \frac{8}{8} + \frac{3}{8}$

$= 5 + $ _____
(2)

Use $5\frac{11}{8}$ in place of $6\frac{3}{8}$.

$6\frac{3}{8} - 2\frac{4}{8} = 5\frac{11}{8} - 2\frac{4}{8} = (5 - 2) + \left(\frac{11}{8} - \frac{4}{8}\right)$

$= 3 + \frac{7}{8} = $ _____
(3)

b. $8\frac{1}{4} - 3\frac{1}{2}$

The LCD is $2 \times 2 = 4$.

$8\frac{1}{4} - 3\frac{1 \times 2}{2 \times 2} = 8\frac{1}{4} - 3\frac{2}{4}$

Rewrite $8\frac{1}{4}$ as

$8\frac{1}{4} = 7 + 1 + \frac{1}{4} = 7 + $ _____ $+ \frac{1}{4}$
(4)

$= 7 + \frac{5}{4}$

Use $7\frac{5}{4}$ in place of $8\frac{1}{4}$.

$8\frac{1}{4} - 3\frac{2}{4} = 7\frac{5}{4} - 3\frac{2}{4} = (7 - 3) + \left(\frac{5}{4} - \frac{2}{4}\right)$

$= 4 + $ _____ $= $ _____
(5)   (6)

---

(1) 8   (2) $\frac{11}{8}$ (3) $3\frac{7}{8}$ (4) $\frac{4}{4}$ (5) $\frac{3}{4}$ (6) $4\frac{3}{4}$

---

29. $9\frac{3}{8} - 2\frac{1}{2}$

30. $8\frac{7}{16} - 4\frac{3}{4}$

31. $10\frac{1}{2} - 6\frac{19}{32}$

32. $5\frac{1}{5} - 4\frac{7}{10}$

33. $7\frac{1}{4} - 3\frac{2}{3}$

34. $12\frac{1}{6} - 8\frac{3}{4}$

35. $7\frac{3}{8} - 1\frac{7}{12}$

36. $9\frac{2}{5} - 6\frac{2}{3}$

37. $16\frac{1}{4} - 9\frac{3}{5}$

38. $6\frac{5}{8} - 2\frac{5}{6}$

39. $11\frac{3}{4} - 2\frac{8}{9}$

40. $5\frac{3}{16} - 1\frac{5}{6}$

3.5 SUBTRACTION

See Example 6.

Complete Example

a. $8 - 4\frac{2}{7}$

Write 8 as $7 + 1 = 7 + \frac{7}{7} = 7\frac{7}{7}$.

Use $7\frac{7}{7}$ in place of 8.

$8 - 4\frac{2}{7} = 7\frac{7}{7} - 4\frac{2}{7}$

$= (7 - 4) + \left(\frac{7}{7} - \frac{2}{7}\right)$

$= \underline{}_{(1)} + \frac{5}{7} = \underline{}_{(2)}$

b. $5 - 1\frac{1}{5}$

Write 5 as $4 + 1 = 4 + \frac{5}{5} = 4\frac{5}{5}$.

Use $4\frac{5}{5}$ in place of 5.

$5 - 1\frac{1}{5} = 4\frac{5}{5} - 1\frac{1}{5}$

$= (4 - \underline{}_{(3)}) + \left(\frac{5}{5} - \frac{1}{5}\right)$

$= 3 + \frac{4}{5} = \underline{}_{(4)}$

(1) 3 (2) $3\frac{5}{7}$ (3) 1 (4) $3\frac{4}{5}$

41. $8 - 3\frac{1}{4}$

42. $6 - 2\frac{2}{3}$

43. $10 - 6\frac{5}{16}$

44. $4 - 1\frac{9}{32}$

45. $7 - 6\frac{3}{5}$

46. $8 - 5\frac{5}{8}$

See Example 7.

Complete Examples

a. $13\frac{3}{10} = 13\frac{3}{10} = 12\frac{13}{10}$
$-7\frac{3}{5} = -7\frac{6}{10} = -7\frac{6}{10}$
$\underline{}_{(1)}$

b. $12 = 11\frac{9}{9}$
$-3\frac{2}{9} = -3\frac{2}{9}$
$\underline{}_{(2)}$

(1) $5\frac{7}{10}$ (2) $8\frac{7}{9}$

47. $17\frac{3}{4}$
$-2\frac{5}{16}$

48. $39\frac{1}{2}$
$-16\frac{2}{3}$

49. $18\frac{2}{5}$
$-3\frac{3}{4}$

50. 48
$-27\frac{7}{16}$

117

FRACTIONS

See Example 8.

Complete Example

$$4\frac{3}{16} - 1\frac{1}{8} + 3\frac{3}{4}$$

The LCD is $2 \times 2 \times 2 \times 2 = \underline{}$.
(1)

$$4\frac{3}{16} - 1\frac{1}{8} + 3\frac{3}{4} = 4\frac{3}{16} - 1\frac{1 \times 2}{8 \times 2} + 3\frac{3 \times 4}{4 \times 4}$$

Change each fractional part into like fractions with denominator 16.

$$= 4\frac{3}{16} - \underline{} + 3\frac{12}{16}$$
(2)

Find the sum or difference of the whole number parts and the fractional parts.

$$= (4 - 1 + 3) + \left(\frac{3}{16} - \frac{2}{16} + \frac{12}{16}\right)$$

Add the two results.

$$= 6 + \underline{} = 6\frac{13}{16}$$
(3)

(1) 16   (2) $1\frac{2}{16}$ (3) $\frac{13}{16}$

51. $2\frac{3}{5} + 4\frac{5}{6} - 3\frac{2}{3}$

52. $6\frac{5}{8} + 3\frac{3}{10} - 4\frac{2}{5}$

53. $5\frac{1}{4} - 2\frac{3}{8} - 1\frac{3}{16} + 7\frac{1}{2}$

54. $8\frac{15}{32} - 2\frac{15}{64} - 3\frac{1}{16} + 2\frac{5}{8}$

55. $9\frac{2}{9} - 3\frac{5}{18} - 2\frac{3}{27} + 4\frac{1}{3}$

56. $16\frac{7}{10} - 7\frac{3}{5} - 3\frac{1}{20} + 2\frac{7}{15}$

Find the missing dimension in Exercises 57 and 58.

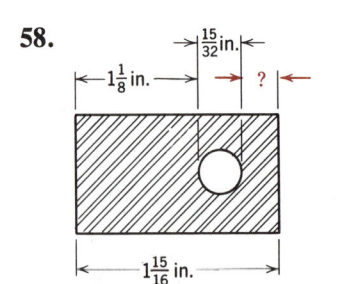

57. (diagram: $1\frac{1}{16}$ in., ?, $\frac{7}{8}$ in., $2\frac{3}{8}$ in.)

58. (diagram: $1\frac{1}{8}$ in., ?, $\frac{15}{32}$ in., $1\frac{15}{16}$ in.)

59. A tank contains 800 gal of oil at the start of a week. The table shows the number of gallons of oil used each day during the week. Determine how many gallons of oil remain in the tank at the end of each of the six days.

	Mon.	Tues.	Wed.	Thurs.	Fri.	Sat.
Gallons	$50\frac{1}{2}$	$60\frac{3}{4}$	$55\frac{3}{8}$	72	$83\frac{1}{2}$	$110\frac{1}{4}$
Remaining	___	___	___	___	___	___

60. A bakery stocked 680 lb of sugar for a week's baking. The table shows the amount of sugar used each day of the week. Determine how much sugar remained in stock at the end of each day.

	Mon.	Tues.	Wed.	Thurs.	Fri.	Sat.
Pounds	$117\frac{1}{2}$	$99\frac{3}{4}$	$74\frac{3}{8}$	$106\frac{7}{8}$	$123\frac{1}{2}$	$139\frac{1}{8}$
Remaining	___	___	___	___	___	___

3.6 MULTIPLICATION AND DIVISION

The product of two or more fractions can be written as a single fraction by the following rule.

> **The product of two or more fractions is a fraction whose numerator is the product of the numerators and whose denominator is the product of the denominators of the given fractions.**

In symbols, the product of two fractions can be represented by

$$\frac{a}{b} \times \frac{c}{d} = \frac{a \times c}{b \times d}$$

where it is understood that b and d are nonzero numbers.

Example 1 a. $\dfrac{2}{3} \times \dfrac{4}{5} = \dfrac{2 \times 4}{3 \times 5}$ b. $\dfrac{4}{15} \times \dfrac{7}{8} = \dfrac{4 \times 7}{15 \times 8}$

$= \dfrac{8}{15}$ $= \dfrac{28}{120} = \dfrac{7}{30}$

In Example 1b., note that *after* the product $\frac{28}{120}$ was obtained, it was reduced to lowest terms. It is usually easier to write all numerators and denominators in completely factored form and "divide out" any factors that are common factors of the numerators and denominators *before* multiplying. Using this method, Example 1b., appears as

$$\frac{4}{15} \times \frac{7}{8} = \frac{\overset{1}{\cancel{2}} \times \overset{1}{\cancel{2}}}{3 \times 5} \times \frac{7}{\underset{1}{\cancel{2}} \times \underset{1}{\cancel{2}} \times 2} = \frac{7}{30}$$

As another alternative, you may prefer to omit the step that shows the completely factored forms and proceed as follows:

$$\frac{\overset{1}{\cancel{4}}}{15} \times \frac{7}{\underset{2}{\cancel{8}}} = \frac{7}{30}$$

Note that in the above example we have shown "1" above 4 and "2" below 8 to emphasize that 1 and 2 are the quotients of dividing 4 and 8 by their common factor, 4. This is a common practice to avoid careless errors.

FRACTIONS

Example 2 a. $\dfrac{2}{3} \times \dfrac{5}{6} \times \dfrac{7}{15}$ b. $\dfrac{5}{6} \times \dfrac{13}{22} \times \dfrac{11}{39}$

$$= \dfrac{\cancel{2}^{1}}{3} \times \dfrac{\cancel{5}^{1}}{\cancel{2} \times 3} \times \dfrac{7}{3 \times \cancel{5}}$$

$$= \dfrac{5}{2 \times 3} \times \dfrac{\cancel{13}^{1}}{2 \times \cancel{11}} \times \dfrac{\cancel{11}^{1}}{3 \times \cancel{13}}$$

$$= \dfrac{7}{3 \times 3 \times 3} = \dfrac{7}{27}$$

$$= \dfrac{5}{2 \times 3 \times 2 \times 3} = \dfrac{5}{36}$$

MULTIPLICATION OF MIXED NUMBERS

To find the product of several factors when one (or more) of the factors is a mixed number, we first *change the mixed numbers to improper fractions*.

Example 3 a. $4\dfrac{3}{5} \times 3\dfrac{3}{4} = \dfrac{(5 \times 4) + 3}{5} \times \dfrac{(4 \times 3) + 3}{4}$

$$= \dfrac{23}{5} \times \dfrac{15}{4}$$

$$= \dfrac{23}{\cancel{5}} \times \dfrac{3 \times \cancel{5}^{1}}{2 \times 2} = \dfrac{69}{4} = 17\dfrac{1}{4}$$

b. $15 \times 16\dfrac{3}{5} = \dfrac{15}{1} \times \dfrac{(5 \times 16) + 3}{5}$

$$= \dfrac{15}{1} \times \dfrac{83}{5}$$

$$= \dfrac{3 \times \cancel{5}^{1}}{1} \times \dfrac{83}{\cancel{5}} = \dfrac{249}{1} = 249$$

The word "of" is often used to indicate multiplication.

Example 4 a. $\dfrac{2}{3}$ of $24 = \dfrac{2}{3} \times 24$ b. $\dfrac{3}{4}$ of $2\dfrac{1}{2} = \dfrac{3}{4} \times \dfrac{(2 \times 2) + 1}{2}$

$$= \dfrac{2}{\cancel{3}} \times \dfrac{\cancel{3}^{1} \times 2 \times 2 \times 2}{1}$$

$$= \dfrac{3}{4} \times \dfrac{5}{2}$$

$$= 16$$

$$= \dfrac{15}{8} = 1\dfrac{7}{8}$$

RECIPROCALS

Consider the products

$$5 \times \dfrac{1}{5} = \dfrac{5}{1} \times \dfrac{1}{5} = 1 \quad \text{and} \quad \dfrac{3}{4} \times \dfrac{4}{3} = 1$$

3.6 MULTIPLICATION AND DIVISION

Observe that in each case the product is 1. We say that 5 and $\frac{1}{5}$ are *reciprocals* of each other, and $\frac{3}{4}$ and $\frac{4}{3}$ are *reciprocals* of each other. In general, if the product of two numbers is 1, then each of the numbers is the **reciprocal** of the other.

It is a simple matter to obtain the reciprocal of a number. For any nonzero number n, the reciprocal is $\frac{1}{n}$. For a fraction $\frac{n}{d}$, where neither n nor d is zero, the reciprocal is $\frac{d}{n}$. In such cases we say that the fraction $\frac{n}{d}$ has been *inverted*.

Example 5 Write the reciprocal of each number.

a. 8   b. $\frac{5}{11}$ c. $2\frac{1}{3}$

Solutions
a. The reciprocal of 8 is $\frac{1}{8}$. ($8 \times \frac{1}{8} = 1$)
b. The reciprocal of $\frac{5}{11}$ is $\frac{11}{5}$. ($\frac{5}{11} \times \frac{11}{5} = 1$)
c. Write $2\frac{1}{3}$ as $\frac{7}{3}$; the reciprocal of $2\frac{1}{3}$ or $\frac{7}{3}$ is $\frac{3}{7}$. ($\frac{7}{3} \times \frac{3}{7} = 1$)

Because the product of zero and *any* number is 0, there is no number whose product with zero is 1. Hence, the number 0 has no reciprocal.

DIVISION

A method for finding the quotient of two fractions can be developed by use of the fundamental principle of fractions together with the concept of reciprocals. As an example, consider the quotient

$$\frac{2}{3} \div \frac{3}{4}$$

which can be written as the fraction

$$\frac{\frac{2}{3}}{\frac{3}{4}}$$

with numerator $\frac{2}{3}$ and denominator $\frac{3}{4}$. Using the fundamental principle of fractions, we can multiply numerator and denominator by $\frac{4}{3}$, the reciprocal of the denominator, and obtain

$$\frac{\frac{2}{3} \times \frac{4}{3}}{\frac{3}{4} \times \frac{4}{3}} = \frac{\frac{2}{3} \times \frac{4}{3}}{1} = \frac{2}{3} \times \frac{4}{3}$$

Thus,

$$\frac{2}{3} \div \frac{3}{4} = \frac{2}{3} \times \frac{4}{3} = \frac{8}{9}$$

This result suggests the following rule.

> **To find the quotient of two fractions, invert the divisor and multiply the resulting fractions.**

FRACTIONS

We represent this rule in symbols by

$$\frac{a}{b} \div \frac{c}{d} = \frac{a}{b} \times \frac{d}{c}$$

where it is understood that b, c, and d are nonzero numbers.

Example 6 a. $\frac{5}{8} \div \frac{7}{12} = \frac{5}{8} \times \frac{12}{7}$ (change to ×, invert divisor)

$= \frac{5}{\cancel{8}} \times \frac{\cancel{12}^3}{7}$
$_2$

$= \frac{15}{14} = 1\frac{1}{14}$

b. $\frac{5}{8} \div 7 = \frac{5}{8} \div \frac{7}{1}$

$\frac{5}{8} \div \frac{7}{1} = \frac{5}{8} \times \frac{1}{7}$ (change to ×, invert divisor)

$= \frac{5}{56}$

Example 7 a. $3\frac{1}{2} \div 2\frac{3}{8}$ b. $2\frac{3}{8} \div 3\frac{1}{2}$

Solutions First write each mixed number as an improper fraction. Then invert the divisor and multiply the fractions.

a. $3\frac{1}{2} \div 2\frac{3}{8}$

$= \frac{7}{2} \div \frac{19}{8}$

$= \frac{7}{\cancel{2}} \times \frac{\cancel{8}^4}{19} = \frac{28}{19} = 1\frac{9}{19}$
$_1$

b. $2\frac{3}{8} \div 3\frac{1}{2}$

$= \frac{19}{8} \div \frac{7}{2}$

$= \frac{19}{\cancel{8}} \times \frac{\cancel{2}^1}{7} = \frac{19}{28}$
$_4$

EXERCISES 3.6

Express fractional answers in lowest terms and improper fractions as mixed numbers.
See Examples 1 and 2.

Complete Examples

a. $\frac{3}{5} \times \frac{2}{7} = \frac{3 \times 2}{5 \times 7}$

$= \underline{\qquad}$
(1)

b. $\frac{2}{5} \times \frac{10}{21} \times \frac{7}{12} = \frac{\cancel{2}^1}{\cancel{5}_1} \times \frac{\cancel{2}^1 \times \cancel{5}^1}{3 \times \cancel{7}_1} \times \frac{\cancel{7}^1}{\cancel{2} \times \cancel{2} \times 3}$

$= \frac{1}{3 \times 3} = \underline{\qquad}$
(2)

---

(1) $\frac{6}{35}$ (2) $\frac{1}{9}$

122

3.6 MULTIPLICATION AND DIVISION

1. $\frac{3}{16} \times \frac{4}{9}$
2. $\frac{2}{3} \times \frac{5}{4}$
3. $\frac{7}{12} \times \frac{3}{14}$
4. $\frac{8}{15} \times \frac{5}{16}$

5. $\frac{6}{25} \times \frac{5}{9}$
6. $\frac{14}{15} \times \frac{3}{4}$
7. $\frac{2}{3} \times \frac{5}{7} \times \frac{3}{4}$
8. $\frac{2}{5} \times \frac{7}{9} \times \frac{5}{12}$

9. $\frac{5}{24} \times \frac{3}{16} \times \frac{8}{15}$
10. $\frac{5}{12} \times \frac{14}{15} \times \frac{8}{21}$
11. $\frac{9}{20} \times \frac{5}{18} \times \frac{11}{12}$
12. $\frac{11}{45} \times \frac{9}{22} \times \frac{2}{15}$

13. $\frac{5}{6} \times \frac{7}{12} \times \frac{24}{35}$
14. $\frac{18}{25} \times \frac{10}{27} \times \frac{3}{4}$
15. $\frac{9}{16} \times \frac{5}{24} \times \frac{8}{27}$
16. $\frac{14}{15} \times \frac{9}{35} \times \frac{10}{21}$

17. $\frac{5}{6} \times \frac{9}{10} \times \frac{4}{25}$
18. $\frac{8}{15} \times \frac{5}{18} \times \frac{9}{16}$

See Example 3.

Complete Examples

a. $4\frac{1}{3} \times 3\frac{3}{5}$

$= \frac{(4 \times 3) + 1}{3} \times \frac{(3 \times 5) + 3}{5}$

$= \frac{13}{3} \times \underline{}$
  (1)

$= \frac{78}{5} = \underline{}$
  (2)

b. $6 \times 8\frac{2}{3}$

$= \frac{6}{1} \times \frac{(8 \times 3) + 2}{3}$

$= \frac{6}{1} \times \underline{}$
  (3)

$= \frac{52}{1} = \underline{}$
  (4)

(1) $\frac{18}{5}$ (2) $15\frac{3}{5}$ (3) $\frac{26}{3}$ (4) 52

19. $7\frac{1}{3} \times 2\frac{1}{4}$
20. $5\frac{2}{5} \times 3\frac{1}{3}$
21. $3\frac{3}{7} \times 2\frac{1}{12}$
22. $2\frac{2}{9} \times 3\frac{3}{5}$

23. $16 \times 6\frac{1}{4}$
24. $24 \times 2\frac{1}{6}$
25. $18 \times 5\frac{3}{4}$
26. $12 \times 2\frac{4}{9}$

FRACTIONS

27. $4\frac{2}{3} \times \frac{6}{7} \times 2\frac{1}{4}$
28. $1\frac{3}{5} \times \frac{3}{8} \times 2\frac{2}{9}$
29. $5\frac{1}{3} \times 3\frac{3}{5} \times \frac{5}{16}$
30. $7\frac{1}{2} \times 4\frac{1}{5} \times \frac{4}{35}$

See Example 4.

Complete Examples

a. $\frac{3}{8}$ of 40

$= \frac{3}{8} \times \frac{40}{1}$

$= \frac{3}{\underset{1}{2} \times \underset{1}{2} \times \underset{1}{2}} \times \frac{5 \times \overset{1}{2} \times \overset{1}{2} \times \overset{1}{2}}{1}$

$= \underline{\qquad}$
   (1)

b. $\frac{2}{5}$ of $4\frac{3}{10}$

$= \frac{2}{5} \times \frac{(4 \times 10) + 3}{10}$

$= \frac{2}{5} \times \underline{\qquad}$
   (2)

$= \frac{43}{25} = \underline{\qquad}$
   (3)

---
(1) 15   (2) $\frac{43}{10}$ (3) $1\frac{18}{25}$
---

31. $\frac{3}{5}$ of 285
32. $\frac{7}{8}$ of 824
33. $\frac{3}{4}$ of $12\frac{1}{3}$
34. $\frac{13}{16}$ of $8\frac{1}{4}$
35. $1\frac{1}{2}$ of $15\frac{2}{9}$
36. $3\frac{3}{10}$ of $18\frac{7}{10}$

Write the reciprocal of each number. See Example 5.

Complete Examples

a. 7

Ans. $\underline{\qquad}$
   (1)

b. $\frac{6}{11}$

Ans. $\underline{\qquad}$
   (2)

c. $3\frac{2}{3}$

$3\frac{2}{3} = \frac{(3 \times 3) + 2}{3} = \underline{\qquad}$
   (3)

Ans. $\underline{\qquad}$
   (4)

---
(1) $\frac{1}{7}$ (2) $\frac{11}{6}$ (3) $\frac{11}{3}$ (4) $\frac{3}{11}$
---

37. 5
38. 9
39. $\frac{3}{5}$
40. $\frac{2}{3}$
41. $3\frac{1}{2}$
42. $4\frac{2}{3}$
43. $11\frac{2}{9}$
44. $25\frac{6}{7}$

3.6 MULTIPLICATION AND DIVISION

Express fractional answers in lowest terms and improper fractions as mixed numbers.

See Example 6.

Complete Examples

a. $\dfrac{5}{6} \div \dfrac{3}{4} = \dfrac{5}{6} \times \underline{\qquad}$ (1) *(change to ×, invert divisor)*

$= \dfrac{5}{\cancel{6}_3} \times \dfrac{\cancel{4}^2}{3}$

$= \dfrac{10}{9} = \underline{\qquad}$ (2)

b. $\dfrac{6}{11} \div 3$

$= \dfrac{6}{11} \div \dfrac{3}{1} = \dfrac{6}{11} \times \underline{\qquad}$ (3) *(change to ×, invert divisor)*

$= \dfrac{\cancel{6}^2}{11} \times \dfrac{1}{\cancel{3}_1} = \underline{\qquad}$ (4)

---

(1) $\dfrac{4}{3}$ (2) $1\dfrac{1}{9}$ (3) $\dfrac{1}{3}$ (4) $\dfrac{2}{11}$

---

45. $\dfrac{3}{4} \div \dfrac{5}{8}$ 46. $\dfrac{15}{32} \div \dfrac{25}{48}$ 47. $\dfrac{4}{5} \div 6$ 48. $\dfrac{5}{6} \div 10$

49. $\dfrac{2}{9} \div 4$ 50. $\dfrac{15}{32} \div 5$ 51. $12 \div \dfrac{8}{13}$ 52. $16 \div \dfrac{12}{25}$

See Example 7.

Complete Examples

a. $3\dfrac{2}{3} \div 4\dfrac{1}{6} = \dfrac{11}{3} \div \underline{\qquad}$ (1)

$= \dfrac{11}{\cancel{3}_1} \times \dfrac{\cancel{6}^2}{25}$

$= \underline{\qquad}$ (2)

b. $4\dfrac{1}{6} \div 3\dfrac{2}{3} = \dfrac{25}{6} \div \dfrac{11}{3}$

$= \dfrac{25}{6} \times \underline{\qquad}$ (3)

$= \dfrac{25}{22} = \underline{\qquad}$ (4)

---

(1) $\dfrac{25}{6}$ (2) $\dfrac{22}{25}$ (3) $\dfrac{3}{11}$ (4) $1\dfrac{3}{22}$

125

FRACTIONS

53. $2\frac{1}{3} \div 5\frac{3}{5}$ 54. $4\frac{1}{2} \div 2\frac{4}{7}$ 55. $3\frac{3}{4} \div 1\frac{7}{8}$ 56. $5\frac{5}{6} \div 2\frac{5}{8}$

57. $2\frac{5}{12} \div 1\frac{5}{24}$ 58. $3\frac{5}{16} \div 2\frac{17}{18}$ 59. $8 \div 3\frac{1}{5}$ 60. $6 \div 4\frac{2}{7}$

61. $5\frac{1}{2} \div \frac{3}{4}$ 62. $3\frac{1}{3} \div \frac{3}{5}$ 63. $7\frac{1}{5} \div 27$ 64. $6\frac{3}{8} \div \frac{17}{20}$

In Exercises 65 and 66, express each answer as either a whole number or a mixed number.

65. Water is poured into an empty tank so that the depth of the water increases by $2\frac{7}{8}$ ft each hour. Compute the depth of the water at the end of each time period.

	a.	b.	c.	d.	e.
Hours	3	$4\frac{1}{2}$	$6\frac{1}{4}$	$8\frac{2}{3}$	$11\frac{5}{12}$
Depth	___	___	___	___	___

66. The following table lists the number of hours worked by each of five employees. If the rate of pay is $6 per hour, find the amount paid to each employee.

Employee	a.	b.	c.	d.	e.
Hours worked	$20\frac{3}{4}$	$38\frac{1}{2}$	$33\frac{1}{4}$	$22\frac{1}{2}$	$29\frac{3}{4}$
Amount paid	___	___	___	___	___

In Exercises 67 and 68, express each answer as either a whole number or a mixed number.

67. A real estate company offers some parcels of land at the prices listed below. Find the cost per acre of each parcel.

	a.	b.	c.	d.
Price (dollars)	1500	868	966	1027
Land (acres)	$2\frac{1}{2}$	$4\frac{2}{3}$	$1\frac{2}{5}$	$\frac{13}{16}$
Cost per acre	___	___	___	___

68. The following table lists the price (in dollars) and the length (in yards) of each of four rolls of cloth. Find the cost per yard of each cloth.

	a.	b.	c.	d.
Price (dollars)	795	782	892	518
Length (yards)	$132\frac{1}{2}$	$97\frac{3}{4}$	$74\frac{1}{3}$	$32\frac{3}{8}$
Cost per yard	___	___	___	___

3.7 DECIMAL EQUIVALENTS OF FRACTIONS

As we have noted, a fraction is a quotient. Hence, a fraction can be written in a decimal form that can be obtained by dividing the numerator by the denominator.

Example 1 Write each fraction in decimal form.

a. $\frac{4}{5}$ b. $\frac{4}{11}$

Solutions Divide the numerator by the denominator.

a. $\frac{4}{5} = 4 \div 5 = 0.8$ b. $\frac{4}{11} = 4 \div 11 = 0.3636\ldots$

3.7 DECIMAL EQUIVALENTS OF FRACTIONS

In Example 1a., we say that 0.8 is the exact decimal equivalent of $\frac{4}{5}$ because no more digits appear to the right of 8. From Example 1b., the quotient of 4 divided by 11 is a *never-ending* decimal, where the symbol ". . ." indicates that there is no last digit. Alternatively, we can use a bar over a group of digits that are repeated.

Example 2 a. $\frac{5}{16} = 5 \div 16 = 0.3125$
$\phantom{\frac{5}{16}}= 0.3$ (to the nearest tenth)
$\phantom{\frac{5}{16}}= 0.31$ (to the nearest hundredth)
$\phantom{\frac{5}{16}}= 0.313$ (to the nearest thousandth)

b. $\frac{5}{12} = 5 \div 12 = 0.416\ldots = 0.41\overline{6}$
$\phantom{\frac{5}{12}}= 0.4$ (to the nearest tenth)
$\phantom{\frac{5}{12}}= 0.42$ (to the nearest hundredth)
$\phantom{\frac{5}{12}}= 0.417$ (to the nearest thousandth)

Example 3 a. $\frac{43}{32} = 43 \div 32 = 1.34375$
$\phantom{\frac{43}{32}}= 1.3$ (to the nearest tenth)
$\phantom{\frac{43}{32}}= 1.34$ (to the nearest hundredth)
$\phantom{\frac{43}{32}}= 1.344$ (to the nearest thousandth)

b. $\frac{38}{33} = 38 \div 33 = 1.1515\ldots = 1.\overline{15}$
$\phantom{\frac{38}{33}}= 1.2$ (to the nearest tenth)
$\phantom{\frac{38}{33}}= 1.15$ (to the nearest hundredth)
$\phantom{\frac{38}{33}}= 1.152$ (to the nearest thousandth)

DECIMAL EQUIVALENTS OF MIXED NUMBERS

The decimal equivalent of a mixed number can be found by adding the whole number part to the decimal equivalent of the fraction.

Example 4 a. $4\frac{3}{8} = 4 + \frac{3}{8} = 4 + 0.375$
$\phantom{4\frac{3}{8}}= 4.375$

b. $16\frac{3}{4} = 16 + \frac{3}{4} = 16 + 0.75$
$\phantom{16\frac{3}{4}}= 16.75$

Sometimes a fraction is part of a decimal. For example, the number $0.06\frac{1}{4}$ (read "six and one-fourth hundredths") means the sum of six hundredths (0.06) and $\frac{1}{4}$ of a hundredth. Because

$$\frac{1}{4} \text{ of } 0.01 = \frac{1}{4} \times 0.01$$
$$= 0.25 \times 0.01 = 0.0025$$

we have that

$$0.06\tfrac{1}{4} = 0.06 + 0.0025 = 0.0625$$

Results such as this suggest a "short cut" procedure. Decimals that include fractional parts can be changed into decimal form simply by replacing the fractional parts by the *digits* (not the decimal point) of their decimal equivalents.

FRACTIONS

For example,

$$0.08\tfrac{3}{4} = 0.0875 \quad \text{(because } \tfrac{3}{4} = 0.75\text{)}$$

$$1.004\tfrac{1}{2} = 1.0045 \quad \text{(because } \tfrac{1}{2} = 0.5\text{)}$$

$$0.83\tfrac{1}{3} = 0.8333 \quad \text{(because } \tfrac{1}{3} = 0.33 \text{ to two places)}$$

COMMON FRACTION–DECIMAL EQUIVALENTS

Fractions with denominators 10, 100, 1000, etc., can be written directly as decimals. For example,

$$\tfrac{1}{10} = 0.1, \quad \tfrac{3}{10} = 0.3, \quad \tfrac{7}{10} = 0.7, \quad \text{and} \quad \tfrac{9}{10} = 0.9$$

$$\tfrac{1}{100} = 0.01, \quad \tfrac{7}{100} = 0.07, \quad \tfrac{23}{100} = 0.23, \quad \text{and} \quad \tfrac{79}{100} = 0.79$$

$$\tfrac{1}{1000} = 0.001, \quad \tfrac{3}{1000} = 0.003, \quad \tfrac{31}{1000} = 0.031, \quad \text{and} \quad \tfrac{271}{1000} = 0.271$$

Also, the following fractions and their decimal equivalents are used so frequently that it is convenient to memorize them:

$$\tfrac{1}{4} = 0.25 \quad \tfrac{1}{2} = 0.5 \quad \tfrac{3}{4} = 0.75 \quad \tfrac{1}{3} = 0.33\tfrac{1}{3} \quad \tfrac{2}{3} = 0.66\tfrac{2}{3}$$

Reversing the reasoning used above, we can write a decimal as a fraction. For example,

$$0.29 = \tfrac{29}{100}, \quad 2.75 = 2\tfrac{3}{4}, \quad \text{and} \quad 4.33\tfrac{1}{3} = 4\tfrac{1}{3}$$

EXERCISES 3.7

Write each fraction in decimal form: a. To the nearest tenth. b. To the nearest hundredth. c. To the nearest thousandth. See Examples 1, 2, and 3.

Complete Examples

a. $\tfrac{5}{8} = 5 \div 8 = $ _____
(1)

(nearest tenth) = 0.6

(nearest hundredth) = _____
(2)

(nearest thousandth) = 0.625

b. $\tfrac{10}{6} = 10 \div 6 = 1.6666\ldots$

(nearest tenth) = _____
(3)

(nearest hundredth) = 1.67

(nearest thousandth) = _____
(4)

---

(1) 0.625   (2) 0.63   (3) 1.7   (4) 1.667

3.7 DECIMAL EQUIVALENTS OF FRACTIONS

1. $\frac{1}{8}$ 2. $\frac{3}{8}$ 3. $\frac{1}{6}$ 4. $\frac{5}{6}$ 5. $\frac{15}{32}$ 6. $\frac{31}{64}$

7. $\frac{7}{9}$ 8. $\frac{5}{18}$ 9. $\frac{17}{16}$ 10. $\frac{11}{8}$ 11. $\frac{15}{13}$ 12. $\frac{23}{19}$

Write each mixed number in decimal form. See Example 4.

Complete Examples

a. $4\frac{4}{5} = 4 + \frac{4}{5}$

$= 4 + \underline{}_{(1)}$

$= \underline{}_{(2)}$

b. $12\frac{3}{4} = 12 + \frac{3}{4}$

$= 12 + \underline{}_{(3)}$

$= \underline{}_{(4)}$

(1) 0.8   (2) 4.8   (3) 0.75   (4) 12.75

13. $5\frac{7}{8}$ 14. $7\frac{1}{8}$ 15. $3\frac{3}{4}$ 16. $6\frac{2}{5}$

17. $12\frac{3}{5}$ 18. $11\frac{1}{4}$ 19. $15\frac{3}{8}$ 20. $19\frac{5}{8}$

Write the decimal equivalent of each fraction by inspection.

21. $\frac{1}{10}$ 22. $\frac{1}{100}$ 23. $\frac{1}{1000}$ 24. $\frac{7}{10}$ 25. $\frac{9}{100}$ 26. $\frac{23}{100}$

27. $\frac{13}{1000}$ 28. $\frac{128}{1000}$ 29. $\frac{1}{2}$ 30. $\frac{1}{4}$ 31. $\frac{3}{4}$ 32. $\frac{1}{3}$

33. $3\frac{1}{4}$ 34. $12\frac{3}{4}$ 35. $5\frac{1}{3}$ 36. $8\frac{2}{3}$

Write the fractional equivalent of each decimal.

37. 0.9 38. 0.03 39. 0.21 40. 0.017 41. 0.127 42. 2.7

43. 3.09 44. 21.19 45. 4.5 46. 5.25 47. 2.75 48. 10.50

49. $0.33\frac{1}{3}$ 50. $0.66\frac{2}{3}$ 51. $1.66\frac{2}{3}$ 52. $5.33\frac{1}{3}$

FRACTIONS

CHAPTER THREE REVIEW EXERCISES

Determine whether the first number is or is not a factor of the second number.

1. 26; 936
2. 48; 884

Write each number in completely factored form.

3. 42
4. 66
5. 72
6. 108

Reduce each fraction to lowest terms.

7. $\frac{22}{72}$
8. $\frac{24}{36}$
9. $\frac{75}{105}$
10. $\frac{72}{126}$

Build each fraction to an equal fraction with the given denominator.

11. $\frac{3}{8} = \frac{}{32}$
12. $\frac{5}{12} = \frac{}{180}$

Write each group of fractions as like fractions with the LCD as the common denominator.

13. $\frac{2}{5}, \frac{3}{7}$
14. $\frac{3}{4}, \frac{5}{12}$
15. $\frac{2}{3}, \frac{5}{6}, \frac{2}{15}$
16. $\frac{4}{1}, \frac{5}{18}, \frac{7}{24}$

Write each set of fractions in order, from the least to the greatest.

17. $\frac{6}{13}, \frac{5}{13}, \frac{8}{13}$
18. $\frac{2}{5}, \frac{1}{7}, \frac{9}{25}$

19. Change $6\frac{3}{8}$ to an improper fraction.
20. Change $4\frac{5}{6}$ to an improper fraction.

Find each sum.

21. $\frac{5}{24} + \frac{7}{24}$
22. $\frac{3}{20} + \frac{5}{12} + \frac{1}{6}$
23. $3\frac{2}{7} + 6\frac{3}{7}$
24. $15\frac{5}{16} + 2\frac{3}{8} + 4\frac{3}{4}$

Compute each of the following.

25. $\frac{25}{36} - \frac{11}{36} - \frac{7}{36}$
26. $\frac{27}{32} - \frac{3}{8} - \frac{1}{4}$
27. $9\frac{25}{32} - 2\frac{5}{32}$
28. $15\frac{5}{6} - 3\frac{1}{4}$
29. $18 - 6\frac{5}{6}$
30. $5\frac{2}{3} + 3\frac{1}{5} - 4\frac{1}{6}$

Find each product.

31. $\frac{3}{4} \times \frac{2}{5}$
32. $\frac{6}{25} \times \frac{5}{12}$

CHAPTER THREE REVIEW EXERCISES

33. $\frac{3}{7} \times \frac{14}{15} \times \frac{5}{8}$

34. $16 \times 3\frac{1}{4}$

35. Compute $\frac{3}{8}$ of 512.

36. Write the reciprocal of $2\frac{3}{5}$.

Compute each of the following.

37. $\frac{15}{28} \div \frac{3}{7}$

38. $\frac{7}{8} \div 14$

39. $3\frac{1}{2} \div 1\frac{3}{4}$

40. $2\frac{3}{16} \div 2\frac{1}{2}$

41. $11\frac{2}{3} \times 3\frac{3}{5} \div 1\frac{1}{5}$

42. $\frac{21}{35} \div \frac{7}{15}$

Write each fraction in decimal form: a. *To the nearest tenth.* b. *To the nearest hundredth.* c. *To the nearest thousandth.*

43. $\frac{3}{7}$

44. $\frac{2}{9}$

45. $\frac{7}{3}$

46. $\frac{13}{11}$

Write each mixed number in decimal form.

47. $2\frac{1}{2}$

48. $5\frac{3}{4}$

49. $12\frac{1}{5}$

50. $9\frac{4}{5}$

51. $6\frac{3}{8}$

52. $12\frac{5}{8}$

Write the fractional equivalent of each decimal.

53. 2.25

54. 5.1

Write each mixed number in decimal form and find each product to the nearest hundredth.

55. $6\frac{1}{2} \times 4.7 \times 3\frac{3}{4}$

56. $12\frac{2}{5} \times 6.31 \times 4\frac{1}{8}$

Miscellaneous Applications (Express answers in fraction form.)

57. If a football team wins 15 games out of 24 games played, what fraction of the games were won?

58. A wooden block has holes that are $\frac{3}{8}$ inch, $\frac{15}{32}$ inch, $\frac{9}{16}$ inch, and $\frac{5}{8}$ inch. Which of these holes are too small for a $\frac{1}{2}$-inch pin?

59. During one week a house painter used $3\frac{1}{2}$ gal of a certain paint on Monday, $2\frac{1}{4}$ gal on Tuesday, and $4\frac{3}{8}$ gal on Friday. How many gallons were used during the week?

60. On Monday morning a store has 310 lb of vegetables. That day $187\frac{1}{2}$ lb are sold, and $96\frac{3}{4}$ lb are sold on Tuesday. How many pounds remain in stock at the close of business on Tuesday?

61. Find the total weight in pounds of $5\frac{1}{3}$ baskets of fruit if each basket weighs $6\frac{3}{4}$ lb.

62. If $3\frac{1}{2}$ acres of land are sold for $4200, find the cost per acre.

4 PERCENT

4.1 PERCENT EQUIVALENTS

In this section we consider relations between percents, decimals, and fractions.

DECIMAL-PERCENT EQUIVALENTS

The percent symbol % is used to mean *hundredths*. For example, the decimal .78 (78 hundredths) can be written in percent form as 78%; the percent form of 6% can be written in decimal form as .06. The following two rules can be used to make these changes.

> **To Change a Decimal to a Percent:**
> 1. Move the decimal point two places to the right.
> 2. Add or drop zeros as needed.
> 3. Add the percent symbol to follow the last digit.

Example 1 a. $0.29 = 29\%$ b. $0.067 = 6.7\%$ c. $1 = 1.00 = 100\%$

> **To Change a Percent to a Decimal:**
> 1. Move the decimal point two places to the left.
> 2. Add or drop zeros as needed.
> 3. Drop the percent symbol.

Example 2 a. $54\% = 0.54$ b. $.02\% = 0.0002$ c. $200\% = 2.00 = 2$

Percents that involve fractional parts can be changed to decimal forms by first changing the fractional part to its decimal equivalent.

PERCENT

Example 3 a. $6\frac{3}{4}\% = 6.75\%$ b. $33\frac{1}{3}\% = 33.3\overline{3}\%$ c. $\frac{3}{4}\% = 0.75\%$
$\qquad\qquad\quad = 0.0675$ $\qquad\qquad\quad = 0.3333$ $\qquad\qquad\quad = 0.0075$

FRACTION–PERCENT EQUIVALENTS

Fractions can be changed to percents by first changing the fraction to its decimal equivalent.

Example 4 a. $\frac{3}{5} = 0.6$ b. $\frac{13}{5} = 2.6$ c. $\frac{7}{8} = 0.875$
$\qquad\qquad\ = 60\%$ $\qquad\qquad = 260\%$ $\qquad\qquad = 87.5\%$

Percents can be changed to fractions by writing an appropriate fraction with the denominator 100 and reducing to lowest terms, if necessary.

Example 5 a. $80\% = \frac{80}{100} = \frac{4}{5}$ b. $112\% = \frac{112}{100} = \frac{28}{25}$

If the numerator of the fraction obtained when changing a percent to a fraction involves a decimal part, we can use the fundamental principle of fractions to rewrite the fraction so that the numerator does not contain a decimal part.

Example 6 a. $0.7\% = \frac{0.7}{100}$ b. $12\frac{1}{2}\% = 12.5\% = \frac{12.5}{100}$
$\qquad\qquad\quad = \frac{0.7 \times 10}{100 \times 10}$ $\qquad\qquad\qquad\quad = \frac{12.5 \times 10}{100 \times 10}$
$\qquad\qquad\quad = \frac{7}{1000}$ $\qquad\qquad\qquad\quad = \frac{125}{1000} = \frac{1}{8}$

EXERCISES 4.1
Write each decimal as a percent. See Example 1.

Complete Examples

a. 0.35 b. 0.028 c. 3

$0.35 = \underline{\quad}\%$ $0.028 = \underline{\quad}\%$ $3.00 = \underline{\quad}\%$
$\quad\ (1)$ $\qquad\quad (2)$ $\qquad\quad (3)$

(1) 35%   (2) 2.8%   (3) 300%

1. 0.33   2. 0.54   3. 0.504   4. 0.686   5. 0.787   6. 0.074
7. 0.021   8. 0.005   9. 0.008   10. 0.0008   11. 5.5   12. 6

134

4.1 PERCENT EQUIVALENTS

Write each percent as a decimal. See Example 2.

Complete Examples

a. 26%
   26% = _____ (1)

b. 0.3%
   00.3% = _____ (2)

c. 150%
   150% = _____ (3)

> (1) 0.26   (2) 0.003   (3) 1.5

13. 15%   **14.** 25%   **15.** 0.4%   **16.** 0.6%   **17.** 6.8%   **18.** 1.1%

19. 2.7%   **20.** 5.8%   **21.** 119%   **22.** 652%   **23.** 259%   **24.** 528%

Write each percent as a decimal. See Example 3.

Complete Examples

a. $5\frac{1}{2}\% = 5.5\%$
   = _____ (1)

b. $25\frac{3}{4}\% = 25.75\%$
   = _____ (2)

c. $\frac{1}{5}\% = 0.2\%$
   = _____ (3)

> (1) 0.055   (2) 0.2575   (3) 0.002

25. $3\frac{1}{4}\%$ **26.** $6\frac{1}{8}\%$ **27.** $2\frac{3}{8}\%$ **28.** $8\frac{3}{4}\%$

29. $8\frac{2}{5}\%$ **30.** $9\frac{3}{5}\%$ **31.** $\frac{1}{2}\%$ **32.** $\frac{1}{4}\%$

Write each fraction as a percent. See Example 4.

Complete Examples

a. $\frac{1}{2} = 0.5$
   = _____ % (1)

b. $\frac{9}{4} = 2.25$
   = _____ % (2)

c. $\frac{5}{8} = 0.625$
   = _____ % (3)

> (1) 50%   (2) 225%   (3) 62.5%

33. $\frac{3}{4}$ **34.** $\frac{1}{5}$ **35.** $\frac{3}{8}$ **36.** $\frac{5}{5}$ **37.** $\frac{3}{5}$ **38.** $\frac{3}{20}$

PERCENT

39. $\dfrac{9}{4}$ 40. $\dfrac{8}{5}$ 41. $\dfrac{12}{5}$ 42. $\dfrac{11}{4}$ 43. $\dfrac{1}{250}$ 44. $\dfrac{1}{500}$

Write each percent as a fraction. See Examples 5 and 6.

Complete Examples

a. $40\% = \dfrac{40}{100}$
   $= \underline{}$
   $\quad (1)$

b. $4\tfrac{1}{2}\% = 4.5\% = \dfrac{4.5 \times 10}{\underline{} \times 10}$
   $\qquad\qquad\qquad\qquad\quad (2)$
   $\qquad = \dfrac{45}{\underline{}} = \dfrac{9}{200}$
   $\qquad\qquad (3)$

(1) $\dfrac{2}{5}$ (2) 100   (3) 1000

45. 25%   46. 10%   47. 125%   48. 150%
49. 0.90%   50. 0.11%   51. $37\tfrac{1}{2}\%$ 52. $87\tfrac{1}{2}\%$

4.2 TYPES OF PERCENT PROBLEMS

CASE I

Of the many problems that involve the use of percent, three kinds are basic. The type most frequently encountered—*finding a percent of a number*—is often referred to as "Case I."

Consider a sentence such as the following.

$$83\% \text{ of } 720 \text{ is what number?}$$

If we let N represent the required number, then we can translate the sentence as follows.

83% of 720 is what number?
↓   ↓   ↓   ↓   ↓
0.83 × 720 =   N

where 83% has been changed to decimal form.

Example 1
a. 83% of 720
$= .83 \times 720$
$= 597.6$

b. 8.8% of 400
$= 0.088\% \times 400$
$= 35.2$

c. $6\tfrac{1}{2}\%$ of 720
$= 6.5\%$ of 720
$= 0.065 \times 720$
$= 46.8$

136

4.2 TYPES OF PERCENT PROBLEMS

CASE II

The second ("Case II") of the three basic percent problems involves finding *what percent one number is of a second number*.

Suppose that you correctly answer 132 questions on a 150-question test. What is your score in percent? This type of question is often stated in either of the following forms.

132 is what percent of 150?   or   What percent of 150 is 132?

As before, we interpret the word "is" to mean "equals" and the word "of" to indicate a multiplication. Then, using P to represent the *decimal* form of the percent we seek, these sentences can be written as

$$132 = P \times 150 \qquad P \times 150 = 132$$

To find P, we have

$$\frac{132}{150} = \frac{P \times 150}{150} \qquad \text{or} \qquad \frac{P \times 150}{150} = \frac{132}{150}$$

$$\frac{132}{150} = P \qquad\qquad P = \frac{132}{150}$$

From either form, $P = 0.88$. Changing 0.88 to percent, we conclude that the score on the test is 88%.

We can use Case I to check the result of a Case II computation. Thus, for the example above, we can verify that

$$88\% \text{ of } 150 = 0.88 \times 150 = 132$$

Example 2   a. 115 is what percent of 250?   b. What percent of 440 is $27\frac{1}{2}$?

Solutions   Let P represent the decimal form of the required percent.

a. $115 = P \times 250$

Divide each side by 250.

$$\frac{115}{250} = \frac{P \times 250}{250}$$

$$P = \frac{115}{250} = 0.46$$

$$P = 46\%$$

b. First change $27\frac{1}{2}$ to 27.5. Then,

$$P \times 440 = 27.5$$

Divide each side by 440.

$$\frac{P \times 440}{440} = \frac{27.5}{440} = 0.0625$$

$$P = \frac{27.5}{440} = 0.0625$$

$$P = 6.25\%$$

Example 3   37 is what percent of 65? Express answer to the nearest tenth of a percent.

Solution   Let P represent the decimal form of the required percent. Then,

$$37 = P \times 65$$

Dividing each side by 65, we have

$$\frac{37}{65} = \frac{P \times 65}{65}$$

PERCENT

In order to obtain the percent to the nearest tenth, we must first obtain a decimal value of the quotient to four decimal places:

$$P = \frac{37}{65} = 0.5692 \quad \text{(to the nearest ten-thousandth)}$$

Thus, $P = 56.92\%$, or 56.9% to the nearest tenth.

CASE III

A third type of basic percent problem ("Case III") involves *finding a number when a percent of the number is known*.

For example, suppose that an advertisement states that $126, the sale price of a camera, is 80% of the regular price of the camera. What is the regular price of the camera? The question can be restated as either

$$126 \text{ is } 80\% \text{ of what number?} \quad \text{or} \quad 80\% \text{ of what number is } 126?$$

As before, we interpret "is" to mean "equals" and "of" to indicate a multiplication. Then, using N to represent the number we seek, and changing 80% to 0.8, we can write these sentences as

$$\underbrace{126}_{126} \; \underbrace{\text{is}}_{=} \; \underbrace{80\%}_{0.8} \; \underbrace{\text{of}}_{\times} \; \underbrace{\text{what number}}_{N} \quad \text{or} \quad \underbrace{80\%}_{0.8} \; \underbrace{\text{of}}_{\times} \; \underbrace{\text{what number}}_{N} \; \underbrace{\text{is}}_{=} \; \underbrace{126?}_{126}$$

Solving for N, we can divide each side of either equation by 0.8 to obtain

$$\frac{126}{0.8} = \frac{\cancel{0.8} \times N}{\cancel{0.8}}$$

To simplify the computations, we can multiply the numerator and denominator by 10 by moving each decimal point one place to the right in order to obtain a whole number in the denominator.

$$N = \frac{126}{0.8} = \frac{126.0}{0.8} = 157.5$$

Hence, the regular price of the camera is $157.50.

Example 4   a.   46.8 is 60% of what number?   b.   40% of what number is 144?

Solutions   Let N represent the required number in each case.

a. $60\% = 0.6$;   hence

$$46.8 = 0.6 \times N$$

Dividing each side by 0.6,

$$\frac{46.8}{0.6} = \frac{\cancel{0.6} \times N}{\cancel{0.6}}$$

$$N = \frac{46.8}{0.6} = \frac{46.8}{0.6} = 78$$

b. $40\% = 0.4$;   hence

$$0.4 \times N = 144$$

Dividing each side by 0.4,

$$\frac{\cancel{0.4} \times N}{\cancel{0.4}} = \frac{144}{0.4}$$

$$N = \frac{144}{0.4} = \frac{144.0}{0.4}$$

$$= 360$$

4.2 TYPES OF PERCENT PROBLEMS

EXERCISES 4.2
Find each of the following. See Example 1.

Complete Examples

a. 24% of 200
   ↓   ↓   ↓
   0.24 × 200
   = _____
     (1)

b. $6\frac{1}{2}$% of 200
   6.5% of 200
   ↓   ↓   ↓
   _____ × 200 = _____
     (2)   (3)

c. 112% of 200
   ↓   ↓   ↓
   1.12 × 200
   = _____
     (4)

---
(1) 48   (2) 0.065   (3) 13   (4) 224
---

1. 25% of 64
2. 10% of 120
3. 30% of 80.5
4. 20% of 95.5
5. 29% of 160
6. 72% of 320
7. 120% of 54
8. 150% of 68
9. 6.5% of 40
10. 8.5% of 60
11. 0.4% of 480
12. 0.6% of 260
13. $\frac{1}{2}$% of 4000
14. $\frac{3}{4}$% of 2000
15. $9\frac{1}{2}$% of 600
16. $8\frac{1}{4}$% of 400

For each of the following: a. Write an equation. b. Solve the equation to find the percent. See Examples 2 and 3

Complete Examples

I. 21 is what percent of 6

II. 35 is what percent of 60? (to the nearest tenth)

Solutions Let P represent the decimal form of the required percent in each case.

Ia. 21 is what percent of 60?
     ↓ ↓   ↓   ↓ ↓
     21 =   P   × 60

Ib. $\frac{21}{60} = \frac{P \times 60}{60}$

   $P = \underline{} = \underline{}$ %
         (1)   (2)

IIa. 35 is what percent of 60?
     ↓ ↓   ↓   ↓ ↓
     35 =   P   × 60

IIb. $\frac{35}{60} = \frac{P \times 60}{60}$

   $P = 0.58\overline{33} = \underline{}$ % (nearest tenth percent)
                               (3)

---
(1) 0.35   (2) 35%   (3) 58.3%

139

PERCENT

17. 4 is what percent of 5?
18. 3 is what percent of 4?
19. What percent of 64 is 16?
20. What percent of 80 is 32?
21. 100 is what percent of 40?
22. 140 is what percent of 40?
23. What percent of 60 is 7.2?
24. What percent of 54 is 8.1?
25. 78 is what percent of 80?
26. What percent of 80 is 68
27. What percent of 20 is 0.5?
28. What percent of 48 is 0.6?
29. What percent of 70 is 2.3? (to the nearest tenth)
30. 80 is what percent of 122? (to the nearest tenth)

For each of the following: a. Write an equation. b. Solve the equation to find the number. See Example 4. Express answers to the nearest tenth.

Complete Examples

I. 32.8 is 40% of what number?

II. $\frac{1}{2}$% of what number is 0.36?

Solutions Let N represent the required number in each case.

Ia. 32.8 is 40% of what number?
$$32.8 = 0.4 \times N$$

Ib. $$\frac{32.8}{0.4} = \frac{0.4 \times N}{0.4}$$

$$N = \frac{32.8}{0.4} = \underline{\quad(1)\quad}$$

IIa. $\frac{1}{2}$% of what number is 0.36?
$$\underline{\quad(2)\quad} \times N = 0.36$$

IIb. $$\frac{0.005 \times N}{0.005} = \frac{0.36}{0.005}$$

$$N = \frac{0.360}{0.005} = \underline{\quad(3)\quad}$$

---

(1) 82   (2) 0.005   (3) 72

---

31. 68 is 20% of what number?
32. 90 is 40% of what number?
33. 60% of what number is 80?
34. 75% of what number is 16?
35. 200 is $7\frac{1}{2}$% of what number?
36. 300 is $8\frac{1}{2}$% of what number?
37. 6.5% of what number is 42?
38. 8.6% of what number is 56?
39. 60 is $8\frac{1}{4}$% of what number?
40. 20 is $5\frac{3}{4}$% of what number?

4.3 APPLICATIONS OF PERCENT

The remaining problems in this set of exercises include Case I, II, and III types of percent. Express answers to the nearest tenth.

41. Find 18% of $95.
42. $16 is what percent of $80?
43. $7.20 is what percent of $57.60?
44. What percent of 8 is 0.4?
45. 9.2 is 62% of what number?
46. Find 120% of $12,700.
47. $1\frac{1}{2}$% of what number is 6?
48. Find $22\frac{1}{2}$% of $18,000.
49. What percent of 560 is 200?
50. 0.092 is what percent of 2.3?
51. Find 7.5% of $238.
52. 36.8 is 8% of what number?

4.3 APPLICATIONS OF PERCENT

The steps on page 32 that we used to solve word problems in previous sections are also applicable to problems that involve percent. Such problems can generally be solved by using, as mathematical models, one of the three cases that we considered in Section 4.2. A number of typical examples are shown in the Exercise set.

EXERCISES 4.3

Solve each problem (Case I).

One guideline for a person renting a house or an apartment is that the monthly rent should not be more than 25% of the monthly income. What is the most a person should plan to spend for each of the following monthly incomes?

Complete Example

Monthly income: $925.

Solution
1. Maximum amount for rent: C
2. C is 25% of $925
   $C = \underline{} \times 925$
   (1)
3. $C = \underline{}$
   (2)
4. A person should pay at most $231.25.

(1) 0.25   (2) $231.25

1. $700
2. $1200
3. $1460
4. $2130

141

PERCENT

A recent report estimates that the average person spends 12% of his income for transportation. Using this statistic, how much is spent by a person with each of the following annual incomes?

5. $16,000   **6.** $24,000   **7.** $42,000   **8.** $54,000

Given the price of an item and the sales tax rate, find the total price of each item to the nearest cent.

Complete Example
   $40; 6%

   Solution 1. Total price: T
   2. T = (item's cost) + (tax on item)
      = 40 + (6% of 40)
      = 40 + (___ × 40)
              (1)
   3. T = 40 + _____
                  (2)
   4. The total price is $_____.
                          (3)

(1) 0.06   (2) 2.4   (3) $42.40

9. $71; 6%   **10.** $120; 6%   **11.** $80; $5\frac{1}{2}$%   **12.** $140; $6\frac{1}{2}$%

The price of an article as specified by a manufacturer is called the **list price.** The amount that a merchant pays for the article is the **cost price.** The cost price of an article is usually determined by subtracting a certain percent of the list price from the list price. The amount to be subtracted is called the **discount;** the percent used to compute the discount is called the **discount rate.**

For each given list price (l.p.) and discount rate (d.r.), find the cost price to the nearest cent.

Complete Example
   l.p. $95.00
   d.r. 30%

      Solution 1. Discount: D;   Cost price: C

142

4.3 APPLICATIONS OF PERCENT

2a. D is 30% of $95.00

3a. $D = \underline{} \times 95 = \underline{}$
        (1)   (2)

2b. $C = (\text{l.p.}) - (D)$

3b. $C = 95.00 - 28.50 = \underline{}$
                                 (3)

4. The cost price is $66.50.

---

(1) 0.3   (2) 28.5   (3) 66.5

---

13. l.p. $73
d.r. 8%

14. l.p. $53
d.r. 6%

15. l.p. $140
d.r. 15%

16. l.p. $160
d.r. 18%

The actual price at which an article is sold to a consumer is referred to as the **selling price**—usually the same as the list price. A business may offer a discounted price to the consumer, in which case the selling price is less than the list price.
For each given l.p. and d.r., find the selling price to the nearest cent.

Complete Example

l.p. $90
d.r. 15%

Solution

1. Discount: D;   Selling price: S

2a. D is 15% of $90

3a. $D = \underline{} \times 90 = \underline{}$
        (1)   (2)

2b. $S = (\text{l.p.}) - (D)$

3b. $S = 90 - 13.50 = \underline{}$
                             (3)

4. The selling price is $76.50.

---

(1) 0.15   (2) 13.50   (3) 76.50

---

17. l.p. $150
d.r. 20%

18. l.p. $80
d.r. 15%

19. l.p. $280
d.r. 8%

20. l.p. $460
d.r. 6%

PERCENT

People who sell goods or services for others are often paid a part of the money that they receive for what they sell. The part of the money that they receive is called **commission.** *The* **rate of commission** *is usually given as a percent. In each of the following, A is the amount of sales and P is the rate of commission. Find the commission.*

Complete Example

$A = \$10,200; \quad P = 4\frac{3}{4}\%$

Solution
1. Commission: C
2. $C = 4\frac{3}{4}\%$ of $10,200$
   $= 4.75\%$ of $10,200$
   $= \underline{}_{(1)} \times 10,200$
3. $C = \underline{}_{(2)}$
4. The commission is $484.50.

(1) 0.0475   (2) 484.5

21. $A = \$12,000$
    $P = 4.2\%$

22. $A = \$16,000$
    $P = 3.6\%$

23. $A = \$4800$
    $P = 6\frac{1}{2}\%$

24. $A = \$6000$
    $P = 8\frac{1}{4}\%$

Solve each problem (Case II). Express each percent answer to the nearest tenth of a percent.

Complete Example

If 28 out of 35 students in a chemistry class own calculators, what percent of the class owns calculators?

Solution
1. Required percent (in decimal form): P
2. 28 is what percent of 35?
   $28 = P \times 35$
3. $\frac{28}{35} = \frac{P \times 35}{35}$
   $P = \underline{}_{(1)} = \underline{}_{(2)}\%$
4. 80% of the class owns calculators.

(1) 0.8   (2) 80%

25. If a student correctly answers 44 out of 70 questions on a test, what percent of the questions are answered correctly?

26. If 14 out of 135 students earn an A grade in a class, what percent of the class earned A grades?

4.3 APPLICATIONS OF PERCENT

27. A large sign uses 1500 light bulbs. If 63 of the bulbs burn out: a. What percent of the bulbs need to be replaced? b. What percent of the bulbs are still in operation?

28. A company spends $24,000 for advertising expenses out of a total sales income of $282,000. What percent of total sales is spent for advertising?

29. If sales tax in the amount $18.20 is added to a charge of $280, find the sales tax rate.

30. A taxpayer pays $3200 federal income tax on an annual income of $18,600. Find her income tax rate.

31. The following table shows the distribution of grades in art history classes in the fall semester. Find the percent of the students that earned each grade. *Hint:* first find the total number of students in the classes.

Grade	A	B	C	D	F	Total
Number of students	52	70	170	26	6	___
%	___	___	___	___	___	___

32. Follow the instructions of Exercise 31 for the spring semester.

Grade	A	B	C	D	F	Total
Number of students	47	62	146	18	9	___
%	___	___	___	___	___	___

Find the percent increase (or decrease) when the first number increases (or decreases) to the second number.

Complete Example

400;   450

Solution
1. Percent of increase: P
2. The amount of change is $450 - 400 = \underline{}_{(1)}$.

$$\begin{pmatrix}\text{amount of} \\ \text{change}\end{pmatrix} = P \times \begin{pmatrix}\text{original} \\ \text{number}\end{pmatrix}$$

$$50 = P \times \underline{}_{(2)}$$

3. $\dfrac{50}{400} = \dfrac{P \times 400}{400} = 0.125$

4. The percent of increase is $\underline{}_{(3)}$ %.

(1) 50   (2) 400   (3) 12.5%

33. 200; 250

34. 300; 360

35. 12,000; 21,000

36. 48,000; 54,000

37. 800; 600

38. 500; 200

39. An employee was given a 15% increase, followed soon after by a 15% decrease in salary. If he began with $360 per week, is he now earning the same salary? More? Less?

40. A rare magazine, at one time valued at $300, appreciated in value by 100%. The publisher then discovered 10,000 copies of the magazine in a warehouse, and the increased value depreciated by 100%. Find the value of the magazine after it depreciated.

145

PERCENT

41. Over a 20-year period, the number of cases of polio in the United States dropped from 58,000 per year to 7 as a result of antipolio vaccines. Find the percent decrease to the nearest hundredth.

42. Over a 5-year period, the number of cases of rubella (German measles) in the United States dropped from 58,000 per year to 12,000 as a result of antirubella vaccines. Find the percent decrease.

In each of the following the present value of an article is given. Find its value: a. If it depreciates by the given percent. b. If it appreciates by the given percent.

Complete Example
$1080; 25%

Solution

1. Amount of depreciation or appreciation: A

2. $\begin{pmatrix} \text{percent of} \\ \text{depreciation or} \\ \text{appreciation} \end{pmatrix} \times \begin{pmatrix} \text{original} \\ \text{amount} \end{pmatrix} = \begin{pmatrix} \text{amount of} \\ \text{depreciation or} \\ \text{appreciation} \end{pmatrix}$

     25% × 1080 = A

3. _____ = A
       (1)

4. a. Subtract the amount of depreciation from the original amount.

   1080 − 270 = _____
                  (2)

   The value depreciates to $810.

   b. Add the amount of appreciation to the original amount.

   1080 + _____ = 1350
             (3)

   The value appreciates to $1350.

(1) 270   (2) 810   (3) 270

43. $940; 10%   **44.** $620; 25%   **45.** $12.50; 9%   **46.** $24.60; 8%

Solve each problem (Case III).
Round off dollars and cents to the nearest cent, percents to the nearest tenth of a percent, and numbers of objects (or people) to the nearest whole number.

Complete Example

If 123 students in a class earned passing grades and 82% of the students passed, how many students were in the class?

Solution

1. Number of students: N

2. 123 is 82% of N

   123 = _____ × N
            (1)

146

4.3 APPLICATIONS OF PERCENT

48. A factory finds that 9% of its employees are absent on Fridays. If 83 employees were absent one particular Friday, how many people are employed in the factory?

50. From a shipment of eggs for a chicken farm, $\frac{4}{5}$% did not hatch. If 48 eggs did not hatch, how many eggs were in the shipment?

Miscellaneous exercises (Cases I, II, and III).

51. The dropout rate for a particular course at Midtown College is 32%. If 560 students enroll in the course, how many can be expected to drop out of the course? (Give answer to the nearest whole number.)

52. In a shipment of 1200 television sets, 37 were damaged in transit. What percent of that shipment was damaged?

53. A used boat was sold for $1450. If this price is 70% of the price when new, what was the price of the new boat?

54. An employee received a salary increase of $4\frac{3}{4}$%. If the increase was $20 a month, what was the original monthly salary?

55. Under certain conditions, $1200 of paid-up insurance has a surrender value of $840. The surrender value is what percent of the paid-up amount?

56. A plant produces 15,000 light bulbs a day. If $\frac{1}{2}$% are defective, how many defective bulbs are produced in one day?

57. An employee's yearly salary is $15,600 If he receives a $5\frac{1}{2}$% increase, how much more will he be earning annually?

58. A store was selling a five-pound sack of sugar for $2.49. A few weeks later the price was $1.39. The new price is what percent of the old price?

59. A college freshman class includes 47% female students. If there are 2052 female students in the group, how many students are in the entire group?

60. A student answered 123 questions correctly on a test and received a score of 82%. How many questions were on the test?

61. A construction company found that $7\frac{1}{2}$% of the 48,650 board-feet of lumber ordered for a building was wasted. a. How many board-feet were wasted? b. How many board-feet were used?

62. An employer contributes $1\frac{3}{4}$% of his payroll to an employee welfare fund. To the nearest cent, find the amount of the contribution on a $428,400 payroll.

PERCENT

63. A student has saved $42 toward the purchase of a $140 typewriter. What percent of the purchase price has been saved?

64. If $135 in sales tax is charged on the purchase of a motorcycle priced at $2700, find the sales tax rate.

4.4 SIMPLE INTEREST

In this section we consider an application of percent that is of great importance to business people and consumers who borrow, lend, or invest money. We will direct our attention primarily to applications of consumer type loans.

The amount of money borrowed is called the **principal** of the loan. The charge paid for the use of the borrowed money is called the **interest,** and the sum of the principal and the interest is called the **amount.**

The interest charged for a loan depends on how much money is borrowed, the interest rate charged by the lender, and the length of time during which the money is kept by the borrower. If the principal does not change during the time of the loan, the interest I is called **simple interest** and is computed by the formula

$$I = P \times r \times t \tag{1}$$

where P is the principal, r is the interest rate given as a decimal form of a *percent* (per year), and t is the length of time (expressed *in years* or as a fraction of a year) over which the interest is to be computed. If the time is a number of months, then t is expressed as a fraction with denominator 12.

For use in Formula (1), it is usually helpful to first change fractions to decimal form, particularly if a calculator is used for the computations.

Example 1 Express each period of time in years or as a fraction of a year in decimal form.

  a. $1\frac{1}{2}$ years
  b. 9 months

Solutions a. $1\frac{1}{2} = 1.5$

  b. 9 out of 12 months can be expressed as $\frac{9}{12} = 0.75$.

Example 2 Find the simple interest and the amount on a loan of $1800 at a $9\frac{1}{2}\%$ interest rate for:

  a. 3 years
  b. 9 months

Solutions In each case substitute into Formula (1).

  a. $P = 1800$; $r = 9\frac{1}{2}\% = 0.095$; $t = 3$.

$$I = P \times r \times t$$
$$I = 1800 \times 0.095 \times 3 = 513$$

The interest is $513. The amount (principal and interest)

$$A = 1800 + 513 = \$2313.$$

4.4 SIMPLE INTEREST

b. $P = 1800$; $r = 9\frac{1}{2}\% = 0.095$; $t = \frac{9}{12} = 0.75$.

$$I = P \times r \times t$$
$$\downarrow \quad \downarrow \quad \downarrow$$
$$I = 1800 \times 0.095 \times 0.75 = 128.25.$$

The interest is $128.25, to the nearest cent. The amount

$$A = 1800 + 128.25 = \$1928.25.$$

Given three of the four quantities I, P, r, or t in the simple interest formula, the fourth quantity can be computed.

Example 3   Find the simple interest rate r on a loan of $2000 for $1\frac{1}{2}$ years if the interest charged is $375.

Solution $P = 2000$; $t = 1.5$; $I = 375$.
Substituting into the formula $I = P \times r \times t$, we obtain

$$375 = 2000 \times r \times 1.5$$
$$375 = 3000 \times r$$

Dividing each side by 3000, we have

$$\frac{375}{3000} = \frac{3000 \times r}{3000};$$
$$r = 0.125$$

Hence, the simple interest rate is 12.5%.

On certain types of loans the simple interest is subtracted from the principal at the time that the loan begins; the amount of money actually given to the borrower is called the **net proceeds.**

Example 4   Find the net proceeds on a loan of $2000 at 10% for 1 year if the simple interest is subtracted from the amount to be borrowed at the time that the loan begins.

Solution $P = 2000$; $r = 10\% = 0.10$; $t = 1$

Substituting into the formula, $I = P \times r \times t$, we obtain

$$I = 2000 \times 0.10 \times 1 = 200$$

Subtracting the interest, 200, from the principal, 2000, we have

$$2000 - 200 = 1800$$

Hence, the net proceeds are $1800.

Example 5   Find the interest rate on the loan in Example 4 based upon the net proceeds rather than upon the amount of the loan.

Solution   The net proceeds were found to be $1800; the interest was $200, and the time was one year. Using the *net proceeds* as P in the formula $I = P \times r \times t$, we have

$$200 = 1800 \times r \times 1$$

149

PERCENT

Dividing each side by 1800, we obtain

$$\frac{200}{1800} = \frac{\cancel{1800} \times r}{\cancel{1800}}$$

$$0.111\overline{1} = r$$

Hence, to the nearest tenth of a percent, $r = 11.1\%$. (Note that this interest rate, which might be called the "true" interest rate, is higher than the 10% rate quoted in the original problem.)

EXERCISES 4.4

Express each time as a decimal for use in the formula $I = P \times r \times t$. See Example 1.

Complete Examples

a. $3\frac{1}{4}$ years

b. 6 months

Solutions

a. $3\frac{1}{4} = \underline{\quad(1)\quad}$

b. 6 months is $\frac{6}{12}$ years; $\frac{6}{12} = \underline{\quad(2)\quad}$

(1) 3.25   (2) 0.5

1. $2\frac{1}{2}$ years
2. $6\frac{3}{4}$ years
3. 3 months
4. 27 months
5. 18 months
6. 15 months

For the following problems, express percents to the nearest tenth of a percent, and money to the nearest cent.

Find: a. The simple interest I. b. The amount A on the given principal at the given rate for the indicated time. See Example 2.

Complete Example

$$\$5250; \quad 12\%; \quad 3 \text{ years}$$

Solutions

a. Substitute into the formula $I = P \times r \times t$.

$$I = P \times r \times t$$
$$\downarrow \quad \downarrow \quad \downarrow$$
$$I = 5250 \times \underline{\quad(1)\quad} \times 3 = \underline{\quad(2)\quad}$$

The interest is $1890.

4.4 SIMPLE INTEREST

b. The amount A is the sum of the principal and interest.

$$A = \text{(principal)} + \text{(interest)}$$

$$A = \underline{\quad\quad}_{(3)} + 1890 = \underline{\quad\quad}_{(4)}$$

The amount is $7140.

---

(1) 0.12   (2) 1890   (3) 5250   (4) 7140

---

7. $400; 14%; 2 years
8. $500; 16%; 6 years
9. $280; 9% 15 years
10. $420; 8%; 8 years
11. $8000; $16\frac{1}{2}$%; 18 months
12. $3000; $14\frac{1}{4}$%; 30 months
13. $2500; 12%; 51 months
14. $7200; 12%; 45 months

Complete each table in Exercises 15–18. See Example 3.

15.

I	P	r	t
$400	$2000	_____	2 years
$ 80	$1600	_____	9 months

16.

I	P	r	t
$30,000	$88,000	_____	4 years
$890	$11,000	_____	18 months

17.

I	P	r	t
$2800	_____	9.25%	15 months
$ 960	_____	8.6%	6 years

18.

I	P	r	t
$6670	$14,500	11.5%	_____
$8670	$ 6800	8.6%	_____

For each of the following loans, find the net proceeds. See Example 4.

Complete Example

$$P: \$850; \quad r: 8.5\%; \quad t: 2 \text{ years}$$

Solution Substitute into the formula $I = P \times r \times t$.

$$I = P \times r \times t$$

$$I = 850 \times \underline{\quad\quad}_{(1)} \times 2 = \underline{\quad\quad}_{(2)}$$

PERCENT

Subtract the interest from the principal.

$$\underline{}_{(3)} - 144.5 = \underline{}_{(4)}$$

The net proceeds are $705.50.

---

(1) 0.085   (2) 144.5   (3) 850   (4) 705.5

---

19. *P:* $900; *r:* 8.6%; time: 6 years

20. *P:* $1200; *r:* 6.9%; time: 10 years

21. *P:* $6400; *r:* 9.8%; time: 15 months

22. *P:* $52,000; *r:* $11\frac{1}{2}$%; time: 3 months

In Exercises 23–26, find the interest rate on the loan based upon the net proceeds rather than upon the amount of the loan.

Complete Example

The loan in the previous example.

Solution The net proceeds were found to be $705.50. Substitute into the formula $I = P \times r \times t$ using 705.50 for P.

$$I = P \times r \times t$$
$$\downarrow = \downarrow \times \downarrow \times \downarrow$$
$$144.5 = 705.5 \times r \times \underline{}_{(1)}$$

$$144.5 = \underline{}_{(2)} \times r$$

$$\frac{144.5}{1411} = \frac{\cancel{1411} \times r}{\cancel{1411}}$$

$$0.1024 = r$$

To the nearest tenth of a percent, $r = \underline{}_{(3)}$%.

---

(1) 2   (2) 1411   (3) 10.2%

---

23. The loan in Exercise 19.

24. The loan in Exercise 20.

25. The loan in Exercise 21.

26. The loan in Exercise 22.

27. A business has a fund of $6700 to pay interest charges on loans.
   a. Find the greatest amount of money that the business can borrow for 6 months at $9\frac{1}{2}$% simple interest.
   b. Find the maximum number of months for which the business can borrow $20,000 at 11% simple interest.

28. If the business of Exercise 27 increases its interest fund to $15,000:
   a. Find the greatest amount of money it can borrow for 18 months at 10.3% simple interest.
   b. Find the greatest number of months for which it can borrow $30,000 at 9.4% simple interest.

CHAPTER FOUR REVIEW EXERCISES

1. Write 0.056 as a percent.
2. Write 18% as a decimal.
3. Write $6\frac{3}{4}$% as a decimal.
4. Write 1.2 as a percent.
5. 25% of 80
6. 64% of 75.6
7. 4.5% of 600
8. 120% of 45
9. 6 is what percent of 8?
10. What percent of 125 is 40?
11. 80 is 40% of what number?
12. 8.6% of what number is 7.74?
13. If a person earns $15,600 a year and spends 9% for transportation, how much is spent on transportation?
14. If a student correctly answers 12 out of 15 questions on a test, what percent of the questions are answered correctly?
15. The sales tax on an item costing $350.00 is $22.75. Find the sales tax rate.
16. The sales tax on a used car was 6%. Find the cost of the car if the tax was $300.
17. Find the cost of an item having a list price of $60 and a discount rate of 8.4%.
18. Find the selling price of an item having a list price of $140 if it is being sold at a discount of 15%.
19. Find the cost price of an item having a list price of $185 and a discount rate of 20%
20. A stereo set has a list price of $220. a. One supplier offers a discount of 24%. b. Another supplier offers a discount of 21%. What is the cost price in each case?

In Exercises 21 and 22, A is the amount of sales, P is the commission rate, and C is the commission.

21. Find C if $A = \$6400$ and $P = 2\%$.
22. Find P if $A = \$14,500$ and $C = \$507.50$.
23. Find the percent increase when 140 increases to 210.
24. Find the percent decrease when 640 decreases to 560.
25. During one season, a basketball player scored 1500 points. During the next season he scored 1620 points. Find the percent increase in points scored.

In Exercises 26–28, find the amount A on the given principal at the given interest rate for the indicated time.

26. $1000; 8.6%; 3 years
27. $2400; 4.9%; 18 months
28. $8600; 9.6%; 6 months
29. Find the yearly rate of interest to the nearest tenth of a percent if $I = \$62$ and $P = \$1400$ for a period of 9 months.
30. How much money would have to be invested at a rate of 8.5% to earn $500 interest in two years?

5 MEASUREMENT

Measurement is the process of comparing a quantity with a standard unit that depends on the *system* of measurement being used. There are two systems of measurement in common use—the United States system and the metric system. In this chapter we consider fundamental relationships in each system, together with techniques for changing units from one system to another.

Numbers that involve units of measure, such as 11 miles and 14 quarts, are called **denominate numbers**. Numbers that do not involve units of measure are **abstract numbers**.

5.1 THE UNITED STATES SYSTEM

The basic unit of length in the U.S. system is the *foot*; the basic unit of weight is the *pound*. These basic units are not always the most convenient for every measurement. For example, we would generally use the *mile* as the unit of measurement for the distance between two cities, the *yard* (or foot) as the unit of measurement for the dimensions of a building, and the *inch* as the unit of measurement for the dimensions of parts of machinery. Relationships between these units, and some standard abbreviations are given below; relationships between these units and others commonly used in the U.S. system are given in the more extensive tables in the Appendix on the inside of the back cover.*

Table A.4

12 inches (in.) = 1 foot (ft)
3 feet = 36 inches = 1 yard (yd)
5280 feet = 1760 yards = 1 mile (mi)

In the above table, note that number 12 is associated with *one* foot for inch-feet conversions, the number 3 is associated with *one* yard for feet-yards conversions,

*All tables labeled with the letter A are to be found in the Appendix.

MEASUREMENT

and so forth. We refer to the numbers associated with one unit, such as 12 and 3, as **conversion numbers**.

CHANGING UNITS

Proportions can be used to convert (change) denominate numbers from one unit to another.

Example 1 Change 108 in. to yards.

Solution Number of yards: Y
Set up a proportion.

First, pair Y (yards) with 108 (inches) as numerators. Next, from Table A.4, note that 36 inches equals 1 yard. Pair 1 (yard) with 36 (inches) as denominators, and solve the resulting proportion.

$$\begin{array}{cc} \textbf{yards} & \textbf{inches} \\ \dfrac{Y}{1} = & \dfrac{108}{36} \end{array}$$

By a direct division,

$$Y = \frac{108}{36} = 3$$

Hence, 108 in. equals 3 yd.

Example 2 Change 6 lb to ounces.

Solution Number of ounces: Z
Set up a proportion.

First, pair Z (ounces) with 6 (pounds) as numerators. Next, from Table A.1, note that 16 oz equals 1 lb. Pair 16 (ounces) with 1 (pound) as denominators, and solve the resulting proportion.

$$\begin{array}{cc} \textbf{ounces} & \textbf{pounds} \\ \dfrac{Z}{16} = & \dfrac{6}{1} \end{array}$$

By the cross-multiplication rule,

$$Z \times 1 = 16 \times 6$$
$$Z = 96$$

Hence, 6 lb equals 96 oz.

In Example 1, note that a smaller unit (inches) is changed to a larger unit (yards) by *dividing* the number of inches by the conversion number (36). In Example 2, note that a larger unit (pounds) is changed to a smaller unit (ounces) by *multiplying* the number of pounds by the conversion number (16). Results such as these suggest the following rules.

5.1 THE UNITED STATES SYSTEM

> **Unit-conversion rules**
> 1. To change a number *S* of smaller units to a number of larger units, divide *S* by the conversion number.
> 2. To change a number *L* of larger units to a number of smaller units, multiply *L* by the conversion number.

Example 3   a. 3660 min = ? hr   b. 54 yd = ? ft

Solution   a. Note that the minutes-hour conversion in Table A.2 is 60. Because the conversion is from smaller to larger units, divide the number of minutes by 60. Thus,

$$3660 \text{ min} = \frac{3660}{60} \text{ hr} = 61 \text{ hr}$$

b. Note that the feet-yards conversion number in Table A.4 is 3. Because the conversion is from larger to smaller units, multiply the number of yards by 3. Thus,

$$54 \text{ yd} = (3 \times 54) \text{ ft} = 162 \text{ ft}$$

DENOMINATE NUMBERS; MORE THAN ONE UNIT

Measurements in the U.S. system are often given in terms of more than one unit. Such numbers can be changed to denominate numbers with only one unit.

Example 4   24 gal 2 qt = ? qt

Solution   Change 24 gallons to quarts, and add 2 quarts to the result.

From Table A.8, the conversion number (gallons-quarts) is 4. Hence,

$$24 \text{ gal } 2 \text{ qt} = (24 \times 4 + 2) \text{ qt} = 98 \text{ qt}$$

DECIMAL FORM OF A DENOMINATE NUMBER

Results of conversions in the U.S. system may be expressed either in fraction or in decimal form. In general, we give answers in decimal form.

Example 5   Write each conversion as a decimal form, to the nearest hundredth:

a. 11,624 ft = ? mi   b. 28 min = ? hr

Solution   a. From Table A.4, the conversion number (feet-miles) is 5280. Hence,

$$11{,}624 \text{ ft} = \frac{11{,}624}{5280} \text{ mi} = 2.20 \text{ mi}$$

to the nearest hundredth.

MEASUREMENT

b. From Table A.2, the conversion number (minutes-hours) is 60. Hence,

$$28 \text{ min} = \frac{28}{60} \text{ hr} = 0.47 \text{ hr}$$

to the nearest hundredth.

Example 6 Write 6 ft 9 in. as a number of feet in decimal form.

Solution Change 9 in. to feet (in decimal form) and add the result to 6 ft. From Table A.4, the conversion number (inches-feet) is 12. Hence,

$$6 \text{ ft } 9 \text{ in.} = \left(6 + \frac{9}{12}\right) \text{ft} = (6 + 0.75) \text{ ft}$$
$$= 6.75 \text{ ft}$$

EXERCISES 5.1

In the following exercises, refer to the appropriate table (1–8) in the Appendix as necessary. If possible, complete Exercises 1–16 without using the tables.

1. 1 ft = ? in.
2. 1 yd = ? in.
3. 1 yd = ? ft
4. 24 in = ? ft
5. 6 ft = ? yd
6. 36 in. = ? yd
7. 1 lb = ? oz
8. 32 oz = ? lb
9. 1 gal = ? qt
10. 1 qt = ? pt
11. 12 pt = ? qt
12. 1 wk = ? da
13. 1 min = ? sec
14. 1 yr = ? mo
15. 1 yr = ? wk
16. 36 mo = ? yr

Convert each measurement. See Examples 1, 2, and 3.

Complete Examples

a. 21,120 ft = ? mi

b. 18 ft = ? in.

Solutions

a. From Table A.4, the conversion number (feet-miles) is _____ . Because the conversion is
(1)
from smaller to larger units, __multiply/divide__ by 5280.
(2)

$$21{,}120 \text{ ft} = \frac{21{,}120}{5280} \text{ mi} = \underline{} \text{ mi}$$
(3)

b. From Table A.4, the conversion number (inches-feet) is _____ .
(4)
Because the conversion is from larger to smaller units, multiply by 12.

$$18 \text{ ft} = (18 \times 12) \text{ in} = \underline{}$$
(5)

---

(1) 5280   (2) divide   (3) 4   (4) 12   (5) 216

5.1 THE UNITED STATES SYSTEM

17. 168 in. = ? ft   **18.** 3 mi = ? yd   **19.** 27 ft = ? yd   **20.** 256 oz = ? lb

21. 20 T = ? lb   **22.** 96 qt = ? gal   **23.** 15 qt = ? pt   **24.** 12,320 yd = ? mi

25. 15 hr = ? min   **26.** 40 pt = ? fl oz   **27.** 24 mi = ? ft   **28.** 31,680 ft = ? mi

Each of the following requires more than one conversion.

Complete Examples

$$256 \text{ pt} = ? \text{ gal}$$

First convert 256 pt to quarts.

$$256 \text{ pt} = \frac{256}{\underline{\quad(1)\quad}} \text{ qt} = \underline{\quad(2)\quad} \text{ qt}$$

Next convert 128 qt to gallons.

$$128 \text{ qt} = \frac{128}{\underline{\quad(3)\quad}} \text{ gal} = \underline{\quad(4)\quad} \text{ gal}$$

(1) 2   (2) 128   (3) 4   (4) 32

29. 2 mi = ? in.   **30.** 1024 fl oz = ? qt   **31.** 640 pt = ? gal   **32.** 15 gal = ? pt

33. 240 min = ? hr   **34.** 1008 hr = ? wk   **35.** 17 pk = ? pt   **36.** 90 qt = ? fl oz

37. 24 bu = ? qt   **38.** 3 wk = ? hr   **39.** 15 da = ? min   **40.** 12 da = ? sec

Convert each measurement. See Example 4.

Complete Examples

a. 15 yd 1 ft = ? ft   b. 3 lb 12 oz = ? oz

Solutions

a. 15 yd 1 ft = (15 × $\underline{\quad(1)\quad}$ + 1) ft   b. 3 lb 12 oz = (3 × $\underline{\quad(3)\quad}$ + 12) oz

       = $\underline{\quad(2)\quad}$ ft   = $\underline{\quad(4)\quad}$ oz

(1) 3   (2) 46   (3) 16   (4) 60

MEASUREMENT

41. 9 lb 14 oz = ? oz

42. 2 T 516 lb = ? lb

43. 17 yd 2 ft = ? ft

44. 5 mi 368 yd = ? yd

45. 8 hr 52 min = ? min

46. 9 wk 4 da = ? da

47. 5 gal 3 qt = ? qt

48. 13 qt 1 pt = ? pt

Express each result as a decimal rounded off to the nearest hundredth. See Example 5.

Complete Examples

a. 75 wk = ? yr

b. 169 min = ? hr

Solutions

a. $75 \text{ wk} = \dfrac{75}{(1)} \text{ yr}$

$= \dfrac{}{(2)} \text{ yr}$

to the nearest hundredth.

b. $169 \text{ min} = \dfrac{169}{(3)} \text{ hr}$

$= \dfrac{}{(4)} \text{ hr}$

to the nearest hundredth.

(1) 52   (2) 1.44   (3) 60   (4) 2.82

49. 67 fl oz = ? pt

50. 71 qt = ? gal

51. 127 oz = ? lb

52. 124 wk = ? yr

53. 3146 yd = ? mi

54. 350 min = ? hr

55. 2700 min = ? da

56. 2700 sec = ? hr

Convert each measurement to decimal form. Round off to the nearest hundredth. See Example 6.

Complete Examples

a. 4 gal 1 qt = ? gal

b. 2 hr 12 min = ? min

Solutions

a. $4 \text{ gal } 1 \text{ qt} = (4 + \frac{1}{4}) \text{ gal}$

$= (4 + \dfrac{}{(1)}) \text{ gal}$

$= \dfrac{}{(2)}$

b. $2 \text{ hr } 12 \text{ min} = (2 + \frac{12}{60}) \text{ hr}$

$= (2 + \dfrac{}{(3)}) \text{ hr}$

$= \dfrac{}{(4)} \text{ hr}$

(1) 0.25   (2) 4.25   (3) 0.2   (4) 2.2

57. 4 ft 7 in. = ? ft

58. 12 ft 5 in. = ? ft

59. 5 mi 700 ft = ? mi

60. 3 mi 960 ft = ? mi

61. 5 gal 3 qt = ? gal

62. 2 gal 1 qt = ? gal

63. Ocean depths are often specified in *fathoms*, 1 fathom being equal to a 6 ft. a. A depth of 4200 fathoms is equivalent to how many feet? b. To how many miles (to the nearest tenth)?

64. To the nearest tenth, a depth of 3179 ft is equivalent to how many fathoms? (See Exercise 63.)

65. In surveying, a *chain* is a distance of 22 yd. a. A distance of 60.5 chains is equivalent to how many yards? b. To how many feet?

66. To the nearest tenth, a distance of 115 yd is equivalent to how many chains? (See Exercise 65.)

67. If a ship is traveling at a speed of 1 *knot*, it travels 6082.2 ft in 1 hr. To the nearest tenth, how many miles is this?

68. A boat traveling with a speed of 15 mi in 1 hr. Express this equivalently in knots. (See Exercise 67.)

69. In the printing trade, 1 *pica* is 0.166044 in. Express 16 picas equivalently in inches (to the nearest hundredth).

70. To the nearest hundredth, express 12 in. in picas. (See Exercise 69.)

5.2 ARITHMETIC OF DENOMINATE NUMBERS

In this section we consider methods for finding sums, differences, products, and quotients of denominate numbers.

ADDITION AND SUBTRACTION

Denominate numbers can be added or subtracted only if they have the same units.

Example 1 a. Compute 8 ft 6 in. + 11 ft 10 in. + 3 ft 9 in.
b. Express the result in decimal form (using the larger unit).

Solutions a. 8 ft 6 in.
11 ft 10 in.
3 ft 9 in.
―――――――――
22 ft 25 in.

Change 25 in. to 2 ft + 1 in. Hence,

22 ft 25 in. = 22 ft + 2 ft + 1 in.
= 24 ft 1 in.

b. 24 ft 1 in. = $\left(24 + \dfrac{1}{12}\right)$ ft = 24.08 ft, to the nearest hundredth.

161

MEASUREMENT

Example 2 a. Compute 9 lb 3 oz − 5 lb 11 oz.
b. Express the result in decimal form (using the larger unit).

Solutions a. Because 11 oz cannot be subtracted from 3 oz, "borrow" 1 lb from the 9 lb and change the 1 lb to 16 oz. Then,

$$9 \text{ lb } 3 \text{ oz} = 8 \text{ lb} + 16 \text{ oz} + 3 \text{ oz} = \quad\quad 8 \text{ lb } 19 \text{ oz}$$
$$\underline{-\ 5 \text{ lb } 11 \text{ oz}} \quad\quad\quad\quad\quad\quad\quad\quad\quad \underline{-\ 5 \text{ lb } 11 \text{ oz}}$$
$$\quad\quad\quad\quad\quad\quad\quad\quad\quad\quad\quad\quad\quad\quad\quad\quad\quad 3 \text{ lb } \ 8 \text{ oz}$$

b. $3 \text{ lb } 8 \text{ oz} = \left(3 + \dfrac{8}{16}\right) \text{ lb} = 3.5 \text{ lb}$

MULTIPLICATION AND DIVISION

The product of an abstract number and a denominate number with more than one unit can be found by multiplying the number associated with each different unit by the abstract number.

Example 3 8 × (5 mi 723 yd)

Solution First multiply the number associated with each unit by 8.

$$\begin{array}{r} 5 \text{ mi } \ 723 \text{ yd} \\ \times\ 8 \\ \hline 40 \text{ mi } 5784 \text{ yd} \end{array}$$

$8 \times 5 \text{ mi}$ $\quad\quad 8 \times 723 \text{ yd}$

Change 5784 yd to 3 mi 504 yd. Hence,

40 mi 5784 yd = 40 mi + 3 mi + 504 yd
= 43 mi 504 yd

In a division, denominate numbers with more than one unit are usually changed to denominate numbers with only one unit.

Example 4 a. (13 ft 6 in.) ÷ 9 b. (13 ft 6 in.) ÷ 9 in.

Solutions In each case,

$$13 \text{ ft } 6 \text{ in.} = (12 \times 13 + 6) \text{ in.} = 162 \text{ in.}$$

a. $\dfrac{13 \text{ ft } 6 \text{ in.}}{9} = \dfrac{162 \text{ in.}}{9} = 18 \text{ in.}$ b. $\dfrac{13 \text{ ft } 6 \text{ in.}}{9 \text{ in.}} = \dfrac{162 \text{ in.}}{9 \text{ in.}} = 18$

In Example 4a. note that the quotient involves units (inches), while the quotient in b. is an abstract number (no units). Example 4a could provide the answer to the problem:

"If a 13 ft 6 in. aluminum rod is cut into 9 pieces (with no waste), how long is each piece?"

The answer is that each piece is 18 in. long. Example 4b. could provide the answer to the problem:

"If a 13 ft 6 in. aluminum rod is cut into 9-in. pieces, how many such pieces can be obtained?"

Assuming no waste, the answer is that 18 such pieces can be obtained.

5.2 ARITHMETIC OF DENOMINATE NUMBERS

EXERCISES 5.2
For each of the following: a. Compute. b. Express the result in decimal form rounded off to the nearest hundredth (using the larger unit). See Examples 1 and 2.

Complete Example

$$2 \text{ hr } 45 \text{ min } + 3 \text{ hr } 17 \text{ min } + 8 \text{ hr } 51 \text{ min}$$

Solutions

a. $\begin{array}{r} 2 \text{ hr } 45 \text{ min} \\ 3 \text{ hr } 17 \text{ min} \\ +8 \text{ hr } 51 \text{ min} \\ \hline \underline{\quad} \text{ hr } 113 \text{ min} \\ (1) \end{array}$ Change 113 min to 1 hr 53 min.
Then 13 hr 113 min = 13 hr + 1 hr 53 min
$$= \underline{\qquad}_{(2)}$$

b. $14 \text{ hr } 53 \text{ min} = \left(14 + \dfrac{53}{60}\right) \text{hr} = \underline{\qquad}_{(3)}$ hr, to the nearest hundredth.

---

(1) 13   (2) 14 hr 53 min   (3) 14.88

---

1. 10 ft + 4 ft + 8 ft 5 in.
2. 15 qt + 3 qt + 6 qt 1 pt
3. 22 lb + 6 lb + 8 lb 7 oz
4. 4 yd + 65 yd + 73 yd 2 ft
5. 39 ft 9 in. + 9 ft 5 in.
6. 27 lb 9 oz + 8 lb 12 oz
7. 6 hr 47 min + 5 hr 33 min + 7 hr
8. 2 bu 3 pk + 9 bu 3 pk + 8 bu 2 pk
9. 34 yd 1 ft + 9 yd 1 ft + 6 yd 2 ft
10. 8 da 12 hr + 4 da 8 hr + 8 da 11 hr

Complete Example

$$4 \text{ ft } 2 \text{ in. } - 1 \text{ ft } 8 \text{ in.}$$

Solutions

a. $\begin{array}{r} 4 \text{ ft } 2 \text{ in.} \\ -1 \text{ ft } 8 \text{ in.} \end{array}$ = 3 ft + 12 in. + 2 in. = $\begin{array}{r} 3 \text{ ft } 14 \text{ in.} \\ -1 \text{ ft } 8 \text{ in.} \\ \hline \underline{\qquad}_{(1)} \end{array}$

b. 2 ft 6 in. = $\left(2 + \dfrac{6}{12}\right)$ ft = $\underline{\qquad}_{(2)}$ ft

---

(1) 2 ft 6 in   (2) 2.5

---

11. 948 mi − 391 mi
12. 89 gal − 63 gal
13. 16 da 5 hr − 8 da 2 hr
14. 65 ft 5 in. − 36 ft 4 in.
15. 92 ft 2 in. − 37 ft 9 in.
16. 75 wk 1 da − 27 wk 4 da

MEASUREMENT

Find each product or quotient. See Examples 3 and 4.

Complete Example

$$7 \times (3 \text{ lb } 5 \text{ oz})$$

Solution

$$\begin{array}{r} 3 \text{ lb } 5 \text{ oz} \\ \times 7 \\ \hline 21 \text{ lb } 35 \text{ oz} \end{array}$$

7×3 lb 7×5 oz

Change 35 oz to _____ lb 3 oz.
                    (1)

21 lb 35 oz = 21 lb + 2 lb + 3 oz
            = _____
                (2)

---

(1) 2   (2) 23 lb 3 oz

---

17. 37 × (31 ft) **18.** 92 × (44 qt) **19.** 8 × (98 ft 2 in.)

20. 5 × (74 lb 6 oz) **21.** 8 × (14 gal 2 qt) **22.** 18 × (19 gal 3 qt)

Complete Examples

a. (3 da 6 hr) ÷ 8   b. (3 da 6 hr) ÷ 8 hr

Solutions

$$3 \text{ da } 6 \text{ hr} = (3 \times \underset{(1)}{\underline{}} + 6) \text{ hr} = \underset{(2)}{\underline{}} \text{ hr}$$

a. $\dfrac{78 \text{ hr}}{8} = \underset{(3)}{\underline{}}$ b. $\dfrac{78 \text{ hr}}{8 \text{ hr}} = \underset{(4)}{\underline{}}$

---

(1) 24   (2) 78   (3) 9.75 hr   (4) 9.75

---

23. (24 da) ÷ 6   **24.** (36 lb) ÷ 12   **25.** (171 hr 36 min) ÷ 9   **26.** (242 gal) ÷ 11

27. (133 lb 8 oz) ÷ 12   **28.** (86 ft 4 in.) ÷ 7   **29.** (8 ft 9 in.) ÷ 15   **30.** (14 yd 2 ft) ÷ 11

31. How many 4-oz portions can a restaurant obtain from 15 lb of ground beef? (Assume that there is no waste.)

32. Assuming no waste, how many 6-oz servings can a restaurant obtain from 27 lb 12 oz of steak?

33. A caterer is to serve ice cream to a group of 256 people, and each person is to be served 4 fl oz.
a. How many fluid ounces of ice cream are needed?
b. If the ice cream is packaged in gallon containers, how many gallons will the caterer need?

34. One can of fruit weighing 1 lb 14 oz is divided equally between 5 people. a. How many ounces did each receive? b. How many cans will be needed to give each of 30 people a 4-oz serving?

35. A board 3 ft 4 in. long is cut from a board that is 8 ft 3 in. long. How long is the remaining piece?

36. A length of pipe measures 16 ft 8 in. Two pieces, one 2 ft 6 in. long and the other 5 ft 9 in. long, are cut from the original pipe. How long is the remaining piece?

37. If a piece of lumber 19 ft 6 in. long is cut into 6 pieces of equal length, how long is each of the resulting pieces?

38. If a bag of potatoes weighing 49 lb 12 oz is separated into four equal parts, what is the weight of each part?

39. Certain concrete blocks are 16 in. long. How many of these blocks, laid end to end, will it take to cover a distance of 53 ft 4 in?

40. How many 9-in.-long pieces of asphalt tile will it take to make a row of tile 17 ft 3 in. long?

5.3   THE METRIC SYSTEM

The metric system of measurement is currently used by most of the countries in the world. The basic unit of length is the **meter** (or **metre**), which is about 3 in. longer than a yard (see Fig. 5.1). The basic unit of weight is the **gram**, which is slightly less than $\frac{1}{28}$ of an ounce (see Fig. 5.2). The basic unit of capacity is the **liter** (or **litre**), which is slightly more than a quart (see Fig. 5.3).

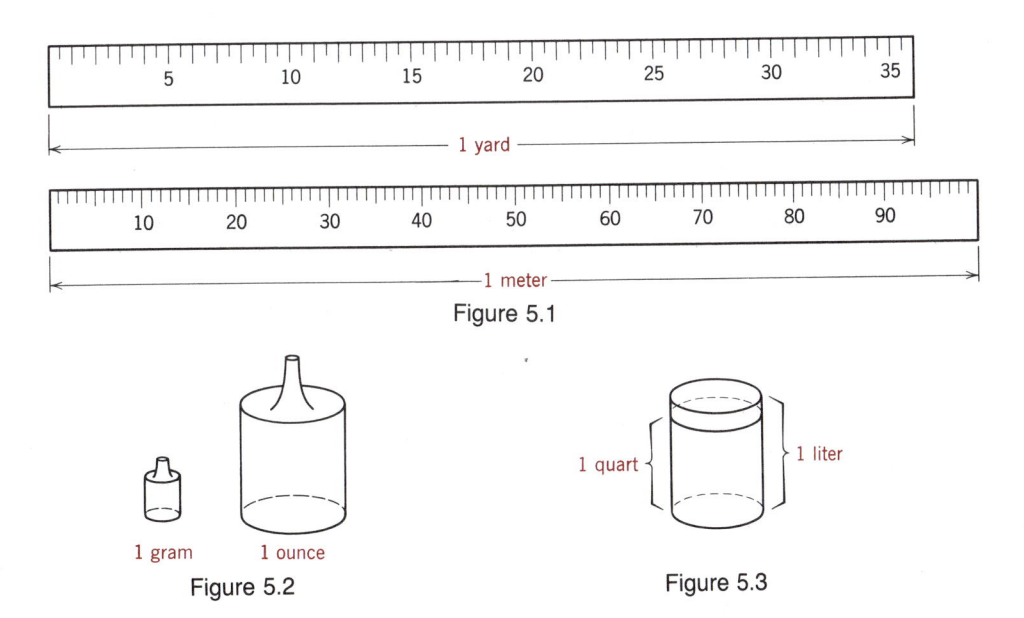

Figure 5.1

1 gram   1 ounce
Figure 5.2

1 quart   1 liter
Figure 5.3

METRIC UNITS AND TABLES

The names of the basic metric units (meter, gram, and liter) are combined with certain prefixes, as listed below, to provide names for other metric units. Each prefix indicates a particular relation between the unit named and the basic unit, as indicated in Table 5.1 on page 166.

MEASUREMENT

Table 5.1
Metric Unit Prefixes

Prefix	Decimal meaning
milli	0.001
centi	0.01
deci	0.1
deka	10
hecto	100
kilo	1000

Example 1
a. "Kilometer" means "one thousand meters."
b. "Kilogram" means "one thousand grams."
c. "Milliliter" means "one thousandth of a liter."
d. "Millimeter" means "one thousandth of a meter."

The metric tables used in this book (see Appendix Tables A.9 to A.13) are arranged so that the smallest unit is listed first; each unit thereafter is larger than the preceding unit. As an example, Table A.9 is reproduced here:

Table A.9
Metric Length

10 millimeters (mm) = 1 centimeter (cm)
10 centimeters   = 1 decimeter (dm)
10 decimeters   = 1 meter (m)
10 meters   = 1 dekameter (dam)
10 dekameters   = 1 hectometer (hm)
10 hectometers   = 1 kilometer (km)

Note that the millimeter is the smallest unit and the kilometer is the largest unit listed.

METRIC CONVERSIONS

A major advantage of the metric system is that a measurement in one unit can be converted to the next smaller unit by multiplying by 10 and can be converted to the next larger unit by dividing by 10. Recall from Section 2.1 that a number can be multiplied by 10 by moving the decimal point one place to the right and can be divided by 10 by moving the decimal point one place to the left. Hence, we have the following rule.

Metric conversions

A measurement in a metric unit can be converted to:

a. The next smaller unit by moving the decimal point one place to the right.
b. The next larger unit by moving the decimal point one place to the left.

Example 2 a. 3.09 cm = ? mm   b. 18.7 hm = ? km

Solutions Refer to Table A.9.

5.3 THE METRIC SYSTEM

a. Because centimeters are to be converted to the next smaller unit (millimeter), move the decimal point one place to the right:

$$3.09 \text{ cm} = 30.9 \text{ mm}$$

b. Because hectometers are to be converted to the next larger unit (kilometers), move the decimal point one place to the left:

$$18.7 \text{ hm} = 1.87 \text{ km}$$

Example 3   a. 623 mg = ? cg   b. 0.54 ℓ = ? dl

Solutions   a. Refer to Table A.10. Because milligrams are to be converted to the next larger unit (centigrams), move the decimal point one place to the left:

$$623 \text{ mg} = 62.3 \text{ cg}$$

b. Refer to Table A.11. Because liters are to be converted to the next smaller unit (deciliters), move the decimal point one place to the right:

$$0.54 \text{ ℓ} = 5.4 \text{ dl}$$

It is sometimes necessary to use more than one conversion to change metric units. For example, multiplying 7.5 by 10, we have

$$7.5 \text{ m} = 75. \text{ dm} = 75 \text{ dm}$$

and, multiplying 75 by 10,

$$75 \text{ dm} = 75.0 \text{ dm} = 750. \text{ cm} = 750 \text{ cm}$$

Thus, 7.5 m equals 750 cm. Note that 7.5 m can be directly converted to centimeters by moving the decimal point two places to the right (multiplying by 100). However, to change 750 cm to meters we would have to divide by 100, which can be done simply by moving the decimal point two places to the left. Results such as these suggest the following procedure:

To change from one metric unit to another

Use the appropriate table in the appendix and count how many conversions (*n*) are needed to convert from one unit to the other. Then:

a. To convert from larger to smaller units, move the decimal point *n* places to the right.
b. To convert from smaller to larger units, move the decimal point *n* places to the left.

Example 4   3.67 km = ? m

Solution   From Table A.9, note that three conversions are needed:

$$\text{kilometers} \longrightarrow \text{hectometers} \longrightarrow \text{dekameters} \longrightarrow \text{meters}$$
$$1 \qquad\qquad\qquad 2 \qquad\qquad\qquad 3$$

Thus, *n* = 3. Because larger units (kilometers) are to be changed to smaller units (meters), move the decimal point three places to the right:

$$3.67 \text{ km} = 3670. \text{ m} = 3670 \text{ m}$$

MEASUREMENT

Example 5   609 mg = ? g

Solution   From Table A.10, note that three conversions are needed:

$$\text{milligrams} \xrightarrow{1} \text{centigrams} \xrightarrow{2} \text{decigrams} \xrightarrow{3} \text{grams}$$

Thus, $n = 3$. Because smaller units (milligrams) are to be changed to larger units (grams), move the decimal point three places to the left:

$$609 \text{ mg} = 0.609 \text{ g}$$

Example 6   14.39 ℓ = ? cl

Solution   From Table A.11 note that two conversions are needed:

$$\text{liters} \xrightarrow{1} \text{deciliters} \xrightarrow{2} \text{centiliters}$$

Thus, $n = 2$. Because larger units (liters) are to be changed to smaller units (centiliters), move the decimal point two places to the right:

$$14.39 \text{ ℓ} = 1439 \text{ cl}$$

EXERCISES 5.3

In the following exercises, use the appropriate table in the Appendix, as necessary.
Write the meaning of each of the following in terms of meter, gram, or liter. See Example 1.

Complete Examples

a.   Centimeter means "one _____(1)_____ of a meter."   b.   Hectometer means "one _____(2)_____ meters."

(1) hundredth   (2) hundred

1. millimeter
2. centiliter
3. centigram
4. decimeter
5. deciliter
6. milligram
7. dekagram
8. kiloliter
9. hectoliter
10. dekameter
11. kilometer
12. hectogram

Convert each measurement as indicated. See Example 2 for Exercises 13–24.

Complete Examples
a.   75.4 dam = ? hm   b.   75.4 dam = ? m

168

5.3 THE METRIC SYSTEM

Solutions Refer to Table A.9.

a. 75.4 dam = __(1)__ hm

b. 75.4 dam = __(2)__ m

(1) 7.54   (2) 754

13. 75 cm = ?mm
14. 21.8 cm = ? dm
15. 17 km = ? hm
16. 1134 m = ? dam
17. 806 dm = ? m
18. 1412 dm = ? cm
19. 93.7 mm = ? cm
20. 205.6 hm = ? km
21. 3.7 hm = ? dam
22. 15.2 cm = ? mm
23. 2700 hm = ? km
24. 977 km = ? hm

See Example 3 for Exercises 25–36.

Complete Examples

a. 82.6 cg = ? mg

b. 82.6 cg = ? dg

Solutions Refer to Table A.10.

a. 82.6 cg = __(1)__ mg

b. 82.6 cg = __(2)__ dg

(1) 826   (2) 8.26

25. 67 cg = ?dg
26. 257 cg = ?mg
27. 60.8 g = ? dag
28. 1023 kg = ?hg
29. 2017 cg = ?mg
30. 6.97 dg = ?g
31. 0.167 dl = ?cl
32. 528 ml = ?cl
33. 1009 hl = ? kl
34. 29.3 hl = ? dal
35. 8192 kl = ? hl
36. 6.04 hl = ? kl

See Examples 4, 5 and 6 for Exercises 37–48.

Complete Examples

a. 5094 cm = ? dam

b. 3.2 ℓ = ? ml

169

MEASUREMENT

Solutions a. Refer to Table A.9.

cm ⟶ dm ⟶ m ⟶ dam
     1   2   3

5094 cm = _____ dam
             (1)

b. Refer to Table A.11.

ℓ ⟶ dl ⟶ cl ⟶ ml
   1   2   3

3.2ℓ = 3.200 ℓ = _____ ml
                  (2)

---

(1) 5.094   (2) 3200

---

37. 39.9 km = ? m
38. 7.38 kg = ? g
39. 888 cg = ? g
40. 313.5 cm = ? m
41. 5.63 ℓ = ? ml
42. 34.5 kl = ? dal
43. 6039 cl = ? kl
44. 9345 dl = ? hl
45. 152.6 mm = ? km
46. 144.8 cm = ? hm
47. 6878 dag = ? mg
48. 4854 hg = ? cg

In each of the following, decide which of the choices is the most reasonable.

Complete Example
The width of a person's hand is about
a. 10 m   b. 10 cm   c. 10 km

Solution 1 m is a little longer than 1 yd or _____ ft, so 10 m is more than 10 yd or
                                                      (1)
_____ ft.; 10 km is 10 × 1000 m, or _____ meters.
   (2)   (3)
The obvious answer is _____ .
                       (4)

---

(1) 3   (2) 30   (3) 10,000   (4) 10 cm

---

49. The width of a thumb is about
    a. 2 cm   b. 2 mm   c. 2 m

50. The height of a door in a house is about
    a. 2 cm   b. 2 mm   c. 2 m

51. A baseball bat is a little longer than
    a. 1 m   b. 10 m   c. 100 m

52. The waist measurement of a beauty-contest winner might be about
    a. .6 m   b. 6 m   c. 60 m

53. An automobile might weigh about
    a. 1300 mg   b. 1300 g   c. 1300 kg

54. The weight of a baseball is about
    a. 14 g   b. 140 g   c. 14,000 g

55. A glass of water contains about
    a. 25 ℓ   b. 2.5 ℓ   c. 0.25 ℓ

56. An automobile gasoline tank might hold about
    a. 68 kl   b. 68 ml   c. 68 ℓ

5.4 UNITED STATES–METRIC CONVERSIONS

At the present time the United States is in the process of changing from the U.S. system to the metric system. During this changeover period, it is useful to be able to convert units from one system to the other. Tables A.14 through A.17 list relationships between commonly used U.S. and metric units. Because the entries in these tables are mainly approximations, in the following examples the word "equals" in phrases such as "0.454 kilogram equals 1 pound" is to be understood to mean "is approximately equal to."

Proportions, together with an appropriate table from the Appendix, can be used to compute U.S.–metric conversions.

Example 1 Convert 6.8 lb to kilograms.

Solution Number of kilograms: K
From Table A.15, note that 1 lb equals 0.454 kg. Set up and solve a proportion.

kilograms   **pounds**
$$\frac{K}{0.454} = \frac{6.8}{1}$$

By the cross-multiplication rule,

$$K \times 1 = 0.454 \times 6.8 = 3.0872$$

To the nearest tenth, 6.8 lb equals 3.1 kg.

Example 2 Convert 500 m to yards (to the nearest tenth).

Solution Number of yards: Y
From Table A.14, note that 1 yd equals 0.914 m. Set up and solve a proportion.

yards   **meters**
$$\frac{Y}{1} = \frac{500}{0.914}$$

By direct division,

$$Y = \frac{500}{0.914} = 547.0 \text{ yd}$$

to the nearest tenth.

We sometimes need to convert measurements that involve different types of units.

Example 3 21.4 mpg = ? kilometers per liter (to the nearest tenth).

Solution First change 21.4 mi to kilometers and then 1 gal to liters.

1. Number of kilometers: K
From Table A.14, note that 1.609 km equals 1 mi. Set up, and solve, a proportion.

kilometers   **miles**
$$\frac{K}{1.609} = \frac{21.4}{1}$$

MEASUREMENT

By the cross-multiplication rule,

$$K \times 1 = 1.609 \times 21.4 = 34.4326$$

Thus, 21.4 mi equals 34.4326 km.

2. From Table A.16, note that 1 gal equals 3.785ℓ. Now, to obtain the number of kilometers per liter, apply the unit-quotient rule and compute the quotient

$$\frac{34.4326}{3.785} = 9.09$$

to the nearest hundredth. Hence, 21.4 mpg equals 9.1 km per liter (to the nearest tenth).

EXERCISES 5.4

In the following exercises, use the appropriate table in the Appendix, as necessary. For Exercises 1–20, see Example 1. Round off to the nearest tenth.

Complete Example

   12.1 pt = ? ℓ

Solution   Number of liters: L

From Table A.16, note that 1 pt = 0.473 ℓ.

$$\underset{\text{liters}}{\frac{L}{0.473}} = \underset{\text{pints}}{\frac{12.1}{1}}$$

$$L \times 1 = 0.473 \times \underline{\qquad}_{(1)} = \underline{\qquad}_{(2)}$$

To the nearest tenth, 12.1 pts equals $\underline{\qquad}_{(3)}$ liters.

---

(1) 12.1   (2) 5.7233   (3) 5.7

---

1. 8.7 lb = ? kg
2. 17.9 lb = ? kg
3. 5.4 lb = ? g
4. 2.5 lb = ? g
5. 9.1 oz = ? g
6. 5.2 oz = ? g
7. 4.7 in = ? cm
8. 7.9 in. = ? cm
9. 14.5 ft = ? m
10. 12.4 ft = ? m
11. 47.6 yd = ? m
12. 16.9 yd = ? m
13. 15.8 mi = ? km
14. 7.9 mi = ? km
15. 27.1 pt = ? ℓ
16. 93.9 pt = ? ℓ
17. 67.3 qt = ? ℓ
18. 49.1 qt = ? ℓ
19. 8.1 gal = ? ℓ
20. 21 gal = ? ℓ

5.4 UNITED STATES–METRIC CONVERSIONS

For Exercises 21–40, see Example 2. Round off to the nearest tenth.

Complete Example

$\qquad$ 5.2 kg = ? lb

Solution $\qquad\qquad\qquad\qquad\qquad\qquad$ Number of pounds: *P*

From Table 1.15, note that 1 lb equals 0.454 kg.

$\qquad\qquad\qquad\qquad\qquad$ pounds $\quad$ kilograms

$$\frac{P}{1} = \frac{5.2}{0.454}$$

$$P = \underline{\qquad}_{(1)}$$

To the nearest tenth, 5.2 kg equals $\underline{\qquad}_{(2)}$ pounds.

---

(1) 11.45 $\qquad$ (2) 11.5

---

21. 8.7 cm = ? in. $\qquad$ 22. 57.2 cm = ? in. $\qquad$ 23. 9.3 m = ? ft $\qquad$ 24. 7.3 m = ? ft

25. 98.9 m = ? yd $\qquad$ 26. 48.1 m = ? yd $\qquad$ 27. 559 km = ? mi $\qquad$ 28. 4.6 km = ? mi

29. 13.8 g = ? oz $\qquad$ 30. 8.5 g = ? oz $\qquad$ 31. 6.5 kg = ? lb $\qquad$ 32. 89 kg = ? lb

33. 11.6 ℓ = ? qt $\qquad$ 34. 21.2 ℓ = ? qt $\qquad$ 35. 8.5 ℓ = ? qt $\qquad$ 36. 17.6 ℓ = ? qt

37. 41 ℓ = ? gal $\qquad$ 38. 87.4 ℓ = ? gal $\qquad$ 39. 596 ℓ = ? gal $\qquad$ 40. 344 ℓ = ? gal

41. A set of spark plugs was used for 10,864 mi. To the nearest tenth, how many kilometers is this?

42. The weight of a rod and piston assembly is 13.07 oz. To the nearest tenth, what is this in grams?

43. Many cameras use film that is 35 mm wide. Convert this width to inches, to the nearest tenth.

44. Motion picture film is made in various widths, two of them being 8 mm and 16 mm. Convert these widths to inches, to the nearest hundredth.

Each of the following exercises gives the rating of various automobiles in miles per gallon. Convert each to kilometers per liter, to the nearest tenth of a kilometer. See Example 3.

Complete Example

$\qquad$ 10.2 mpg = ? kilometers per liter

Solution $\qquad\qquad\qquad\qquad\qquad\qquad$ Number of kilometers: *K*

From Table A.14, note that 1.609 km equals 1 mile.

173

MEASUREMENT

$$\frac{K}{1.609} = \frac{10.2}{1}$$

kilometers   miles

$$K \times 1 = 1.609 \times \underline{}_{(1)} = \underline{}_{(2)}$$

Thus, 10.2 miles equals 16.4118 km. From Table A.16, note that 1 gal equals 3.785 ℓ. Apply the unit-quotient rule.

$$\frac{16.4118}{3.785} = \underline{}_{(3)}$$

Thus, 10.2 mpg equals $\underline{}_{(4)}$ km per liter to the nearest tenth.

---

(1) 10.2   (2) 16.4118   (3) 4.33   (4) 4.3

---

45. 18.3 mpg   **46.** 15.1 mpg   **47.** 14.6 mpg   **48.** 23.5 mpg

Convert each of the following to miles per gallon, to the nearest tenth of a mile.

49. 8.7 km per liter   **50.** 9.3 km per liter   **51.** 7.9 km per liter   **52.** 10.1 km per liter

5.5 APPLICATIONS

In this section, we consider several applications of measurements and averages that occur in our daily activities.

COMPARISON SHOPPING

In today's marketplace, the same product is usually found in many different size packages. Hence, in order to make the best choice when selecting a particular size package, a shopper needs to compare costs. For this purpose, the concept of *unit cost*, as introduced in Section 2.3 is applicable. The unit cost is the cost of a single item or of a single unit of a product. If a particular product appears in different size packages, the size with the lowest unit cost is usually considered to be the best buy.

Example 1   The following table lists the price of each of four different size packages of cheese. Determine the best buy.

5.5 APPLICATIONS

Weight	14 oz	1 lb	1 lb 11 oz	3 lb 2 oz
Price	$.79	$.87	$1.42	$2.75

Solution When more than one unit is involved, it is convenient to work entirely with the smallest unit. Thus, we first change each weight that involves pounds to ounces, and we then compute unit costs in terms of "cost per ounce." Because 1 pound equals 16 ounces, we have

$$1 \text{ lb } 11 \text{ oz} = (16 + 11) \text{ oz} = 27 \text{ oz}$$

$$3 \text{ lb } 2 \text{ oz} = (3 \times 16 + 2) \text{ oz} = 50 \text{ oz}$$

Now we use the unit-quotient rule and the fact that "cost per ounce" suggests "money divided by ounces."

Weight	14 oz	16 oz	27 oz	50 oz
Price	$.79	$.87	$1.42	$2.75
Calculation	$\dfrac{\$0.79}{14}$	$\dfrac{\$0.87}{16}$	$\dfrac{\$1.42}{27}$	$\dfrac{\$2.75}{50}$
Unit cost	$0.056	$0.054	$0.053	$0.055

By inspection, we see that $0.053 is the lowest cost per ounce. Hence, the 27 oz, or 1 lb 11 oz size, is the best buy.

Example 2 The following table lists the prices (including all taxes) of four brands of tires, each with a different mileage guarantee. Determine the best buy.

Brand	A	B	C	D
Price	$30.48	$43.12	$48.41	$56.79
Guarantee	18,000 mi	24,000 mi	26,000 mi	35,000 mi

Solution By the unit-quotient and the fact that "cost per mile" suggests "money divided by miles," we have

Brand	A	B	C	D
Calculation	$\dfrac{30.48}{18{,}000}$	$\dfrac{43.12}{24{,}000}$	$\dfrac{48.41}{26{,}000}$	$\dfrac{56.79}{35{,}000}$
Cost per mile	$0.0017	$0.0018	$0.0019	$0.0016

Because $0.0016 is the lowest cost per mile, brand D is the best buy.

REGULATING BODY WEIGHT

The food that you eat provides the energy needed by your body to survive. The amount of energy provided by a particular food is measured by a unit called a *Calorie* (a measure of heat). The number of Calories that you need each day to maintain your body weight—neither losing nor gaining—depends on such factors as your age, how active you are, the climate where you live, and so on. The process

MEASUREMENT

of losing weight can often be helped by increasing physical activity, because all activity uses up Calories. Table 5.2 lists the approximate number of Calories used each hour for certain physical activities.

Table 5.2
Activity Chart*

Activity	Calories Per Hour
Walking (3 mi per hr)	270
Bicycling (10 mi per hr)	420
Swimming (30 yd per min)	500
Jogging	500
Skiing	600

Example 3 How many Calories are used in bicycling (at 10 mi per hr) for 45 min?

Solution Number of Calories: C

From Table 5.2, bicycling for 1 hour (60 minutes) uses 420 Calories. Thus, we set up the proportion.

$$\begin{array}{cc} \text{calories} & \text{minutes} \\ \dfrac{C}{420} = & \dfrac{45}{60} \end{array}$$

By the cross-multiplication rule,

$$C \times 60 = 420 \times 45$$

Dividing each side by 60, we have

$$\frac{C \times 60}{60} = \frac{420 \times 45}{60}$$

$$C = 315$$

Hence, 315 Calories are used.

Example 4 If a student bicycles 45 minutes each day for 180 days, how many Calories does she use?

Solution From Example 3 we have that 45 minutes of bicycling in one day uses 315 Calories. For 180 days, we apply the unit-product rule:

$$315 \times 180 = 56{,}700$$

Hence, the student uses 56,700 Calories in 180 days.

AVERAGES

The notion of the *average* of a set of numbers is defined as follows:

> **The average (A) of N numbers is equal to the sum of the numbers divided by N:**
>
> $$A = \frac{\text{sum of } N \text{ numbers}}{N}$$

*These figures are from an article by Dr. Jean Mayer (Department of Nutrition, Harvard University)

5.5 APPLICATIONS

Example 5 A student receives scores of 70%, 82%, 73%, 81%, and 83%, on five tests. Find the average score to the nearest tenth of a percent.

Solution Because there are five tests, $N = 5$, and

$$A = \frac{70 + 82 + 73 + 81 + 83}{5} = 77.8$$

To the nearest tenth, the average score is 77.8%.

Example 6 If a score of 80% (or higher) is needed for a B grade, what is the minimum grade the student of Example 5 must score on the sixth test in order to earn a B grade?

Solution Score on the sixth test: G
We have that the average (A) must be 80% and $N = 6$. Hence,

$$80 = \frac{70 + 82 + 73 + 81 + 83 + G}{6}$$

$$80 = \frac{389 + G}{6}$$

Multiplying each side by 6, we obtain

$$80 \times 6 = \frac{(389 + G)}{\cancel{6}} \times \cancel{6}$$

$$480 = 389 + G$$

Subtracting 389 from each side, we have

$$480 - 389 = 389 + G - 389$$

$$91 = G$$

The student must receive 91% on the sixth test to earn a B grade.

Sometimes a set of measurements for which we want to compute an average includes the same measurement more than once. For example, in many colleges the "grade point average" (GPA) of a student is computed on the basis of the total number of completed units of study. First, 4.0 points is assigned to each unit with an A grade, 3.0 points to each unit with a B grade, 2.0 points to each unit with a C grade, and 1.0 point to each unit with a D grade. Next the total number of grade points earned is divided by the total number of completed units to determine the GPA. For problems of this type it is convenient to use the method illustrated in the next example.

Example 7 The following table lists the number of completed units together with the grades earned for a certain student.
Compute the grade point average of the student to the nearest tenth. Use the 4.0–3.0–2.0–1.0 system described in the preceding paragraph.

Grade	A	B	C	D
Units	12	7	10	3

Solution First, we find the total number of units completed:

$$12 + 7 + 10 + 3 = 32$$

Next, we use the fact that there are 12 units at 4.0 points each, 7 units at 3.0 points

MEASUREMENT

each, 10 units at 2.0 points each, 3 units at 1.0 point each and determine the GPA as follows.

$$\text{GPA} = \frac{(12 \times 4.0) + (7 \times 3.0) + (10 \times 2.0) + (3 \times 1.0)}{32}$$

$$= \frac{48.0 + 21.0 + 20.0 + 3.0}{32} = 2.875$$

The grade point average is 2.9, to the nearest tenth.

AVERAGE DAILY BALANCE

Many department stores keep records of how much each charge account customer owes per day. The amount owed each day is called the *daily balance*. The *average daily balance* is the average amount of money owed each day over a period of 30 days.

Example 8 Find the average daily balance for the following account:

Number of Days	Daily Balance
8	$ 42.27
6	$ 85.91
10	$126.55
6	$294.17

Solution From the table we see that the customer had a daily balance of $42.27 for 8 days, followed by a daily balance of $85.91 for the next 6 days, etc. Note also that 8 + 6 + 10 + 6 equals 30 days. We compute the average daily balance as follows:

$$\frac{(8 \times 42.27) + (6 \times 85.91) + (10 \times 126.55) + (6 \times 294.17)}{30}$$

$$= \frac{338.16 + 515.46 + 1265.50 + 1765.02}{30} = 129.47133$$

The average daily balance is $129.47, to the nearest cent.

Customers who use charge accounts are billed monthly for the amount owed. If the amount owed is not paid before a date specified on the monthly bill, a finance charge based upon the average daily balance may be charged.

Example 9 If a finance charge at the rate of $1\frac{1}{2}$% of the average daily balance is charged each month for late payment, find the finance charge on the average daily balance computed in Example 8.

Solution From Example 8, the average daily balance is $129.47. Hence, the finance charge equals

$$1\tfrac{1}{2}\% \text{ of } \$129.47 = 0.015 \times 129.47$$

$$= 1.94205$$

The finance charge is $1.94, to the nearest cent.

5.5 APPLICATIONS

EXERCISES 5.5

Each of Exercises 1–6 gives the price and package size of an item. Find the unit cost of each item to the nearest cent. See Example 1.

Complete Examples

a. $1.75 for 6 oz
b. $3.25 for 1 gal 2 qt

Solutions

a. Divide cost by ounces.

$$\frac{1.75}{6} = \$ \underline{}_{(1)} \text{ (to the nearest cent)}$$

b. Change 1 gal 2 qt to quarts; then divide cost by quarts.

$$1 \text{ gal } 2 \text{ qt} = (1 \times 4 + 2) \text{ qt} = \underline{}_{(2)} \text{ qt}$$

$$\frac{3.25}{6} = \$ \underline{}_{(3)} \text{ (to the nearest cent)}$$

(1) $0.29   (2) 6   (3) $0.54

1. 78¢ for 14 oz
2. $5.15 for 3 lb
3. $16.29 for 4 qt
4. $25.82 for 5 gal
5. $1.18 for 2 lb 8 oz
6. 89¢ for 1 lb 4 oz

Determine the best buy of each of the following items. See Example 1.

7. Liquid detergent:

Size	28 fl oz	63 fl oz	3 qt 11 fl oz
Price	95¢	$2.22	$3.64
Unit Cost			

8. Powdered milk:

Size	16 oz	2 lb	44.8 oz
Price	$1.09	$2.17	$2.81
Unit Cost			

9. Paint:

Size	1 qt	2 qt	1 gal
Price	$4.39	$7.89	$11.69
Unit Cost			

10. Paint:

Size	1 qt	2 qt	1 gal
Price	$4.79	$8.99	$15.99
Unit Cost			

MEASUREMENT

In Exercises 11-12, determine the best tire buy. See Example 2.

Complete Examples

Brand	A	B	C
Price	$30.00	$50.00	$60.00
Guarantee	18,000 mi	25,000 mi	30,000 mi
Calculation	$\frac{30}{18,000}$	$\frac{50}{25,000}$	$\frac{60}{30,000}$
Cost per mile	$ _____ (1)	$ _____ (2)	$ _____ (3)

Brand _____ is the best buy.
    (4)

---

(1) $0.0017   (2) $0.002   (3) $0.002   (4) A

11.

Brand	A	B	C
Price	$14.72	$17.95	$27.65
Guarantee	12,000 mi	18,000 mi	24,000 mi
Cost per mile	_____	_____	_____

12.

Brand	A	B	C
Price	$51.51	$53.55	$75.55
Guarantee	34,000 mi	36,000 mi	40,000 mi
Cost per mile	_____	_____	_____

13. A Model X refrigerator sells for $430. If the cost of the electricity needed to operate the refrigerator for 14 years is $1077, find: a. The total cost of buying and operating the refrigerator for 14 years. b. The cost per year of owning the refrigerator for 14 years.

14. A Model Y refrigerator sells for $465. If the cost of the electricity needed to operate the refrigerator for 14 years is $921, find: a. The total cost of buying and operating the refrigerator for 14 years. b. The cost per year of owning the refrigerator for 14 years.

The tables of Exercises 15 and 16 list the prices of different brands of refrigerators with similar features, together with the total operating costs over a period of 14 years. In each problem, complete the table and determine the better buy.

15.

Brand	Price	14-Yr Operating Cost	Total Cost	Cost Per Year
A	$435	$938	_____	_____
B	$390	$1088	_____	_____

5.5 APPLICATIONS

16.

Brand	Price	14-Yr Operating Cost	Total Cost	Cost Per Year
C	$525	$994	_____	_____
D	$540	$1020	_____	_____

The persons in the following exercises are assumed to be eating just enough to maintain their weight. If, in addition to their normal activities, they engage in the indicated type of exercise, find: a. The number of Calories used each day. b. The expected weight loss at the end of the stated time if 3500 Calories equals one pound. Round off answers to the nearest tenth. Refer to Table 5.2. See Examples 3 and 4.

Complete Example
Skiing 4 hours a day for 3 days.

Solution

a.
From Table 5.2, set up the proportion.

Number of Calories: C

Calories   hours

$$\frac{C}{600} = \frac{4}{1}$$

$$C \times 1 = 4 \times \underline{\quad}_{(1)}$$

$$C = \underline{\quad}_{(2)} \text{ Calories in one day.}$$

b. Apply the unit-product rule.

$$2400 \times \underline{\quad}_{(3)} = 7200 \text{ Calories in 3 days.}$$

Expected weight loss: P

Set up the proportion.

pounds   Calories

$$\frac{P}{1} = \frac{7200}{3500}$$

$$P = \underline{\quad}_{(4)} \text{ (to the nearest tenth of a pound)}$$

The person can expect to lose 2.1 pounds in 3 days.

(1) 600   (2) 2400   (3) 3   (4) 2.1

17. Walking 1 hour a day for 30 days.

18. Swimming $\frac{1}{2}$ hour a day for 60 days.

MEASUREMENT

19. Jogging 45 minutes a day for 45 days.

20. Skiing 2 hours a day for 20 days.

21. Bicycling 40 minutes a day for 120 days

22. Walking 90 minutes a day for 75 days.

In each of the following exercises, determine which of the two choices would result in the greater loss of weight.

23. Walk 1 hour a day for 30 days, or jog $\frac{1}{2}$ hour a day for 35 days?

24. Bicycle 1 hour a day for 30 days, or swim $\frac{1}{2}$ hour a day for 45 days?

25. Ski 2 hours a day for 40 days, or bicycle 90 minutes a day for 30 days?

26. Walk 2 hours a day for 60 days, or ski 90 minutes a day for 25 days?

27. Bicycle 45 minutes a day for 20 days, or eat 5000 fewer Calories?

28. Jog 75 minutes a day for 40 days, or eat 20,000 fewer Calories?

29. Swim 2 hours a day for 15 days, or eat 18,000 fewer Calories?

30. Walk 3 hours a day for 12 days, or eat 10,000 fewer Calories?

The following tables list the points scored by certain players in each of five basketball games. Find the average number of points (to the nearest tenth of a point) scored by each player. See Example 5.

Complete Example

Player	1	2	3	4	5	Average
Jack	22	30	16	18	15	_____ (1)
Connie	28	30	32	30	26	_____ (2)

Solution N is 5 in both cases.

$$\text{Jack's average is: } \frac{22 + 30 + 16 + 18 + 15}{5} = \underline{}_{(1)}$$

$$\text{Connie's average is: } \frac{28 + 30 + 32 + 30 + 26}{5} = \underline{}_{(2)}$$

(1) 20.2   (2) 29.2

31.

Player	1	2	3	4	5	Average
Pat	26	23	16	15	20	_____
Fran	27	33	26	19	15	_____
Chris	20	29	19	24	33	_____

5.5 APPLICATIONS

32.
Player	1	2	3	4	5	Average
Jon	11	11	14	17	19	_____
Jan	29	17	21	19	32	_____
Terry	24	35	31	21	25	_____

33.
Student	1	2	3	4	5	6	Average
E. Adams	73	68	84	79	81	81	_____
M. Lopez	75	77	71	83	80	82	_____
L. Frye	62	61	59	68	71	69	_____

34.
Student	1	2	3	4	5	6	Average
K. Shindo	93	95	87	88	81	89	_____
R. Cary	65	58	66	70	64	68	_____
C. Ricci	75	85	62	91	58	77	_____

For Exercises 35–40: a 90% average is needed for a grade of A, 80% for a grade of B, and 70% for a grade of C.

What minimum score must each of the following named students (of Exercises 33 and 34) receive on the seventh test in order to earn the indicated grade? See Example 6.

Complete Example

E. Garcia; C grade.

Student	1	2	3	4	5	6
E. Garcia	42	86	92	76	80	76

Solution   Score on seventh test: G

$$70 = \frac{42 + 86 + 92 + 76 + 80 + 76 + G}{7}$$

$$70 = \frac{452 + G}{7}$$

$$7 \times 70 = \frac{452 + G}{7} \times 7$$

$$\underline{}_{(1)} = 452 + G$$

$$490 - 452 = 452 + G - \underline{}_{(2)}$$

$$\underline{}_{(3)} = G$$

E. Garcia must receive 38% on the seventh test to earn a C grade.

(1) 490   (2) 452   (3) 38

183

MEASUREMENT

35. E. Adams; B grade
36. K. Shindo; A grade
37. M. Lopez; B grade
38. R. Cary; C grade
39. L. Frye; C grade
40. C. Ricci; C grade

The following tables list the number of completed units together with the grades earned for each of two students. Use the 4.0–3.0–2.0–1.0 system to compute the grade point average for each student to the nearest tenth. See Example 7.

Complete Example

Grade	A	B	C	D
Units	30	28	12	5

Solution

Total number of units: $30 + 28 + 12 + 5 =$ _____ (1)

Total number of grade points: $(30 \times 4) + (28 \times 3) + (12 \times 2) + (5 \times 1) =$ _____ (2)

$$\text{GPA} = \frac{233}{75} = \underline{\qquad} \text{ (to the nearest tenth)} \quad (3)$$

The grade point average is 3.1 to the nearest tenth.

(1) 75   (2) 233   (3) 3.1

41.
Grade	A	B	C	D
Units	21	19	15	5

42.
Grade	A	B	C	D
Units	35	30	45	10

Find the average daily balance for each of the following accounts. See Example 8.

Complete Example

No. of Days	Daily Balance
10	$39.00
15	$75.00
5	$126.00

184

5.5 APPLICATIONS

Solution

$$\text{Average daily balance} = \frac{(10 \times 39) + (15 \times 75) + (5 \times 126)}{\underline{\qquad(1)\qquad}}$$

$$= \frac{390 + 1125 + 630}{30} = \underline{\qquad(2)\qquad}$$

The average daily balance is $71.50.

(1) 30   (2) 71.5

43.
No. of Days	Daily Balance
11	$ 19.24
5	84.19
3	275.49
11	304.98

44.
No. of Days	Daily Balance
14	$ 0.00
5	215.92
9	429.17
2	512.43

45.
No. of Days	Daily Balance
8	$ 42.50
6	173.28
6	209.44
11	473.82

46.
No. of Days	Daily Balance
9	$ 25.50
5	273.74
10	562.18
7	611.95

For late payments, a department store computes finance charges at the monthly rate of $1\frac{1}{2}$% of the average daily balance. Compute the finance charges on each of the average daily balances of Exercises 47 and 48. See Example 9.

Complete Example
a. $71.50

b. $1,492.25

Solutions

a. $1\frac{1}{2}$% of $71.50

$= 0.015 \times 71.50$

$= \underline{\$\qquad}$ (to the nearest cent)
  (1)

b. $1\frac{1}{2}$% of 1,492.25

$= 0.015 \times 1492.25$

$= \underline{\$\qquad}$ (to the nearest cent)
  (2)

(1) $1.07   (2) $22.38

MEASUREMENT

47. **a.** $208.22   **b.** $346.91   **c.** $412.74   **d.** $692.08

48. **a.** $835.49   **b.** $1,024.65   **c.** $1,815.75   **d.** $2,309.76

In Exercises 49 and 50: a. Find the average daily balance for each of the following accounts.
b. Compute the $1\frac{1}{2}$% monthly finance charge to be charged if the monthly payment is not made on time.

49.	No. of Days	Daily Balance		50.	No. of Days	Daily Balance
	9	$ 8.44			9	$ 36.10
	7	53.18			8	421.00
	6	179.23			13	1002.17
	8	256.88				

CHAPTER FIVE REVIEW EXERCISES

Convert each measurement.

1. 156 in. = ? ft
2. 216 qt = ? gal
3. 3 mi = ? in.
4. 5 da = ? sec
5. 525 oz = ? lb (round off answer to the nearest tenth)
6. The height of a horse is sometimes measured in *hands*, one hand being equal to 4 in. If the height of a horse is 5 ft 8 in., find its height in hands.
7. Find the sum: 5 ft 6 in. + 8 ft 7 in. + 4 ft 9 in.
8. Find the difference: 28 lb 6 oz − 10 lb 12 oz.
9. Find the product: 7 × (4 gal 3 qt).   10. Find the quotient: (16 yd 2 ft) ÷ 5.
11. Two pieces are cut from a piece of lumber 16 ft long. The first piece is 4 ft 8 in. long and the second is 3 ft 10 in. long. What is the length of the remaining piece?

Convert each measurement.

12. 87 cm = ? mm
13. 307 hm = ? km
14. 88.2 g = ? dag
15. 2031 kg = ? hg
16. 94.3 hm = ? m
17. 8.37 kg = ? g

Convert each measurement. Express answers to the nearest tenth.

18. 7.4 lb = ? kg
19. 16.3 mi = ? km
20. 12.4 cm = ? in.
21. 864.3 g = ? oz
22. A certain dog sled race in Alaska covers a distance of 1049 miles. Convert this distance to kilometers to the nearest kilometer.

CHAPTER FIVE REVIEW EXERCISES

The persons in Exercises 23 and 24 are assumed to be eating just enough to maintain their weight under normal conditions. Use Table 5.2 and round off answers to the nearest day.

23. If a person walks 60 minutes a day, how long will it take to lose 10 pounds?

24. If a person bicycles 45 minutes a day, how long will it take to lose 8 pounds?

Find the unit cost of each item, given the price and package size.

25. 90¢ for 18 oz

26. $42.35 for 5 gal

Determine the best buy of the listed items.

27. A tire guaranteed for 20,000 miles at a price of $25.98, and a tire guaranteed for 30,000 miles selling for $36.84.

28. A battery guaranteed for 36 months at a price of $23.55, and a battery guaranteed for 5 years selling for $38.00.

29. In five basketball games a player scored 20, 18, 15, 17, and 24 points. Find his average score per game.

30. On six tests a student received scores of 87, 89, 88, 90, 84, and 92. If an average of 90 is necessary to receive a grade of A, how many points must be earned on the seventh test for the student to receive an A?

MEASUREMENT

SUMMARY PART I

References to appropriate sections are shown in parentheses.

PROPERTIES OF OPERATIONS

1. The order in which numbers are added (or multiplied) does not change the sum (or product). (1.3, 2.1)
2. *Roles of 0 and 1*
   (a) The sum of any number and 0 is the number. (1.3)
   (b) The product of any number and 1 is the number. (2.1)
   (c) The product of any number and 0 is 0. (2.1)
   (d) Division by 0 is not possible. (2.3)
3. If a number is added to and subtracted from a given number, the result is the given number. (1.4)
4. If a given number is multiplied and divided by a given nonzero number, the result is the given number. (2.4)

PROPERTIES USED TO SOLVE EQUATIONS

1. If a number is added to, or subtracted from, each side of an equation, the resulting equation is equivalent to the original equation. (1.5)
2. If each side of an equation is multiplied by, or divided by, the same nonzero number, the resulting equation is equivalent to the original equation. (2.6)
3. In a proportion, the product of the extremes is equal to the product of the means. (2.7)

$$\text{If } \frac{a}{b} = \frac{c}{d} \quad \text{then} \quad a \times d = b \times c$$

SOLVING WORD PROBLEMS

1. *Suggested procedure*
   (a) Read the problem and note what is asked for. Write a short word phrase to describe each quantity.
   (b) Choose a letter (the unknown) to represent a number that you are trying to find. Any letter can be used.
   (c) Write an equation that involves the unknown and is a "model" for the word sentence that involves the condition on the unknown.
   (d) Solve the equation.
   (e) Interpret the solution in terms of what is asked for in the problem. (1.6)
2. *Some Basic Relationships*
   (a) If a given number is associated with one unit, *the number associated with more than one such unit* is equal to the product

   $$(\text{number of units}) \times (\text{number associated with one unit}) \quad (2.5)$$

   (b) If a given number is associated with more than one unit, *the number associated with one such unit* is equal to the quotient

   $$\frac{\text{number associated with more than one unit}}{\text{number of units}} \quad (2.5)$$

PROPERTIES OF FRACTIONS

1. *Fundamental principle of fractions*
   If both the numerator and the denominator of a given fraction are multiplied or divided by the same nonzero number, the resulting fraction is equal to the given fraction. (3.2)

   $$\frac{n}{d} = \frac{n \times k}{d \times k} \quad \text{or} \quad \frac{n}{d} = \frac{n \div k}{d \div k}$$

2. *To reduce a fraction to lowest terms*
   (a) Factor the numerator and the denominator
   (b) Then, "divide out" common factors. (3.2)

   $$\frac{a \times c}{b \times c} = \frac{a}{b}$$

SUMMARY PART 1

3. *To build a fraction to higher terms*
   (a) Obtain the building factor by inspection or by dividing the new denominator by the given denominator.
   (b) Then, multiply the numerator and denominator of the given fraction by the building factor. (3.2)

   $$\frac{a}{b} = \frac{a \times c}{b \times c}$$

4. *To find the LCD of a set of unlike fractions*
   (a) Express each denominator in completely factored form, aligning common factors where possible.
   (b) Then, write a product whose factors are each of the different prime factors that occur in any of the denominators, and include each of these factors the greatest number of times that it appears in any one of the given denominators. (3.3)

5. *To find the sum (or difference) of two or more like fractions*
   Add or subtract the numerators. Use the common denominator of the original fractions as the denominator of the sum (or difference). (3.4, 3.5)

   $$\frac{a}{d} + \frac{b}{d} = \frac{a+b}{d}$$

6. *To find the sum (or difference) of two or more unlike fractions*
   (a) Find the lowest common denominator LCD of the given fractions.
   (b) Then, change the given fractions into like fractions with the LCD as the common denominator and find the sum (or difference) of the resulting fractions. (3.4, 3.5)

7. *To find the product of two or more fractions*
   Write a fraction whose numerator is the product of the numerators and whose denominator is the product of the denominators of the given fractions. (3.6)

   $$\frac{a}{b} \times \frac{c}{d} = \frac{a \times c}{b \times d}$$

8. *To find the quotient of two fractions*
   Invert the divisor and multiply the resulting fractions. (3.6)

   $$\frac{a}{b} \div \frac{c}{d} = \frac{a}{b} \times \frac{d}{c}$$

CONVERSIONS OF METRIC UNITS

1. *To change a measurement in one metric unit to the next smaller unit,* move the decimal point one place to the right.

2. *To change a measurement in one metric unit to the next larger unit,* move the decimal point one place to the left.

MENTAL ARITHMETIC

1. To multiply by: Move the decimal point:
   0.1   One place to the left
   0.01   Two places to the left
   0.001, etc.   Three places to the left, etc.
                       Add or drop zeros as needed   (2.2)

2. To multiply by: Move the decimal point:
   10   One place to the right
   100   Two places to the right
   1000, etc.   Three places to the right, etc.
                       Add or drop zeros as needed   (2.2)

3. To divide by: Move the decimal point:
   10   One place to the left
   100   Two places to the left
   1000, etc.   Three places to the left, etc.
                       Add or drop zeros as needed   (2.4)

4. *Common fraction–decimal conversions* (3.7)

   $\frac{1}{4} = 0.25$ $\frac{1}{2} = 0.5$ $\frac{3}{4} = 0.75$

   $\frac{1}{3} = 0.33\frac{1}{3}$ $\frac{2}{3} = 0.66\frac{2}{3}$

5. *To change a decimal to a percent*
   Move the decimal point two places to the right. Add or drop zeros as needed. Add the percent symbol to follow the last digit. (4.1)

6. *To change a percent to a decimal*
   Move the decimal point two places to the left. Add or drop zeros as needed. Drop the percent symbol. (4.1)

MEASUREMENT

CUMULATIVE REVIEW FOR PART I

1. Find 6% of $240.
2. Compute 17.1 × 4.3.
3. Convert 204 inches to feet.
4. Add: $2\frac{3}{8} + 1\frac{1}{4}$.
5. Solve: 9% × N = 657.
6. Divide 46.23 by 7.4 and round off the result to the nearest tenth.
7. Find the difference: 7 gal 1 qt − 2 gal 3 qt.
8. Compute: 1.703 − 0.942 + 2.74.
9. Reduce $\frac{12}{40}$ to lowest terms.
10. Multiply: $8 \times 3\frac{3}{4}$.
11. Write 234.07 in words.
12. Convert 20 pounds to kilograms. Express the result to the nearest tenth.
13. What percent of 64 is 24?
14. If 3 pounds of grapes cost $1.32, find the cost of 7 pounds.
15. Divide: $4\frac{2}{3} \div \frac{4}{3}$.
16. What number added to 612.3 equals 792?
17. Write $\frac{5}{4}$ as an equal fraction with a denominator of 24.
18. Convert 28 kilometers to miles (nearest tenth).
19. A company finds that there are 4 rejects out of each 1000 spark plugs manufactured. What is the percent of rejects?
20. A student's test scores were 74, 83, 91, 64, and 72. Find her average.
21. Find the percent increase when 40 increases to 60.
22. Find the percent decrease when 60 decreases to 40.
23. One inch on a map represents 50 miles. How many miles does $3\frac{3}{4}$ inches represent?
24. The sales tax on an item costing $6400 is $400. Find the sales tax rate.
25. One 18-oz box of laundry soap costs $1.50. A larger 44-oz box costs $3.40. Which box is the best buy?
26. Write the number "twenty-three thousand, forty-five" using numerals.
27. Find the sum $\frac{3}{5} + \frac{2}{3}$.
28. Find the quotient 2.25 ÷ 1.3, rounded off to the nearest tenth.
29. What is 12% of 210?
30. What percent of 140 is 35?
31. Factor 120 completely.
32. Solve: $\frac{N}{1.4} = 12.3$
33. Find the difference 121.34 − 68.72.
34. Given the formula $A = l \times w$, find w if $l = 1.7$ and $A = 5.44$.
35. If a student correctly answers 32 out of 40 questions on a test, what percent of the questions does she answer correctly?
36. Find the simple interest on $120 at 9% interest for 2 years.
37. Find the least common denominator of $\frac{5}{12}$ and $\frac{1}{15}$.
38. Solve: $\frac{N}{4.2} = \frac{3.7}{2}$
39. Round off $5.936 to the nearest cent.

CUMULATIVE REVIEW FOR PART I

40. Find the quotient $6\frac{1}{4} \div 2\frac{1}{2}$.

41. A typist can type 68 words a minute. At the same rate, how many words can he type in 3 minutes?

42. A car travels 195.3 miles on 12 gallons of gasoline. What is the distance travelled per gallon?

43. Write the number 215.35 in words.

44. Reduce $\frac{24}{40}$

45. Solve: $11.51 + N = 78.32$

46. A woman's weight decreased from 150 pounds to 120 pounds. What was the percent of decrease?

47. What is the total cost of a guitar that sold for $220 if 6% sales tax is added?

48. Compute: $1.3 \times 4.5 \times 7$.

49. Write $\frac{2}{5}$ as a percent.

50. Find 15% of $160.

DO EACH OF THE FOLLOWING PROBLEMS MENTALLY.

51. $\frac{12 \times 132}{12}$

52. 240×0.01

53. $62.1 + 23.4 - 62.1$

54. $\frac{32}{10}$

55. Write $\frac{3}{4}$ as a percent.

56. Write 25% as a fraction.

57. Add: $6\frac{1}{5} + 3\frac{2}{5}$.

58. Write 0.45 as a percent.

59. Convert 2500 kilograms to grams.

60. Find the product: 1.6 qt $\times 100$.

61. Write $\frac{1}{3}$ as a percent.

62. Convert 9 yards to feet.

63. Round off 26.95 to the nearest tenth.

64. Compute: 68.3×0.01

65. Compute: $8 \div \frac{1}{2}$.

66. $\frac{342}{100}$

67. Write 0.7 as a fraction

Solve.

68. $N + 50 = 50$

69. $50 \times N = 50$

70. $N - 50 = 50$

71. $50 \div N = 50$

72. $\frac{N}{6} = 4$

73. $3 + N = 15$

74. $3 \times N = 21$

75. $12 - N = 7$

Use rounded-off numbers to estimate each of the following.

76. $112 \div 22$

77. 41×298

78. $798 + 416 + 286$

79. $6493 - 2487$

80. 204×0.98

PART II
ELEMENTARY ALGEBRA

In this part of the book, we continue to focus our attention on numbers. We will use the same procedures and symbols we used in arithmetic, together with certain new symbols. The vocabulary used in arithmetic will apply in algebra. We will be studying arithmetic, but from a more general point of view.

You may want to refer to the Summary at the end of Part I before starting a particular chapter where a review of the prerequisite material would be helpful.

6

INTEGERS

6.1 INTEGERS AND THEIR GRAPHS

The numbers we use to count things are called **natural numbers.** The numbers

$$1, 2, 3, 4, 5, 6, 7, \ldots$$

↑—Read "and so on."

are natural numbers, whereas $\frac{2}{3}$, 3.141, and $\sqrt{2}$ are not.

When the number 0 is included with the natural numbers, the numbers in the enlarged collection

$$0, 1, 2, 3, 4, 5, \ldots$$

(as we noted in Section 1.1) are called **whole numbers.** Thus, we can refer to numbers such as 2, 3, and 6 as natural numbers or whole numbers.

INTEGERS

On occasion, we use natural numbers to represent physical quantities such as money (5 dollars), temperature (20 degrees), and distance (10 miles). Since this representation does not differentiate between gains and losses, degrees above or below zero, or distances in opposite directions from a starting point, mathematicians have found it convenient to represent these ideas symbolically by the use of plus (+) and minus (−) signs. For example, we may represent

A loss of five dollars as −$5.
A gain of five dollars as +$5.
Ten degrees below zero as −10°.
Ten degrees above zero as +10°.
Ten miles to the west of a starting point as −10 miles.
Ten miles to the east of a starting point as +10 miles.

195

INTEGERS

Nonzero numbers that are preceded by a minus sign are called **negative numbers.** Nonzero numbers that are preceded by a plus sign are called **positive numbers.** Together these numbers are called **signed numbers.** Note that zero is not considered positive or negative.

The signed numbers, together with zero,

$$\ldots, -3, -2, -1, 0, +1, +2, +3, \ldots$$

are called **integers.** When a numeral is written without a sign, for example, 3, 5, 9, we assume that a plus sign is intended.

We refer to the numbers

$$\ldots, -6, -4, -2, 0, 2, 4, 6, \ldots$$

as **even integers.** And we refer to the numbers

$$\ldots, -5, -3, -1, 1, 3, 5, \ldots$$

as **odd integers.**

Note that even integers are multiples of 2. An integer is odd if it leaves a remainder of 1 when it is divided by 2.

NUMBER LINE

The integers are ordered. That is, we can always say that a particular integer is *greater than, equal to,* or *less than* another. We can use a **number line** to represent the relative order of a set of integers.

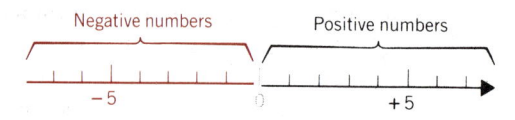

Figure 6.1

To construct a number line (see Figure 6.1):

1. Draw a straight line.
2. Mark a point on the line to represent 0. This point is called the *origin*.
3. Decide on a convenient unit of scale and mark off units of this length on the line on both sides of the origin.
4. On the bottom side of the line, label enough of these units to establish the scale, usually two or three points.
5. Add a small arrow pointing to the right to indicate that numbers are larger to the right.
6. On the top side of the line, label the numbers to be graphed. Graph the numbers by placing dots at the appropriate places on the line.

Example 1   Graph the numbers on a number line.

a. $-5, -3, -2, -1, 0, 3$

b. The odd integers between -5 and 7.

Solutions

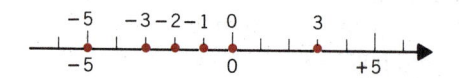

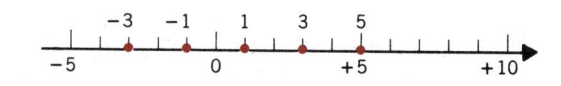

6.1 INTEGERS AND THEIR GRAPHS

ORDER OF INTEGERS

Given any two numbers, the number whose graph on a number line is to the left is *less than* the number whose graph is to the right. In Figure 6.2, the graph of 3 is *to the left* of the graph of 7. Therefore, 3 *is less than* 7. We could also state this relationship as "7 *is greater than* 3."

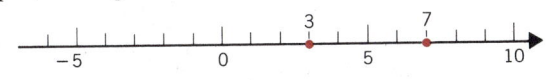

Figure 6.2

We use special symbols to indicate the order relationship between two numbers:

$<$ means "is less than";

$>$ means "is greater than."

For example,

$2 < 5$ is read "2 is less than 5,"

and

$5 > 2$ is read "5 is greater than 2."

Note that the point of the symbols $<$ and $>$ always points to the smaller number.

$2 < 5$ $\quad$ $5 > 2$

Points to smaller number. $\quad$ Points to smaller number.

Negative integers are ordered in the same way. The number line is particularly useful in visualizing the relative order of any two integers.

Example 2   For the graph shown, replace the comma in each pair with the proper symbol: $<$, $>$, or $=$.

a. p, n b. m, p c. r, r

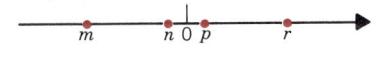

Solutions   a. $p > n$ (the graph of p is to the right of the graph of n)
b. $m < p$ (the graph of m is to the left of the graph of p)
c. $r = r$

The notion of the order of integers is consistent with our physical experiences. A temperature of $-5°$ is lower or less than one of $-2°$, $-3°$ is less than $3°$, and $-1°$ is less than $0°$.

Example 3   Replace the comma in each pair with the proper symbol: $<$ or $>$.

a. $-3, 3$ b. $-1, 0$ c. $-2, -5$ d. $0, -2$

Solutions   Referring to the number line,

a. $-3 < 3$, read "negative three *is less than* three";
b. $-1 < 0$, read "negative 1 *is less than* zero";
c. $-2 > -5$, read "negative two *is greater than* negative five";
d. $0 > -2$, read "zero *is greater than* negative two."

INTEGERS

THE SYMBOL "−"

We have now used the symbol "−" to indicate a sign of operation (subtraction) and to indicate negative numbers. We also use the symbol "−" to mean the "opposite" or "negative" of a number.

> **−x means the "opposite" of x**

If x is positive, then $-x$ is negative. For example,
$$-(5) = -5 \quad \text{and} \quad -(10) = -10$$

If x is negative, then $-x$ is positive. For example,
$$-(-5) = +5 \quad \text{and} \quad -(-10) = +10$$

For any a, positive or negative, the opposite of the opposite of a is a. In symbols we write
$$-(-a) = a$$

ABSOLUTE VALUE

The **absolute value** of a signed number is the value of the number without regard to its sign. We can think of a signed number as having two parts: the sign of the number and the absolute value of the number. When no sign is written, we assume that the sign is positive. For example,

−4 — sign ↑ ↑ absolute value: 4

+7 — sign ↑ ↑ absolute value: 7

5 — sign understood to be + ↑ ↑ absolute value: 5

In the case of 0, we define the absolute value of zero to be zero. Thus, the absolute value of any integer is never negative.

We designate absolute value by vertical bars.

Example 4 Simplify.

a. $|-5|$ b. $|7|$ c. $|0|$

Solution
a. $|-5| = 5$, read "the absolute value of negative five is equal to five";
b. $|7| = 7$, read "the absolute value of seven is equal to 7";
c. $|0| = 0$, read "the absolute value of zero is equal to zero."

We define the absolute value of x more formally as
$$|x| = \begin{cases} x & \text{if } x \text{ is greater than or equal to 0.} \\ -x & \text{if } x \text{ is less than 0.} \end{cases}$$

Example 5
a. $|3| = 3$, $|5| = 5$, and $|0| = 0$. In this case, 3, 5, and 0 are greater than or equal to 0. Thus, the absolute value is the number itself.

b. $|-3| = -(-3) = 3$ and $|-5| = -(-5) = 5$. In this case, -3 and -5 are less than 0. So we take the negative of -3 and the negative of -5 to find the absolute value.

6.1 INTEGERS AND THEIR GRAPHS

EXERCISES 6.1
Graph the numbers on a number line. See Example 1.

Complete Examples

a. $-8, -6, -2, 1, 5,$ and 9.

b. The integers between -2 and 4.

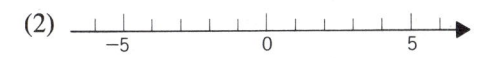

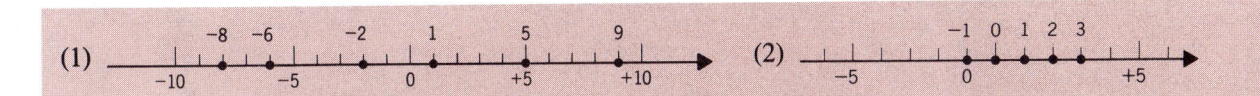

(1) [number line showing $-8, -6, -2, 1, 5, 9$ plotted from -10 to $+10$]

(2) [number line showing $-1, 0, 1, 2, 3$ plotted from -5 to $+5$]

1. $0, -3, 5, -5, 2,$ and -1.
2. $-4, 3, -2, -1, 6,$ and 9.
3. $-2, 4, 7, -5, -3,$ and 3.
4. $-4, -7, 0, -2, 4,$ and 8.
5. The natural numbers less than 10.
6. The integers between -3 and 7.
7. The even integers between -3 and 11.
8. The odd integers between -5 and 5.
9. The odd integers between -5 and -1.
10. The odd integers between -10 and 0.

Let a, b, c, d represent whole numbers. Their graphs are shown on the number line below.

[number line with points a, b to the left of 0, and c, d to the right of 0]

Replace the comma in each pair with the proper symbol: $<, >,$ or $=$. See Example 3.

Complete Examples

a. $a, 0$ b. a, c c. c, c

$a \underline{\quad} 0$ (1) $a \underline{\quad} c$ (2) $c \underline{\quad} c$ (3)

(1) $<$ (2) $<$ (3) $=$

11. b, c
12. c, a
13. a, d
14. b, d
15. b, b
16. a, a
17. $0, b$
18. $c, 0$
19. a, b
20. c, d
21. d, a
22. d, b

199

INTEGERS

Replace the comma in each pair with the proper symbol: $<$ or $>$. See Examples 2 and 3.

Complete Examples

a. 1, −3

The graph of 1 is to the right of the graph of −3.

$$1 \underset{(1)}{\_\_\_\_} -3$$

b. −9, −6

The graph of −9 is to the left of the graph of −6.

$$-9 \underset{(2)}{\_\_\_\_} -6$$

(1) $>$ (2) $<$

23. 4, 8
24. 7, 2
25. 0, 4
26. 0, −4
27. 3, −7

28. −4, 5
29. −5, −9
30. −2, −8
31. −5, −2
32. −8, −1

33. −5, 7
34. 4, −5
35. 0, −4
36. 8, 0

Simplify. See Examples 4 and 5.

Complete Examples

a. $|-7| = \underset{(1)}{\_\_\_\_}$

b. $-|-5|$
$= -(\underset{(2)}{\_\_\_\_})$
$= \underset{(4)}{\_\_\_\_}$

c. $|-2|^3$
$= (\underset{(3)}{\_\_\_\_})^3$
$= -5$

(1) 7   (2) 5   (3) 2   (4) 8

37. $|6|$
38. $|-2|$
39. $|-10|$
40. $|5|$

41. $-|9|$
42. $-|-4|$
43. $-|-6|$
44. $-|7|$

200

Replace the comma in each pair with the proper symbol: <, >, or = .

45. $|-7|, 8$
46. $|-2|, -2$
47. $|-5|, |5|$
48. $|-3|, |3|$

49. $-|-4|, -5$
50. $-|-8|, -6$
51. $|6|, 0$
52. $|-12|, 0$

53. $|0|, 6$
54. $|0|, -3$
55. $|-7|, |-6|$
56. $|-2|, |4|$

Choose the correct word to make a true statement.

57. If x is a positive number, then $|x|$ is (*positive/negative*).

58. If x is a negative number, then $|x|$ is (*positive/negative*).

59. If x is a positive number, then $-x$ is (*positive/negative*).

60. If x is a negative number, then $-x$ is (*positive/negative*).

6.2 SUMS OF INTEGERS

To see what is meant by the sum of two signed numbers, we can think of such numbers as representing gains and losses, (+) numbers denoting gains and (−) numbers denoting losses.

	+5 gain	−5 loss	−5 loss	+5 gain
	+3 gain	−3 loss	+3 gain	−3 loss
Sum:	+8 gain	−8 loss	−2 loss	+2 gain

In algebra, we usually write sums horizontally. Thus, we would write the sum of +5 and +3 as

$$+5 + (+3) = +8 \quad \text{or} \quad 5 + 3 = 8$$

And we would write the sum of −5 and −3 as

$$-5 + (-3) = -8$$

For the sum of −5 and +3, we would have

$$-5 + (+3) = -2 \quad \text{or} \quad -5 + 3 = -2$$

and for the sum of +5 and −3,

$$+5 + (-3) = 2 \quad \text{or} \quad 5 + (-3) = 2$$

The above examples suggest a rule for the addition of integers.

INTEGERS

> To add integers with like signs:
> 1. Add the absolute values of the numbers.
> 2. The sum has the same sign as the numbers.
>
> To add integers with unlike signs:
> 1. Find the absolute value of each and subtract the lesser absolute value from the greater absolute value.
> 2. The sum has the same sign as the number with the greater absolute value.

Example 1 Add

a. $\begin{array}{r}-4\\-7\\\hline\end{array}$

b. $\begin{array}{r}-5\\+2\\\hline\end{array}$

Solutions a. Since the signs are alike, add the absolute values.

$$|-4| + |-7| = 4 + 7 = 11$$

The sum has the same sign as the numbers, -4 and -7.

Ans. -11

b. Since the signs are unlike, subtract the lesser absolute value from the greater absolute value.

$$|-5| = 5; \quad |+2| = 2; \quad 5 - 2 = 3$$

Since $|-5|$ is greater than $|+2|$, the sum is negative.

Ans. -3

Example 2 Add

a. $(-5) + (-7)$

b. $(+6) + (-2)$

Solutions a. Since the signs are alike, add the absolute values.

$$|-5| + |-7| = 5 + 7 = 12$$

The sum has the same sign as the numbers, -5 and -7.

Ans. -12

b. Since the signs are unlike, subtract the lesser absolute value from the greater absolute value.

$$|+6| = 6; \quad |-2| = 2; \quad 6 - 2 = 4$$

Since $|+6|$ is greater than $|-2|$, the sum is positive.

Ans. $+4$ or 4

Example 3 a. $(-5) + (+7) + (-3)$
$= (+2) + (-3)$
$= -1$

b. $(+8) + (-6) + (-4)$
$= (+2) + (-4)$
$= -2$

6.2 SUMS OF INTEGERS

PROPERTIES OF ADDITION

In Section 1.3 we noted that for all whole numbers a and b,

$$a + b = b + a$$

This property was called the **commutative law of addition.** A second important property of addition, called the **associative law of addition,** states that the way in which three numbers are grouped for addition does not change the sum. That is, for all numbers a, b, and c,

$$a + (b + c) = (a + b) + c$$

In fact, the commutative and associative laws of addition hold for all integers. For example,

$$2 + (-3) = -3 + 2$$

and

$$[2 + (-3)] + 4 = 2 + [(-3) + 4]$$

EXERCISES 6.2
Add. See Example 1.

Complete Examples

a. $+5$
    $+7$
    ―――
    (1)

b. -5
    $+7$
    ―――
    (2)

c. $+5$
    -7
    ―――
    (3)

d. -5
    -7
    ―――
    (4)

(1) 12   (2) 2   (3) −2   (4) −12

1. $+3$
   $+4$

2. $+2$
   $+8$

3. -6
   -3

4. -4
   -9

5. $+7$
   -3

6. -4
   $+8$

7. -8
   $+2$

8. $+4$
   -6

9. $+8$
   0

10. -6
    0

11. -7
    $+7$

12. $+8$
    -8

Add. See Example 2.

Complete Examples

a. $(-14) + (-12)$
   $= \underline{}$
   (1)

b. $-15 + (+9)$
   $= \underline{}$
   (2)

c. $(+15) + (-9)$
   $= \underline{}$
   (3)

(1) −26   (2) −6   (3) 6

203

INTEGERS

13. $(+7) + (+3)$	14. $(+7) + (+2)$	15. $(-3) + (-7)$	16. $(-6) + (3)$
17. $(+8) + (-3)$	18. $(-8) + (+3)$	19. $(-6) + (+2)$	20. $(-15) + (+1)$
21. $(+5) + 0$	22. $0 + (-5)$	23. $(-6) + (+6)$	24. $(-8) + (+8)$
25. $(+5) + (+3)$	26. $(-6) + (-2)$	27. $(-8) + (0)$	28. $(-5) + (+3)$
29. $(+7) + (-2)$	30. $(-5) + (-6)$		

Add. See Example 3.

Complete Examples

a. $(-7) + (+3) + (-5)$

$$= (\underline{\qquad}_{(1)}) + (-5)$$

$$= \underline{\qquad}_{(2)}$$

b. $(+8) + (-12) + (+3)$

$$= (\underline{\qquad}_{(3)}) + (+3)$$

$$= \underline{\qquad}_{(4)}$$

(1) -4 (2) -9 (3) -4 (4) -1

31. $(-5) + (+8) + (+3)$	32. $(+7) + (-7) + (+2)$	33. $(-4) + (-5) + (+6)$
34. $(-5) + (-3) + (-4)$	35. $(+4) + (0) + (-7)$	36. $(-9) + (+9) + (0)$
37. $(+6) + (0) + (-8) + (+5)$	38. $(+4) + (0) + (-7) + (-6)$	39. $(-7) + (0) + (+7) + (+3)$
40. $(-5) + (+6) + (+6) + (-3)$		

6.3 DIFFERENCES OF INTEGERS

When we discussed the difference of two natural numbers in Section 1.2, we saw that $5 - 3$ is a number 2 that, when added to 3, gives 5. In general, $a - b$ is a number that, when added to b, gives a. The same idea holds for the difference of two integers. For example,

$$(+5) - (+2) = (+3) \quad \text{because} \quad (+2) + (+3) = (+5)$$

Now, since

$$(+5) + (-2) = 3$$

we have

$$(+5) - (+2) = (+5) + (-2)$$

This and similar examples suggest the following rule.

6.3 DIFFERENCES OF INTEGERS

> To subtract an integer *b* from an integer *a*:
>
> 1. Change the subtraction to addition.
> 2. Change the sign of *b*.
> 3. Proceed as in addition.

In symbols,

$$a - b = a + (-b)$$

We can now view an expression such as $5 - 2$ as the difference $(+5) - (+2)$ or as the sum $(+5) + (-2)$. In either case, the expression equals 3. Similarly, $2 - 5$ can be viewed as the difference $(+2) - (+5)$ or as the sum $(+2) + (-5)$, where both expressions equal -3. Because it is easier to work with sums, we usually rewrite a difference as a sum, and then add.

Example 1 Find each difference.
  a. $7 - 3$
  b. $(+7) - (-3)$
  c. $(-7) - (+3)$
  d. $-7 - (-3)$

Solutions
  a. $7 - 3 = 7 + (-3) = 4$
  b. $(+7) - (-3) = (+7) + (+3) = 10$
  c. $(-7) - (+3) = (-7) + (-3) = -10$
  d. $-7 - (-3) = -7 + (+3) = -4$

We can also write a difference as a sum in expressions that contain more than two terms.

Example 2
  a. $5 - 7 + 3 - 2 = 5 + (-7) + 3 + (-2)$
  $= -2 + 3 + (-2)$
  $= 1 + (-2)$
  $= -1$

  b. $-2 - 4 + 6 - 8 = (-2) + (-4) + 6 + (-8)$
  $= -6 + 6 + (-8)$
  $= 0 + (-8)$
  $= -8$

INTEGERS

When more than two terms are involved, we can use the commutative and associative laws to help us rewrite sums or differences by grouping the positive terms and grouping the negative terms. Thus, the examples above could be simplified as follows.

Example 3 a. $5 - 7 + 3 - 2 = 5 + (-7) + (3) + (-2)$
$= (5 + 3) + [(-7) + (-2)]$
$= 8 + (-9)$
$= -1$

b. $-2 - 4 + 6 - 8 = (-2) + (-4) + 6 + (-8)$
$= 6 + [(-2) + (-4) + (-8)]$
$= 6 + (-14)$
$= -8$

Note that each minus sign in the expression $5 - 7 + 3 - 2$ can be viewed as the sign associated with the number that follows, while the operation is understood to be addition.

We can also find the difference between two integers written in vertical form by changing the sign on the number being subtracted and then adding the two numbers.

Example 4 Subtract the bottom number from the top number.
  a. $+ 8$ b. $+ 8$
      $+11$ -11

Solutions a. subtract $\begin{array}{r} + 8 \\ +11 \end{array}$ is changed to the sum   add $\begin{array}{r} + 8 \\ -11 \\ \hline - 3 \end{array}$

b. subtract $\begin{array}{r} + 8 \\ -11 \end{array}$ is changed to the sum   add $\begin{array}{r} + 8 \\ +11 \\ \hline +19 \end{array}$

EXERCISES 6.3
Subtract by first changing each problem to an equivalent addition problem. See Example 1.

Complete Examples
a. $(-6) - (-4)$
  $= (-6) + (\underline{\quad\quad})$
  (1)

  $= \underline{\quad\quad}$
  (2)

b. $(+5) - (+2)$
  $= (+5) + (\underline{\quad\quad})$
  (3)

  $= \underline{\quad\quad}$
  (4)

c. $3 - 8$
  $= 3 + (\underline{\quad\quad})$
  (5)

  $= \underline{\quad\quad}$
  (6)

(1) 4   (2) -2 (3) -2 (4) 3   (5) -8 (6) -5

6.3 DIFFERENCES OF INTEGERS

1. $(+8) - (+2)$
2. $(+7) - (+1)$
3. $(+2) - (-6)$
4. $(+3) - (-9)$
5. $(-4) - (-7)$
6. $(-8) - (-6)$
7. $(-5) - (-2)$
8. $(-8) - (-1)$
9. $(7) - (2)$
10. $(8) - (5)$
11. $(3) - (8)$
12. $(1) - (7)$
13. $6 - 3$
14. $9 - 5$
15. $4 - 5$
16. $5 - 12$
17. $-4 - 7$
18. $-5 - 8$

Subtract. See Examples 2 and 3.

Complete Examples

a. $(-5) + (-3) - (-4)$
  $= (-5) + (-3) + (\underline{}_{(1)})$
  $= \underline{}_{(2)} + 4 = \underline{}_{(3)}$

b. $4 - 8 + 3 - 5$
  $= 4 + (-8) + 3 + (\underline{}_{(4)})$
  $= [4 + 3] + [(-8) + (-5)]$
  $= 7 + (\underline{}_{(5)}) = \underline{}_{(6)}$

(1) 4   (2) −8   (3) −4   (4) −5   (5) −13   (6) −6

19. $(-5) + (-4) - (+3)$
20. $(-6) - (-2) + (-4)$
21. $(+9) - (+3) + (-4)$
22. $(-2) + (+6) - (-2)$
23. $(+6) - (+2) - (-4)$
24. $(+6) - (-2) - (-3)$
25. $(4) - (2) + (-2)$
26. $(-2) + (6) + (2)$
27. $(-2) + (-2) - (-2) + (2)$
28. $(4) - (-4) + (4) - (4)$
29. $(-4) - (-2) + (5) + (1)$
30. $(7) - (4) + (-3) + (-1)$
31. $6 - 3 + 4$
32. $8 + 5 - 3$
33. $-3 + 2 - 9$
34. $-6 + 9 - 7$
35. $8 + 0 + 3 - 8$
36. $-5 + 4 + 0 - 4$

Subtract the bottom number from the top number. See Example 4.

Complete Examples

a. $\begin{array}{r} +6 \\ +8 \\ \hline \end{array}$ Change to and add. $\begin{array}{r} +6 \\ -8 \\ \hline \underline{}_{(1)} \end{array}$

b. $\begin{array}{r} -5 \\ -8 \\ \hline \end{array}$ Change to and add. $\begin{array}{r} -5 \\ +8 \\ \hline \underline{}_{(2)} \end{array}$

(1) −2   (2) 3

INTEGERS

37. +4 +4	38. +8 +8	39. +7 +2	40. +5 +3	41. +6 +9	42. +3 +8
43. −6 0	44. 0 −6	45. −8 −2	46. −3 −1	47. −4 −8	48. −2 −4

6.4 PRODUCTS AND QUOTIENTS OF INTEGERS; FORMS OF PRODUCTS

Recall that when we multiply two numbers a and b, the result is called the product of a and b, and a and b are the **factors** of the product. In arithmetic, we used the symbol × to represent multiplication. But in algebra, the symbol × may sometimes be confused with the variable x, which we use so frequently. So in algebra we usually indicate multiplication either by a dot between the numbers or by parentheses around one or both of the numbers.

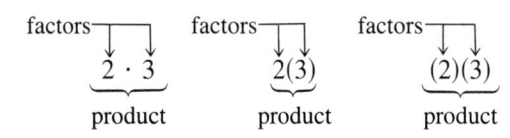

$$\underbrace{2 \cdot 3}_{\text{product}} \quad \underbrace{2(3)}_{\text{product}} \quad \underbrace{(2)(3)}_{\text{product}}$$

(factors indicated above each expression)

Multiplication of variables may be written the same way or may be written with the symbols side by side. For example,

$$\underbrace{ab}_{\text{product}} \quad \underbrace{3x}_{\text{product}}$$

(factors indicated above each expression)

where

ab means "the number a times the number b," and

$3x$ means "the number 3 times the number x."

Since multiplication is a form of addition we can think of $3(2)$ as the sum of three 2's $(2 + 2 + 2)$, and we can think of $3(-2)$ as the sum of three -2's $[(-2) + (-2) + (-2)]$, which is -6. Thus, it would appear that the product of a positive and negative number should be a negative number.

Now let us investigate the product of two negative numbers. First, consider the following sequence of products:

$$4(-2) = -8$$
$$3(-2) = -6$$
$$2(-2) = -4$$
$$1(-2) = -2$$
$$0(-2) = 0$$

If we continue the sequence with the product $-1(-2)$, it seems plausible (from the sequence above) that the number 2 is the next number in the sequence following the equals sign. That is, the sequence continues with

208

6.4 PRODUCTS AND QUOTIENTS OF INTEGERS; FORMS OF PRODUCTS

$$-1(-2) = 2$$
$$-2(-2) = 4$$
$$-3(-2) = 6$$

and so on. It appears (at least intuitively) that the product of two negative numbers is a positive number. In fact, this is the case and we adopt the following rule.

To find the product of two signed numbers:

1. Multiply the absolute value of the numbers.
2. Determine the sign of the product:
   if the factors have like signs, the product is positive;
   if the factors have unlike signs, the product is negative.

In symbols,

$$(+a)(+b) = +(ab), \qquad (-a)(-b) = +(ab)$$
$$(+a)(-b) = -(ab) \quad \text{and} \quad (-a)(+b) = -(ab)$$

Also, the product of zero and any integer is zero.

Example 1
a. $(+5)(+9) = +45$ b. $(+5)(-9) = -45$ c. $(-5)(+9) = -45$
d. $(-5)(-9) = +45$ e. $(+5) \cdot 0 = 0$ f. $(-5) \cdot 0 = 0$

PROPERTIES OF MULTIPLICATION

In Section 2.1, we noted that for all whole numbers a and b

$$a \cdot b = b \cdot a$$

which was called the **commutative law of multiplication.** Another basic property of multiplication states that the way three factors in a product are grouped for multiplication does not change the product. That is, for all numbers a, b, and c

$$(a \cdot b) \cdot c = a \cdot (b \cdot c)$$

This property is called the **associative law of multiplication.**

In fact, the commutative and associative laws of multiplication hold for signed numbers. For example, by the commutative law,

$$(-2)(3) = 3(-2)$$

And by the associative law,

$$[(-2)(3)](4) = (-2)[(3)(4)]$$

SIGNS OF PRODUCTS

We can determine the sign of the product of any number of factors. If we have an *odd* number of negative factors, the product is negative; if we have an *even* number of negative factors, the product is positive.

INTEGERS

Example 2

a. one negative factor → $(\overset{\downarrow}{-}2)(3)(4) = (-6)(4) = \overset{\downarrow}{-}24$ ← negative product

b. three negative factors → $(\overset{\downarrow}{-}2)(\overset{\downarrow}{-}3)(\overset{\downarrow}{-}4) = 6(-4) = \overset{\downarrow}{-}24$ ← negative product

c. two negative factors → $(\overset{\downarrow}{-}2)(3)(\overset{\downarrow}{-}4) = (-6)(-4) = +24$ ← positive product

d. four negative factors → $(\overset{\downarrow}{-}2)(\overset{\downarrow}{-}3)(\overset{\downarrow}{-}2)(\overset{\downarrow}{-}4) = 6(-2)(-4)$
$= (-12)(-4) = +48$ ← positive product

EXPONENTIAL NOTATION

We introduced exponential notation in Section 2.2 as an efficient way of writing a product of repeated factors. For example,

$$2^{\overset{\text{exponent}}{3}} = \underbrace{2 \cdot 2 \cdot 2}_{\text{3 factors of 2}}$$

base ↑

In general,

$$a^n = a \cdot a \cdot a \cdot \ldots \cdot a \quad (n \text{ factors})$$

Thus,

5^2 means $(5)(5)$, read "five squared" or "5 to the second power;"

2^4 means $(2)(2)(2)(2)$, read "two to the fourth power;"

Example 3   a. $4 \cdot 4 \cdot 4 = 4^3$ b. $(-3)(-3) = (-3)^2$ c. $(-2)(-2)(-2)(-2) = (-2)^4$

We can view a negative number as the product of -1 and a natural number. For example,

$$-2 = -1 \cdot 2, \quad -3 = -1 \cdot 3, \quad \text{and} \quad -3^2 = -1 \cdot 3^2$$

Thus, we can also use exponential notation when we write signed numbers in prime factored form.

Example 4   a. $24 = 2 \cdot 2 \cdot 2 \cdot 3$ b. $-36 = -1 \cdot 2 \cdot 2 \cdot 3 \cdot 3$ c. $-16 = -1 \cdot 2 \cdot 2 \cdot 2 \cdot 2$
$= 2^3 \cdot 3$ $= -1 \cdot 2^2 \cdot 3^2$ $= -1 \cdot 2^4$

Common Error   An exponent only applies to the number directly preceding it. For example, consider the two expressions -3^2 and $(-3)^2$. From our above discussion,

$$-3^2 = -1 \cdot 3^2 = -9$$

On the other hand,

$$(-3)^2 = (-3)(-3) = 9$$

Thus, $-3^2 \neq (-3)^2$.

6.4 PRODUCTS AND QUOTIENTS OF INTEGERS; FORMS OF PRODUCTS

Example 5   a. $-5^2 = -1 \cdot 5 \cdot 5$ b. $(-5)^2 = (-5)(-5)$ c. $-2 \cdot 3^2 = -2 \cdot 3 \cdot 3$
$ = -25$ $ = 25$ $ = -18$

QUOTIENTS

We define the quotient of two integers the same way we define the quotient of two natural numbers (Section 1.3). However, the sign of the quotient has to be consistent with the rule of signs for the multiplication of signed numbers. Recall that for natural numbers, the quotient a/b is the number q such that $b \cdot q = a$. Let us examine the quotient of two signed numbers by considering some numerical cases.

$$\frac{+6}{+3} = +2 \quad because \quad (+3)(+2) = +6$$

$$\frac{+6}{-3} = -2 \quad because \quad (-3)(-2) = +6$$

$$\frac{-6}{+3} = -2 \quad because \quad (+3)(-2) = -6$$

$$\frac{-6}{-3} = +2 \quad because \quad (-3)(+2) = -6$$

These examples suggest the following rule.

To find the quotient of two signed numbers:
1. Find the quotient of the absolute values of the numbers.
2. Determine the sign of the quotient:
    if the dividend and divisor have like signs, the quotient is positive;
    if the dividend and divisor have unlike signs, the quotient is negative.

In symbols:

$$\frac{+a}{+b} = +\left(\frac{a}{b}\right), \qquad \frac{-a}{-b} = +\left(\frac{a}{b}\right)$$

$$\frac{+a}{-b} = -\left(\frac{a}{b}\right) \quad \text{and} \quad \frac{-a}{+b} = -\left(\frac{a}{b}\right)$$

QUOTIENTS INVOLVING ZERO

As for whole numbers, a quotient equals zero if the dividend is zero and the divisor is not zero. And as always, *division by 0 is undefined*. For example,

$$\frac{0}{3} = 0 \quad \text{and} \quad \frac{0}{-3} = 0$$

whereas

$$\frac{3}{0}, \quad \frac{-3}{0} \quad \text{and} \quad \frac{0}{0} \quad \text{are undefined}$$

INTEGERS

Example 6 a. $\dfrac{+12}{+3} = +4$ b. $\dfrac{+12}{-3} = -4$ c. $\dfrac{-12}{-3} = +4$

d. $\dfrac{-12}{+3} = -4$ e. $\dfrac{0}{+3} = 0$ f. $\dfrac{+12}{0}$ is undefined

EXERCISES 6.4
Multiply. See Examples 1 and 2.

Complete Examples
a. $(-3)(-7)$
 $= \underline{}$
 (1)

b. $(-4)(7)$
 $= \underline{}$
 (2)

c. $(-3)(5)(0) = (-15)(0)$
 $= \underline{}$
 (3)

(1) 21   (2) −28   (3) 0

1. $(5)(-3)$
2. $(-2)(4)$
3. $(-5)(-6)$
4. $(4)(8)$
5. $(5)(0)$
6. $0(-4)$
7. $(4)(-2)$
8. $(-3)(-3)$
9. $(-5)(3)$
10. $(-8)(-4)$
11. $(2)(-7)$
12. $(6)(7)$
13. $(-6)(0)$
14. $(0)(8)$
15. $(2)(-3)(4)$
16. $(-2)(-3)(-4)$
17. $(-6)(-1)(3)$
18. $(4)(-1)(-7)$
19. $(-4)(0)(6)$
20. $(-5)(4)(0)$
21. $(-2)(5)(4)(-3)$
22. $(3)(-5)(-2)(-2)$
23. $(-4)(-2)(-1)(-1)$
24. $(-3)(-2)(1)(-2)$

Write in exponential form. See Example 3.

Complete Examples
a. $8 \cdot 8 \cdot 8 \cdot 8 = \underline{}$
 (1)

b. $(-1)(-1)(-1) = \underline{}$
 (2)

(1) 8^4 (2) $(-1)^3$

25. $3 \cdot 3 \cdot 3$
26. $5 \cdot 5 \cdot 5$
27. $(-4)(-4)$
28. $(-6)(-6)$
29. $(-2)(-2)(-2)$
30. $(-7)(-7)(-7)$
31. $2 \cdot 2 \cdot 2 \cdot 2 \cdot 2$
32. $4 \cdot 4 \cdot 4 \cdot 4 \cdot 4$

6.4 PRODUCTS AND QUOTIENTS OF INTEGERS; FORMS OF PRODUCTS

Write in completely factored form using exponents. Check the factorization by using the table of prime factors inside the back cover. See Example 4.

Complete Examples

a. $81 = 3 \cdot 3 \cdot 3 \cdot \underline{\quad}_{(1)}$
  $= \underline{\quad}_{(2)}$

b. $100 = 2 \cdot 2 \cdot 5 \cdot \underline{\quad}_{(3)}$
  $= \underline{\quad}_{(4)}$

c. $-27 = -1 \cdot 3 \cdot 3 \cdot \underline{\quad}_{(5)}$
  $= \underline{\quad}_{(6)}$

(1) 3   (2) 3^4 (3) 5   (4) $2^2 \cdot 5^2$ (5) 3   (6) $-1 \cdot 3^3$

33. 45 **34.** 18 **35.** -28 **36.** -8
37. 27 **38.** 54 **39.** -64 **40.** -32

Simplify. See Example 5.

Complete Examples

a. $-2^2 = -1 \cdot 2 \underline{\quad}_{(1)}$
  $= \underline{\quad}_{(2)}$

b. $(-2)^2 = (-2)(\underline{\quad}_{(3)})$
  $= \underline{\quad}_{(4)}$

(1) 2   (2) -4 (3) -2 (4) 4

41. -4^2 **42.** -6^2 **43.** $(-4)^2$ **44.** $(-6)^2$
45. $-3 \cdot 2^2$ **46.** $-2 \cdot 3^2$ **47.** $(-3 \cdot 2)^2$ **48.** $(-2 \cdot 3)^2$
49. $(-1)^2 \cdot 2$ **50.** $(-1)^2 \cdot 5$ **51.** $(-1)(-4)^2$ **52.** $(-1)(-5)^2$

Simplify. See Example 6.

Complete Examples

a. $\dfrac{-12}{-2} = \underline{\quad}_{(1)}$

b. $\dfrac{15}{-3} = \underline{\quad}_{(2)}$

c. $\dfrac{0}{-2} = \underline{\quad}_{(3)}$

d. $\dfrac{-2}{0}$ is $\underline{\quad}_{(4)}$

(1) 6   (2) -5 (3) 0   (4) undefined

213

INTEGERS

53. $\dfrac{-25}{-5}$ 54. $\dfrac{-32}{-8}$ 55. $\dfrac{-27}{9}$ 56. $\dfrac{12}{-4}$

57. $\dfrac{0}{7}$ 58. $\dfrac{0}{-4}$ 59. $\dfrac{8}{0}$ 60. $\dfrac{-5}{0}$

61. $-12 \div 2$ 62. $-20 \div (-5)$ 63. $15 \div (-3)$ 64. $-35 \div 7$

65. $16 \div (-8)$ 66. $-24 \div (-3)$ 67. $0 \div (-2)$ 68. $0 \div (-5)$

69. $-4 \div 0$ 70. $-8 \div 0$

6.5 ORDER OF OPERATIONS

The expression $4 + 6 \cdot 2$ can be interpreted two ways. We can see it as meaning either

$$(4 + 6) \cdot 2 \quad \text{or} \quad 4 + (6 \cdot 2)$$

in which case the result is either

$$10 \cdot 2 = 20 \quad \text{or} \quad 4 + 12 = 16$$

To avoid such confusion, we make certain assumptions about parentheses and fraction bars and about the order of performing mathematical operations.

> Order of Operations
> 1. Perform any operations inside parentheses, or above or below a fraction bar.
> 2. Compute all indicated powers.
> 3. Perform all other multiplication operations and any division operations in the order in which they occur from left to right.
> 4. Perform additions and subtractions in order from left to right.

Example 1 Simplify $4 + 6 \cdot 2$.

Solution We first multiply to get

$$4 + 12$$

and then add to obtain

$$16$$

Expressions that involve powers require special attention, as shown in Examples 2–4.

Example 2 Simplify $2^3 + 3(4 + 1)$.

Solution First, we simplify the quantity in the parentheses

$$2^3 + 3(5)$$

We then compute the power and multiply to get

$$8 + 15$$

Last, we add to obtain

$$23$$

Example 3 Simplify $\dfrac{8-4}{2} + 2 \cdot 3^2$.

Solution We first simplify above the fraction bar, then compute the power to obtain

$$\frac{4}{2} + 2 \cdot 9$$

Then, we divide and multiply to get

$$2 + 18$$

And finally, we add to obtain

$$20$$

Example 4 Simplify $\dfrac{2+7}{5-2} - \dfrac{2^2 + 2^2}{4}$.

Solution We first simplify above and below the fraction bars

$$\frac{9}{3} - \frac{8}{4}$$

Then, we divide as indicated

$$3 - 2$$

and last, we subtract to obtain

$$1$$

All the numbers in the above examples were whole numbers. The order of operations that are listed on page 214 are, in fact, applicable to all signed numbers.

Example 5 Simplify $\dfrac{5 + (-3)^2}{-2} + 3 \cdot (-2)^2$.

Solution We first compute powers to obtain

$$\frac{5 + 9}{-2} + 3 \cdot 4$$

Then we simplify above the fraction bar to get

$$\frac{14}{-2} + 3 \cdot 4$$

Then, we divide and multiply to obtain

$$-7 + 12$$

And finally we add to obtain

$$5$$

INTEGERS

EXERCISES 6.5
Simplify. See Examples 1 and 2.

Complete Examples
a. $35 - (4)(3)$
$= 35 - \underline{\quad\text{(1)}\quad}$
$= \underline{\quad\text{(2)}\quad}$

b. $3^2 + 4(3 + 2)$
$= 3^2 + 4(5)$
$= \underline{\quad\text{(3)}\quad} + 20$
$= \underline{\quad\text{(4)}\quad}$

(1) 12 (2) 23 (3) 9 (4) 29

1. $(2)(4) - 3$
2. $(4)(3) - 5$
3. $2 + 4 \cdot 3$
4. $9 - 3 \cdot 2$
5. $4(0) + 5$
6. $5 - 3(0)$
7. $(2)(3) - 2^2$
8. $(3)(5) - 2^3$
9. $14 - 3^2$
10. $18 - 4^2$
11. $4(3 + 5)$
12. $3(7 - 4)$
13. $(5 + 4)(3)$
14. $(6 - 1)(4)$
15. $(3 + 2)(5 - 1)$
16. $(5 - 3)(3 + 1)$
17. $3(5 - 3)^2$
18. $4(4 + 1)^2$
19. $2 \cdot 3^2 + 4$
20. $3 \cdot 2^2 - 5$
21. $(2 \cdot 3)^2 + 4$
22. $(3 \cdot 2)^2 - 5$
23. $(2 \cdot 2)^3 - 15$
24. $(2 \cdot 2)^3 - 25$

Simplify. See Examples 3 and 4.

Complete Example
a. $\dfrac{5(4)}{2} - 10$
$= \dfrac{20}{2} - 10$
$= \underline{\quad\text{(1)}\quad} - 10$
$= \underline{\quad\text{(2)}\quad}$

b. $\dfrac{8 + 4}{6 - 4} - \dfrac{3^2 + 2^4}{3 + 2}$
$= \dfrac{12}{2} - \dfrac{25}{5}$
$= 6 - \underline{\quad\text{(3)}\quad}$
$= \underline{\quad\text{(4)}\quad}$

(1) 10 (2) 0 (3) 5 (4) 1

25. $\dfrac{4(3)}{2} - 2$
26. $10 - \dfrac{12(3)}{9}$
27. $\dfrac{5 + 7}{4} - 1$
28. $\dfrac{16 - 4}{6} + 5$
29. $\dfrac{3(8)}{12} - \dfrac{6 + 4}{5}$
30. $\dfrac{5 + 9}{7} + \dfrac{2(8)}{4}$
31. $\dfrac{3^2 + 5}{2} + \dfrac{5^2 - 4}{3}$
32. $\dfrac{2^3}{4} + \dfrac{5 + 3^3}{8}$
33. $\dfrac{5^2 - 1}{6} + \dfrac{2(3^2)}{6}$

34. $\dfrac{2^3 - 1}{7} + \dfrac{3(2^3)}{4}$

35. $\dfrac{(2^2)(3^2)}{5 + 1} - \dfrac{7^2 - 6^2}{5 + 8}$

36. $\dfrac{4^2 - 2^2}{(4 - 2)^2} + \dfrac{3^3 - 2^4}{4 + 7}$

37. $\dfrac{2(3) + 4}{6 - 1} - \dfrac{8(3)}{3(4)}$

38. $\dfrac{3^3 + 3}{5(2)} + \dfrac{2 + 2^3}{5(2)} - \dfrac{8^2}{16}$

39. $\dfrac{4^2 - 3^2}{7} + \dfrac{2(3^2) + 2}{2(5)} - \dfrac{26}{3(5) - 2}$

40. $\dfrac{26 - 2(3)^2}{4^2 - 3(4)} + \dfrac{4 + 6^2}{3(5) - 7} - \dfrac{5 + 5^2}{6 + 3^2}$

41. $3\left(\dfrac{5^3 - 100}{3 + 2}\right)\left(\dfrac{2^5 + 4}{15 - 3^2}\right)$

42. $4\left(\dfrac{8^2 - 2(3)^2}{5^2 - 2}\right)\left(\dfrac{6^3 - 4(5)^2}{5^2 + 4}\right)$

Simplify. See Example 5.

Complete Examples

a. $4 - 2(-3)^2$

$= 4 - 2 \cdot \underline{}_{(1)}$

$= 4 - 18$

$= \underline{}_{(2)}$

b. $\dfrac{3 + (-2)^2}{-7} - \dfrac{4 - (-2)^3}{3}$

$= \dfrac{3 + 4}{-7} - \dfrac{4 - (-8)}{3}$

$= \dfrac{7}{-7} - \dfrac{12}{3}$

$= \underline{}_{(3)} - \underline{}_{(4)}$

$= \underline{}_{(5)}$

(1) 9   (2) −14   (3) −1   (4) 4   (5) −5

43. $2 + (-2)(3)^2$

44. $4 - (3)(-2)^2$

45. $\dfrac{(-2)^2 + 6}{-5} + 2$

46. $\dfrac{(-3)^2 + 6}{-3} - 4$

47. $\dfrac{4(-3)^2}{-2} + \dfrac{3(-2)^2}{6}$

48. $\dfrac{2(-4)^2}{-4} - \dfrac{2(-3)^2}{2}$

49. $\dfrac{3 - 2 \cdot 3^2}{-5} - \dfrac{-2 \cdot 4^2}{8}$

50. $\dfrac{4 + (-2)(-3)^2}{4 - (-3)} - \dfrac{6 - (-2)^2}{-4 - (-2)}$

CHAPTER SIX REVIEW EXERCISES

1. Graph the integers −6, −2, −1, 2, 5, 7 on a number line.
2. Graph the odd integers between −6 and 4 on a number line.
3. Arrange the integers 2, −3, 5, −4, 0 in order from the smallest to the largest.
4. Replace the comma in each pair with the proper symbol: <, >, or =.
    a. −5, 2
    b. $|-3|$, 3
    c. $|-2|$, 0
5. Add:
    a. $+3$
       $\underline{-5}$
    b. -2
       $\underline{-4}$
    c. -6
       $\underline{7}$

217

INTEGERS

Simplify.

6. a. $(-2) + (-1)$ b. $(-3) + (0)$ c. $(-8) + (5)$
7. a. $4 + 3 - 2$ b. $6 - 5 - 7$ c. $-4 - 3 - 8$

8. Subtract the bottom number from the top number.
   a. $\underline{6}\atop 8$ b. $\underline{-7}\atop -3$ c. $\underline{0}\atop -2$

9. a. $(-3)(-2)$ b. $4(-3)$ c. $-2 \cdot 5$
10. a. $4(0)(-2)$ b. $5(-1)(-3)$ c. $-2(4)(-2)$
11. a. $\dfrac{-15}{-3}$ b. $\dfrac{-8}{2}$ c. $\dfrac{12}{-4}$
12. a. $-16 \div 2$ b. $8 \div (-8)$ c. $0 \div (-6)$

13. Write each product in exponential form.
    a. $4 \cdot 4$ b. $5 \cdot 5 \cdot 5$ c. $-2(-2)(-2)(-2)$

14. Write each product in completely factored form.
    a. 20 b. -36 c. -56

Simplify.

15. a. -8^2 b. $(-8)^2$ c. $-5 \cdot 2^2$
16. a. $(-2)^2 \cdot 3^2$ b. $(-2)^3 \cdot (-3)^2$ c. $-2^3(-3)^2$
17. a. $2 + (3)(4)$ b. $4 - 2(3)$ c. $4 \cdot 3 - 2$
18. a. $\dfrac{12 + 3}{2 + 3} - \dfrac{10}{2}$ b. $\dfrac{5 \cdot 4}{10} + \dfrac{2 + 8}{7 - 2}$ c. $\dfrac{6 - 3}{3 - 2} - \dfrac{6 \cdot 3}{9}$
19. a. $\dfrac{(-2)^2 - 1}{3} - \dfrac{1 + 2^2}{5}$ b. $\dfrac{4 - (-2)^2}{3} + \dfrac{2^2 + 1}{5}$ c. $\dfrac{3^2 - 2^2}{5} - \dfrac{5 - 3^2}{2}$
20. a. $3(-2)^2 - 4 \cdot 3^2$ b. $-2(3)^2 + 4(-2)^2$ c. $\dfrac{(-2)^2 + 2^2}{2} + (-3)^2$

7 POLYNOMIALS

7.1 ALGEBRAIC EXPRESSIONS

In Chapter 6 we considered integers and operations on integers. In this chapter, we consider results of such operations that involve variables that represent integers.

EXPONENTIAL NOTATION

In Section 6.4 we defined

$$a^n = a \cdot a \cdot a \cdot \ldots \cdot a \quad (n \text{ factors})$$

For example,

$$3^2 = 3 \cdot 3 \quad \text{and} \quad (-2)^3 = (-2)(-2)(-2)$$

It is also convenient to use exponential notation for products that involve variables. For example,

$$x^2 = x \cdot x \quad \text{and} \quad y^5 = y \cdot y \cdot y \cdot y \cdot y$$

Example 1
a. $3 \cdot 3 \cdot 3 \cdot xxyyy = 3^3 x^2 y^3$
b. $3xyy - 2 \cdot 2 \cdot xxxy = 3xy^2 - 2^2 x^3 y$
c. $(5b)(5b) - 5bb = (5b)^2 - 5b^2$
d. $5(x - 2)(x - 2) = 5(x - 2)^2$

When we write a variable such as x with no exponent indicated, it is understood that the exponent is 1. That is,

$$x = x^1$$

In a product that involves a power as a factor, it is understood that the exponent is attached only to the **base** of the power and not to other factors in the product. Thus,

$3x^2$, read "three x squared," means $3xx$,

$5x^2y^3$, read "five x squared y cubed," means $5xxyyy$,

$2x^3$, read "two x cubed," means $2xxx$,

$(2x)^3$, read "the quantity $2x$ cubed," means $(2x)(2x)(2x)$.

219

POLYNOMIALS

Example 2 a. $12(3x)^2 = 2 \cdot 2 \cdot 3(3x)(3x)$ b. $4a^2(7a)^2 = 2 \cdot 2 \cdot aa(7a)(7a)$
c. $(2x + 1)^2$
  $= (2x + 1)(2x + 1)$
d. $y^2(y - 2)^3$
  $= yy(y - 2)(y - 2)(y - 2)$

ALGEBRAIC EXPRESSIONS

An **algebraic expression,** or simply, an **expression,** is any meaningful collection of numbers, variables, and signs of operation. For example,

$$4x, \quad 3x - y, \quad \text{and} \quad -3x^2y + xy^2 + z$$

are algebraic expressions.

TERM OF AN EXPRESSION

Any single collection of factors in an algebraic expression, such as

$$-3x^2y, \; xy^2, \quad \text{or} \quad 2$$

is called a **term.** If the term does not contain variables, such as 2, then the term is called a **constant.**

POLYNOMIALS

A **polynomial** is the sum or difference of terms, where the exponents on the variables are natural numbers. For example,

$$4x^5 + 2x^2 - 7x - 5, \quad 3xy^2 - 2x - 3x^4 + 4,$$
$$4xy^2, \quad 5x^2y - 3x - 7, \quad \text{and} \quad 2x + 2y$$

are polynomials. Notice that a polynomial can have any number of terms (*poly* is the Greek prefix for "many").

If a polynomial has only one term, we call it a **monomial** (*mono* is the Greek prefix for "one").

If a polynomial has exactly two terms, we call it a **binomial** (*bi* is the Greek prefix for "two").

If a polynomial has exactly three terms, we call it a **trinomial** (*tri* is the Greek prefix for "three").

Example 3 7, $4x$, y^2z, $3x^2y^3$, and $2x^2y$ are monomials;
$2x + y$, $3xy^2 + 4$, and $x^2 - y^2$ are binomials;
$x + y - z$, $2a + 3b + 5c$, and $x^2 - 3x + 4$ are trinomials.

Polynomials in one variable are generally written in descending powers of the variable.

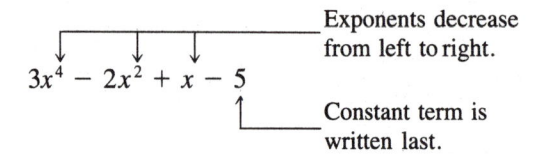

$3x^4 - 2x^2 + x - 5$ — Exponents decrease from left to right.
— Constant term is written last.

220

7.1 ALGEBRAIC EXPRESSIONS

COEFFICIENTS

Any collection of factors in a term is called the **coefficient** of the remaining factors in the term. Of particular interest is the numerical coefficient of a term. For example, in $3xy$, 3 is the numerical coefficient. In terms such as x^2 and $-x^2$, the numerical coefficients are understood to be 1 and -1, respectively.

Example 4 $3x^2$ has a numerical coefficient of 3;
$-2xy^2$ has a numerical coefficient of -2;
x^4 has a numerical coefficient of 1;
$-y^2$ has a numerical coefficient of -1.

In Section 6.3 we noted that any difference could be expressed as a sum. For example,

$$7 - 2 - 3 = 7 + (-2) + (-3)$$

Similarly, we can write the polynomial

$$7x^3 - 2x^2 - 3x$$

as

$$7x^3 + (-2x^2) + (-3x)$$

and view the coefficients of x^2 and x to be -2 and -3, respectively. Hence we shall consider the signs in a polynomial to be part of the numerical coefficient of the terms and the operation of addition is understood.

When we refer to the coefficient of a term, it will be understood that we mean the numerical coefficient unless otherwise stated.

Example 5 a. In the polynomial $2x - 3y - 4z$, the coefficient of x is 2, the coefficient of y is -3, and the coefficient of z is -4.

b. In the polynomial $3x^2 - x + 4$, the coefficient of x^2 is 3 and the coefficient of x is -1.

DEGREE OF A POLYNOMIAL

In a term containing only one variable, the exponent on the variable is called the **degree** of the term. The degree of a constant term is considered 0.

Example 6 $3x^2$ is of second degree;
$-2y^3$ is of third degree;
$4z$ is of first degree (the exponent on z is understood to be 1);
7 has degree 0.

The degree of a polynomial in one variable is the degree of the term of highest degree.

Example 7 $2x + 1$ is of first degree;
$3y^2 - 2y + 4$ is of second degree;
$y^5 - 3y^2 + y$ is of fifth degree.

POLYNOMIALS

EQUIVALENT EXPRESSIONS

We say that two expressions are **equivalent** if they name the same number for *all* replacements of the variable. Although expressions may name the same number for *some* replacements of the variable, they are not necessarily equivalent—they must name the same number for *all* replacements. For example,

$$2x^3 \quad \text{and} \quad (2x)^3$$

name the same number for $x = 0$, but when any other value is used for x, say, $x = 2$, we get

$$2x^3 = 2(2)^3 = 2(2 \cdot 2 \cdot 2)$$
$$= 2 \cdot 8 = 16;$$

and

$$(2x)^3 = (2 \cdot 2)^3 = 4^3$$
$$= 4 \cdot 4 \cdot 4 = 64$$

Thus, $2x^3$ and $(2x)^3$ are not equivalent.

Common Errors Note that in the expression $2x^3$, the exponent applies only to the factor x and not to the product $2x$. In general,

$$ab^n \neq (ab)^n$$

Also note that 2^3 does not mean $3 \cdot 2$. In general,

$$a^n \neq n \cdot a$$

Example 8 Use a numerical example to show that $4x^2$ and $(4x)^2$ are not equivalent.

Solution Let x be 2. Hence,

$$4x^2 = 4 \cdot 2^2 \quad \text{whereas} \quad (4x)^2 = (4 \cdot 2)^2$$
$$= 4 \cdot 4 = 16 \quad \quad \quad \quad \quad = 8^2 = 64.$$

Thus, $4x^2$ is not equivalent to $(4x)^2$.

EXERCISES 7.1
Write in exponential form. See Example 1.

Complete Examples

a. $5 \cdot 5 \cdot 5 \cdot yyxxx = $ _____ (1)

b. $3(x + 5)(x + 5) = $ _____ (2)

c. $2xxy - 3 \cdot 3xyyy = $ _____ (3)

d. $(2z)(2z) - 2zz = $ _____ (4)

---

(1) $5^3 y^2 x^3$ (2) $3(x + 5)^2$ (3) $2x^2 y - 3^2 xy^3$ (4) $(2z)^2 - 2z^2$

7.1 ALGEBRAIC EXPRESSIONS

1. $6 \cdot 6$
2. $5 \cdot 5 \cdot 5$
3. xxx
4. $yyyy$
5. $3 \cdot 3yy$
6. $3 \cdot 3 \cdot 3xx$
7. $2xxyyy$
8. $5 \cdot 5xxyyz$
9. $2aabccc$
10. $5aaabbc$
11. $(x-3)(x-3)$
12. $y(y+2)(y+2)$
13. $3 \cdot 3 + 2 \cdot 2 \cdot 2$
14. $3 \cdot 3 \cdot 3 + 5 \cdot 5$
15. $3xx + 5yyy$
16. $2xxx - 4yzz$
17. $(3a)(3a) - bb$
18. $a(3a)(3a) - (2b)(2b)$
19. $xxx + xxyy$
20. $xxy - xyyy$
21. $3aa - (3a)(3a)$
22. $(2a)(2a) + 2aa$
23. $(x-y)(x-y) + yy$
24. $xxy - (x+y)(x+y)$

Write in completely factored form without exponents. See Example 2.

Complete Examples

a. $3x^2(2y)^3$
   $= 3 \cdot x \cdot \underline{\quad} (2y)(2y)(\underline{\quad})$
   $\qquad\quad$ **(1)** $\qquad\qquad$ **(2)**

b. $x^3(x+2)^2$
   $= x \cdot x \cdot \underline{\quad} (x+2)(\underline{\quad})$
   $\qquad\quad$ **(3)** $\qquad\qquad$ **(4)**

> **(1)** x **(2)** $2y$ **(3)** x **(4)** $x+2$

25. $9x^2y^3$
26. $6x^4y^3$
27. $20a^2b^2c$
28. $10ab^2c^2$
29. $y^2(3x)^2$
30. $6x(2y)^3$
31. $(5a)^2(2b)^3$
32. $(3a)^3(3b)^2$
33. $3x^2$
34. $5x^2$
35. $(3x)^2$
36. $(5x)^2$
37. $(a-4)^3$
38. $(2a+1)^2$
39. $y^3(3y+4)^2$
40. $x^2(2x-1)^3$

a. *Identify each expression as monomial, binomial, or trinomial. See Example 3.*
b. *Write each term separately and state its numerical coefficient. See Examples 4 and 5.*

Complete Examples

a. $4y^3 - 2y^2 + y$ is a $\underline{\qquad}$.
   $\qquad\qquad\qquad\qquad\quad$ **(1)**
   $4y^3$ has a numerical coefficient of 4.
   $-2y$ has a numerical coefficient of -2.
   y has a numerical coefficient of $\underline{\qquad}$.
   $\qquad\qquad\qquad\qquad\qquad\qquad\quad$ **(2)**

b. $3x^2 + 2y^2$ is a $\underline{\qquad}$.
   $\qquad\qquad\qquad\qquad$ **(3)**
   $3x^2$ has a numerical coefficient of 3.
   $2y^2$ has a numerical coefficient of $\underline{\qquad}$.
   $\qquad\qquad\qquad\qquad\qquad\qquad$ **(4)**

> **(1)** trinomial   **(2)** 1   **(3)** binomial   **(4)** 2

223

POLYNOMIALS

41. $2y^2$ **42.** $3x^3$ **43.** $5x^2 + 3x$ **44.** $4y^2 + 2y$

45. $x^2 + 5x$ **46.** $4x^2 + x$ **47.** $-y^5$ **48.** $-x^4$

49. $3y^4 - 2x^2$ **50.** $x^3 - y$ **51.** $x + y$ **52.** $y^2 + x^3$

53. $3x^4 + 3y$ **54.** $2y^3 + 4x$ **55.** $3x^2 + 3y - 4z$ **56.** $3y^3 - 4y + x$

Write each term separately and state the degree of the term. See Example 6.

Complete Example

$4x^2 + 2x^3 + x + 5$

$4x^2$ has degree ____(1)____ , $2x^3$ has degree ____(2)____ ,

x has degree ____(3)____ , 5 has degree ____(4)____ .

(1) 2   (2) 3   (3) 1   (4) 0

57. $5y^3 + y^2$ **58.** $x^3 + 3x$ **59.** $y^4 - 3y + 4$

60. $2x^2 - x + 1$ **61.** $4y^2 - 2y + 1$ **62.** $y^4 - 2y^3 + y$

63. $4x^3 + 2x^2 + 2x$ **64.** $z^5 + 2z^2 + z$

Give the degree of each polynomial. See Example 7.

Complete Examples

a. $3y^4 + 2y^2 + 1$ has degree ____(1)____ since the term of highest degree, $3y^4$, has degree 4.

b. $5x^6 + 3x^3 + x$ has degree ____(2)____ , since the term of highest degree, ____(3)____ , has degree 6.

(1) 4   (2) 6   (3) $5x^6$

65. $2x^4 + x^2 + 4x$ **66.** $4y^5 + 2y^2 + y$ **67.** $y^5 + 2y^3 + y$

68. $4x^3 + 2x^2 + 2x$ **69.** $z^4 + 2z^2 - z$ **70.** $z^5 + 3z^3 - z$

Use a numerical example to show that the following pairs of expressions are not equivalent. See Example 8.

71. $-x^2$; $(-x^3)$
72. $-y^2$; $(-y)^2$
73. $-2x^2$; $(-2x)^2$
74. $-3x^2$; $(-3x)^2$
75. $-2y^3$; $-(2y)^3$
76. $-3y^3$; $(-3y)^3$

7.2 NUMERICAL EVALUATION

The process of substituting given numbers for variables and simplifying the arithmetic expression according to the order of operations given in Section 6.5 is called *numerical evaluation*. It is helpful to use parentheses (as in the following examples) each time a substitution is made.

Example 1   Evaluate $3x^2 + 2x - 3$ for $x = -3$.

Solution   We first substitute -3 for x, using parentheses to avoid sign mistakes, and obtain

$$3(-3)^2 + 2(-3) - 3$$

Now, we evaluate powers and multiply to get

$$3(9) + 2(-3) - 3 = 27 + (-6) - 3$$

Last, we add or subtract to obtain

$$27 + (-6) - 3 = 18$$

Example 2   Evaluate $3x^2 - xy + 2y^2$ for $x = 2$, $y = -1$.

Solution   We first substitute 2 for x and -1 for y to obtain

$$3(2)^2 - (2)(-1) + 2(-1)^2$$

Now, we evaluate powers and multiply to get

$$3(4) - (2)(-1) + 2(1) = 12 - (-2) + 2$$

Last, we add or subtract to obtain

$$12 - (-2) + 2 = 16$$

SPECIAL NOTATION

In Section 7.1 we defined polynomials. Polynomials are frequently represented by expressions such as $P(x)$, $D(y)$, and $Q(z)$, where the symbol in parentheses designates the variable. These symbols are read "*P* of *x*," "*D* of *y*," and "*Q* of *z*," respectively. For example, we might write

$$P(x) = x^2 - 2x + 1$$
$$D(y) = y^6 - 2y^2 + 3y - 2$$
$$Q(z) = 8z^4 + 3z^3 - 2z^2 + z - 1$$

POLYNOMIALS

We can use this notation to represent values of the polynomial for specific values of the variable. Thus, we write $P(2)$ to represent the value of $P(x)$ when x is replaced by 2. Therefore, to evaluate $P(2)$, we first replace x with 2 and then follow the proper order of operations.

Example 3 For $P(x) = x^2 - 2x + 1$, find a. $P(2)$ b. $P(-4)$

Solutions a. First, we replace x by 2 to obtain

$$P(2) = (2)^2 - 2(2) + 1$$

Now, following our order of operations yields

$$P(2) = 4 - 4 + 1 = 1$$

b. First, we replace x by -4 to obtain

$$P(-4) = (-4)^2 - 2(-4) + 1$$

Now, following our order of operations yields

$$P(-4) = 16 + 8 + 1 = 25$$

EXERCISES 7.2
Find the value of each of the following expressions, given $x = -2$. See Example 1.

Complete Example
$2x^2 - 2x + 1$

$= 2(\underline{})^2 - 2(-2) + 1$ Substitute   for x.
     (1)

$= 2(\underline{}) - 2(-2) + 1$ Evaluate powers.
     (2)

$= 8 - (\underline{}) + 1$ Multiply.
     (3)

$= \underline{}$ Add or subtract.
  (4)

(1) -2 (2) 4 (3) -4 (4) 13

1. $4x$
2. $-4x$
3. $3x^2$
4. $-3x^2$
5. $-x^2$
6. $(-x)^2$
7. $1 - 4x$
8. $5 + 3x$
9. $6x^3$
10. $-x^3$
11. x^4
12. x^5
13. $3x^2 + x$
14. $x^2 - x$
15. $-x^2 + 5x$
16. $2x^2 + 4x$
17. $\dfrac{2x + 2}{2} + 1$
18. $\dfrac{2x + 1}{3} - 2$
19. $\dfrac{x^2}{2} - \dfrac{2}{x}$
20. $\dfrac{6}{x} - \dfrac{4}{x^2}$

7.2 NUMERICAL EVALUATION

Find the value of each of the following expressions, given $x = 1$, $y = -2$. See Example 2.

Complete Example

$2x^2 - 3xy + y^2$

$= 2(1)^2 - 3(1)(\underline{\quad(1)\quad}) + (\underline{\quad(2)\quad})^2$ Substitute 1 for x and -2 for y.

$= 2(1) - 3(1)(-2) + \underline{\quad(3)\quad}$ Evaluate powers.

$= 2 - (\underline{\quad(4)\quad}) + 4$ Multiply.

$= \underline{\quad(5)\quad}$ Add or subtract.

(1) -2 (2) -2 (3) 4 (4) -6 (5) 12

21. xy
22. $-4xy$
23. $x + y$
24. $x - y$

25. $3x + y$
26. $x^2 y$
27. $-x^2 y^2$
28. $5x^2 y^3$

29. $-3x^3 y^2$
30. $x^2 + y$
31. $3x^2 + y$
32. $x^2 - 2y^2$

33. $x^2 + xy + y^2$
34. $2x^2 - 4xy + y^2$
35. $-x^2 - y - y^2$
36. $3x^2 - (3y)^2$

37. $\dfrac{x - y}{3} + y^2$
38. $\dfrac{4x - 2y}{4} - 3y$

Given $a = 1$, $b = -2$, $c = -3$, $d = 0$.

39. $a + bc$
40. $a + b + c + d$
41. $3a + b - d$
42. $2a - 2b + 2c$

43. $-abc$
44. $ab^2 c$
45. $a^2 b^2 c^2 d^2$
46. $-abc^2$

47. $-4a^2 bc$
48. $6abcd$
49. $a^2 + b^2$
50. $a^2 - b^2 - c$

51. $ab^2 - cd^2$
52. $a^2 + ac - b^2$
53. $3a^2 - 2bc + 2d^2$
54. $\dfrac{3ab}{c}$

55. $\dfrac{a - b}{-c}$
56. $\dfrac{a + cd - b}{bc - 3a}$
57. $\dfrac{bd}{ac} + \dfrac{bc^2}{a^2}$
58. $\dfrac{cb^2}{a} - \dfrac{cd}{b}$

POLYNOMIALS

Find the values of each polynomial for the specified values of the variable. See Example 3.

Complete Examples

If $P(x) = x^2 + 2x - 4$, find $P(-2)$ and $P(0)$.

a. $P(-2) = (\underline{})^2 + 2(\underline{}) - 4$ Substitute -2 for x.
            (1)   (2)

$ = 4 + (-4) - 4 = \underline{}$
                        (3)

b. $P(0) = (0)^2 + 2(0) - 4$ Substitute 0 for x.

$ = 0 + \underline{} - 4 = \underline{}$
           (4)   (5)

---

(1) -2 (2) -2 (3) -4 (4) 0 (5) -4

---

59. If $P(x) = x^2 + 3$, find $P(-1)$ and $P(2)$.

60. If $P(x) = x^2 - 4$, find $P(-2)$ and $P(1)$.

61. If $Q(x) = x^2 - 2x + 1$, find $Q(-2)$ and $Q(0)$.

62. If $Q(x) = x^2 + 2x - 1$, find $Q(-1)$ and $Q(0)$

63. If $D(y) = y^2 + 3y - 6$, find $D(-3)$ and $D(3)$.

64. If $D(y) = y^2 - 3y + 4$, find $D(-3)$ and $D(3)$.

7.3 SUMS INVOLVING VARIABLES

We can add natural numbers by using a counting procedure. If we wish to add 3 to 5, we can first count out five units, then starting with the next unit, count out three more, giving us the number 8 as the sum. Now suppose we wish to add three 2's to five 2's, that is, $5(2) + 3(2)$. We can add these *like quantities* by counting five 2's, arriving at the number 10, and then counting three more 2's, to make a total of eight 2's or 16. This addition is shown on the number line in Figure 7.1

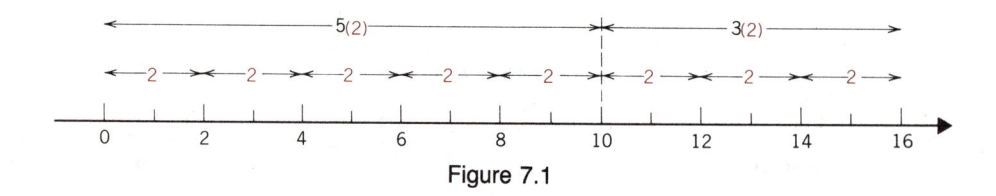

Figure 7.1

LIKE TERMS

In algebra, where terms are usually made up of both numerals and variables, we have to decide what constitutes *like quantities* so that we can apply the idea of addition just developed. We could add 2's as we did because they represented a common unit in each number. For variables, we are sure that $x = x$ and $ab = ab$,

228

7.3 SUMS INVOLVING VARIABLES

regardless of the numbers these letters represent. It is also clear that, in general $x \neq x^2$, $a^3 \neq a^2$, $x \neq xy$, and so forth. We therefore define **like terms** to be any terms that are *exactly alike* in their variable factors. Like terms may differ only in numerical coefficients. Thus,

$$2x \text{ and } 3x, \qquad 4x^2 \text{ and } 7x^2$$
$$5xy \text{ and } 3xy, \qquad 2x^2y \text{ and } 4x^2y$$

are like terms, whereas

$$2x \text{ and } 3x^2$$
$$2x^2y \text{ and } 2xy$$

are not like terms because the variable factors are different.

Note that, by the commutative property, expressions such as xy and yx are equivalent. Hence, $5yx$ and $3xy$ are like terms. Similarly, $2x^2y$ and $4yx^2$ are like terms.

ADDING LIKE TERMS

In view of our definition for like terms and the discussion above, we state the following rule:

> **To add like terms, add their numerical coefficients**

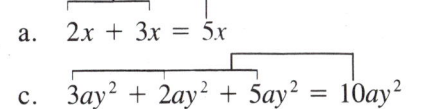

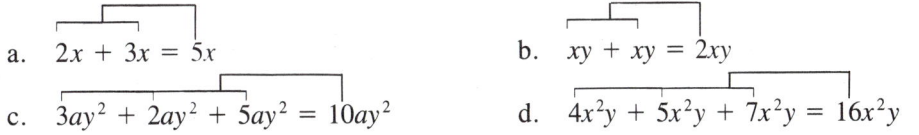

Example 1
a. $2x + 3x = 5x$
b. $xy + xy = 2xy$
c. $3ay^2 + 2ay^2 + 5ay^2 = 10ay^2$
d. $4x^2y + 5x^2y + 7x^2y = 16x^2y$

In Example 1b above, it is understood that the numerical coefficient of xy is 1.

We can illustrate the addition of like terms on a number line, by considering the unit of distance to be equal to the variable part for each term. Thus, we can represent $2x + 3x = 5x$ as shown in Figure 7.2.

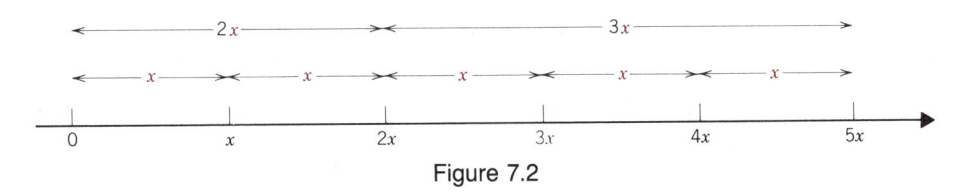

Figure 7.2

The examples above illustrate a basic property of numbers called the **distributive law**:

$$b \cdot a + c \cdot a = (b + c) \cdot a$$

Using the property, we can write

$$4x + 3x = (4 + 3)x = 7x$$
$$2x^2 + 5x^2 = (2 + 5)x^2 = 7x^2$$
$$3x^2y + 5x^2y = (3 + 5)x^2y = 8x^2y$$

We sometimes refer to this addition process as **combining like terms.**

229

POLYNOMIALS

Many expressions contain both like terms and unlike terms. In such expressions we can combine only the like terms.

Example 2 Simplify.

a. $3x^2 + 2x + 5x^2 + x$
b. $3x^2y + 2xy^2 + x^2y^2 + 4x^2y^2 + 5xy^2$

Solutions Add the coefficients of like terms.

a. $3x^2 + 2x + 5x^2 + x$
    $= 8x^2 + 3x$

b. $3x^2y + 2xy^2 + x^2y^2 + 4x^2y^2 + 5xy^2$
    $= 3x^2y + 7xy^2 + 5x^2y^2$

Common Errors Notice that

$$2x + 3y \neq 5xy$$

since $2x$ and $3y$ are *not* like terms, and

$$2x^2 + 3x^2 \neq 5x^4$$

since we add the *numerical* coefficients only; the exponents are *not added*.

COEFFICIENTS WITH SIGNED NUMBERS

We can rewrite sums of terms with positive or negative coefficients according to the laws of signs in Section 6.2.

Example 3 Simplify.

a. $(+5x) + (+3x)$
b. $(-5x) + (-3x)$
c. $(+5x) + (-3x)$
d. $(-5x) + (+3x)$

Solutions Add numerical coefficients.

a. $(+5x) + (+3x) = 8x$
b. $(-5x) + (-3x) = -8x$

Add numerical coefficients.

c. $(+5x) + (-3x) = 2x$
d. $(-5x) + (+3x) = -2x$

Recall that if a variable is preceded by a minus sign, the coefficient -1 is understood. Thus,

$$-x = -1 \cdot x$$
$$-x^2 = -1 \cdot x^2$$

and so on.

7.3 SUMS INVOLVING VARIABLES

Example 4 Add numerical coefficients.

a. $4x^2 + (-x^2) = [4 + (-1)]x^2$
$= 3x^2$

b. $5x^2 + 6x + (-x)$
$= 5x^2 + [6 + (-1)]x$
$= 5x^2 + 5x$

Add numerical coefficients.

c. $3y^2 + 6y + (-y^2)$
$= 3y^2 + (-y^2) + 6y$
$= [3 + (-1)]y^2 + 6y$
$= 2y^2 + 6y$

d. $7x^2 + 2y^2 + (-x^2)$
$= 7x^2 + (-x^2) + 2y^2$
$= [7 + (-1)]x^2 + 2y^2$
$= 6x^2 + 2y^2$

It is convenient to rewrite sums that involve parentheses without using parentheses. We can therefore write $(+5x) + (+3x)$ simply as $5x + 3x$. In general,

$$a + (+b) = a + b$$

We can extend this idea to expressions in which two or more terms are grouped by parentheses.

> **In expressions involving only addition, parentheses that are preceded by a plus sign may be dropped; each term within the parentheses keeps its original sign.**

Thus,

$$x + (y + z) = x + y + z$$

Example 5

a. $4x + (-2x + 6)$
$= 4x + (-2x) + 6$
$= 2x + 6$

b. $(2x^2 + x) + (3x^2 + 2x)$
$= 2x^2 + x + 3x^2 + 2x$
$= 5x^2 + 3x$

EXERCISES 7.3
Simplify. See Example 1.

Complete Examples

a. $5b + 7b = $ _____
   (1)

b. $6a^2 + 3a^2 + 4a^2 = $ _____
   (2)

(1) $12b$ (2) $13a^2$

231

POLYNOMIALS

1. $4y + 2y$
2. $3x + 4x$
3. $6y^2 + 3y^2$
4. $2x^2 + 6x^2$
5. $2x^2 + 4x^2 + x^2$
6. $3y^2 + y^2 + 3y^2$
7. $b^3 + 3b^3 + 2b^3$
8. $2b^3 + 5b^3 + b^3$

Simplify. See Example 2.

Complete Examples

a. $4x^2 + 2x + 6x^2 + 3x = $ _____ (1)

b. $a^3 + b^2 + c^3 + 4b^2 + 2c^3 = $ _____ (2)

(1) $10x^2 + 5x$ (2) $a^3 + 5b^2 + 3c^3$

9. $4xy + 3xy + xy$
10. $3xy + xy + 7xy$
11. $xy^3 + 3xy^3 + 6xy^3$
12. $2x^2y + 3x^2y + 7x^2y$
13. $x^2y + 4x^2y + 2xy^2$
14. $5xy^2 + 4x^2y + x^2y$
15. $3x + 2xy + 4xy + 4y$
16. $6x^2y + 3xy + 3x^2y + xy^2$

Simplify. See Example 3.

Complete Examples

a. $(+6x) + (-8x)$
   $= $ _____ (1)

b. $(-4x^2) + (9x^2)$
   $= $ _____ (2)

c. $(-3xy) + (+6xy)$
   $= $ _____ (3)

(1) $-2x$ (2) $5x^2$ (3) $3xy$

17. $(+4x) + (-2x)$
18. $(+x) + (+8x)$
19. $(-6y) + (+3y)$
20. $(-7a) + (-4a)$
21. $(+3x) + (-3x)$
22. $(-5a) + (-5a)$
23. $(+6hk) + (8hk)$
24. $(-6xz) + (+4xz)$

7.3 SUMS INVOLVING VARIABLES

Simplify. See Example 4.

Complete Examples

a. $2x^2 + 5x^2 + x$
    $= [2 + 5]x^2 + x$ Add numerical coefficients of like terms.
    $= \underline{\qquad}$
       (1)

b. $-4a^2 + 2b^2 + (-a^2)$
    $= -4a^2 + (-a^2) + 2b^2$
    $= [-4 + (\underline{\qquad})]a^2 + 2b^2$ Add numerical coefficients of like terms.
              (2)
    $= \underline{\qquad}$
       (3)

---

(1) $7x^2 + x$ (2) -1 (3) $-5a^2 + 2b^2$

---

25. $5x^2 + 2x^2 + x$
26. $4y^2 + 2y^2 + 7y$
27. $3x^2 + 9c + (-c)$
28. $2b + 9b^2 + (-4b^2)$

29. $6y + 4y^2 + (-y)$
30. $-2x + 5x^2 + 7x$
31. $7x + 3y + (-2x)$
32. $-2y^2 + 4x + 6y^2$

33. $3y + 4x + (-x)$
34. $6x + (-3x) + 2y$
35. $-xy + 3x + 8xy$
36. $7xy + (-4xy) + y$

37. $x^2y + xy + xy^2 + (-xy^2) + (-xy)$
38. $3x^2y + (-4x) + (-2x)$

39. $2x + 3x + (-4x^2) + (-x^3) + 2x^3$
40. $-x + (-2x^2) + x^3 + 2x + x^2$

41. $x^2yz + xy^2z + 2xyz^2 + (-3xy^2z) + (-xyz^2)$
42. $abc + a^3bc + (-2a^2bc) + (-a^3bc) + 5abc$

43. $8x^3yz + xyz + 5x^3yz + (-10x^2yz) + (-3xyz)$
44. $15xyz + (-3wxy) + (-4wxz) + 3wxy + 7wxz$

Simplify. See Example 5.

Complete Examples

a. $x + (x + 3)$
    $= x + x + 3$
    $= \underline{\qquad}$
       (1)

b. $(2x + 3y) + (-x + 4y)$
    $= 2x + 3y + (-x) + 4y$
    $= \underline{\qquad}$
       (2)

---

(1) $2x + 3$ (2) $x + 7y$

233

POLYNOMIALS

45. $4x + (5x + 2)$ **46.** $6y + (4y + 1)$ **47.** $-3y + (x + 3y)$

48. $-7x + (5x + 4y)$ **49.** $(x^2 + x) + (3x^2 + 5x)$ **50.** $(4y^2 + 2y) + (y^2 + 3y)$

51. $(ab^2 + 4b) + (4ab^2 + 2b)$ **52.** $(5a^2 + 2ab) + (-a^2 + ab)$ **53.** $(a^2b^2 + a^2) + (-7a^2b^2 + b^2)$

54. $(4ab + 3b) + (7a + 6b)$

7.4 DIFFERENCES INVOLVING VARIABLES

In Section 7.3 we rewrote sums of algebraic expressions by adding the numerical coefficients of like terms. Using a similar method, we can also rewrite differences of algebraic expressions by first using the fact that a difference $a - b$ can be written as the sum $a + (-b)$ (see Section 6.2).

Example 1
a. $(+7x) - (+2x) = (+7x) + (-2x) = 5x$ (Change to addition. Change sign.)

b. $(+7x) - (-2x) = (+7x) + (+2x) = 9x$ (Change to addition. Change sign.)

c. $(-7x) - (+2x) = (-7x) + (-2x) = -9x$ (Change to addition. Change sign.)

d. $(-7x) - (-2x) = (-7x) + (+2x) = -5x$ (Change to addition. Change sign.)

We can also rewrite as sums those differences that do not involve parentheses.

Example 2
a. $6x^2 - 3x^2 = 6x^2 + (-3x^2) = 3x^2$
b. $4xy - 7xy = 4xy + (-7xy) = -3xy$
c. $-5a^2b - 2a^2b = -5a^2b + (-2a^2b) = -7a^2b$

In any expression that contains some like terms and some unlike terms, we can also rewrite the expression so that all the operations involved are additions.

Example 3
a. $4x^2 + 3x - 2x^2 = 4x^2 + 3x + (-2x^2)$
   $= 2x^2 + 3x$

b. $6a + 3b - 5b - 4a = 6a + 3b + (-5b) + (-4a)$
   $= 2a + (-2b)$
   $= 2a - 2b$

7.4 DIFFERENCES INVOLVING VARIABLES

We must be careful when removing parentheses preceded by a minus sign. For example, to rewrite

$$(+5x - 2y) - (+7x - 4y)$$

without parentheses, we must remember that the minus sign applies to the entire binomial $(+7x - 4y)$. Therefore, we must change the sign of *each term* in $(+7x - 4y)$. Since $+7x - 4y$ is the same as $+7x + (-4y)$, changing the signs yields $-7x + (+4y)$ and we get

$$(+5x - 2y) - (+7x - 4y) = 5x - 2y - 7x + 4y$$
$$= 5x + (-2y) + (-7x) + 4y$$
$$= -2x + 2y$$

Each sign is changed.

As a general rule,

> **If an expression is inside parentheses preceded by a minus sign, the sign of each term is changed when the expression is written without parentheses.**

Example 4 a. $(a^2 + 2b^2) - (4a^2 - b^2 + c^2)$

Each sign is changed.

$$= a^2 + 2b^2 - 4a^2 + b^2 - c^2$$
$$= a^2 + 2b^2 + (-4a^2) + b^2 + (-c^2)$$
$$= -3a^2 + 3b^2 - c^2$$

b. $(4x^2 + 2x - 5) - (2x^2 - 3x - 2)$

Each sign is changed.

$$= 4x^2 + 2x - 5 - 2x^2 + 3x + 2$$
$$= 4x^2 + 2x + (-5) + (-2x^2) + 3x + 2$$
$$= 2x^2 + 5x - 3$$

We can also use a vertical format to subtract polynomials. Again, we must be careful to properly change signs.

Example 5 Subtract the bottom polynomial from the top polynomial.

a. $+5x - 2y$
    $\underline{+7x - 4y}$

b. $4x^2 + 2x - 5$
    $\underline{2x^2 - 3x - 2}$

Solutions a. Subtract. $\quad \begin{array}{r} +5x - 2y \\ \underline{+7x - 4y} \end{array}$ Change signs of each term and add. $\begin{array}{r} +5x - 2y \\ \underline{-7x + 4y} \\ -2x + 2y \end{array}$

235

POLYMIALS

b.
$$\text{Subtract.} \quad \begin{array}{r} 4x^2 + 2x - 5 \\ 2x^2 - 3x - 2 \end{array} \longrightarrow \begin{array}{c} \text{Change signs of each} \\ \text{term and add.} \end{array} \quad \begin{array}{r} 4x^2 + 2x - 5 \\ -2x^2 + 3x + 2 \\ \hline 2x^2 + 5x - 3 \end{array}$$

Common Errors Many errors occur when differences are involved. You must be careful to change signs properly. For example,

$$(3x - 4y) - (x + 2y) \neq 3x - 4y - x + 2y$$
Sign has *not* been changed. ⬆

and

$$(4x^2 + 5x) - (-2x^2 - 3x) \neq 4x^2 + 5x + 2x^2 - 3x$$
Sign has *not* been changed. ⬆

EXERCISES 7.4
Simplify. See Example 1.

Complete Examples

a. $(6x) - (-2x)$
$= 6x + (\underline{\quad}_{(1)})$
$= \underline{\quad}_{(2)}$

b. $(2y^2) - (4y^2) - (-5y^2)$
$= 2y^2 + (-4y^2) + (\underline{\quad}_{(3)})$
$= \underline{\quad}_{(4)}$

(1) $2x$ (2) $8x$ (3) $5y^2$ (4) $3y^2$

1. $(x) - (4x)$
2. $(5x) - (2x)$
3. $(-3x) - (-x)$
4. $(6x) - (3x)$
5. $(3y) - (-3y)$
6. $(-2y) - (2y)$
7. $(6xy) - 2(xy)$
8. $(4xy) - (-xy)$
9. $(-2x^3y) - (5x^3y)$
10. $(-2xz) - (-3xz)$
11. $(5y^2) - (-y^2)$
12. $(-4y^2) - (y^2)$
13. $(2x) + (3x) - (-x)$
14. $(2y) - (-5y) - (3y)$
15. $(-8g) + (-3g) - (-7g)$
16. $(2rx) - (-2rx) + (-3rx)$
17. $(5ab^2) - (-ab^2) + (-3ab^2)$
18. $(7a) - (a) - (-2a)$

Simplify. See Examples 2 and 3.

Complete Examples

a. $3x^2 - 7x^2$
$= 3x^2 + (\underline{\quad}_{(1)})$
$= \underline{\quad}_{(2)}$

b. $4a - 6b + 2b$
$= 4a + (\underline{\quad}_{(3)}) + 2b$
$= \underline{\quad}_{(4)}$

(1) $-7x^2$ (2) $-4x^2$ (3) $-6b$ (4) $4a - 4b$

7.4 DIFFERENCES INVOLVING VARIABLES

19. $2x - 5x$ **20.** $4x - 9x$ **21.** $-8y^2 - 3y^2$ **22.** $-y^2 - 6y^2$
23. $7a^2b - 4a^2b$ **24.** $3ab^2 - ab^2$ **25.** $3ab^2 - 2ab^2 + a^2b$ **26.** $a^2b + 4ab^2 - 5a^2b$
27. $xy^2 + 3x^2y - 8xy^2$ **28.** $4x^2y - 6x^2y + xy^2$

Simplify. See Example 4.

Complete Example

$(3x^2 - 2y + z) - (-2x^2 + 3y - z)$

$= 3x^2 - 2y + z + \underline{\qquad}$ (1)   Remove ( ); change signs.

$= 3x^2 + (\underline{\qquad}) + z + 2x^2 + (\underline{\qquad}) + z$ Express as a sum.
                (2)   (3)

$= \underline{\qquad}$ (4)   Add like terms.

---

(1) $2x^2 - 3y + z$ (2) $-2y$ (3) $-3y$ (4) $5x^2 - 5y + 2z$

---

29. $(3x^2 + 2x - 1) - (3x^2 - 2x + 3)$ **30.** $(5x^2 + 2x - 3) - (x^2 + 4x + 1)$
31. $(2y^2 - y + 1) - (6y^2 + 2y + 1)$ **32.** $(4x^2 - 3x - 1) - (5x^2 + x - 1)$
33. $(z^2 - 4z + 1) - (2z^2 + z + 1)$ **34.** $(y^2 - 3y + 5) - (y^2 + 4y - 3)$
35. $(2p^2 - 3p + 1) - (2p^2 - 3p + 1)$ **36.** $(y^2 - 3y + 1) - (2y^2 - 6y + 2)$
37. $(x^2y - xy + xy^2) - (2x^2y + 3xy)$ **38.** $(a^2b^2 + 2ab + 1) - (3 - ab)$
39. $(x^2y^2 - 2xy + 5) - (xy + 2)$ **40.** $(2g^2h + h - g) - (2g^2h + h)$
41. $(2xy^2 + 6xy - x) - (2xy + x)$ **42.** $(5ax^2 + 3ax + 4) - (2ax^2 - 3)$
43. $2x - y + (x + y)$ **44.** $3a - 2b + (2a + b)$
45. $2x + 3 - (x - 4)$ **46.** $(2x + 3) - x - 4$
47. $6a + 5b - (2a - 5b) - 2a$ **48.** $3c - 2d + 1 - (2 + 2c - d) + c - 1$
49. $(x + y - z) + (x + y + 2z) - (x + y + z) + (3x - y + 2z)$
50. $(2x + y - z) + (x - 2y + z) - (x + y + 2z) - (x - 3y - 4z)$
51. $(a - b - c) + (a - b - c) - (a - b - c) + (a + b + c)$
52. $(2g + 3h - k) + (2g + h + k) - (2g + 2h + 2k) - (3g - h + k)$
53. $(2x + 2y - z) - (x + 2y - z) - (3x + 2y - z) + (x + 4y + 5z)$
54. $(a - b + c) - (2a + b - 2c) + (-a + b + c) - (a - 2b - 3c)$

237

POLYNOMIALS

Subtract the bottom polynomial from the top polynomial. See Example 5.

Complete Examples

a. Subtract. $\begin{array}{r} 4x^2 + 2x - 7 \\ -3x^2 - 5x + 1 \end{array}$ Change signs of each term and add. $\begin{array}{r} 4x^2 + 2x - 7 \\ +3x^2 + 5x - 1 \\ \hline (1) \end{array}$

b. Subtract. $\begin{array}{r} -2x^2 - 3x + 5 \\ 4x^2 - x + 3 \end{array}$ Change signs of each term and add. $\begin{array}{r} -2x^2 - 3x + 5 \\ -4x^2 + x - 3 \\ \hline (2) \end{array}$

(1) $7x^2 + 7x - 8$ (2) $-6x^2 - 2x + 2$

55. $\begin{array}{r} 3x - 7 \\ 2x + 4 \end{array}$ 56. $\begin{array}{r} 5x^2 + 6x \\ -2x^2 + x \end{array}$ 57. $\begin{array}{r} 3y^2 - 2 \\ 4y^2 - 5 \end{array}$ 58. $\begin{array}{r} 7x^2 + 3 \\ -2x^2 - 2 \end{array}$

59. $\begin{array}{r} 3y^2 + 2y - 1 \\ -2y^2 - 2y - 1 \end{array}$ 60. $\begin{array}{r} -4x^2 - 3x + 7 \\ 2x^2 - 2x + 5 \end{array}$ 61. $\begin{array}{r} 10y^2 + 2y + 1 \\ -4y^2 - 4y + 5 \end{array}$ 62. $\begin{array}{r} -3x^2 + 4x - 2 \\ 4x^2 - 3x - 1 \end{array}$

7.5 PRODUCTS INVOLVING VARIABLES

We have used exponents to indicate the number of times a given factor occurs in a product. For example, $x^3 = (x)(x)(x)$. Exponential notation provides us with a simple way to multiply expressions that contain powers with the same base. Consider the product $(x^2)(x^3)$, which in completely factored form appears as

$$(x)(x) \cdot (x)(x)(x)$$

This, in turn, may be written x^5, since it contains x as a factor five times. Again, $(y^5)(y^2)$ is equivalent to

$$(y)(y)(y)(y)(y) \cdot (y)(y)$$

which may be written as y^7.

By applying both the commutative and associative properties for multiplication mentioned in Section 1.2, we can arrange the factors in a product in any order we wish. For example, we can write the product $(2x^2y)(5xy^2)$ in completely factored form as

$$(2)(x)(x)(y)(5)(x)(y)(y)$$

Then, by the associative and commutative laws, we can write

$$(2)(5)(x)(x)(x)(y)(y)(y)$$

which is equivalent to $10x^3y^3$.

7.5 PRODUCTS INVOLVING VARIABLES

Example 1   a. $x^2 \cdot x^3 \cdot x$ b. $(-5x)(2x^3)$ c. $(2xy)(5xy^2)(x^2y)$

$= xx \cdot xxx \cdot x$ $= (-5)(2)x \cdot xxx$ $= 2 \cdot 5(xy)(xyy)(xxy)$

$= x^6$ $= -10x^4$ $= 2 \cdot 5xxxxyyyy$

$ = 10x^4y^4$

FIRST LAW OF EXPONENTS

In the examples above, we can obtain the product of two expressions with the same base by adding the exponents of the powers to be multiplied. We can make a more general statement by considering the product $(a^m)(a^n)$. In completely factored form, $(a^m)(a^n)$ appears as

$$\underbrace{(aaa \cdots a)}_{m \text{ factors}} \underbrace{(aaa \cdots a)}_{n \text{ factors}} = \underbrace{aaa \cdots a}_{(m+n) \text{ factors}}$$

which in exponential notation is written as a^{m+n}. Thus,

> **To multiply two powers with the same base, add the exponents.**

This property is called the **first law of exponents.** In symbols,

$$a^m \cdot a^n = a^{m+n}$$

For example,

$3^4 \cdot 3^3 = 3^{4+3} = 3^7$ — Add exponents.
Same base

$y^3 \cdot y^2 = y^{3+2} = y^5$ — Add exponents.
Same base

Example 2   a. $(-4x^3)(3x^2)$ b. $(2xy^3)(5x^2y^2)(xy)$

$ = -4 \cdot 3x^{3+2}$ $ = -4 \cdot 3x^{3+2}$

$ = -12x^5$ $ = 10x^4y^6$

To simplify an expression involving products and sums, we follow the proper order of operations (see Section 6.5).

Example 3   a. $a^3 + 3a^2(a)$ b. $(3x)(xy^3) - xy + 4x^2y^3$ Multiply factors.

$ = a^3 + 3a^3$ $ = 3x^2y^3 - xy + 4x^2y^3$ Combine like terms.

$ = 4a^3$ $ = 7x^2y^3 - xy$

239

POLYNOMIALS

Common Errors   Notice that

$$x^3 \cdot x^2 \neq x^{3 \cdot 2} \quad \text{or} \quad x^6$$

If $x = 2$, then

$$2^3 \cdot 2^2 = 8 \cdot 4 = 32$$

whereas

$$2^6 = 64$$

By the first law of exponents, we must *add* the exponents. Thus,

$$x^3 \cdot x^2 = x^{3+2} = x^5$$

EXERCISES 7.5

Find each of the following products by first expressing in a rearranged, completely factored form, then simplifying. See Example 1.

Complete Examples

a. $a^3 \cdot a^4 = aaa \cdot aaaa$
$= \underline{\qquad}$
$\quad$ (1)

b. $(5x^2y^2)(3xy^3)$
$= 5 \cdot 3(xxyy)(xyyy)$
$= 5 \cdot 3xxx \cdot yyyyy$
$= \underline{\qquad}$
$\quad$ (2)

c. $(-3xy^2)(-2x^2)$
$= -3(-2)(xyy)(xx)$
$= -3(-2)(xxx \cdot yy)$
$= \underline{\qquad}$
$\quad$ (3)

---

(1) a^7 (2) $15x^3y^5$ (3) $6x^3y^2$

---

1. $x^3 \cdot x^2$
2. $x^2 \cdot x^6$
3. $y^3 \cdot y^7$
4. $y \cdot y^5$
5. $(-4y^3)(-3y^2)$
6. $(-6y)(-3y^3)$
7. $(3a)(a^2)(4a^3)$
8. $7a^2(a)(a^3)$
9. $(3a^3b)(4ab^2)$
10. $(6ab^3)(2a^2b^2)$
11. $(-5x^4y^2)(2x^2y^2)$
12. $(3xy^2)(-2x^3y^2)$

Find the products directly by using the First Law of Exponents. See Example 2.

Complete Examples

a. $3c^4 \cdot 5c^2$
$= 3 \cdot 5c^{4+2}$
$= \underline{\qquad}$
$\quad$ (1)

b. $(3x^2y)(3xy^2)(2xy)$
$= 3 \cdot 3 \cdot 2x^{2+1+1}y^{1+2+1}$
$= \underline{\qquad}$
$\quad$ (2)

c. $(-3ab^2)(4a^4b)$
$= -3 \cdot 4a^{1+4}b^{2+1}$
$= \underline{\qquad}$
$\quad$ (3)

---

(1) $15c^6$ (2) $18x^4y^4$ (3) $-12a^5b^3$

13. $(2x)(3x^2)$ 14. $(6y^2)(3y^2)$ 15. $(-y^4)(-y^3)$ 16. $(-x)(-4x^4)$
17. $(2a^2b)(8ab^3)$ 18. $(a^3b^3)(b^2c)$ 19. $(ab)(ab)(ab)$ 20. $(a^2b)(a^3b^2)(b^3)$
21. $(a^2b)(b^2c)(a^3c^2)$ 22. $(ac^2)(a^2b^3)(b^2c^2)$ 23. $(-2abc)(5b^2c^2)(ab^2c)$ 24. $(7a^2bc)(2bc)(-abc)$

Simplify. See Example 3.

Complete Examples

a. $3b^3 + 2b^2(b)$
$= 3b^3 + \underline{}_{(1)}$
$= \underline{}_{(2)}$

b. $(4a)(ax^2) - ax + 2a^2x^2$
$= \underline{}_{(3)} - ax + 2a^2x^2$
$= \underline{}_{(4)} - ax$

c. $(-3x)(xy) + xy - x^2y$
$= -3x^2y + xy - x^2y$
$= \underline{}_{(5)} + xy$

(1) $2b^3$ (2) $5b^3$ (3) $4a^2x^2$ (4) $6a^2x^2$ (5) $-4x^2y$

25. $4x^3 - x(3x^2)$ 26. $3x^2(3x) - 2x^3$ 27. $x^5 + 2x(x^4)$
28. $3b(a^2) + 3a^2(b) - a^2b$ 29. $(-x)(-x^2) + (y^2)(2y) - x^3$ 30. $x - x(-xy) + x^2y$
31. $6a^2(b^2) - b^2$ 32. $a^2 + 4a^2(3b)$ 33. $2y(y^3) - 2y^4(3y)$
34. $6t^2(2t) + 4t(3)$ 35. $2a(-a^2) - 3a^2(-a) - a^2$ 36. $b^2b^3 - 3b(-b^2) - b^2b$
37. $5a(2b) - 2a(3b) + 3b$ 38. $a(3b) + 4a(2b^2) - 3ab(b)$
39. $4a^2(ab) + a(a^2b) + 3a^3b$ 40. $8a(ab^2) + 3ab(ab) - 2a^2b^2$

7.6 QUOTIENTS INVOLVING VARIABLES

In Section 6.4 we rewrote quotients of integers. In this section we will rewrite quotients involving variables with integer coefficients. In our work we will assume that *no divisor is equal to zero*. Thus, in a quotient such as $\dfrac{5}{x}$, x will *not* represent zero.

FUNDAMENTAL PRINCIPLE OF FRACTIONS

In algebra, we can use the following **fundamental principle** of fractions to rewrite a quotient in which the denominator is a factor of the numerator

$$\dfrac{a \cdot c}{c} = a$$

Thus, we can simplify quotients by expressing the dividend and divisor in completely factored form and dividing out common factors.

241

POLYNOMIALS

Example 1 a. $\dfrac{15}{5} = \dfrac{3 \cdot \cancel{5}}{\cancel{5}} = 3$ b. $\dfrac{-36}{6} = \dfrac{-1 \cdot \cancel{2} \cdot 2 \cdot \cancel{3} \cdot 3}{\cancel{2} \cdot \cancel{3}} = -6$

We can also use this method for quotients involving variable factors.

Example 2 a. $\dfrac{x^5}{x^2} = \dfrac{\cancel{x} \cdot \cancel{x} \cdot x \cdot x \cdot x}{\cancel{x} \cdot \cancel{x}} = x^3$

b. $\dfrac{12x^3y^4}{4x^2y} = \dfrac{\cancel{2} \cdot \cancel{2} \cdot 3\cancel{x}\cancel{x}x\cancel{y}yyy}{\cancel{2} \cdot \cancel{2}\cancel{x}\cancel{x}\cancel{y}} = 3xy^3$

Slash bars can be used on the original quotient.

Example 3 a. $\dfrac{-\cancel{12}\cancel{x^3}\cancel{y^4}}{\cancel{4}\cancel{x^2}\cancel{y}} = -3xy^3$ b. $12a^3b^2 \div (3ab) = \dfrac{\cancel{12}\cancel{a^3}\cancel{b^2}}{\cancel{3}\cancel{a}\cancel{b}} = 4a^2b$

If the exponents on the same variable in the dividend and divisor are the same, the quotient of these two powers is 1. Thus,

$$\dfrac{x^2}{x^2} = 1, \quad \dfrac{y}{y} = 1, \quad \text{and} \quad \dfrac{z^5}{z^5} = 1$$

To simplify expressions involving sums, differences, products, and quotients, we follow the proper order of operations (see page 214).

Example 4 a. $\dfrac{3x^2y + 5x^2y}{2xy}$ b. $\dfrac{8y^2 - 4y^2}{2y}$ Combine like terms above the fraction bar.

$= \dfrac{8x^2y}{2xy}$ $= \dfrac{4y^2}{2y}$ Divide out common factors.

$= \dfrac{\cancel{8}\cancel{x^2}\cancel{y}}{\cancel{2}\cancel{x}\cancel{y}} = 4x$ $= \dfrac{\cancel{4}\cancel{y^2}}{\cancel{2}\cancel{y}} = 2y$

Common Errors

Notice that when we apply the fundamental principle of fractions, we can only divide out *common factors*. We *cannot* divide out *terms*. That is,

Terms *cannot* be divided out. $\dfrac{\cancel{2} + 4}{\cancel{2}} \neq 4$

Terms *cannot* be divided out. $\dfrac{\cancel{x} + 3}{\cancel{x}} \neq 3$

2 is not a factor of the entire numerator. $\dfrac{\cancel{2}x + 3}{\cancel{2}} \neq x + 3$

242

7.6 QUOTIENTS INVOLVING VARIABLES

EXERCISES 7.6
Find each of the following quotients by first writing the expression in completely factored form, then simplify. If the expression is undefined, so state. See Examples 1 and 2.

Complete Examples

a. $\dfrac{36a^3b}{9a^2} = \dfrac{2 \cdot 2 \cdot 3 \cdot 3 \cdot aaab}{3 \cdot 3a \underline{\quad\quad}}$

$\qquad (1)$

$= \underline{\quad\quad}$
$\qquad\quad (2)$

b. $6a^3b^4c^2 \div (ab^3c) = \dfrac{3 \cdot 2aaabbbbcc}{a \underline{\quad\quad} c}$

$\qquad\qquad\qquad\qquad\qquad\qquad\quad (3)$

$= \underline{\quad\quad}$
$\qquad\quad (4)$

(1) a (2) $4ab$ (3) bbb (4) $6a^2bc$

1. $\dfrac{9x}{3}$
2. $\dfrac{16x}{4}$
3. $\dfrac{-12y}{3y}$
4. $\dfrac{-18x}{6x}$
5. $\dfrac{35y}{0}$
6. $\dfrac{5x^2}{0}$
7. $\dfrac{0}{9x}$
8. $\dfrac{0}{5y^3}$
9. $\dfrac{-x^5}{x^2}$
10. $\dfrac{-y^7}{y^3}$
11. $y^4 \div y$
12. $x^5 \div x^4$
13. $\dfrac{x^2y^4}{xy^2}$
14. $\dfrac{x^5y^3}{x^3y}$
15. $x^4y^2 \div (xy)$
16. $x^3y^5 \div (x^2y^2)$
17. $\dfrac{-8x^2y^5}{4xy}$
18. $\dfrac{-21x^4y^2}{3x^2y}$
19. $21x^2y^4 \div (7xy)$
20. $12x^3y^3 \div (4xy^2)$

Find the quotients directly by finding the quotient of the numerical coefficients and the quotient of the variable factors. If the expression is undefined, so state. See Example 3.

Complete Examples

a. $\dfrac{-24x^3y^3}{6xy^2} = \dfrac{\overset{-4x^2y}{-24\cancel{x^3}\cancel{y^3}}}{\underset{111}{\cancel{6}\cancel{x}\cancel{y^2}}}$

$= \underline{\quad\quad}$
$\qquad (1)$

b. $\dfrac{-20x^2y^5}{-4xy^3} = \dfrac{\overset{5x\ y^2}{-20\cancel{x^2}\cancel{y^5}}}{\underset{111}{-4\cancel{x}\cancel{y^3}}}$

$= \underline{\quad\quad}$
$\qquad (2)$

(1) $-4x^2y$ (2) $5xy^2$

21. $\dfrac{3x}{-x}$
22. $\dfrac{-8x^2}{2x}$
23. $\dfrac{-x^2y^3}{-xy}$
24. $\dfrac{-12x^3}{-4x}$
25. $\dfrac{-6x}{6x}$
26. $\dfrac{24y^3}{-12}$
27. $\dfrac{20x^3}{-4x}$
28. $\dfrac{-2xy}{y}$

243

POLYNOMIALS

29. $\dfrac{-6x^2y^3}{-xy}$ 30. $\dfrac{-x^3y^2}{-xy}$ 31. $\dfrac{6x^3}{-6x^2}$ 32. $\dfrac{3xy^3}{xy^2}$

33. $\dfrac{ax^2}{ax}$ 34. $\dfrac{8bc}{4c}$ 35. $\dfrac{-9xy^2}{3xy}$ 36. $\dfrac{-12c^2d^2}{6c^2d}$

37. $\dfrac{24x^3y}{-6xy}$ 38. $\dfrac{-36x^3y^3z^3}{18xy^2z^3}$ 39. $\dfrac{x^2y^2z^2}{x^2y^2z^2}$ 40. $\dfrac{26abc^2}{13ab}$

41. $\dfrac{-8x^2y^2}{-2xy}$ 42. $\dfrac{-15a^3b^2}{-5a^3b^2}$ 43. $\dfrac{30g^2h^3y^4}{15g^2h^2y^2}$ 44. $\dfrac{24mn^2}{12mn}$

45. $\dfrac{-33x^2y}{11x^2y}$ 46. $\dfrac{-26cd}{26c}$ 47. $\dfrac{18xy^2z}{-xz}$ 48. $\dfrac{-56x^2y^3z}{7x^2y^2z}$

Simplify. (Review Section 6.5 for order of operations.) See Example 4.

Complete Example

$y - \dfrac{3y + 5y}{4}$

$= y - \dfrac{8y}{4}$ Combine like terms in the numerator.

Divide.

$= y - \underline{}$
(1)

Combine like terms.

$= \underline{}$
(2)

(1) $2y$ (2) $-y$

49. $3x - \dfrac{x + 2x}{x}$ 50. $\dfrac{x^2 + 3x^2}{-2} - \dfrac{x^2}{x}$ 51. $\dfrac{-3x + x}{-2} + x$ 52. $5x^2 + \dfrac{2x^2 - x^2}{x}$

53. $\dfrac{-xy}{x} + 6y$ 54. $xy^2 + \dfrac{2x^3 - x^3}{x}$ 55. $\dfrac{x^2y}{-y} + x^2$ 56. $\dfrac{x^3y^3}{xy} - 3x^2$

57. $\dfrac{-xy^3}{y^3} - 4x$ 58. $\dfrac{-y}{-y} - 8$ 59. $\dfrac{x^2y^2}{-y} + \dfrac{-x^3y}{-x}$ 60. $\dfrac{-y^2}{y} + \dfrac{-xy}{-x}$

61. $\dfrac{3x^2 - x^2}{-x^2} + \dfrac{4x^3 + 2x^3}{3x^3}$ 62. $\dfrac{4y^3 + 8y^3}{6y^3 - 3y^3} - \dfrac{3x^3 + 7x^3}{5x^3 - 3x^3}$

63. $\dfrac{6xy^2 - 2xy^2}{2xy} + \dfrac{7x^2y + 8x^2y}{6x^2 - x^2}$ 64. $\dfrac{13a^2b^2 - a^2b^2}{7ab - ab} - \dfrac{ab^3 + 19ab^3}{(2b)^2}$

65. $\dfrac{24a(2b^3 - b^3)}{6b^2 - 2b^2} - \dfrac{15a^2b^2 + 3a^2b^2}{3a(4b - b)}$ 66. $\left[\dfrac{(6a)^2 - (5a)^2}{10a + a}\right]^2 + \dfrac{7a^2b - a^2b}{4b - 10b}$

244

CHAPTER SEVEN REVIEW EXERCISES

1. Write in exponential form.
   a. $4aabbb$
   b. $-xyyzzz$
   c. $3 \cdot 3ccd$

2. Write in completely factored form without exponents.
   a. $6xy^2$
   b. a^3b^2
   c. $27cd^2$

3. If $a = 1$, $b = 0$, $c = -2$, find the value of each expression.
   a. $a^2 + c$
   b. $4a + 3b + c^2$
   c. $\dfrac{c^2 - b}{2a}$

4. If $a = 2$ and $b = -3$, find the value of each expression.
   a. $2(a + b)^2$
   b. $2a^2 + 2b^2$
   c. $(2a)^2 + (2b)^2$

5. If $P(x) = 2x^2 - x + 3$, find the value of each expression.
   a. $P(-2)$
   b. $P(0)$
   c. $P(3)$

6. If $x = -1$, $y = -2$, $z = 1$, find the value of each expression.
   a. $\dfrac{-x^2 y}{-z}$
   b. $\dfrac{x^2 - z^2}{-2y}$
   c. $\dfrac{x^2 - y}{z}$

Simplify.

7. a. $3xy + 2y + 3xy$
   b. $6a^2 - a^2 - 3a$
   c. $3r + 5s - r - s$

8. a. $-2x^2 + 3x + x^2$
   b. $6x^2 + 2xy + 2x^2 - 3xy$
   c. $2xy^2 + 3xy - xy^2 + xy$

9. a. $(3x^2 - 2x) + (x^2 - x)$
   b. $(3x^2 - 1) - (2x^2 + 2)$
   c. $(4y^2 - 2y) - (y - 1)$

10. a. $(x + y + 2z) - (3x^2 + z)$
    b. $(2a + 3b - 4c) - (a + b + c)$
    c. $(x + y - 2z) - (2x + y - z)$

11. a. $(xy^2)(x^2y)$
    b. $(3b^3)(2a)(2b)$
    c. $(r^3)(s^2)(rs)$

12. a. $3x^2 - x^2(2x) + x^2$
    b. $ab(b^2) - b^2$
    c. $2r(rs^2) - r^2s^2$

13. a. $\dfrac{4a^2b}{2a}$
    b. $\dfrac{3a^3b}{3a^3b}$
    c. $\dfrac{12xy^3}{4y^2}$

14. a. $\dfrac{xy}{-x}$
    b. $\dfrac{-xyz}{xyz}$
    c. $\dfrac{-4x}{-4}$

15. a. $\dfrac{3x - x}{2x} + 4$
    b. $\dfrac{7x - 4x}{3x} - 1$
    c. $\dfrac{x^2}{x} - \dfrac{2x^3 + 4x^3}{2x^2}$

16. a. $\dfrac{3x^2 - 5x^2}{x} + 6x$
    b. $\dfrac{-x^3}{x^2} - \dfrac{12x}{3}$
    c. $\dfrac{3x^2 - 4x^2 + x^2}{7} + 1$

17. What is an algebraic expression consisting of two terms called?
18. In the expression $4x^3$, the number 4 is called the _?_ of x^3.
19. What is the numerical coefficient in the expression x^2?
20. What is the degtree of the trinomial $2x^4 - x^2 - 6x$?

245

8 First-Degree Equations and Inequalities

8.1 SOLVING EQUATIONS

Recall from Section 1.5 that we can determine whether or not a given number is a solution of a given equation by substituting the number in place of the variable and determining the truth or falsity of the result.

Example 1 Determine if the value 3 is a solution of the equation

$$4x - 2 = 3x + 1$$

Solution We substitute the value 3 for x in the equation and see if the left-hand member equals the right-hand member.

$$4(3) - 2 \stackrel{?}{=} 3(3) + 1$$
$$12 - 2 \stackrel{?}{=} 9 + 1$$
$$10 = 10$$

Ans. 3 is a solution.

FINDING SOLUTIONS BY INSPECTION

The variable in the equations that we consider in this chapter have an exponent of one and are called **first-degree equations.** Such equations have at most one solution. The solution to many such equations can be determined by inspection.

Example 2 Find the solution of each equation by inspection.

a. $x + 5 = 12$ b. $4 \cdot x = -20$

Solutions a. 7 is the solution since $7 + 5 = 12$.

b. -5 is the solution since $4(-5) = -20$.

FIRST-DEGREE EQUATIONS AND INEQUALITIES

The solutions of most equations are not immediately evident by inspection. Hence, we shall use the mathematical "tools" for solving equations that were introduced in Chapters 1 and 2.

EQUIVALENT EQUATIONS

Recall that equivalent equations are equations that have identical solutions. Thus,

$$3x + 3 = x + 13, \quad 3x = x + 10, \quad 2x = 10, \quad \text{and} \quad x = 5$$

are equivalent equations, because 5 is the only solution of each of them. In solving any equation, we transform a given equation whose solution may not be obvious to an equivalent equation whose solution is easily noted.

In Chapters 1 and 2 we introduced two properties that were used to generate equivalent equations. The following property, sometimes called the **addition-subtraction property** is one way.

> **If the same quantity is added to or subtracted from both members of an equation, the resulting equation is equivalent to the original equation.**

In symbols,

$$a = b, \quad a + c = b + c, \quad \text{and} \quad a - c = b - c$$

are equivalent equations.

Example 3 Write an equation equivalent to

$$x + 3 = 7$$

by subtracting 3 from each member.

Solution Subtracting 3 from each member yields

$$x + 3 - 3 = 7 - 3$$

or

$$x = 4$$

Notice that $x + 3 = 7$ and $x = 4$ are equivalent equations since the solution is the same for both, namely 4. The next example shows how we can generate equivalent equations by first simplifying one or both members of an equation.

Example 4 Write an equation equivalent to

$$4x - 2 - 3x = 4 + 6$$

by combining like terms and then by adding 2 to each member.

Solution Combining like terms yields

$$x - 2 = 10$$

Adding 2 to each member yields

248

8.1 SOLVING EQUATIONS

$$x - 2 + 2 = 10 + 2$$
$$x = 12$$

To solve an equation, we use the addition-subtraction property to transform a given equation to an equivalent equation of the form $x = a$, from which we can find the solution by inspection.

Example 5 Solve $2x + 1 = x - 2$.

Solution We want to obtain an equivalent equation in which all terms containing x are in one member and all terms not containing x are in the other. If we first add -1 to (or subtract 1 from) each member, we get

$$2x + 1 - 1 = x - 2 - 1$$
$$2x = x - 3$$

If we now add $-x$ to (or subtract x from) each member, we get

$$2x - x = x - 3 - x$$
$$x = -3$$

where the solution -3 is obvious.*

Since each equation obtained in the process is equivalent to the original equation, -3 is also a solution of $2x + 1 = x - 2$. We can check the solution by substituting -3 for x in the original equation

$$2(-3) + 1 \stackrel{?}{=} (-3) - 2$$
$$-5 = -5$$

The *symmetric property of equality* introduced in Section 1.5 states

If $a = b$ then $b = a$

This enables us to interchange the members of an equation whenever we please without having to be concerned with any changes of sign. Thus,

If $4 = x + 2$ then $x + 2 = 4$

If $x + 3 = 2x - 5$ then $2x - 5 = x + 3$

If $d = rt$ then $rt = d$

There may be several different ways to apply the addition property above. Sometimes one method is better than another, and in some cases, the symmetric property of equality is also helpful.

Example 6 Solve $2x = 3x - 9$.   (1)

Solution If we first add $-3x$ to each member, we get

$$2x - 3x = 3x - 9 - 3x$$
$$-x = -9$$

---

*The solution of the original equation is the number -3; however, the answer is often displayed in the form of the equation $x = -3$.

FIRST-DEGREE EQUATIONS AND INEQUALITIES

where the variable has a negative coefficient. Although we can see by inspection that the solution is 9, because $-(9) = -9$, we can avoid the negative coefficient by adding $-2x$ and $+9$ to each member of Equation (1). In this case, we get

$$2x - 2x + 9 = 3x - 9 - 2x + 9$$
$$9 = x$$

from which the solution 9 is obvious. If we wish, we can write the last equation as

$$x = 9$$

by the symmetric property of equality.

A second property introduced in Section 2.6, sometimes called the **multiplication and division property** is also helpful to generate equivalent equations.

> **If both members of an equation are multiplied by, or divided by, the same (nonzero) quantity, the resulting equation is equivalent to the original equation.**

In symbols,

$$a = b, \quad a \cdot c = b \cdot c, \quad \text{and} \quad \frac{a}{c} = \frac{b}{c} \quad (c \neq 0)$$

are equivalent equations.

Example 7 Write an equation equivalent to

$$-4x = 12$$

by dividing each member by -4.

Solution Dividing both members by -4 yields

$$\frac{-4x}{-4} = \frac{12}{-4} \quad \text{or} \quad x = -3$$

In solving equations, we use the above property to produce equivalent equations in which the variable has a coefficient of 1.

Example 8 Solve $3y + 2y = 20$.

Solution We first combine like terms to get

$$5y = 20$$

Then, dividing each member by 5, we obtain

$$\frac{5y}{5} = \frac{20}{5} \quad \text{or} \quad y = 4$$

8.1 SOLVING EQUATIONS

In the next example, we use the addition-subtraction property and the division property to solve an equation.

Example 9   Solve $4x + 7 = x - 2$.

Solution   First, we add $-x$ and -7 to each member to get

$$4x + 7 - x - 7 = x - 2 - x - 7$$

Next, combining like terms yields

$$3x = -9$$

Last, we divide each member by 3 to obtain

$$\frac{3x}{3} = \frac{-9}{3} \quad \text{or} \quad x = -3$$

The following examples show how the multiplication property can be used to produce equivalent equations that are free of fractions.

Example 10   Write an equivalent equation to

$$\frac{x}{6} = -3$$

by multiplying each member by 6.

Solution   Multiplying each member by 6 yields

$$(6)\frac{x}{6} = (6)(-3) \quad \text{or} \quad x = -18$$

Example 11   Solve $\frac{3x}{5} = 9$.

Solution   First, multiply each member by 5 to get

$$(5)\frac{3x}{5} = (5)9$$

$$3x = 45$$

Now, divide each member by 3,

$$\frac{3x}{3} = \frac{45}{3} \quad \text{or} \quad x = 15$$

251

FIRST-DEGREE EQUATIONS AND INEQUALITIES

Example 12   Solve $\dfrac{8y - 3y}{3} = -10$.

Solution   First, simplify above the fraction bar to get

$$\frac{5y}{3} = -10$$

Next, multiply each member by 3 to obtain

$$(\overset{1}{\cancel{3}}) \frac{5y}{\underset{1}{\cancel{3}}} = (3)(-10) \quad \text{or} \quad 5y = -30$$

Last, dividing each member by 5 yields

$$\frac{\overset{1}{\cancel{5}} y}{\underset{1}{\cancel{5}}} = \frac{\overset{-6}{\cancel{-30}}}{\underset{1}{\cancel{5}}} \quad \text{or} \quad y = -6$$

EXERCISES 8.1
Determine whether each equation is true for the indicated value of the variable; that is, determine whether the given number is, or is not, the solution of the given equation. See Example 1.

Complete Examples

a. $3x + 4 = 10$, for $x = 2$

$3(\underline{\quad}) + 4 \overset{?}{=} 10$
$ (1)$

$\phantom{3(\_) + 4 \overset{?}{=}} 10 = 10$

2 $\underline{\text{is / is not}}$ the solution.
$ (2)$

b. $y - 4 = 3y + 1$, for $y = -2$

$(-2) - 4 \overset{?}{=} 3(\underline{\quad}) + 1$
$\phantom{(-2) - 4 \overset{?}{=} 3()} (3)$

$\phantom{(-2) - 4 \overset{?}{=}} -6 \neq -5$

-2 $\underline{\text{is / is not}}$ the solution
$ (4)$

---

(1) 2   (2) is   (3) -2 (4) is not

---

1. $\dfrac{3x + 2}{8} = 4$ for $x = -6$

2. $\dfrac{4x - 5}{5} = -5$, for $x = -5$

3. $2y + 1 = 3y + 3$ for $y = -2$

4. $4y - 2 = 3y + 1$, for $y = 3$

5. $\dfrac{x}{4} - 2 = \dfrac{x}{2}$, for $x = -8$

6. $\dfrac{x}{6} + 2 = \dfrac{x}{3}$, for $x = 12$

8.1 SOLVING EQUATIONS

Find the solution of each equation by inspection. See Example 2.

Complete Examples
a. $x + 6 = 4$

Ans. ____(1)____ (because $-2 + 6 = 4$)

b. $5x = -15$

Ans. ____(2)____ (because $5(-3) = -15$)

(1) -2 (2) -3

7. $x + 4 = -1$
8. $x + 7 = -2$
9. $x - 5 = -4$
10. $x - 10 = -5$
11. $8 - x = -5$
12. $4 - x = -7$
13. $-5x = -25$
14. $-4x = -32$
15. $\dfrac{x}{5} = -35$
16. $\dfrac{x}{-8} = 4$
17. $\dfrac{x}{-3} = 0$
18. $\dfrac{x}{12} = 0$

Write an equation equivalent to the given equation. See Examples 3 and 4.

Complete Examples

a. $x + 6 = 15$,
by adding -6 to each member.

$x + 6 - 6 = 15 - \underline{\quad(1)\quad}$

$x = \underline{\quad(2)\quad}$

b. $3x + 4 - 2x = 5 + 7$,
by combining like terms.

$x + 4 = \underline{\quad(3)\quad}$

by adding -4 to each member.

$x + 4 - 4 = 12 - \underline{\quad(4)\quad}$

$x = \underline{\quad(5)\quad}$

(1) 6   (2) 9   (3) 12   (4) 4   (5) 8

19. $2x - x + 5 = 8$, by a. combining like terms and
b. adding -5 to each member.

20. $3y - 4 + 2 - 2y = 8$, by a. combining like terms and
b. adding 2 to each member.

253

FIRST-DEGREE EQUATIONS AND INEQUALITIES

21. $2z = 9 + z$,   by   a. adding $-z$ to each member and
                                     b. combining like terms.

22. $3x + 4 - 2x = 8 + 3$, by   a. combining like terms and
                                           b. adding -4 to each member.

Solve each equation. See Examples 5 and 6.

Complete Example

$3x - x + 7 = 12 + x$ Combine like terms.

$\underline{}_{(1)} + 7 = 12 + x$ Add -7 to each member.

$2x + 7 - 7 = 12 + x - \underline{}_{(2)}$ Combine like terms.

$2x = 5 + x$ Add $-x$ to each member.

$2x - x = 5 + x - \underline{}_{(3)}$ Combine like terms.

$x = \underline{}_{(4)}$

(1) $2x$ (2) 7 (3) x (4) 5

23. $4x = 3x + 5$ **24.** $5x = 4x - 6$ **25.** $6x = -7 + 7x$

26. $8x = -3 + 9x$ **27.** $0 = 4x + 5 - 3x$ **28.** $0 = 5x - 4x - 6$

29. $2r + 3r = 4r + 1$ **30.** $6s - 4s = s - 2$ **31.** $8 + 3t - t = 6 + t$

32. $u + 2u + 3u = 5u - 1$ **33.** $5z + 3 + 6z - 1 = 10z + 5$ **34.** $6t + 7 - 3t - 2 = 4 + 2t$

8.1 SOLVING EQUATIONS

Complete Example

$$4(3x - x) = 3(5x - 3x) + x + 6$$

$$4(\underline{}) = 3(2x) + x + 6$$
(1)

Combine like terms in parentheses.

$$8x = \underline{} + x + 6$$
(2)

Perform the multiplications.

$$8x = \underline{} + 6$$
(3)

Combine like terms $6x$ and x.

$$8x - 7x = 7x + 6 - \underline{}$$
(4)

Add $-7x$ to each member.

$$x = \underline{}$$
(5)

Combine like terms.

(1) $2x$ (2) $6x$ (3) $7x$ (4) $7x$ (5) 6

---

35. $2(4t - t) + 6 = 2(2t + t) + 8 - t$

36. $-3(x - 3x) + 5 = -4(3x - x) + 7 + 13x$

37. $\dfrac{4x - 2x}{2} + 3(x + 2x) = 2(3x + x) + x$

38. $5(2x + x) - \dfrac{3(2x + x)}{9} = 2(3x + 4x) + x$

39. $\dfrac{3(6y - y)}{5 - 2} + \dfrac{8y - 2y}{3} = 2(5y - 2y) + 4$

40. $\dfrac{3(5y - y)}{4 + 2} - \dfrac{4(2y - y)}{2} = 8 + y$

Solve each equation. See Examples 7 and 8.

Complete Example

$$2x + 4x = -18$$

$$\underline{} = -18$$
(1)

Combine like terms.

$$\dfrac{6x}{6} = \dfrac{-18}{\underline{}}$$
(2)

Divide each member by 6.

$$x = \underline{}$$
(3)

(1) $6x$ (2) 6 (3) -3

255

FIRST-DEGREE EQUATIONS AND INEQUALITIES

41. $5x = 15$ **42.** $6y = 24$ **43.** $12 = 3y$ **44.** $35 = 7x$
45. $-12 = 3y$ **46.** $-25 = 5z$ **47.** $-x = 6$ **48.** $-y = 3$
49. $8z - 2z = 12$ **50.** $7x - 2x = 20$ **51.** $2y + 4y = -6$ **52.** $2z + z = -18$

Complete Example (see Example 9)

$$3y - 6 = 6y + 3$$
$$3y - 6 - 3y - 3 = 6y + 3 - \underline{\quad\text{(1)}\quad} - 3$$

Add $-3y$ and -3 to each member.

$$-9 = \underline{\quad\text{(2)}\quad}$$

Combine like terms.

$$\frac{-9}{3} = \frac{3y}{\underline{\quad\text{(3)}\quad}}$$

Divide each member by 3.

$$y = \underline{\quad\text{(4)}\quad}$$

Simplify and exchange members.

(1) $3y$ (2) $3y$ (3) 3 (4) -3

53. $5 = 6p - 13$ **54.** $8 = 3y + 2$ **55.** $-5 = -2 - 3x$ **56.** $-20 = 1 - 7t$
57. $-6t = 3t$ **58.** $7r = 5r$ **59.** $7x = 14 + 5x$ **60.** $5y + 3 = 13 - 5y$
61. $3d + 2 - 4d = 6$ **62.** $3x - 4 = 4x + 2$ **63.** $3 = 6x - 3 - 3x$ **64.** $30 = 6r - 24 + 3r$

Solve each equation. See Examples 10 and 11.

Complete Example

$$\frac{2x}{3} = 6$$

Multiply each member by 3.

$$\overset{1}{\cancel{3}}\left(\frac{2x}{\cancel{3}}\right) = \underline{\quad\text{(1)}\quad} (6)$$

$$2x = \underline{\quad\text{(2)}\quad}$$

Divide each member by 2.

$$\frac{2x}{2} = \frac{18}{2}$$

$$x = \underline{\quad\text{(3)}\quad}$$

(1) 3 (2) 18 (3) 9

65. $\dfrac{x}{2} = 10$ 66. $\dfrac{x}{3} = 6$ 67. $\dfrac{z}{3} = -4$ 68. $\dfrac{z}{7} = -4$

69. $\dfrac{2a}{5} = 8$ 70. $\dfrac{4a}{5} = -12$ 71. $8 = \dfrac{2b}{3}$ 72. $4 = \dfrac{2b}{5}$

73. $\dfrac{2c}{4} = -10$ 74. $\dfrac{4c}{2} = -8$ 75. $-8 = \dfrac{4a}{5}$ 76. $-10 = \dfrac{5a}{3}$

77. $\dfrac{-b}{3} = 12$ 78. $\dfrac{-b}{5} = 6$ 79. $\dfrac{-2c}{3} = -10$ 80. $\dfrac{-3c}{4} = -12$

Complete Example (see Example 12)

$\dfrac{5x - 2x}{4} = -6$

Simplify above the fraction bar. Multiply each member by 4.

$(\cancel{4})\dfrac{3x}{\cancel{4}} = \underline{}(-6)$

$3x = \underline{}$

Divide each member by 3.

$\dfrac{\cancel{3}x}{\cancel{3}} = \dfrac{-24}{3}$

$x = \underline{}$

(1) 4   (2) −24   (3) −8

81. $\dfrac{4t - t}{6} = 5$ 82. $\dfrac{7t - t}{18} = -2$ 83. $\dfrac{5x + 2x}{4} = 7$ 84. $\dfrac{8x - 3x}{4} = -10$

85. $\dfrac{4y + 6y}{20} = \dfrac{7 + 2}{3}$ 86. $\dfrac{y - 3y}{3} = \dfrac{2 - 18}{4}$ 87. $\dfrac{3^2 + 4^2}{5} = \dfrac{3x + 5x}{24}$ 88. $\dfrac{7^2 - 5^2}{-6} = \dfrac{2y + 4y}{9}$

8.2 FURTHER SOLUTIONS OF EQUATIONS

Now we know all the techniques needed to solve most first-degree equations. There is no specific order in which the properties should be applied. Any one or more of the following steps listed on page 258 may be appropriate.

FIRST-DEGREE EQUATIONS AND INEQUALITIES

> **Steps to solve first-degree equations:**
> 1. Combine like terms in each member of an equation.
> 2. Using the addition or subtraction property, write the equation with all terms containing the unknown in one member and all terms not containing the unknown in the other.
> 3. Combine like terms in each member.
> 4. Use the multiplication property to remove fractions.
> 5. Use the division property to obtain a coefficient of 1 for the variable.

Example 1 Solve $5x - 7 = 2x - 4x + 14$.

Solution First, we combine like terms, $2x - 4x$, to yield

$$5x - 7 = -2x + 14$$

Next, we add $+2x$ and $+7$ to each member and combine like terms to get

$$5x - 7 + 2x + 7 = -2x + 14 + 2x + 7$$

$$7x = 21$$

Finally, we divide each member by 7 to obtain

$$\frac{\cancel{7}x}{\cancel{7}} = \frac{\cancel{21}}{\cancel{7}} \quad \text{or} \quad x = 3$$

In the next example, we simplify above the fraction bar before applying the properties that we have been studying.

Example 2 Solve $\dfrac{4x - 2x}{3} + 3 = 5$.

Solution First, we combine like terms, $4x - 2x$, to get

$$\frac{2x}{3} + 3 = 5$$

Then we add -3 to each member and simplify

$$\frac{2x}{3} + 3 - 3 = 5 - 3$$

$$\frac{2x}{3} = 2$$

Next, we multiply each member by 3 to obtain

$$\cancel{3}\left(\frac{2x}{\cancel{3}}\right) = 3(2) \quad \text{or} \quad 2x = 6$$

Finally, we divide each member by 2 to get

$$\frac{\cancel{2}x}{\cancel{2}} = \frac{\cancel{6}}{\cancel{2}} \quad \text{or} \quad x = 3$$

8.2 FURTHER SOLUTIONS OF EQUATIONS

EXERCISES 8.2
Solve.

Complete Example (see Example 1)

$$6x + 3 = 6x - 2x - 7$$
$$6x + 3 = \underline{} - 7 \qquad \text{Combine like terms, } 6x - 2x.$$
$$\hspace{2em}{}_{(1)}$$
$$6x + 3 - 4x - 3 = 4x - 7 - 4x - 3 \qquad \text{Add } -4x \text{ and } -3 \text{ to each member and combine like terms.}$$
$$\underline{} = \underline{}$$
$$\hspace{1em}{}_{(2)} \hspace{2em} {}_{(3)}$$

$$\frac{\overset{1}{\cancel{2}}x}{\underset{1}{\cancel{2}}} = \frac{-10}{2} \qquad \text{Divide each member by 2.}$$

$$x = \underline{}$$
$$\hspace{2em}{}_{(4)}$$

| (1) $4x$ | (2) $2x$ | (3) -10 | (4) -5 |

1. $7 - 2c = c + 1$
2. $c + 2 = 6 - 3c$
3. $6 - x = 6 + 2x$

4. $8 + x = 8 - 5x$
5. $4x - 3 = 2x + 5$
6. $6x - 5 = 2x + 7$

7. $3x - 14 = 5x - 4x + 2$
8. $2x - 3 + 2x = 4 - x + 8$
9. $2y - 3 + 3y = 4y + 2$

10. $6z + 2 - 7z = 10 - 2z$
11. $6a - 4 + 2 = 3a + 1$
12. $5y + 1 - y = 10 + y$

13. $3x + 4 - 5x + 2 = 0$
14. $5x + 7 - 2x - 16 = 0$
15. $0 = 7 - 2x + 3 - 3x$

16. $0 = 3x + 5 - 7x + 3$
17. $-x(2 + 5) = 2x + 16 - x$
18. $3y(7 - 2) + 17 = 16y + y - 1$

259

FIRST-DEGREE EQUATIONS AND INEQUALITIES

Complete Example (see Example 2)

$\dfrac{6y - 3y}{4} + 2 = 8$ Combine like terms, $6y - 3y$.

$\dfrac{3y}{4} + 2 = 8$ Add -2 to each member; simplify.

$\dfrac{3y}{4} + 2 - 2 = 8 - \underline{\quad(1)\quad}$

$\dfrac{3y}{4} = \underline{\quad(2)\quad}$

Multiply each member by 4; simplify.

$(\cancel{4}) \dfrac{3y}{\cancel{4}} = 4(6)$

$3y = \underline{\quad(3)\quad}$

Divide each member by 3; simplify.

$\dfrac{\cancel{3}y}{\cancel{3}} = \dfrac{24}{3}$ or $y = \underline{\quad(4)\quad}$

(1) 2   (2) 6   (3) 24   (4) 8

19. $4 = 1 - \dfrac{3x}{7}$

20. $-4 = 2 + \dfrac{3y}{5}$

21. $0 = \dfrac{5x}{2} + 10$

22. $0 = 6 - \dfrac{2y}{3}$

23. $7 = -5 + \dfrac{2y}{3}$

24. $\dfrac{3x}{2} - 4 = 2$

25. $\dfrac{4x - 2x}{2} + 4 = 12 - 5$

26. $\dfrac{x - 3x}{4} + 2 = 11 - 2$

27. $\dfrac{3x + 2x}{2} + 2 = \dfrac{10 + 4}{2}$

28. $\dfrac{5x - 3x}{4} + 2 = \dfrac{8 + 2}{5}$

29. $\dfrac{4x + x}{3} - 3 = 4 + 3$

30. $\dfrac{3z - z}{6} + 3 = 2 + 9$

31. $18 + \dfrac{3(5x - 4x)}{2} = 15$

32. $-3 + \dfrac{7(4x - 3x)}{2} = 25$

33. $14 + \dfrac{3(5x - 3x)}{9} = 4 - 2^3$

34. $\dfrac{2(7x - 5x)}{5} - 6 = 4^2 + 6$

35. $\dfrac{3(8x - 6x)}{5} + 3 = 2^3 + 7$

36. $\dfrac{2(5x - 4x)}{3} - 5 = 4^2 - 3$

37. $\dfrac{5(4x - x)}{3} + x(3^2 - 1) = x - 36$

38. $\dfrac{4(5x + x)}{2} + x(2^3 - 4) = 4x - 24$

39. $\dfrac{8(2t + 5t)}{3^2 - 2} + \dfrac{2(3t + t)}{2^2} = 8(2t + t) + 28$

40. $\dfrac{6(5u - u)}{2^3} - \dfrac{5(u + 5u)}{3} = 3(2u - u) + 30$

8.3 SOLVING FORMULAS

Equations that involve variables for the measures of two or more physical quantities are called **formulas.** We can solve for any one of the variables in a formula if the values of the other variables are known. We substitute the known values in the formula and solve for the unknown variable by the methods we used in the preceding sections.

Example 1 In the formula $d = rt$, find t if $d = 24$ and $r = 3$.

Solution We can solve for t by substituting 24 for d and 3 for r. That is,

$$d = rt$$
$$(24) = (3)t$$
$$8 = t$$

It is often necessary to solve formulas or equations in which there is more than one variable for one of the variables in terms of the others. We use the same methods demonstrated in the preceding sections.

Example 2 In the formula $d = rt$, solve for t in terms of r and d.

Solution We may solve for t in terms of r and d by dividing both members by r to yield

$$\frac{d}{r} = \frac{\cancel{r}t}{\cancel{r}}$$

$$\frac{d}{r} = t$$

from which, by the symmetric law,

$$t = \frac{d}{r}$$

In the above example, we solved for t by applying the division property to generate an equivalent equation. Sometimes, it is necessary to apply more than one such property.

Example 3 In the equation $ax + b = c$, solve for x in terms of a, b, and c.

Solution We can solve for x by first adding $-b$ to each member to get

$$ax + b - b = c - b$$
$$ax = c - b$$

Then dividing each member by a, we have

$$\frac{\cancel{a}x}{\cancel{a}} = \frac{c - b}{a}$$

$$x = \frac{c - b}{a}$$

FIRST-DEGREE EQUATIONS AND INEQUALITIES

EXERCISES 8.3

Evaluate each formula for the specified symbol. See Example 1.

Complete Example

In the formula $f = ma$, find m if $f = -64$ and $a = -32$.

$(\underline{}_{(1)}) = m(\underline{}_{(2)})$ Substitute for f and a.

Divide each member by -32.

$$\frac{\overset{2}{\cancel{-64}}}{\underset{1}{\cancel{-32}}} = \frac{m(\overset{1}{\cancel{-32}})}{\underset{1}{\cancel{-32}}}$$

$\underline{}_{(3)} = m$ or $m = 2$

---

(1) -64 (2) -32 (3) 2

---

1. $f = ma$ Find a if $m = 3$ and $f = -27$.

2. $f = ma$ Find m if $a = -32$ and $f = -96$.

3. $d = rt$ Find r if $d = 80$ and $t = 16$.

4. $d = rt$ Find t if $r = 60$ and $d = 240$.

5. $v = lwh$ Find h if $v = 60$, $l = 15$, and $w = 2$.

6. $v = lwh$ Find w if $v = 45$, $l = 5$, and $h = 3$.

7. $I = prt$ Find t if $I = 100$, $p = 25$, and $r = 2$.

8. $I = prt$ Find t if $I = 70$, $p = 7$, and $r = 2$.

9. $s = \dfrac{at^2}{2}$ Find a if $s = -144$ and $t = 3$.

10. $s = \dfrac{at^2}{2}$ Find a if $s = -64$ and $t = 2$.

11. $v = k + gt$ Find g if $v = 32$, $k = 20$, and $t = 4$.

12. $v = k + gt$ Find t if $v = 35$, $k = 15$, and $g = 4$.

13. $F = \dfrac{kmM}{d^2}$ Find M if $F = 10$, $k = 5$, $m = 4$, and $d = 2$.

14. $F = \dfrac{kmM}{d^2}$ Find k if $F = 8$, $m = 2$, $M = 12$, and $d = 3$.

8.3 SOLVING FORMULAS

Solve each of the following formulas for the symbol in color. See Example 2.

Complete Examples

a. $c = 2\pi r$

$2\pi r = c$

$$\frac{\overset{1}{\cancel{2}}\overset{1}{\cancel{\pi}} r}{\underset{1}{\cancel{2}}\underset{1}{\cancel{\pi}}} = \frac{c}{\underline{\quad(1)\quad}}$$

$r = \underline{\quad(2)\quad}$

b. $P = 4s$

$4s = P$

$$\frac{\overset{1}{\cancel{4}} s}{\underset{1}{\cancel{4}}} = \frac{P}{\underline{\quad(3)\quad}}$$

$s = \underline{\quad(4)\quad}$

Exchange members.

Divide each member by the appropriate expression to obtain the symbol in color by itself.

(1) 2π (2) $\dfrac{c}{2\pi}$ (3) 4 (4) $\dfrac{P}{4}$

15. $d = rt$
16. $v = k + gt$
17. $v = lwh$
18. $f = ma$
19. $c = \pi d$
20. $I = prt$
21. $d = rt$
22. $v = lwh$
23. $v = lwh$
24. $f = ma$
25. $I = prt$
26. $I = prt$
27. $v = k + gt$
28. $s = \dfrac{at^2}{2}$
29. $F = \dfrac{kmM}{d^2}$
30. $F = \dfrac{kmM}{d^2}$

Solve for x. See Example 3.

Complete Example

$5ax - 2c = 2ax$

$5ax - 2c - 2ax + 2c = 2ax - 2ax + \underline{\quad(1)\quad}$

$3ax = \underline{\quad(2)\quad}$

$\dfrac{\overset{1}{\cancel{3}}\overset{1}{\cancel{a}} x}{\underset{1}{\cancel{3}}\underset{1}{\cancel{a}}} = \dfrac{2c}{3a}$ or $x = \underline{\quad(3)\quad}$

Add $-2ax$ and $+2c$ to each member.

Combine like terms.

Divide each member by $3a$.

(1) $2c$ (2) $2c$ (3) $\dfrac{2c}{3a}$

31. $x - a = 0$
32. $x - 3a = 0$
33. $3x - 3a = x - a$
34. $2x + 3a = 9a - x$
35. $5a - 2x = 2a - x$
36. $5a + x = 2x - a$

263

FIRST-DEGREE EQUATIONS AND INEQUALITIES

Solve for y.

37. $ay - b = 0$
38. $2ay + 2b = 4b + ay$
39. $3a - 3by = -9a + by$
40. $2aby + 6a = aby$
41. $5aby - 3b = 7b - aby$
42. $3ay - 4ab + ay = 0$

Solve for x or y.

43. $cx - a^2 = 0$
44. $dy - a^2 - 3dy = 0$
45. $bx + 2b = 5b - bx + b$
46. $ax - 2ar^2 = ar^2$
47. $\frac{a}{b}x - c = 0$
48. $0 = \frac{b}{c}y + a$

8.4 APPLICATIONS

In Chapters 1 and 2 we solved a variety of simple word problems by first translating word sentences into equations. Because this is not always an easy thing to do as word problems become more difficult, we suggested a series of steps to follow that we repeat here. Furthermore, to highlight the importance of writing word phrases for the quantity or quantities you want to find in order to "get started" on a problem, we have renumbered steps 1a. and 1b. in the four steps shown in Section 1.6 as steps 1 and 2. With algebraic word problems, a sketch or table is often very helpful when analyzing the information and we have added this idea as step 3.

> To solve a word problem:
>
> 1. Represent each quantity you want to find as a word phrase.
> 2. Represent each quantity from step 1 in terms of a single variable.
> 3. When applicable, make a sketch or a table and indicate all quantities.
> 4. Write an equation that represents a word sentence relating the known and unknown quantities.
> 5. Solve the resulting equation.
> 6. Using the result from step 5, answer the original question.

In practice, we usually combine steps 1 and 2.

Example 1 Follow the six steps listed above to solve the following word problem.

> The sum of a certain number and 9 is equal to four times the number. What is the number?

Solution **Steps 1–2** We first write what we want to find (a number) as a word phrase. Then, we use a variable to represent this number.

The number: x

8.4 APPLICATIONS

Step 3   A sketch is not applicable in this problem.

Step 4   We write an equation relating the known and unknown quantities.
$$x + 9 = 4x$$

Step 5   We now solve the resulting equation.
$$x + 9 - x = 4x - x$$
$$9 = 3x$$
$$3 = x$$

Step 6   The number is 3.

CONSECUTIVE INTEGERS

Some algebraic word problems involve the concepts of consecutive integers.

Consecutive integers are integers that differ by 1 or -1. For example, 3 and 4 are consecutive integers and -2 and -1 are consecutive integers. If x represents an integer, then the next consecutive integer is represented by $x + 1$.

Consecutive even integers are even integers that differ by 2 or -2. For example, 6 and 8 are consecutive even integers and -4 and -2 are consecutive even integers. If x represents an even integer, then the next consecutive even integer is represented by $x + 2$.

Consecutive odd integers are odd integers that differ by 2 or -2. For example, 5 and 7 are consecutive odd integers and -3 and -1 are consecutive odd integers. If x represents an odd integer, then the next consecutive odd integer is represented by $x + 2$.

Example 2   Follow the six steps listed on page 264 to solve the following word problem.

The sum of two consecutive odd integers is 12. What are the integers?

Solution   *Steps 1–2*   We first write what we want to find (the integers) as two word phrases. Then, we use a variable to represent the integers.

The smaller odd integer: x

The next consecutive odd integer: $x + 2$

Step 3   A sketch is not applicable in this problem.

Step 4   Next, we write an equation relating known and unknown quantities.
$$x + (x + 2) = 12$$

Step 5   We now solve the resulting equation.
$$x + x + 2 = 12$$
$$2x + 2 = 12$$
$$2x = 10$$
$$x = 5$$

Step 6   The integers are 5 and 5 + 2, or 7.

FIRST-DEGREE EQUATIONS AND INEQUALITIES

In the following example, a sketch is helpful.

Example 3 Follow the six steps listed on page 264 to solve the following word problem.

A board 186 centimeters long is cut into two pieces so that one piece is twice as long as the other. How long is each of the two pieces?

Solution   *Steps 1–2*   We first write what we want to find (the lengths) as word phrases. Then, we use a variable to represent the lengths.

The smaller piece: x

The larger piece: $2x$

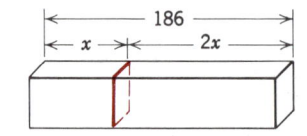

Step 3   We now make a sketch and label dimensions.

Step 4   Writing an equation relating the known and the unknown quantities yields

$$x + 2x = 186$$

Step 5   We now solve the equation.

$$3x = 186$$
$$x = 62$$

Step 6   The smaller piece is 62 centimeters and the larger piece is 2(62) or 124 centimeters.

Common Error

When translating a phrase such as "*a* subtracted from *b*," the order is important. We must write

$$b - a, \quad \text{not} \quad a - b$$

since *a* is *subtracted from b*. For example, to translate

"If seven is subtracted from 4 times an integer, the result is 17,"

we represent the integer by a symbol, say *x*, and then write the sentence as follows.

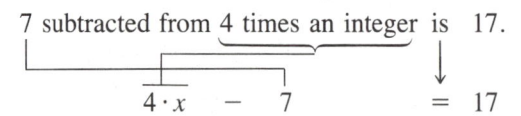

7 subtracted from 4 times an integer is 17.

$$4 \cdot x - 7 = 17$$

EXERCISES 8.4

For Exercises 1–42, follow the six steps outlined on page 264 and solve the word problems. Omit step 3 if a sketch is not applicable.

Complete Example (see Example 1)

The sum of a certain number and 12 is equal to three times the number. What is the number?

8.4 APPLICATIONS

Steps 1–2   Represent the quantity you want to find as a word phrase.
Represent the quantity in terms of a variable.

$$\text{The number: } x$$

Step 3   A sketch is not applicable.

Step 4   Write an equation relating known and unknown quantities.

$$x + \underline{\quad\quad}_{(1)} = \underline{\quad\quad}_{(2)}$$

Step 5   Solve the equation.

$$12 = \underline{\quad\quad}_{(3)} \;\; ; \;\; \underline{\quad\quad}_{(4)} = x$$

Step 6   The required number is 6.

---

(1) 12   (2) 3x   (3) 2x   (4) 6

---

1. If 5 if added to twice a certain number, the result is 19. What is the number?

2. If 4 is added to three times a number, the result is 25. What is the number?

3. Two times an integer added to four times the integer gives -30. Find the integer.

4. Five times an integer added to three times the integer gives -48. Find the integer.

5. Twice a number subtracted from five times the number gives 24. Find the number.

6. Three times a number subtracted from six times the number gives 45. Find the number.

7. A certain integer increased by eight is equal to twice the integer. Find the integer.

8. A certain integer increased by fifteen is equal to four times the integer. Find the integer.

9. What number subtracted from four times itself gives 12?

10. What number subtracted from three times itself gives 14?

Complete Example (see Example 2)

The sum of two consecutive even integers is 46. What are the integers?

Steps 1–2   Represent the two quantities you want to find as two word phrases. Represent the quantities in terms of a variable.

$$\text{The smaller even integer: } x$$

$$\text{The next consecutive even integer: } x + 2$$

Step 3   A sketch is not applicable.

Step 4   Write an equation relating known and unknown quantities.

$$x + (\underline{\quad\quad}_{(1)}) = \underline{\quad\quad}_{(2)}$$

267

FIRST-DEGREE EQUATIONS AND INEQUALITIES

Step 5 Solve the equation.

$$\underline{}_{(3)} + 2 = 46$$

$$2x = 44; \qquad x = \underline{}_{(4)}$$

Step 6 The integers are 22 and 22 + 2, or 24.

---

(1) $x + 2$ (2) 46   (3) $2x$ (4) 22

---

11. The sum of two consecutive even integers is 26. Find the integers.

12. The sum of two consecutive even integers is 86. Find the integers.

13. The sum of two consecutive odd integers is 32. Find the integers.

14. The sum of two consecutive odd integers is 28. Find the integers.

15. Find three consecutive integers whose sum is -33.

16. The sum of three consecutive integers is 57. Find the integers.

17. Find three consecutive odd integers whose sum is -21.

18. Find three consecutive even integers whose sum is 84.

19. The sum of three consecutive even integers equals four times the smallest integer. What are the integers?

20. The sum of three consecutive odd integers equals 1 less than four times the smallest integer. What are the integers?

Complete Example (see Example 3)

A board 36 centimeters is cut into two pieces so that one piece is five times as long as the other. How long is each of the two pieces?

Steps 1–2 Represent the two quantities you want to find as two word phrases. Represent the quantities in terms of a variable.

The smaller piece: x

The larger piece: $\underline{}_{(1)}$

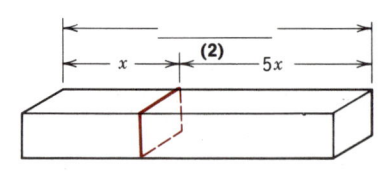

Step 3 Draw a sketch and label quantities.

Step 4 Write an equation relating known and unknown quantities.

$$x + 5x = \underline{}_{(3)}$$

Step 5 Solve the equation.

$$\underline{}_{(4)} = 36$$

$$x = \underline{}_{(5)}$$

268

8.4 APPLICATIONS

Step 6   The smaller piece is 6 centimeters and the larger piece is 5(6) or 30 centimeters.

---

(1) $5x$ (2)   36   (3)   36   (4) $6x$ (5)   6

---

21. A board 144 centimeters long is cut into two pieces so that one piece is 24 centimeters longer than the other piece. How long is each of the two pieces?

22. A board 112 centimeters long is cut into two pieces so that one piece is three times as long as the other. How long are the two pieces?

23. A board 76 feet long is cut into two pieces so that one piece is 12 feet longer than the other piece. How long is each of the two pieces?

24. A board 39 centimeters long is cut into two pieces so that one piece is 9 centimeters longer than the other piece. How long is each of the two pieces?

25. A board 24 feet long is cut into three pieces of which the second is three times as long as the first, and the third is 4 feet longer than the first. How long are the three pieces?

26. A board 12 feet long is cut into three pieces so that the second is three times as long as the first, and the third is 2 feet longer than the first. How long are the three pieces?

27. A board 51 feet long is cut into three pieces so that the second piece is 3 feet longer than the first, and the third is twice as long as the first. How long are the three pieces?

28. A board 63 meters long is cut into three pieces so that the longest piece is three times the length of the shortest piece, and the other piece is 3 meters longer than the shortest piece. How long are the three pieces?

Complete Example

There were 2480 votes cast in an election. The winning candidate received 142 votes more than the losing candidate. How many votes did each candidate receive?

Steps 1–2   Represent the two quantities you want to find as two word phrases. Represent the quantities in terms of a variable.

$$\text{The number of votes of the winner:} \quad x$$

$$\text{The number of votes of the loser:} \quad \underline{}_{(1)}$$

Step 3   A sketch is not applicable.

Step 4   Write an equation relating the known and unknown quantities.

$$x + (x - 142) = \underline{}_{(2)}$$

Step 5   Solve the equation.

$$x + x - 142 = 2480$$

$$\underline{}_{(3)} = 2622$$

$$x = \underline{}_{(4)}$$

Step 6   The winner received 1311 votes and the loser received $1311 - 142$, or 1169 votes.

---

(1) $x - 142$ (2)   2480   (3) $2x$ (4)   1311

269

FIRST-DEGREE EQUATIONS AND INEQUALITIES

29. At a recent election, the winning candidate received 50 votes more than his opponent. If there were 4376 votes cast in all, how many votes did each candidate receive?

30. There were 12,822 votes cast in a recent election. The winning candidate received 132 votes more than his opponent. How many votes did each candidate receive?

31. Two keypunch operators compared their output over a period of time and found that one had punched 24 more cards than the other. But after combining their cards, neither could remember how many he had himself punched. If together they had produced 212 cards, how many cards had each operator punched?

32. An oil well begins to pump 70 barrels a day into an empty tank. After three days the tank begins to leak. After a total of seven days of pumping, the tank contains 410 barrels of oil. How many barrels per day were leaking from the tank?

33. A person weighing 78 kilograms steps on a scale while carrying a briefcase. If the scale reads 88 kilograms, what is the weight of the briefcase?

34. A car 72 meters from a fallen tree skidded to a stop 5 meters from the tree. How far did the car skid?

35. How many minutes can be devoted to the entertainment portion of a half-hour television show if 9 minutes are taken up by commercials?

36. If the flying time from Los Angeles to San Francisco is 47 minutes, how many minutes from San Francisco is a plane that is 19 minutes from Los Angeles?

37. A 340-centimeter board is cut into four pieces of equal length and a 24-centimeter piece remains. How long is each of the four pieces?

38. If a 36-ounce solution fills four glass containers of the same size with 4 ounces of the solution left over, how many ounces will each container hold?

39. A man drove from town A to town B and then returned to town A. Leaving town A again to return to town B, the man found that after 5 kilometers on the road, he had traveled a total of 19 kilometers since he first left town A. How far is it from town A to town B?

40. Two ships leave port at the same time, traveling in the same direction. In one hour, one ship sales three times as far as the other. If the ships are then 14 kilometers apart, how far has the slower ship sailed?

41. Two cars leave town A at the same time, traveling in the same direction. In one hour, one car travels four times as far as the other. If the cars are then 54 miles apart, how far has the slower car traveled?

42. A woman drove from town A to town B. Leaving town B to return to town A, the woman found that after 13 kilometers on the road, she had traveled a total of 37 kilometers since she first left town A. How far is town A from town B?

8.5 INEQUALITIES

ORDER RELATIONSHIPS

In Sections 6.1 we saw that of two different numbers, the graph of the lesser number lies to the left of the graph of the greater number on a number line. These order relationships can be expressed by using the following symbols:

$<$ means "is less than,"

$>$ means "is greater than."

$\leq$ means "is less than or equal to,"

$\geq$ means "is greater than or equal to."

For example,

"1 is less than 3" can be written as $1 < 3$.

"7 is less than 9" can be written as $7 < 9$.

"-3 is greater than -5" can be written as $-3 > -5$.

"2 is less than or equal to x" can be written as $2 \leq x$.

"4 is greater than or equal to y" can be written as $4 \geq y$.

Statements that involve any of the above symbols are called **inequalities.** Inequalities such as

$$1 < 3 \quad \text{and} \quad 7 < 9$$

are said to be of the *same order* or *opposite sense* because the left-hand member is less than the right-hand member in each case. Inequalities such as

$$7 < 9 \quad \text{and} \quad -3 > -5$$

are said to be of *opposite order* or *opposite sense* because in one case the left-hand member is less than the right-hand member and in the other case the left-hand member is greater than the right-hand member.

PROPERTIES OF INEQUALITIES

In Section 8.1, we saw that a first-degree equation in one variable has only one solution. But a first-degree inequality has an infinite number of solutions. For example, the graphs of the infinite number of integer solutions of the inequality $x > 3$ are shown in Figure 8.1.

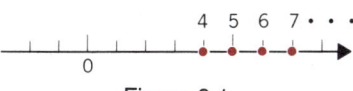

Figure 8.1

Sometimes it is not possible to determine the solutions of a given inequality simply by inspection. But using the following properties, we can form equivalent inequalities (inequalities with the same solutions) in which the solution is evident by inspection.

> 1. **If the same expression is added to or subtracted from each member of an inequality, the result is an equivalent inequality in the *same* order.**

In symbols,

$$a < b, \quad a + c < b + c, \quad \text{and} \quad a - c < b - c$$

are equivalent inequalities.

FIRST-DEGREE EQUATIONS AND INEQUALITIES

Example 1 a. Because $3 < 5$, then $3 + 4 < 5 + 4$ or $7 < 9$

b. If $x < 7$, then $x + 2 < 7 + 2$

c. If $4 < y$, then $4 + (-2) < y + (-2)$

> **2. If each member of an inequality is multiplied or divided by the same positive number, the result is an equivalent inequality in the *same* order.**

In symbols,

$$\text{If } c > 0, \text{ then } a < b, \quad ac < bc, \quad \text{and} \quad \frac{a}{c} < \frac{b}{c}$$

are equivalent inequalities.

Example 2 a. Because $2 < 3$ and $5 > 0$,

$$2(5) < 3(5) \quad \text{or} \quad 10 < 15$$

b. If $3x < 12$, then

$$\frac{3x}{3} < \frac{12}{3} \quad \text{or} \quad x < 4$$

c. If $5 < y$ and $z > 0$, then

$$5(z) < y(z) \quad \text{or} \quad 5z < yz$$

> **3. If each member of an inequality is multiplied or divided by the same negative number, the result is an equivalent inequality in the *opposite* order.**

In symbols,

$$\text{If } c < 0, \text{ then } a < b, \quad ac > bc, \quad \text{and} \quad \frac{a}{c} > \frac{b}{c}$$

are equivalent inequalities.

Example 3 a. Because $3 < 5$ and $-2 < 0$,

$$3(-2) > 5(-2) \quad \text{or} \quad -6 > -10$$

b. If $-3x < 12$, then

$$\frac{-3x}{-3} > \frac{12}{-3} \quad \text{or} \quad x > -4$$

c. If $2 < x$ and $y < 0$, then

$$2(y) > x(y) \quad \text{or} \quad 2y > xy$$

The three properties above also apply to inequalities of the form $a > b$, as well as $a < b$.

SOLVING INEQUALITIES

Now let us see how the three properties can help us solve inequalities.

Example 4 Solve $\frac{3x}{2} < 3$, where x is an integer.

Solution Multiplying each member by 2 (a positive number), we have

$$\overset{1}{\cancel{2}}\left(\frac{3x}{\cancel{2}}\right) < 2(3)$$

$$3x < 6$$

Then dividing each member by 3, we get

$$\frac{\cancel{3}x}{\cancel{3}} < \frac{6}{3}$$

$$x < 2$$

The graph of this inequality is

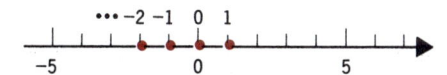

In the above example, all the inequalities were in the same order because we only applied Property 2 above. Now consider the following inequality.

Example 5 Solve $-3x + 1 > 7$, where x is an integer.

Solution Adding -1 to each member, we get

$$-3x + 1 + (-1) > 7 + (-1)$$

$$-3x > 6$$

Now we apply Property 3 and divide each member by -3. In this case we have to *reverse* the order of the inequality.

$$\frac{-\cancel{3}x}{-\cancel{3}} < \frac{\cancel{6}x}{-\cancel{3}}$$

$$x < -2$$

The graph of the inequality is

When solving word problems involving inequalities, we follow the six steps outlined on page 264 except the word *equation* will be replaced by the word *inequality*.

FIRST-DEGREE EQUATIONS AND INEQUALITIES

EXERCISES 8.5
Complete each statement. See Examples 1–3.

Complete Examples
a. If $y < 7$, then $y - 2 \underset{(1)}{\_\_\_} 7 - 2$.
b. If $3 < x$, then $3 + 5 \underset{(2)}{\_\_\_} x + 5$.
c. If $x < y$, then $5x \underset{(3)}{\_\_\_} 5y$.
d. If $x < y$, then $-5x \underset{(4)}{\_\_\_} -5y$.

(1) <   (2) <   (3) <   (4) >

1. If $x < 5$, then $x + 3$? $5 + 3$.
2. If $y > 4$, then $y + (-2)$? $4 + (-2)$.
3. If $x > y$, then $7x$? $7y$.
4. If $x < y$, then $-3x$? $-3y$.
5. If $\frac{x+1}{2} < 5$, then $2\left(\frac{x+1}{2}\right)$? $2(5)$.
6. If $-5x < 30$, then $\frac{-5x}{-5}$? $\frac{30}{-5}$.

Solve each inequality. see Examples 4 and 5.

Complete Examples
a. $x - 6 > 2$
    $x - 6 + 6 > 2 + \underset{(1)}{\_\_\_}$ Add 6 to each member.
    $x > \underset{(2)}{\_\_\_}$ Simplify.

b. $-3y < 21$
    $\frac{-3y}{-3} \underset{(3)}{\_\_\_} \frac{21}{-3}$ Divide each member by -3; reverse inequality.
    $y > \underset{(4)}{\_\_\_}$ Simplify.

(1) 6   (2) 8   (3) >   (4) −7

7. $x + 3 > 7$
8. $x - 5 < 8$
9. $y + 10 < -5$
10. $y - 4 < -3$
11. $x + 2 \geq 3^2 - 4$
12. $x + 5 \geq 4^2 - 3^2$
13. $x - 4^2 > 3(2 + 1)$
14. $x + 3^2 < 3(4 - 2)$
15. $-3y < 15$
16. $-5y > 20$
17. $\frac{x}{3} \leq 4$
18. $\frac{x}{5} \geq 2$

8.5 INEQUALITIES

19. $4y > 12$ **20.** $6y < 24$ **21.** $\dfrac{x}{-3} > 7$

22. $\dfrac{x}{-4} > 8$

Solve each inequality and graph the solutions. All variables are integers. See Examples 4 and 5.

Complete Example

$\dfrac{2x - 6x}{8} > -2$ Combine like terms, $2x - 6x$.

$\dfrac{-4x}{8} > \underline{}_{(1)}$

Multiply each member by 8.

$\overset{1}{\cancel{8}}\left(\dfrac{-4x}{\underset{1}{\cancel{8}}}\right) > 8(-2)$

$-4x > \underline{}_{(2)}$

Divide each member by -4 and reverse the inequality.

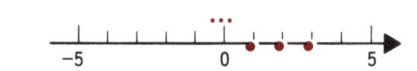

$\dfrac{\overset{1}{-\cancel{4}x}}{\underset{1}{-\cancel{4}}} < \dfrac{\overset{4}{-\cancel{16}}}{\underset{1}{-\cancel{4}}}$

$x < \underline{}_{(3)}$;

(1) -2 (2) -16 (3) 4

23. $2x + 3 > 7$ **24.** $3x - 4 < 11$ **25.** $-3x + 2 < 11$

26. $-4x - 3 > -11$ **27.** $5x \leq 4x - 6$ **28.** $4x \geq 3x + 5$

29. $2 + 3x < 4x - 1$ **30.** $3x + 5 > 2x + 3$ **31.** $0 < 4y + 5 - 3y$

32. $0 > 8y + 6 - 7y$ **33.** $-6 > \dfrac{3y}{5}$ **34.** $12 < \dfrac{2y}{3}$

35. $\dfrac{2x}{3} + 1 < 3$ **36.** $\dfrac{-3x}{4} - 2 > 4$ **37.** $\dfrac{6x - 4x}{3} < 2$

38. $\dfrac{-3z + z}{6} > 4$ **39.** $\dfrac{3x + 2x}{3} - 3 \geq 7$ **40.** $\dfrac{7x - 9x}{4} + 2 \leq 2$

FIRST-DEGREE EQUATIONS AND INEQUALITIES

In Exercises 41–46, follow the six steps outlined on page 264 and solve the following word problems.

Complete Example

The sum of three consecutive integers is greater than 108. What value could the smallest integer be?

Steps 1–2   Represent the unknown quantities you want to find as word phrases. Represent the quantities in terms of a variable.

$$\text{Smallest integer: } x$$
$$\text{Next consecutive integer: } x + 1$$
$$\text{Next consecutive integer: } \underline{\quad(1)\quad}$$

Step 3   A sketch is not applicable.

Step 4   Write an inequality relating known and unknown quantities.

$$x + (x + 1) + (x + 2) > \underline{\quad(2)\quad}$$

Step 5   Solve the inequality.

$$\underline{\quad(3)\quad} + 3 > 108$$
$$3x > 105$$
$$x > \underline{\quad(4)\quad}$$

Step 6   The smallest integer must be greater than 35.

---

(1) $x + 2$ (2) 108   (3) $3x$ (4) 35

---

41. The sum of three consecutive integers is greater than 93. What value could the smallest integer be?

42. The sum of three consecutive integers is less than 126. What value could the smallest integer be?

43. Four times an integer is at least 10 more than three times the integer. What can the integer be?

44. Six times an integer is at least 3 more than three times the integer. What can the integer be?

45. A 40-foot board is cut into two pieces so that one piece is three times the length of the other piece. The carpenter wants at least 4 feet of the board to remain. How long can the shortest piece be?

46. A 31-foot board is cut into two pieces so that one piece is twice the length of the other. The carpenter wants at least 4 feet of the board to remain. How long can the shortest piece be?

CHAPTER EIGHT REVIEW EXERCISES

1. Write an equation expressing the following:
   a. A number added to 3 equals two less than twice the same number.
   b. A number subtracted from 12 equals twice the same number.
   c. Three times a number divided by 4 equals 21 less 6.

Solve.

2. a. $2 + x = 8$ b. $5y = 2 + 4y$ c. $\dfrac{2a + 4a}{3} = 5a + 3$

3. a. $4x + 3x = 35$ b. $4x - 4 = 2x - 4$ c. $8z + 6z = 2z - 12$

4. a. $\dfrac{2a}{3} = -12$ b. $\dfrac{b + 4b}{3} = 15$ c. $\dfrac{6x - 2x}{3} = -4$

5. a. $\dfrac{-9a - a}{2} = 10$ b. $\dfrac{3x + 5x}{2} = 6 + 3x$ c. $\dfrac{8x - 4x}{2} = \dfrac{8 + 10}{3}$

6. a. $-5y + 1 < 26$ b. $3x + 2 > x - 10$ c. $\dfrac{4x - 5x}{4} \geq 2$

7. Show by direct substitution that the solutions you obtained in Exercises 2a, 2b, and 2c are correct.

8. Solve each of the following formulas for the symbol in color.
   a. $f = ma$ b. $v = k + gt$ c. $M = \dfrac{a + b}{2}$

9. If an odd integer is represented by x, how may the next consecutive odd integer be represented in terms of x?

10. If an even integer is represented by x, how may the next consecutive even integer be represented in terms of x?

11. If an integer is represented by x, how may the next four consecutive integers be represented in terms of x?

12. If the width of a rectangular garden is represented by x, how may the length, which is three times this width, be represented in terms of x?

13. If a man's height is represented by x, how may the height of a second man be represented who is 7 centimeters taller? 7 centimeters shorter?

14. If a person's weight is represented by x, how may the weight of a second person be represented who weighs 18 kilograms more? 12 kilograms less?

15. The sum of four consecutive integers is 54. Find the integers.

16. The sum of three consecutive odd integers is five times the smallest integer. Find the integers.

17. A delivery man delivered 431 papers to a location where two girls were to pick them up for house delivery. The man could not remember how many customers each girl had on her route but did remember that one girl had 27 more customers than the other. How should he have divided the papers?

18. A 32-foot board is cut into three pieces, so that one of the pieces is 3 feet longer than a second piece, and the third is 5 feet longer than the second. How long is each piece?

19. A businesswoman is allowed $50 per day to rent a car. If the daily route is $25 plus $.05 per mile, how far can she drive so that she does not exceed $50?

20. A businesswoman is allowed $91 for food expenses. She has three dinners out and each one costs twice as much as the previous meal. How much could the first meal cost if she is to stay within her budget?

9 PRODUCTS AND FACTORS

In Section 7.5 we wrote the products of two or more monomials in simpler forms. We also wrote such products in factored forms. In this chapter we will continue with products and factors; however, now we will extend our work to include polynomials that consist of more than one term.

9.1 THE DISTRIBUTIVE LAW

When we multiply whole numbers or decimals that contain more than one digit (see Section 2.1) we use a procedure that involves adding partial products. For example, consider a simple case of the product 7×14 that can be obtained as follows:

$$\begin{array}{r} 14 \\ \times\ 7 \\ \hline 28 \leftarrow (7 \cdot 4) \\ 70 \leftarrow (7 \cdot 10) \\ \hline 98 \leftarrow (7 \cdot 4) + (7 \cdot 10) \end{array}$$

or

$$\begin{aligned} 7(4 + 10) &= 7 \cdot 4 + 7 \cdot 10 \\ &= 28 + 70 \\ &= 98 \end{aligned}$$

This important property that we used so often in computing products of whole numbers and decimals is also important when multiplying algebraic expressions in which at least one of the factors contains more than one term. This property, called the **distributive property,** can be represented symbolically as

$$a(b + c) = ab + ac \quad \text{or} \quad (b + c)a = ba + ca$$

By applying the distributive law to algebraic expressions that involve parentheses, we can obtain equivalent expressions without parentheses.

PRODUCTS AND FACTORS

Example 1 Write $2x(x - 3)$ without parentheses.

Solution We think of $2x(x - 3)$ as $2x[x + (-3)]$ and then apply the distributive law to obtain

$$2x(x - 3) = 2x(x) + 2x(-3)$$
$$= 2x^2 - 6x$$

The above method works equally as well with the product of a monomial and trinomial.

Example 2 Write $-y(y^2 + 3y - 4)$ without parentheses.

Solution Applying the distributive property yields

$$-y(y^2 + 3y - 4) = (-y)(y^2) + (-y)(3y) + (-y)(-4)$$
$$= -y^3 - 3y^2 + 4y$$

When simplifying expressions involving parentheses, we first remove the parentheses and then combine like terms.

Example 3 Simplify $a(3 - a) - 2(a + a^2)$.

Solution We begin by removing parentheses to obtain

$$a(3 - a) - 2(a + a^2) = 3a - a^2 - 2a - 2a^2$$

Now, combining like terms yields

$$a - 3a^2$$

We can use the distributive property to rewrite expressions in which the coefficient of an expression in parentheses is $+1$ or -1.

Example 4 Write each expression without parentheses.

a. $+(3a - 2b)$ b. $-(2a - 3b)$

Solution a. Since $+(3a - 2b) = +1(3a - 2b)$, we have
$$+1(3a - 2b) = 3a - 2b.$$

b. Since $-(2a - 3b) = -1(2a - 3b)$, we have
$$-1(2a - 3b) = -2a + 3b.$$

Notice that in Example 4b, the sign of each term is changed when the expression is written without parentheses. This is the same result that we would have obtained if we used the procedures that we introduced in Section 7.4 to simplify expressions.

EXERCISES 9.1
Write each expression without parentheses.

Complete Examples (See Example 1)

a. $3x(x-2) = 3x(\underline{}_{(1)}) + 3x(\underline{}_{(2)})$
   $= \underline{}_{(3)} - 6x$

b. $-4a(a-3) = (-4a)(a) + (-4a)(\underline{}_{(4)})$
   $= \underline{}_{(5)} + 12a$

(1) x (2) -2 (3) $3x^2$ (4) -3 (5) $-4a^2$

1. $3(x-4)$
2. $4(x+1)$
3. $5(2y-2)$
4. $2(3y+6)$

5. $-2(x+4)$
6. $-3(x-7)$
7. $2a(5a+3)$
8. $3a(3a-1)$

9. $-b(b-2)$
10. $-3b(2b+1)$
11. $xy(x+y)$
12. $xy(x-2y)$

13. $-x^2(2x+3y)$
14. $-y^2(x-2y^2)$

Complete Examples (See Example 2)

a. $x(2x^2 - x + 1)$
   $= x(2x^2) + x(\underline{}_{(1)}) + x(1)$
   $= \underline{}_{(2)}$

b. $-y(y^2 - 2y + 3)$
   $= -y(\underline{}_{(3)}) + (-y)(\underline{}_{(4)}) + (-y)(3)$
   $= \underline{}_{(5)}$

(1) $-x$ (2) $2x^3 - x^2 + x$ (3) y^2 (4) $-2y$ (5) $-y^3 + 2y^2 - 3y$

15. $x(x^2 - 2x + 1)$
16. $y(y^2 + 3y + 4)$
17. $-y(y^2 - y + 2)$
18. $-x(x^2 + 3x - 2)$

19. $4x^3(x^2 - 3x + 4)$
20. $2y^3(y^2 + y - 2)$
21. $-y^3(y^3 - y + 1)$
22. $-x^3(y^3 + 2y^2 - 1)$

23. $-xy(x^2 + xy + y^2)$
24. $-xy(2x^2 - xy + 3y^2)$

PRODUCTS AND FACTORS

Simplify. See Example 3.

Complete Example

$-4x(x + 1) + x(x - 3)$
$= -4x^2 - \underline{}_{(1)} + \underline{}_{(2)} - 3x$ Remove parentheses.

$= \underline{}_{(3)}$ Combine like terms.

---

(1) $4x$ (2) x^2 (3) $-3x^2 - 7x$

---

25. $-a(x + 1) + ax$ **26.** $by - b(1 - y)$ **27.** $a(x + 1) + x(a + 1)$

28. $2a(x + 3) - 3a(x - 3)$ **29.** $a(x + y) - 2(ax + y)$ **30.** $4x(3 + 2y) - 3(2x + y)$

31. $3(x^2 + 2x - 1) - 2(x^2 + x - 2)$ **32.** $ax(x^2 + 2x - 3) - a(x^3 + 2x^2)$

33. $3(x - 2y) - 2(x + 3y) + 2x$ **34.** $2x(3 - x) + 2(x^2 + 1) - 2$

35. $3y(2y - 5) + 2y^2 - 5(y^2 + 2y)$ **36.** $3(y^2 - 2y + 1) + 3(1 - y^2)$

37. $3(ax^2 + ax - a) - 2a(x^2 - 1)$ **38.** $3x^2(a - b + c) - 2x(ax - bx + cx)$

39. $-3ab(x + y - 2) - 2a(bx - by + 2b) + b(ax + 1)$ **40.** $3ab^2(2 + 3a) - 2ab(3ab + 2b) - 2b^2(a^2 - 2a)$

Write each expression without parentheses and simplify. See Example 4.

Complete Examples

a. $+ (4x - 3y) = \underline{}_{(1)} (4x - 3y)$ b. $-(3x - 4y) = \underline{}_{(3)} (3x - 4y)$

       $= 4x - \underline{}_{(2)}$ $= \underline{}_{(4)} + 4y$

---

(1) $+1$ (2) $3y$ (3) -1 (4) $-3x$

---

41. $-(a + c)$ **42.** $-(2 - x)$ **43.** $+(a - 2b + c)$

44. $+(2a + b - c)$ **45.** $-(3x + 2y - z)$ **46.** $-(2r - s - t)$

47. $-(1 - 3x + x^2)$ **48.** $-(3 + 3x - 2x^2)$ **50.** $(a - 2b) - (a - b)$

51. $-(x - y) - (x + y)$ **52.** $-(x^2 - x) + (x^2 + x)$

9.2 FACTORING MONOMIALS FROM POLYNOMIALS

From the symmetric property of equality, we know that if

$$a(b + c) = ab + ac, \quad \text{then} \quad ab + ac = a(b + c)$$

Thus, if there is a monomial factor common to all terms in a polynomial, we can write the polynomial as the product of the common factor and another polynomial. For instance, since each term in $x^2 + 3x$ contains x as a factor, we can write the expression as the product $x(x + 3)$. Rewriting a polynomial in this way is called **factoring**, and the number x is said to be factored "from" or "out of" the polynomial $x^2 + 3x$.

To factor a monomial from a polynomial:

1. Write a set of parentheses preceded by the monomial common to each term in the polynomial.
2. Divide the monomial factor into each term in the polynomial and write the quotient in the parentheses.

Generally, we can find the common monomial factor by inspection.

Example 1 a. $4x + 4y = 4(x + y)$ b. $3xy - 6y = 3y(x - 2)$

We can check that we factored correctly by multiplying the factors and verifying that the product is the original polynomial. Using Example 1, we get

$$4(x + y) = 4x + 4y \quad \text{and} \quad 3y(x - 2) = 3xy - 6y$$

If the common monomial is hard to find, we can write each term in prime factored form and note the common factors.

Example 2 Factor $4x^3 - 6x^2 + 2x$.

Solution We can write

$$4x^3 - 6x^2 + 2x = 2 \cdot 2xxx - 2 \cdot 3xx + 2x$$

We now see that $2x$ is a common monomial factor to all three terms. Then we factor $2x$ out of the polynomial

$$2x(\qquad)$$

Now, we divide each term in the polynomial by $2x$

$$\frac{4x^3}{2x} = 2x^2; \quad \frac{-6x^2}{2x} = -3x; \quad \frac{2x}{2x} = 1$$

and write the quotients inside the parentheses to get

$$2x(2x^2 - 3x + 1)$$

We can check our answer in Example 2 by multiplying the factors to obtain

$$2x(2x^2 - 3x + 1) = 4x^3 - 6x^2 + 2x$$

PRODUCTS AND FACTORS

In this book, we will restrict the common factors to monomials consisting of numerical coefficients that are integers and to integral powers of the variables. The choice of sign for the monomial factor is a matter of convenience. Thus,

$$-3x^2 - 6x$$

can be factored either as

$$-3x(x + 2) \quad \text{or as} \quad 3x(-x - 2)$$

The first form is usually more convenient.

Example 3 Factor out the common monomial, including -1.

    a. $-3x^2 - 3xy$ b. $-x^3 - x^2 + x$

Solutions a. $-3x^2 - 3xy$ b. $-x^3 - x^2 + x$

            $= -3xx + (-3)xy$ $= -xxx + (-x)x - (-x)$

            $= -3x(x + y)$ $= -x(x^2 + x - 1)$

Sometimes it is convenient to write formulas in factored form.

Example 4 a. $A = P + PRT$ b. $S = 4kR^2 - 4kr^2$

               $= P(1 + RT)$ $= 4k(R^2 - r^2)$

EXERCISES 9.2
Factor

Complete Examples (See Example 1)

a. $3a + 3b = 3(\underline{})$ b. $2xy - 4x = 2x(\underline{})$
              (1)   (2)

(1) $a + b$ (2) $y - 2$

1. $3x + 6$ 2. $4x - 8$ 3. $2x - 6y$ 4. $10x + 5y$ 5. $2y^2 - 2y$

6. $3y^2 + 6y$ 7. $ay^2 + y$ 8. $by^2 - b$ 9. $9ay^2 + 6y$ 10. $4bx^2 - 12x$

Complete Example (See Example 2)

$3y^3 - 6y^2 + 9y$

     $= 3yyy - 3 \cdot 2yy + 3 \cdot 3y$ Write in prime factored form.

     $= \underline{}(\underline{} - \underline{} + \underline{})$ Factor out the common monomial.
        (1)   (2)   (3)   (4)

(1) $3y$ (2) y^2 (3) $2y$ (4) 3

9.3 BINOMIAL PRODUCTS I

11. $3y^2 - 3y + 3$ **12.** $2y^2 - 4y - 2$ **13.** $ax + ay - az$ **14.** $2bx - 6by + 4bz$

15. $x^2 - 3x + xy$ **16.** $y^2 + 2y - 4xy$ **17.** $4y^3 - 2y^2 + 2y$ **18.** $6x^3 + 6x^2 - 9x$

19. $6ax^2y - 18axy^2 + 24axy$ **20.** $3a^2x^2y - 12ax^2y^2 + 9ax^2y$

Factor out the common monomial, including -1. *See Example 3.*

Complete Examples
a. $-4x^3 - 4xy = -4x(\underline{}_{(1)} + \underline{}_{(2)})$ b. $-2x^2 - x + 5 = -(\underline{}_{(3)})$

(1) x^2 (2) y (3) $2x^2 + x - 5$

21. $-a^2 - ab$ **22.** $-a^2 - a$ **23.** $-x - x^2$ **24.** $-ab - ac$

25. $-abc - ab - bc$ **26.** $-b^2 - bc - ab$ **27.** $-6y^3 - 3y^2 - 3y$ **28.** $-2x^2 - 4x - 2$

29. $-x + x^2 - x^3$ **30.** $-3x^2 + 3xy - 3x$ **31.** $-xy^5 - xy^4 + xy^2$ **32.** $-x^2y + xy^2 - 3xy$

Factor the right-hand member of each of the following formulas. See Example 4.

Complete Examples
a. $P = 2l + 2w$

$P = 2(\underline{}_{(1)})$

b. $V = 2\pi R^2 - 2\pi r^2$

$V = 2\pi(\underline{}_{(2)})$

(1) $l + w$ (2) $R^2 - r^2$

33. $d = k + kat$ **34.** $A = kR^2 + 2kr^2$ **35.** $S = kr^2h + kr^2$ **36.** $R = r + rat$

37. $V = 2ga^2D - 2ga^2d$ **38.** $L = 2an + n^2 - nd$ **39.** $A = ar^2 + br^2 + cr^2$ **40.** $S = 2kr^2 + 2krh$

9.3 BINOMIAL PRODUCTS I

We can use the distributive law to multiply two binomials. Although there is little need to multiply binomials in arithmetic as shown in the following example, the distributive law also applies to expressions containing variables.

285

PRODUCTS AND FACTORS

$$(10 + 4)(10 + 2) = 10 \cdot 10 + 10 \cdot 2 + 4 \cdot 10 + 4 \cdot 2$$
$$= 100 + 20 + 40 + 8$$
$$= 168$$

We will now apply the above procedure for an expression containing variables.

Example 1 Write $(x - 2)(x + 3)$ without parentheses.

Solution First, apply the distributive property to get

$$(x - 2)(x + 3) = x^2 + 3x - 2x - 6$$

Now, combine like terms to obtain

$$x^2 + x - 6$$

With practice, you will be able to mentally add the second and third products. The above process is sometimes called the FOIL method. F, O, I, and L stand for:

1. The product of the First terms.
2. The product of the Outer terms.
3. The product of the Inner terms.
4. The product of the Last terms.

The FOIL method can also be used to square binomials.

Example 2 Write $(x + 3)^2$ without parentheses.

Solution First, rewrite $(x + 3)^2$ as $(x + 3)(x + 3)$. Next, apply the FOIL method to get

$$(x + 3)^2 = (x + 3)(x + 3) = x^2 + 3x + 3x + 9$$

Combining like terms yields

$$x^2 + 6x + 9$$

When we have a monomial factor and two binomial factors, it is easiest to first multiply the binomials.

9.3 BINOMIAL PRODUCTS I

Example 3 Write $3x(x-2)(x+3)$ without parentheses.

Solution First, multiply the binomials to obtain

$$3x(x^2 + 3x - 2x - 6) = 3x(x^2 + x - 6)$$

Now, apply the distributive law to get

$$3x(x^2 + x - 6) = 3x^3 + 3x^2 - 18x$$

Common Errors Notice in Example 2,

$$(x+3)^2 \neq x^2 + 3^2$$

Similarly,

$$(x-3)^2 \neq x^2 - 3^2$$

In general,

$$(a+b)^2 \neq a^2 + b^2, \text{ and } (a-b)^2 \neq a^2 - b^2$$

EXERCISES 9.3
Write each product as an expression without parentheses.

Complete Examples (see Example 1)

a. $(x+2)(x-3)$

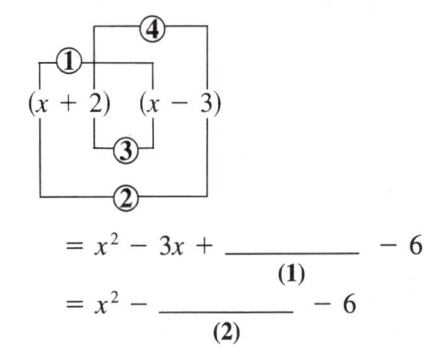

$(x+2)\ (x-3)$

$= x^2 - 3x + \underline{\qquad}_{(1)} - 6$

$= x^2 - \underline{\qquad}_{(2)} - 6$

b. $(y-a)(y+a)$

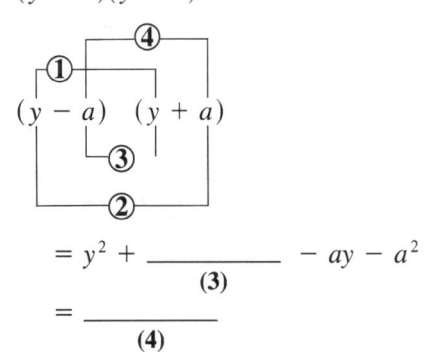

$(y-a)\ (y+a)$

$= y^2 + \underline{\qquad}_{(3)} - ay - a^2$

$= \underline{\qquad}_{(4)}$

(1) $2x$ (2) x (3) ay (4) $y^2 - a^2$

1. $(x+3)(x+4)$
2. $(x-2)(x-3)$
3. $(y-3)(y+1)$
4. $(y+4)(y-2)$
5. $(a+5)(a+2)$
6. $(a-5)(a-2)$
7. $(b-4)(b+2)$
8. $(b+5)(b-3)$
9. $(x+1)(x+8)$
10. $(x-2)(x-9)$
11. $(y-1)(y-7)$
12. $(y+1)(y-6)$
13. $(a+4)(a+4)$
14. $(a-3)(a-3)$
15. $(b-5)(b+5)$
16. $(b+7)(b-7)$

287

PRODUCTS AND FACTORS

17. $(x + 1)(x + 1)$ **18.** $(x - 9)(x - 9)$ **19.** $(y - 1)(y + 1)$ **20.** $(y + 4)(y - 4)$

21. $(2 + x)(2 - x)$ **22.** $(5 - x)(5 - x)$ **23.** $(6 + y)(6 - y)$ **24.** $(9 - y)(9 + y)$

25. $(x - 3b)(x - b)$ **26.** $(x - a)(x - 2a)$ **27.** $(x + 2y)(x - y)$ **28.** $(x + b)(x + 2b)$

29. $(x + 2a)(x + 2a)$ **30.** $(x + 3b)(x + 3b)$ **31.** $(y - 6a)(y + 6a)$ **32.** $(x + 3z)(x - 3z)$

33. $(x - t)(x + t)$ **34.** $(y - c)(y + c)$

Complete Examples (see Example 2)

a. $(x + 6)^2 = (x + 6)(x + 6)$
$= x^2 + \underline{}_{(1)} + 6x + 36$
$= x^2 + \underline{}_{(2)} + 36$

b. $(y - 3)^2 = (y - 3)(y - 3)$
$= y^2 - 3y - \underline{}_{(3)} + 9$
$= y^2 - \underline{}_{(4)} + 9$

(1) $6x$ (2) $12x$ (3) $3y$ (4) $6y$

35. $(x + 4)^2$ **36.** $(y - 5)^2$ **37.** $(x - 7)^2$ **38.** $(y + 1)^2$

39. $(x - 1)^2$ **40.** $(y + 8)^2$ **41.** $(x + 2)^2$ **42.** $(y - 10)^2$

43. $(a - b)^2$ **44.** $(x - y)^2$

Write without parentheses. See Example 3.

Complete Example

$4(x + 4)(x - 1)$
$= 4(\underline{}_{(1)} - x + \underline{}_{(2)} - \underline{}_{(3)})$ Multiply binomial factors.
$= 4(x^2 + 3x - 4)$ Simplify.
$= \underline{}_{(4)}$ Multiply by monomial.

(1) x^2 (2) $4x$ (3) 4 (4) $4x^2 + 12x - 16$

45. $2(x + 1)(x + 2)$ **46.** $4(x - 3)(x + 2)$ **47.** $6(y + 5)(y + 5)$

48. $4(y - 7)(y + 1)$ **49.** $6(x - 1)^2$ **50.** $3(y + 3)^2$

51. $a(a - 1)(a + 5)$ **52.** $b(b - 2)(b + 7)$ **53.** $a(a - 2)(a + 2)$

54. $b(b + 3)(b - 3)$ **55.** $x(y - 3)^2$ **56.** $x^2(y - 4)^2$

9.4 FACTORING TRINOMIALS I

In Section 4.3, we saw how to find the product of two binomials. Now we will reverse this process. That is, given the product of two binomials, we will find the binomial factors. The process involved is another example of factoring. As before, we will only consider factors in which the terms have integral numerical coefficients. Such factors do not always exist, but we will study the cases where they do.

Consider the following product.

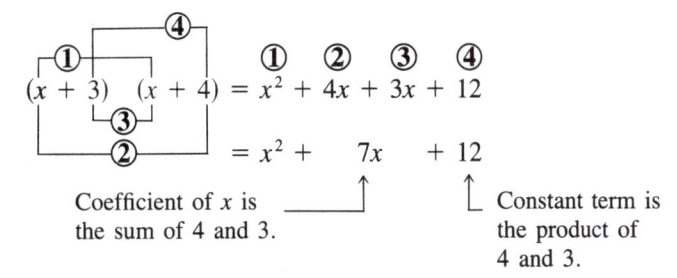

$$(x + 3)(x + 4) = x^2 + 4x + 3x + 12$$
$$= x^2 + 7x + 12$$

Coefficient of x is the sum of 4 and 3. Constant term is the product of 4 and 3.

Notice that the first term in the trinomial, x^2, is product ①; the last term in the trinomial, 12, is product ④; and the middle term in the trinomial, $7x$, is the *sum* of products ② and ③.

In general,

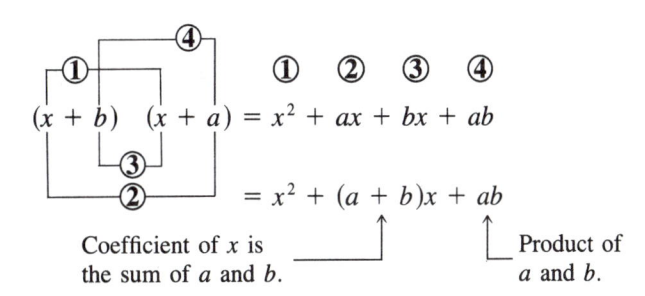

$$(x + b)(x + a) = x^2 + ax + bx + ab$$
$$= x^2 + (a + b)x + ab$$

Coefficient of x is the sum of a and b. Product of a and b.

We use this equation (from right to left) to factor any trinomial of the form $x^2 + Bx + C$. We find two numbers whose product is C and whose sum is B.

Example 1 Factor $x^2 + 7x + 12$.

Solution We look for two integers whose *product* is 12 and whose *sum* is 7. Consider the following pairs of factors whose product is 12.

Pairs of factors	Product of factors	Sum of factors
12, 1	12	13
−12, −1	12	−13
3, 4	12	7
−3, −4	12	−7
2, 6	12	8
−2, −6	12	−8

We see that the only pair of factors whose product is 12 and whose sum is 7 is 3 and 4. Thus,

$$x^2 + 7x + 12 = (x + 3)(x + 4)$$

PRODUCTS AND FACTORS

Note that when all terms of a trinomial are positive, we need only consider pairs of *positive* factors because we are looking for a pair of factors whose product and sum are positive. That is, the factored term of

$$x^2 + 7x + 12$$

would be of the form

$$(\ +\ )(\ +\ )$$

When the first and third terms of a trinomial are positive but the middle term is negative, we need only consider pairs of *negative* factors because we are looking for a pair of factors whose product is positive but whose sum is negative. That is, the factored form of

$$x^2 - 5x + 6$$

would be of the form

$$(\ -\ )(\ -\ )$$

Example 2 Factor $x^2 - 5x + 6$.

Solution Because the third term is positive and the middle term is negative, we find two negative integers whose product is 6 and whose sum is -5. We list the possibilities.

Pairs of factors	Product	Sum
$-6, -1$	6	-7
$-3, -2$	6	-5

We see that the only pair of factors whose product is 6 and whose sum is -5 is -3 and -2. Thus,

$$x^2 - 5x + 6 = (x - 3)(x - 2)$$

When the first term of a trinomial is positive and the third term is negative, the signs in the factored form are opposite. That is, the factored form of

$$x^2 - x - 12$$

would be of the form

$$(\ +\ )(\ -\ )\ \text{or}\ (\ -\ )(\ +\ )$$

Example 3 Factor $x^2 - x - 12$.

Solution We must find two integers whose product is -12 and whose sum is -1. We list the possibilities.

Pairs of factors	Product	Sum
12, -1	-12	11
-12, 1	-12	-11
6, -2	-12	4
-6, 2	-12	-4
4, -3	-12	1
-4, 3	-12	-1

9.4 FACTORING TRINOMIALS I

We see that the only pair of factors whose product is -12 and whose sum is -1 is -4 and 3. Thus,

$$x^2 - x - 12 = (x - 4)(x + 3)$$

It is easier to factor a trinomial completely if any monimial factor common to each term of the trinomial is factored first. For example, we can factor

$$12x^2 + 36x + 24$$

as

$$(12x + 24)(x + 1); \quad (12x + 12)(x + 2); \quad (6x + 12)(2x + 2)$$
$$(2x + 4)(6x + 6); \quad (4x + 8)(3x + 3) \quad \text{or} \quad (3x + 6)(4x + 4)$$

A monomial can then be factored from these binomial factors. However, first factoring the common factor 12 from the original expression yields

$$12(x^2 + 3x + 2)$$

Factoring again, we have

$$12(x + 2)(x + 1)$$

which is said to be in **completely factored form.** In such cases, it is not necessary to factor the numerical factor itself, that is, we do not write 12 as $2 \cdot 2 \cdot 3$.

Example 4 Factor $3x^2 + 12x + 12$ completely.

Solution First we factor out the 3 from the trinomial to get

$$3(x^2 + 4x + 4)$$

Now, we factor the trinomial and obtain

$$3(x + 2)(x + 2)$$

The techniques we have developed are also valid for a trinomial such as $x^2 + 5xy + 6y^2$.

Example 5 Factor $x^2 + 5xy + 6y^2$.

Solution We find two positive factors whose product is $6y^2$ and whose sum is $5y$ (the coefficient of x). The two factors are $3y$ and $2y$. Thus,

$$x^2 + 5xy + 6y^2 = (x + 3y)(x + 2y)$$

When factoring, it is best to write the trinomial in descending powers of x. If the coefficient of the x^2-term is negative, factor out a negative before proceeding.

Example 6 Factor $8 + 2x - x^2$.

Solution We first rewrite the trinomial in descending powers of x to get

$$-x^2 + 2x + 8$$

Now, we can factor out the -1 to obtain

$$-(x^2 - 2x - 8)$$

291

PRODUCTS AND FACTORS

Finally, we factor the trinomial to yield
$$-(x - 4)(x + 2)$$

Sometimes, trinomials are not factorable.

Example 7 Factor $x^2 + 5x + 12$.

Solution We look for two integers whose product is 12 and whose sum is 5. From the table in Example 1 on page 289, we see that there is no pair of factors whose product is 12 and whose sum is 5. In this case, the trinomial *is not* factorable.

Skill at factoring is usually the result of extensive practice. If possible, do the factoring process mentally, writing your answer directly.

You can check the results of a factorization by multiplying the binomial factors and verifying that the product is equal to the given trinomial.

EXERCISES 9.4
Factor completely.

Complete Example (see Examples 1 and 2)

$$x^2 - 14x + 45$$

Find two negative integers whose product is _____ and whose sum is _____ .
$\quad\quad\quad\quad\quad\quad\quad\quad\quad\quad\quad\quad\quad\quad\quad\quad\quad\quad$ (1) $\quad\quad\quad\quad\quad\quad\quad\quad\quad\quad$ (2)

The two integers are -5 and _____ .
$\quad\quad\quad\quad\quad\quad\quad\quad\quad\quad\quad\quad$ (3)

$$x^2 - 14x + 45 = (x - 5)(\underline{\quad\quad\quad})$$
$\quad\quad\quad\quad\quad\quad\quad\quad\quad\quad\quad\quad\quad\quad\quad\quad\quad\quad$ (4)

---

(1) 45   (2) -14 (3) -9 (4) $x - 9$

---

1. $x^2 + 5x + 6$
2. $x^2 + 9x + 20$
3. $x^2 + 11x + 30$
4. $x^2 + 20x + 100$
5. $x^2 + 14x + 45$
6. $y^2 - 8y + 15$
7. $y^2 - 3y + 2$
8. $y^2 - 14y + 13$
9. $y^2 - 16y + 63$
10. $x^2 - 46x + 45$

Complete Example (see Example 3)

$$x^2 - 2x - 3$$

Find two integers whose product is _____ and whose sum is _____ .
$\quad\quad\quad\quad\quad\quad\quad\quad\quad\quad\quad\quad\quad\quad\quad\quad$ (1) $\quad\quad\quad\quad\quad\quad\quad\quad\quad\quad$ (2)

9.4 FACTORING TRINOMIALS I

The two integers are -3 and ___(3)___ .

$$x^2 - 2x - 3 = (x - \underline{})(x + \underline{})$$

---

(1) -3 (2) -2 (3) 1 (4) 3 (5) 1

---

11. $x^2 - x - 12$ **12.** $x^2 - 3x - 10$ **13.** $y^2 + y - 20$ **14.** $y^2 + 2y - 8$

15. $a^2 + 2a - 35$ **16.** $a^2 + 8a - 20$ **17.** $b^2 - 19b - 20$ **18.** $b^2 - 4b - 12$

19. $a^2 - 5a - 50$ **20.** $a^2 - a - 72$ **21.** $b^2 - 4b - 45$ **22.** $b^2 - 12b - 45$

23. $y^2 - 44y - 45$ **24.** $y^2 - 14y - 51$

Complete Example (see Example 4)

$2b^3 - 8b^2 - 10b$

$= \underline{} (b^2 - \underline{} - 5)$ Factor out the common monomial.

$= 2b(b - 5)(\underline{})$ Factor the trinomial.

---

(1) $2b$ (2) $4b$ (3) $b + 1$

---

25. $2x^2 + 10x + 12$ **26.** $3a^2 - 3a - 18$ **27.** $y^3 - 2y^2 - 3y$

28. $b^3 + 2b^2 + b$ **29.** $5c^2 - 25c + 30$ **30.** $2a^2 - 38a - 40$

31. $4a^2b + 12ab - 72b$ **32.** $3x^2y - 6xy - 105y$

Complete Example (see Example 5)

$$x^2 - 5xy + 6y^2$$

Find two negative factors whose product is ___(1)___ and whose sum is ___(2)___ .

The two factors are $-3y$ and ___(3)___ .

$$x^2 - 5xy + 6y^2 = (x - 3y)(\underline{})$$

---

(1) $6y^2$ (2) $-5y$ (3) $-2y$ (4) $x - 2y$

PRODUCTS AND FACTORS

33. $x^2 + 4ax + 4a^2$
34. $x^2 - 3xy + 2y^2$
35. $a^2 - 3ab + 2b^2$
36. $r^2 + 4rx + 3x^2$
37. $s^2 + 5as + 6a^2$
38. $x^2 + 15xy + 36y^2$

Complete Example (see Example 6)

$6 - 5x - x^2$

$= -x^2 - \underline{\quad(1)\quad} + 6$ Write in decreasing powers of x.

$= -(x^2 + 5x - \underline{\quad(2)\quad})$ Since the coefficient of x^2 is -1, factor out -1 from each term.

$= -(x + 6)(\underline{\quad(3)\quad})$ Factor the trinomial.

(1) $5x$ (2) 6 (3) $x - 1$

39. $10 + 7y + y^2$
40. $30 - 11y + y^2$
41. $8 - 9x + x^2$

42. $81 + 18x + x^2$
43. $32 - 12z + z^2$
44. $56 - 15y + y^2$

45. $21 - 4x - x^2$
46. $6 - x - x^2$
47. $24 + 10z - z^2$

48. $63 - 2y - y^2$
49. $18 + 7y - y^2$
50. $54 + 3y - y^2$

Factor completely; if the trinomial is not factorable, so state. See Example 7.

51. $x^2 + 4x + 2$
52. $x^2 + 3x + 1$
53. $y^2 - 4y + 2$

54. $y^2 - 2y - 2$
55. $x^2 + 12x + 30$
56. $x^2 + 11x + 20$

57. $x^2 + 5xy + 4y^2$
58. $x^2 - 2xy + 4y^2$
59. $7 - 3y + y^2$

60. $5 - 2y + y^2$

9.5 BINOMIAL PRODUCTS II

In this section, we use the procedure developed in Section 9.3 to multiply binomial factors whose first-degree terms have numerical coefficients other than 1 or -1.

Example 1 Write as a polynomial.

    a. $(2x - 3)(x + 1)$ b. $(3x - 2y)(3x + y)$

Solutions We first apply the FOIL method and then combine like terms.

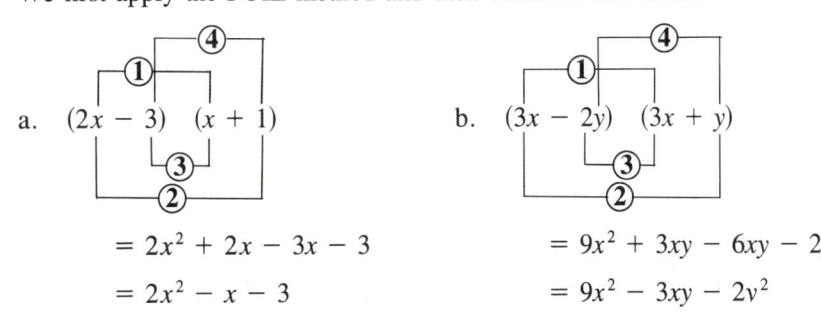

a. $(2x - 3)(x + 1)$

$= 2x^2 + 2x - 3x - 3$
$= 2x^2 - x - 3$

b. $(3x - 2y)(3x + y)$

$= 9x^2 + 3xy - 6xy - 2y^2$
$= 9x^2 - 3xy - 2y^2$

As before, if we have a squared binomial, we first rewrite it as a product, then apply the FOIL method.

Example 2   a. $(3x + 2)^2$ b. $(2x - y)^2$

$= (3x + 2)(3x + 2)$ $= (2x - y)(2x - y)$

$= 9x^2 + 6x + 6x + 4$ $= 4x^2 - 2xy - 2xy + y^2$

$= 9x^2 + 12x + 4$ $= 4x^2 - 4xy + y^2$

As you may have seen in Section 4.3, the product of two bionimals may have no first-degree term in the answer.

Example 3   a. $(2x - 3)(2x + 3)$ b. $(3x - y)(3x + y)$

$= 4x^2 + 6x - 6x - 9$ $= 9x^2 + 3xy - 3xy - y^2$

$= 4x^2 - 9$ $= 9x^2 - y^2$

When a monomial factor and two binomial factors are being multiplied, it is easiest to multiply the binomials first.

Example 4 Write $3x(2x - 1)(x + 2)$ as a polynomial.

Solution We first multiply the binomials to get

$$3x(2x^2 + 4x - x - 2) = 3x(2x^2 + 3x - 2)$$

Now multiplying by the monomial yields

$$3x(2x^2) + 3x(3x) + 3x(-2) = 6x^3 + 9x^2 - 6x$$

PRODUCTS AND FACTORS

EXERCISES 9.5
Write as a polynomial.

Complete Example (see Example 1)

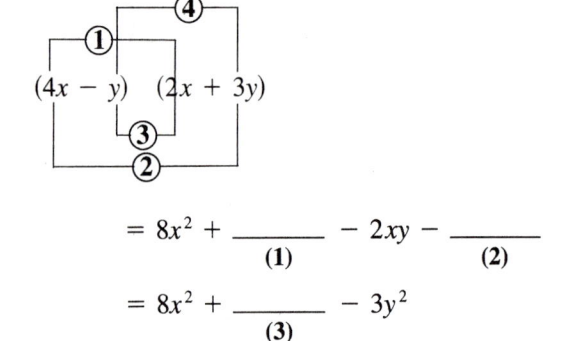

(4x − y) (2x + 3y) with connections ①, ②, ③, ④

$= 8x^2 + \underline{} - 2xy - \underline{}$ Apply the FOIL method.

$= 8x^2 + \underline{} - 3y^2$ Combine like terms.

(1) $12xy$ (2) $3y^2$ (3) $10xy$

1. $(2x + 1)(x + 3)$
2. $(4x − 2)(x − 1)$
3. $(3y − 2)(y + 1)$
4. $(y + 5)(5y − 2)$
5. $(3y + 1)(2y + 3)$
6. $(4y − 1)(3y + 2)$
7. $(5x − 2)(4x + 3)$
8. $(6x − y)(2x + 5y)$
9. $(2x + 3y)(2x − 5y)$
10. $(3x − y)(2x + y)$
11. $(4y − 3x)(2y + 3x)$
12. $(6y − 5x)(6y + 3x)$

Complete Example (see Example 2)

$(4x − y)^2$

$= (4x − y)(\underline{})$ Rewrite as the product of binomials.

$= 16x^2 − 4xy − \underline{} + \underline{}$ Apply the FOIL method.

$= 16x^2 − \underline{} + y^2$ Combine like terms.

(1) $4x − y$ (2) $4xy$ (3) y^2 (4) $8xy$

13. $(2x + 1)^2$
14. $(3x + 1)^2$
15. $(5x + 2)^2$
16. $(2x − 3)^2$
17. $(4y + 5)^2$
18. $(3y + 7)^2$
19. $(x − 2y)^2$
20. $(2x − 3y)^2$

9.5 BINOMIAL PRODUCTS II

21. $(3x - y)^2$ **22.** $(3x - 2y)^2$ **23.** $(8x + 3y)^2$ **24.** $(2x + y)^2$

25. $(2x + 3y)^2$ **26.** $(3x - 4y)^2$

Complete Example (see Example 3)

$(4x - 2)(4x + 2)$

$= 16x^2 + \underline{} - 8x - \underline{}$ Apply the FOIL method.
 (1) (2)

$= \underline{}$ Combine like terms.
 (3)

(1) $8x$ (2) 4 (3) $16x^2 - 4$

27. $(2x + 3)(2x - 3)$ **28.** $(3x + 1)(3x - 1)$ **29.** $(6y - 5)(6y + 5)$

30. $(4y - 3)(4y + 3)$ **31.** $(2x + a)(2x - a)$ **32.** $(3x - a)(3x + a)$

33. $(3x - 2y)(3x + 2y)$ **34.** $(2x - 5y)(2x + 5y)$ **35.** $(4x + 7y)(4x - 7y)$

36. $(5x + 9y)(5x - 9y)$

Complete Example (see Example 4)

$x(x + 2)(3x - 5)$

$= x(3x^2 - 5x + \underline{} - \underline{})$ Multiply binomials.
 (1) (2)

$= x(3x^2 + \underline{} - 10)$ Simplify.
 (3)

$= 3x^3 + \underline{} - 10x$ Multiply by the monomial.
 (4)

(1) $6x$ (2) 10 (3) x (4) x^2

37. $2(3x + 1)(x - 3)$ **38.** $4(x - 2)(2x - 3)$ **39.** $3(2y + 1)(2y - 1)$

40. $6(3y + 2)(3y - 2)$ **41.** $3(2x - 5)^2$ **42.** $3(x + 1)^2$

43. $x(x - 2)(2x + 5)$ **44.** $y(y + 2)(y - 1)$ **45.** $x(2x - 1)^2$

46. $y(y + 1)^2$ **47.** $r(3r - 1)(3r + 1)$ **48.** $s(2s - 3)(2s + 3)$

PRODUCTS AND FACTORS

9.6 FACTORING TRINOMIALS II

In Section 9.4 we factored trinomials of the form $x^2 + Bx + C$ where the second-degree term had a coefficient of 1. Now we want to extend our factoring techniques to trinomials of the form $Ax^2 + Bx + C$, where the second-degree term has a coefficient other than 1 or -1.

First, we consider a test to determine if a trinomial is factorable. A trinomial of the form $Ax^2 + Bx + C$ is factorable if we can find two integers whose product is $A \cdot C$ and whose sum is B.

Example 1 Determine if $4x^2 + 8x + 3$ is factorable.

Solution We check to see if there are two integers whose product is $(4)(3) = 12$ and whose sum is 8 (the coefficient of x). Consider the following possibilities.

Pairs of Factors with Product 12	Sum of Factors
12, 1	13
$-12, -1$	-13
6, 2	8
$-6, -2$	-8
4, 3	7
$-4, -3$	-7

Since the factors 6 and 2 have a sum of 8, the value of B in the trinomial $Ax^2 + Bx + C$, the trinomial is factorable.

Example 2 The trinomial $4x^2 - 5x + 3$ is not factorable, since the above table shows that there is no pair of factors whose product is 12 and whose sum is -5. The test to see if the trinomial is factorable can usually be done mentally.

Once we have determined that a trinomial of the form $Ax^2 + Bx + C$ is factorable, we proceed to find a pair of factors whose product is A, a pair of factors whose product is C, and an arrangement that yields the proper middle term. We illustrate by examples.

Example 3 Factor $4x^2 + 8x + 3$.

Solution In Example 1 we determined that this polynomial is factorable.

need a product of $4x^2$ → $4x^2$

$(4x)(x)$
$(2x)(2x)$

need a product of 3 → 3

$(4x + 3)\ (x + 1)$
$(4x + 1)\ (x + 3)$
$(2x + 3)\ (2x + 1)$
$(2x + 1)\ (2x + 3)$

1. We consider all pairs of factors whose product is 4. Since 4 is positive, only positive integers need to be considered. The possibilities are 4, 1 and 2, 2.

2. We consider all pairs of factors whose product is 3. Since the middle term is positive, consider positive pairs of factors only. The possibilities are 3, 1. We write all possible arrangements of the factors as shown.

298

9.6 FACTORING TRINOMIALS II

$(2x + 1)\ (2x + 3)$
    $\lfloor 2x \rfloor$
    $6x$
    $8x$

3. We select the arrangement in which the sum of products ② and ③ yields a middle term of $8x$.

Now, we consider the factorization of a trinomial in which the constant term is negative.

Example 4   Factor $6x^2 + x - 2$.

Solution   First, we test to see if $6x^2 + x - 2$ is factorable. We look for two integers that have a product of $6(-2) = -12$ and a sum of 1 (the coefficient of x). The integers 4 and -3 have a product of -12 and a sum of 1, so the trinomial is factorable. We now proceed.

need a product $\longrightarrow 6x^2$
of $6x^2$

$(6x\quad)(x\quad)$
$(2x\quad)(3x\quad)$

1. We consider all pairs of factors whose product is 6. Since 6 is positive, only positive integers need to be considered. The possibilities are 6, 1 and 2, 3.

need a product $\longrightarrow -2$
of -2

$(6x + 2)\quad (x - 1)$
$(6x - 1)\quad (x + 2)$
$(6x - 2)\quad (x + 1)$
$(6x + 1)\quad (x - 2)$
$(2x + 2)\quad (3x - 1)$
$(2x - 1)\quad (3x + 2)$
$(2x - 2)\quad (3x + 1)$
$(2x + 1)\quad (3x - 2)$

2. We consider all pairs of factors whose product is -2. The possibilities are 2, -1 and -2, 1. We write all possible arrangements of the factors as shown.

$(2x - 1)\ (3x + 2)$
    $\lfloor -3x \rfloor$
    $4x$
    x

3. We select the arrangement in which the sum of products ② and ③ yields a middle term of x.

With practice, you will be able to mentally check the combinations and will not need to write out all the possibilities. Paying attention to the signs in the trinomial is particularly helpful for mentally eliminating possible combinations.

It is easiest to factor a trinomial written in descending powers of the variable.

Example 5   Factor.

   a. $3 + 4x^2 + 8x$ b. $x - 2 + 6x^2$

Solutions   Rewrite each trinomial in descending powers of x and then follow the solutions of Examples 3 and 4.

   a. $4x^2 + 8x + 3$ b. $6x^2 + x - 2$

As we said in Section 9.4, if a polynomial contains a common monomial factor in each of its terms, we should factor this monomial from the polynomial before looking for other factors.

299

PRODUCTS AND FACTORS

Example 6 Factor $24x^2 - 44x - 40$.

Solution We first factor 4 from each term to get
$$4(6x^2 - 11x - 10)$$
We then factor the trinomial, to obtain
$$4(3x + 2)(2x - 5)$$

ALTERNATIVE METHOD OF FACTORING TRINOMIALS

If the above "trial and error" method of factoring does not yield quick results, an alternative method, which we will now demonstrate using the earlier example $4x^2 + 8x + 3$, may be helpful.

We know that the trinomial is factorable because we found two numbers whose product is 12 and whose sum is 8. Those numbers are 2 and 6. We now proceed and use these numbers to rewrite $8x$ as $2x + 6x$.

Rewrite the middle term as $2x + 6x$.

$$4x^2 + 2x + 6x + 3$$

$$2x(2x + 1) + 3(2x + 1)$$

$$(2x + 1)(2x + 3)$$

We now factor the first two terms, $4x^2 + 2x$ and the last two terms, $6x + 3$.

A common factor, $2x + 1$, is in each term, so we can factor again.

This is the same result that we obtained before.

EXERCISES 9.6

Determine if the trinomial is factorable. See Examples 1 and 2.

Complete Example

$$3x^2 + 6x - 2$$

We check to see if there are two numbers whose product is

$3(-2) = $ _____ and whose sum is _____
   (1)   (2)

Pairs of factors whose product is −6	Sum of factors
−6, 1	−5
6, −1	5
3, −2	1
−3, ____	____
(3)	(4)

300

9.6 FACTORING TRINOMIALS II

Because there are no two numbers whose product is -6 and whose sum is 6, the trinomial __is/is not__ factorable.
 (5)

---

(1) -6 (2) 6 (3) 2 (4) -1 (5) is not

---

1. $5x^2 - 4x + 3$
2. $2x^2 - 11x + 4$
3. $6y^2 - 11y + 3$
4. $4y^2 + 3y - 1$
5. $3x^2 + 5xy - y^2$
6. $4x^2 - 5xy - 2y^2$

Factor. Assume the trinomials are factorable.

Complete Example (see Examples 3 and 4)

$6x^2 - 11x + 4$

$6x^2$

$(6x)(x)$
$(3x)(2x)$ Consider all pairs of positive factors whose product is $6x^2$.

4

Since the middle term is negative, consider all pairs of negative factors whose product is 4.

$(6x - 2)(x - 2)$
$(3x - 4)(\underline{})$
 (1)
$(3x - 1)(\underline{})$
 (2)

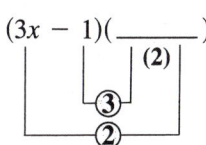

Select the arrangement in which the sum of products ② and ③ yields a middle term of _____.
 (3)

$(3x - 4)\ (2x - 1)$
$\llcorner{-8x}\lrcorner$
$\llcorner{-3x}\lrcorner$
$\overline{-11x}$

---

(1) $2x - 1$ (2) $2x - 4$ (3) $-11x$

---

7. $3a^2 + 4a + 1$
8. $2r^2 + 3r + 1$
9. $2x^2 - 3x + 1$
10. $2y^2 + 5y + 3$
11. $9b^2 - 6b + 1$
12. $4a^2 + 4a + 1$
13. $2x^2 - 7x + 3$
14. $4y^2 - 4y + 1$
15. $4y^2 - 5y + 1$
16. $4a^2 - 11a + 6$
17. $64x^2 + 64x + 15$
18. $4y^2 + 16y + 15$
19. $4y^2 - 3y - 1$
20. $4y^2 + 3y - 1$
21. $4a^2 + a - 5$
22. $16x^2 - 2x - 5$

PRODUCTS AND FACTORS

23. $16x^2 - 38x - 5$ **24.** $16x^2 - 11x - 5$ **25.** $16x^2 + 79x - 5$ **26.** $9x^2 - 21x - 8$

27. $2t^2 - 5st - 3s^2$ **28.** $2a^2 + 5ab - 3b^2$ **29.** $3x^2 - 7ax + 2a^2$ **30.** $9y^2 - 3yz - 2z^2$

Complete Example (see Examples 3, 4, and 5)

$9x^2 - 8 - 21x$

$= 9x^2 - \underline{}_{(1)} - 8$ Arrange in descending powers of x.

$= (3x - \underline{}_{(2)})(\underline{}_{(3)})$ Factor.

---

(1) $21x$ (2) 8 (3) $3x + 1$

---

31. $2x^2 - 3 + x$ **32.** $2x^2 + 3 + 7x$ **33.** $2x^2 - 3 - x$ **34.** $6a^2 - 1 - a$

35. $1 + 6a^2 + 5a$ **36.** $23a + 4a^2 - 6$ **37.** $16x^2 - 5 - 16x$ **38.** $10y - 8 + 25y^2$

Complete Example (see Examples 3, 4, and 5)

$13xy + 3x^2 + 4y^2$

$= 3x^2 + \underline{}_{(1)} + 4y^2$ Arrange in descending powers of one variable.

$= (3x + \underline{}_{(2)})(\underline{}_{(3)})$ Factor.

---

(1) $13xy$ (2) y (3) $x + 4y$

---

39. $5by + 4y^2 + b^2$ **40.** $9ab + 9a^2 - 4b^2$

41. $4a^2 + 15b^2 + 16ab$ **42.** $9x^2 - 2y^2 + 3xy$

Complete Example (see Example 6)

$4a^2b + 10ab + 6b$

$= 2b(2a^2 + \underline{}_{(1)} + \underline{}_{(2)})$ Factor out monomial factor.

$= 2b(2a + 3)(\underline{}_{(3)})$ Factor trinomial.

---

(1) $5a$ (2) 3 (3) $a + 1$

302

43. $6x^2 + 8x + 2$

44. $6x^3 + 21x^2 + 9x$

45. $8y^2 - 6y - 2$

46. $18x^2 - 3x - 3$

47. $18x^2 - 9x - 27$

48. $4x^3 - 10x^2 - 6x$

49. $27y^3 - 9y^2 - 6y$

50. $4y^3 + 10y^2 + 6y$

51. $12ab^2 + 15a^2b + 3a^3$

52. $27a^2b + 27ab^2 - 12b^3$

53. $50xy^3 + 20x^2y^2 - 16x^3y$

54. $4a^2bx^2 - 2abx - 12b$

9.7 FACTORING THE DIFFERENCE OF TWO SQUARES

Some polynomials occur so frequently that it is helpful to recognize these special forms, which in turn enables us to directly write their factored form. Observe that

$$(a + b)(a - b) = a^2 - ab + ab - b^2$$
$$= a^2 - b^2$$

In this section we are interested in viewing this relationship from right to left, from the polynomial $a^2 - b^2$ to its factored form $(a + b)(a - b)$.

> **The difference of two squares, $a^2 - b^2$, equals the product of the sum $a + b$ and the difference $a - b$.**

Example 1 a. $x^2 - 9 = x^2 - 3^2$
$\qquad\qquad\quad = (x + 3)(x - 3)$

b. $x^2 - 16 = x^2 - 4^2$
$\qquad\qquad = (x + 4)(x - 4)$

Since

$$(3x)(3x) = 9x^2$$

we can view a binomial such as $9x^2 - 4$ as $(3x)^2 - 2^2$ and use the above method to factor.

Example 2 a. $9x^2 - 4 = (3x)^2 - 2^2$
$\qquad\qquad\quad = (3x + 2)(3x - 2)$

b. $4y^2 - 25x^2 = (2y)^2 - (5x)^2$
$\qquad\qquad\quad = (2y + 5x)(2y - 5x)$

As before, we always factor out a common monomial first whenever possible.

Example 3 a. $x^3 - x^5 = x^3(1 - x^2)$
$\qquad\qquad\quad = x^3(1 + x)(1 - x)$

b. $a^2x^2y - 16y = y(a^2x^2 - 16)$
$\qquad\qquad\quad = y[(ax)^2 - 4^2]$
$\qquad\qquad\quad = y(ax - 4)(ax + 4)$

303

PRODUCTS AND FACTORS

EXERCISES 9.7
Factor.

Complete Example (see Example 1)

$x^2 - 36 = x^2 - (\underline{})^2$
$ = (x + \underline{})(x - \underline{})$

Write as the difference of squares.

Factor.

---

(1) 6   (2) 6   (3) 6

---

1. $x^2 - 9$
2. $y^2 - 25$
3. $x^2 - 1$
4. $x^2 - 81$

5. $x^2 - z^2$
6. $x^2 - 9y^2$
7. $x^2y^2 - 16$
8. $x^2y^2 - 36$

9. $a^2x^2 - 49b^2$
10. $x^2 - 100a^2b^2$
11. $36 - x^2$
12. $b^2 - y^2$

Complete Example (see Example 2)

$100 - 81b^2 = (\underline{})^2 - (\underline{})^2$
$ = (10 + \underline{})(\underline{})$

Write as the difference of squares.

Factor.

---

(1) 10   (2) 9b   (3) 9b   (4) 10 − 9b

---

13. $4b^2 - 9$
14. $9b^2 - 1$
15. $25x^2 - 16$
16. $4y^2 - 25$

17. $9 - 4x^2$
18. $25 - 9y^2$
19. $81 - 4x^2$
20. $9 - 64y^2$

21. $4a^2 - 121b^2$
22. $64x^2 - 9y^2$
23. $25y^2 - 49x^2$
24. $100x^2 - 81y^2$

25. $49a^2x^2 - 144b^2y^2$
26. $49a^2x^2 - 36b^2y^2$
27. $4x^2y^2 - 81$
28. $121 - 49x^2y^2$

29. $36a^2b^2 - 1$
30. $1 - 100a^2b^2$

9.8 EQUATIONS INVOLVING PARENTHESES

Complete Example (see Example 3)

$4ax^2y^2 - 16a = 4a(\underline{\qquad}_{(1)})$ Factor out common monomial.

$ = 4a[(\underline{\qquad}_{(2)})^2 - (\underline{\qquad}_{(3)})^2]$ Write as the difference of squares.

$ = 4a(xy + \underline{\qquad}_{(4)})(\underline{\qquad}_{(5)})$ Factor.

---

(1) $x^2y^2 - 4$ (2) xy (3) 2 (4) 2 (5) $xy - 2$

---

31. $5x^2 - 5$
32. $2x^2 - 8$
33. $3x^3 - 3x$
34. $3a^2 - 75$
35. $2x^2 - 8y^2$
36. $3xy^2 - 12xb^2$
37. $3a^2b^2 - 12c^2d^2$
38. $8x^2y^2z^2 - 18$
39. $4x^2y^2 - 16x^2$
40. $3x^2y^2 - 12y^2$

9.8 EQUATIONS INVOLVING PARENTHESES

Often we must solve equations in which the variable occurs within parentheses. We can solve these equations in the usual manner after we have simplified them by applying the distributive law to remove the parentheses.

Example 1 Solve $4(5 - y) + 3(2y - 1) = 3$.

Solution We first apply the distributive law to get

$$20 - 4y + 6y - 3 = 3$$

Now combining like terms and solving for y yields

$$2y + 17 = 3$$
$$2y = -14$$
$$y = -7$$

The same method can be applied to equations involving binomial products.

Example 2 Solve $(x + 5)(x + 3) - x = x^2 + 1$.

Solution First, we apply the FOIL method to remove parentheses and obtain

$$x^2 + 8x + 15 - x = x^2 + 1$$

Now, combining like terms and solving for x yields

$$x^2 + 7x + 15 = x^2 + 1$$
$$7x = -14$$
$$x = -2$$

PRODUCTS AND FACTORS

EXERCISES 9.8
Solve.

Complete Example (see Example 1)

$$3(2 - y) + 4(3y + 1) = 19$$

$$6 - \underline{\quad(1)\quad} + 12y + \underline{\quad(2)\quad} = 19$$

$$\underline{\quad(3)\quad} + 10 = 19$$

$$9y = \underline{\quad(4)\quad}$$

$$y = \underline{\quad(5)\quad}$$

Apply distributive law.

Solve for y.

---

(1) $3y$ (2) 4 (3) $9y$ (4) 9 (5) 1

---

1. $3(x - 5) = 6$
2. $5(3x - 2) = 35$
3. $2(4y + 5) = 2$
4. $6(3y - 4) = -60$
5. $28 = 4(1 + 2x)$
6. $0 = 7(8 - 2x)$
7. $-3(2x + 1) - 4 = -1$
8. $-5(2x - 3) + 2 = 47$
9. $7(y - 3) = 2y - 31$
10. $4(5 - y) = 10 - 6y$
11. $3y + 35 = 4(2y + 5)$
12. $5x - 64 = -2(3x - 1)$
13. $-x - (8 + x) = 2$
14. $3(7 + 2x) = 30 + 7(x - 1)$
15. $5x - (x + 2) = 7 + (x + 3)$
16. $4(y - 1) = 5(y - 2)$
17. $-2y + 5(y + 1) = 25 + 7y$
18. $25 + 5y = -2(y - 4) - 18$
19. $(a - 1) - (a + 2) = 2a - 3$
20. $3(2a - 1) + 2(a + 5) = 15$
21. $(b + 5) - (b - 1) = 3b$
22. $5a - 4(1 - a) = 41 - 6a$
23. $b = 4(b + 6) + 3b$
24. $b = (b + 1) - (b - 5)$

Complete Example (see Example 2)

$$(x + 3)(x - 4) - x = x^2 - 14$$

$$x^2 - 4x + 3x - \underline{\quad(1)\quad} - x = x^2 - 14$$

$$-2x = \underline{\quad(2)\quad}$$

$$x = \underline{\quad(3)\quad}$$

Apply the FOIL method.

Solve for x.

---

(1) 12 (2) -2 (3) 1

306

9.9 WORD PROBLEMS INVOLVING NUMBERS

25. $(x - 1)(x + 2) = x^2 + 1$
26. $(x - 4)(x - 1) = 9 + x^2$
27. $(x + 2)(x - 2) = x^2 - 4x$
28. $(y + 3)(y + 1) = y^2 - 5$
29. $(y + 2)(y + 4) = y(y + 8)$
30. $(y - 4)(y + 4) = y^2 + 4y$
31. $(a + 3)(a - 2) + 1 = a^2$
32. $(a - 1)(a - 1) = a^2 - 11$
33. $(a - 2)(a + 2) + 3 = a^2 - a$
34. $2(a - 1)(a + 1) = 2a^2 + 2a$
35. $(a - 3)^2 = a^2 - 15$
36. $(a + 2)^2 = a^2 + 5a - 2$
37. $(3 - 2x)(x + 1) + 2x^2 = 0$
38. $(x - 3)(x + 4) = -2 + x^2$
39. $5 - (x - 4)(x + 3) = 12 - x^2$
40. $3 - (x + 1)(3x - 2) = -3x^2$

9.9 WORD PROBLEMS INVOLVING NUMBERS

Parentheses are useful in representing products in which the variable is contained in one or more terms in any factor.

Example 1 One integer is three more than another. If x represents the smaller integer, represent in terms of x

    a. The larger integer.
    b. Five times the smaller integer.
    c. Five times the larger integer.

Solutions   a. $x + 3$ b. $5x$ c. $5(x + 3)$

Let us say we know the sum of two numbers is 10. If we represent one number by x, then the second number must be $10 - x$ as suggested by the following table.

One number	Second number	Sum
3	7 or 10 − 3	10
5	5 or 10 − 5	10
7	3 or 10 − 7	10
x	10 − x	10

In general, if we know the sum of two numbers is S and x represents one number, the other number must be $S - x$.

Example 2 The sum of two integers is 13. If x represents the smaller integer, represent in terms of x

    a. The larger integer.
    b. Five times the smaller integer.
    c. Five times the larger integer.

Solution   a. $13 - x$ b. $5x$ c. $5(13 - x)$

Example 3 The difference of the squares of two consecutive odd integers is 24. If x represents the smaller integer, represent in terms of x

    a. The larger integer.
    b. The square of the smaller integer.
    c. The square of the larger integer.

Solution   a. $x + 2$ b. x^2 c. $(x + 2)^2$

PRODUCTS AND FACTORS

Sometimes, the mathematical models (equations) for word problems involve parentheses. We can use the approach outlined on page 264 to obtain the equation. Then, we proceed to solve the equation by first writing equivalently the equation without parentheses.

Example 4 One integer is five more than a second integer. Three times the smaller integer plus twice the larger equals 45. Find the integers.

Solution **Steps 1–2** First, we write what we want to find (the integers) as word phrases. Then, we represent the integers in terms of a variable.

The smaller integer: x

The larger integer: $x + 5$

Step 3 A sketch is not applicable.

Step 4 Now, we write an equation that represents the condition in the problem and get

$$3x + 2(x + 5) = 45$$

Step 5 Applying the distributive law to remove parentheses yields

$$3x + 2x + 10 = 45$$
$$5x = 35$$
$$x = 7$$

Step 6 The integers are 7 and 7 + 5 or 12.

EXERCISES 9.9

In each of the following statements, assign a variable to one of the unknowns and present all other unknowns in terms of this variable.

Complete Example (see Example 1)

One integer is seven less than another. If x represents the larger integer, represent in terms of x

a. The smaller integer
b. Three times the larger integer
c. Three times the smaller integer.

_____ _____ _____
 (1)   (2)   (3)

(1) $x - 7$ (2) $3x$ (3) $3(x - 7)$

---

1. One integer is four more than a second integer. If x represents the smaller integer, represent in terms of x
   a. The larger integer.
   b. Five times the smaller integer.
   c. Five times the larger integer.

2. One integer is six less than a second integer. If n represents the larger integer, represent in terms of n
   a. The smaller integer.
   b. Two times the smaller integer.
   c. Two times the larger integer.

9.9 WORD PROBLEMS INVOLVING NUMBERS

3. One integer is six more than another. If n represents the smaller integer, represent in terms of n, three times the larger integer.

4. One integer is five less than another. If n represents the larger integer, represent in terms of n, six times the smaller integer.

5. One integer is eight less than another. If n represents the larger integer, represent in terms of n, five times the smaller.

6. One integer is three more than another. If n represents the smaller integer, represent in terms of n, seven times the larger.

Complete Example (see Example 2)

The sum of two integers is 15. If x represents the smaller integer, represent in terms of x

a. The larger integer.   b. Four times the smaller integer.   c. Four times the larger integer.

_____   _____   _____
  (1)   (2)   (3)

(1) $15 - x$ (2) $4x$ (3) $4(15 - x)$

7. The sum of two integers is 27. If n represents the smaller integer, represent in terms of n
   a. The larger integer.
   b. Three times the smaller integer.
   c. Three times the larger integer.

8. The sum of two integers is 39. If n represents the larger integer, represent in terms of n
   a. The smaller integer.
   b. Six times the smaller integer.
   c. Four times the larger integer.

9. The difference of two integers is 16. If n represents the smaller integer, represent in terms of n
   a. The larger integer.
   b. Five times the smaller integer.
   c. Two times the larger integer.

10. The difference of two integers is 21. If n represents the larger integer, represent in terms of n
    a. The smaller integer.
    b. Two times the larger integer.
    c. Three times the smaller integer.

11. The sum of two integers is 42. If n represents the larger integer, represent in terms of n
    a. The smaller integer.
    b. Four times the smaller integer.
    c. Five times the larger integer.

12. The sum of two integers is 19. If n represents the smaller integer, represent in terms of n
    a. The larger integer.
    b. Three times the larger integer.
    c. Six times the smaller integer.

Complete Example (see Example 3)

The difference of the squares of two consecutive positive integers is 36. If x represents the smaller integer, represent in terms of x

a. The larger integer.   b. The square of the smaller integer.   c. The square of the larger integer.

_____   _____   _____
  (1)   (2)   (3)

(1) $x + 2$ (2) x^2 (3) $(x + 2)^2$

309

PRODUCTS AND FACTORS

13. The difference of the squares of two consecutive positive odd integers is 48. If x represents the smaller integer, represent in terms of x
   a. The larger integer.
   b. The square of the smaller integer.
   c. The square of the larger integer.

14. The difference of the squares of two consecutive negative even integers is 28. If x represents the smaller integer, represent in terms of x
   a. The larger integer.
   b. The square of the smaller integer.
   c. The square of the larger integer.

15. Three consecutive integers are such that the sum of the first plus three times the second, less the third is equal to 16. If x represents the smaller integer, represent in terms of x
   a. The second integer.
   b. Three times the second integer.
   c. The third integer.

16. Three consecutive integers are such that the sum of twice the first plus the second, minus the third is equal to 13. If x represents the smaller integer, represent in terms of x
   a. Twice the first integer.
   b. The second integer.
   c. The third integer.

17. The product of two consecutive integers is 5 less than the square of the second one. If x represents the smaller integer, represent in terms of x
   a. The larger integer.
   b. The square of the larger integer.
   c. The product of the integers.

18. The product of two consecutive even integers is 12 less than the square of the larger one. If x represents the smaller integer, represent in terms of x
   a. The larger integer.
   b. The square of the larger integer.
   c. The product of the integers.

Solve each of the following problems completely. Follow the six steps outlined on page 264.

Complete Example (see Example 4)

One integer is four more than a second integer. Twice the smaller plus three times the larger equals 22. Find the integers.

Steps 1–2 The smaller integer: x
The larger integer: _____
(1)

Step 3 A sketch is not applicable.

Step 4 $2x + 3(\underline{}) = 22$
(2)

Step 5 $2x + 3x + \underline{} = 22$
(3)

$5x = \underline{}$; $x = 2$
(4)

Step 6 The smaller integer is 2 and the larger integer is _____ .
(5)

---

(1) $x + 4$ (2) $x + 4$ (3) 12   (4) 10   (5) 6

---

19. One integer is two less than a second integer. The larger integer plus four times the smaller equals 17. Find the integers.

20. One integer is three more than a second integer. Four times the second integer plus twice the first equals 42. Find the integers.

21. One positive integer is three less than a second integer. The larger plus four times the smaller equals 123. Find the integers.

22. One integer is four more than a second integer. Five times the second integer plus twice the first equals 113. Find the integers.

23. The sum of two integers is 25. Twice the larger integer is one more than 5 times the smaller. Find the integers.

24. The sum of two integers is 24. Twice the larger is two less than three times the smaller. Find the integers.

25. The difference of the squares of two consecutive positive *even* integers is 52. Find the integers.

26. The difference of the squares of two consecutive integers is 17. Find the integers.

27. One positive integer is three times another. If 5 is added to the smaller, the sum is equal to 9 subtracted from the larger. Find the integers.

28. One positive integer is four times another. A third integer is six more than the smaller of the other two. The sum of the three integers is equal to the smallest integer plus 46. What are the integers?

29. Find three consecutive even integers such that twice the first plus three times the second, minus the third equals 26.

30. Find three consecutive *odd* integers such that twice the smallest, plus three times the second, minus three times the third, equals 8.

31. The product of two consecutive even integers is 22 less than the square of the second one. Find the integers.

32. The product of two consecutive odd integers is 14 less than the square of the first. Find the integers.

33. The sum of the squares of three consecutive integers is equal to three times the square of the first integer plus 59. Find the integers.

34. The sum of the squares of three consecutive even integers is equal to three times the square of the third integer minus 76. Find the integers.

9.10 APPLICATIONS

In this section, we will examine several applications of word problems that lead to equations that involve parentheses. Once again, we will follow the six steps outlined on page 264 when we solve the problems.

COIN PROBLEMS

The basic idea of problems involving coins (or bills) is that the value of a number of coins of the same denomination is equal to the product of the value of a single coin and the total number of coins.

$$\begin{bmatrix} \text{value of} \\ n \\ \text{coins} \end{bmatrix} = \begin{bmatrix} \text{value of} \\ 1 \\ \text{coin} \end{bmatrix} \times \begin{bmatrix} \text{number} \\ \text{of} \\ \text{coins} \end{bmatrix}$$

A table like the one shown in the next example is helpful in solving coin problems.

Example 1 A collection of coins consisting of dimes and quarters has a value of $5.80. There are 16 more dimes than quarters. How many dimes and quarters are in the collection?

311

PRODUCTS AND FACTORS

Solution *Steps 1–2* We first write what we want to find as word phrases. Then, we represent each phrase in terms of a variable.

The number of quarters: x
The number of dimes: $x + 16$

Step 3 Next, we make a table showing the number of coins and their value.

Denomination	Value of one coin in cents	Number of coins	Value of coins in cents
Quarters	25	x	$25x$
Dimes	10	$x + 16$	$10(x + 16)$

Step 4 Now we can write an equation.

$$\begin{bmatrix}\text{value of}\\ \text{quarters}\\ \text{in cents}\end{bmatrix} + \begin{bmatrix}\text{value of}\\ \text{dimes}\\ \text{in cents}\end{bmatrix} = \begin{bmatrix}\text{value of}\\ \text{collection}\\ \text{in cents}\end{bmatrix}$$

$$25x + 10(x + 16) = 580$$

Step 5 Solving the equation yields

$$25x + 10x + 160 = 580$$
$$35x = 420$$
$$x = 12$$

Step 6 There are 12 quarters and $12 + 16$ or 28 dimes in the collection.

INTEREST PROBLEMS

The basic idea of solving interest problems is that the amount of interest i earned in one year at simple interest equals the product of the rate of interest r and the amount of money p invested ($i = r \cdot p$). For example, $1000 invested for one year at 9% yields $i = (0.09)(1000) = \$90$.

A table like the one shown in the next example is helpful in solving interest problems.

Example 2 Two investments produce an annual interest of $320. $1000 more is invested at 11% than at 10%. How much is invested at each rate?

Solution *Steps 1–2* We first write what we want to find as word phrases. Then, we represent each phrase in terms of a variable.

Amount invested at 10%: x
Amount invested at 11%: $x + 100$

Step 3 Next, we make a table showing the amount of money invested, the rates of interest, and the amounts of interest.

Amount of money	Rate of interest	Amount of interest
x	0.10	$0.10x$
$x + 1000$	0.11	$0.11(x + 1000)$

9.10 APPLICATIONS

Step 4 Now, we can write an equation relating the interest from each investment and the total interest received.

$$0.10x + 0.11(x + 1000) = 320$$

Step 5 To solve for x, first multiply each member by 100 to obtain

$$10x + 11(x + 1000) = 32{,}000$$
$$10x + 11x + 11{,}000 = 32{,}000$$
$$21x = 21{,}000$$
$$x = 1000$$

Step 6 $1000 invested at 10%; $1000 + $1000, or $2000, invested at 11%.

MIXTURE PROBLEMS

The basic idea of solving mixture problems is that the amount (or value) of the substances being mixed must equal the amount (or value) of the final mixture.

Example 3 How much candy worth 80¢ a kilogram (kg) must a grocer blend with 60 kg of candy worth $1 a kilogram to make a mixture worth 90¢ a kilogram?

Solution *Steps 1–2* We first write what we want to find as a word phrase. Then, we represent the phrase in terms of a variable.

Kilograms of 80¢ candy: x

Step 3 Next, we make a table showing the types of candy, the amount of each, and the total values of each.

	Value of 1 kg in cents	Number of kilograms	Value of candy in cents
0.80 candy	80	x	$80x$
1.00 candy	100	60	$100(60)$
0.90 candy	90	$x + 60$	$90(x + 60)$

Step 4 We can now write an equation.

$$\begin{bmatrix} \text{value of} \\ x \text{ kg} \\ \text{of 80¢ candy} \end{bmatrix} + \begin{bmatrix} \text{value of} \\ 60 \text{ kg} \\ \text{of \$1 candy} \end{bmatrix} = \begin{bmatrix} \text{value of} \\ (x + 60) \text{ kg} \\ \text{of 90¢ candy} \end{bmatrix}$$

$$80x \quad + \quad 60(100) \quad = \quad 90(60 + x)$$

Step 5 Solving the equation yields

$$80x + 6000 = 5400 + 90x$$
$$600 = 10x$$
$$60 = x$$

Step 6 The grocer should use 60 kg of the 80¢ candy.

313

PRODUCTS AND FACTORS

Another type of mixture problem is one that involves the mixture of the two liquids.

Example 4 How many quarts of a 20% solution of acid should be added to 10 quarts of a 30% solution of acid to obtain a 25% solution?

Solution *Steps 1–2* We first write what we want to find as a word phrase. Then, we represent the phrase in terms of a variable.

Number of quarts of 20% solution to be added: x

Step 3 Next, we make a table or drawing showing the percent of each solution, the amount of each solution, and the amount of pure acid in each solution.

Percent of acid	Amount of solution	Amount of acid in solution
20%	x	$0.20x$
30%	10	$0.30(10)$
25%	$(x + 10)$	$0.25(x + 10)$

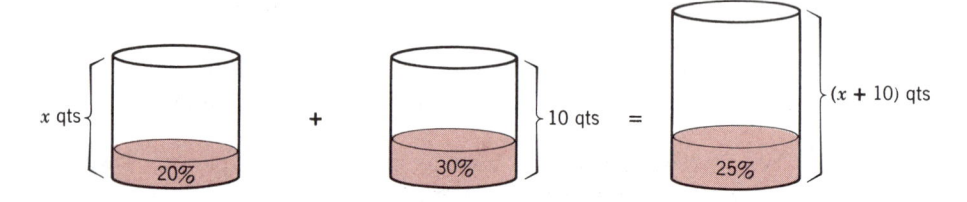

Step 4 We can now write an equation relating the *amounts of pure acid* before and after combining the solutions.

$$\begin{bmatrix} \text{Pure acid in} \\ \text{20\% solution} \end{bmatrix} + \begin{bmatrix} \text{Pure acid in} \\ \text{30\% solution} \end{bmatrix} = \begin{bmatrix} \text{Pure acid in} \\ \text{25\% solution} \end{bmatrix}$$

$$0.20x \quad + \quad 0.30(10) \quad = \quad 0.25(x + 10)$$

Step 5 To solve for x, first multiply each member by 100 to obtain

$$20x + 30(10) = 25(x + 10)$$
$$20x + 300 = 25x + 250$$
$$50 = 5x$$
$$10 = x$$

Step 6 Add 10 quarts of 20% solution to produce the desired solution.

EXERCISES 9.10
Solve each of the following problems completely. Follow the six steps outlined on page 264.

Complete Example (see Example 1)

A collection of coins consisting of nickels and dimes has a value of $1.80. There are 3 more dimes than nickels. How many nickels and dimes are in the collection?

9.10 APPLICATIONS

Steps 1–2 Number of nickels: x
Number of dimes: _____(1)_____

Step 3

Denomination	Value of one coin in cents	Number of coins	Value of coins in cents
Nickels	5	x	$5x$
Dimes	10	$x + 3$	_____(2)_____

Step 4

$$\begin{bmatrix} \text{value of} \\ \text{nickels} \\ \text{in cents} \end{bmatrix} + \begin{bmatrix} \text{value of} \\ \text{dimes} \\ \text{in cents} \end{bmatrix} = \begin{bmatrix} \text{value of} \\ \text{collection} \\ \text{in cents} \end{bmatrix}$$

_____(3)_____ $+ \; 10(x + 3) \; = \; 180$

Step 5

$5x + 10x + 30 = 180$
$15x = $ _____(4)_____ ; $x = 10$

Step 6 There are 10 nickels and _____(5)_____ dimes in the collection.

---

(1) $x + 3$ (2) $10(x + 3)$ (3) $5x$ (4) 150   (5) 13

---

1. A woman had $1.80 in change. The change was entirely in the form of dimes and nickels. If she had three more dimes than nickels, how many of each coin did she have?

2. A man had $1.45 in change. The money consisted of quarters and dimes only. If he had four fewer quarters than he had dimes, how many of each coin did he have?

3. A woman had $1.14 in change consisting of pennies, nickels, and dimes. She had six more nickels than pennies and six fewer dimes than pennies. How many of each coin did she have?

4. A man had $1.47 in change consisting of pennies, nickels, and quarters. He had three more pennies than quarters and one more nickel than pennies. How many of each coin did he have?

5. One thousand tickets were sold at a football game. Adults paid $4.25 each for their tickets, and children paid $1.50 each. If the total receipts for the game were $3150, how many tickets of each kind were sold?

6. Three hundred tickets were sold at a baseball game. Adults paid $3.50 each for their tickets, and chldren paid $1.50 each. If the total receipts for the game were $750, how many tickets of each kind were sold?

PRODUCTS AND FACTORS

Complete Example (see Example 2)

Two investments produce an annual interest of $350. $2000 more is invested at 12% than at 10%. How much is invested at each rate?

Steps 1–2 Amount invested at 10%: x
Amount invested at 12%: _____
(1)

Step 3

Amount of money	Rate of interest	Amount of interest
x	0.10	$0.10x$
$x + 2000$	0.12	_____ (2)

Step 4 $0.10x + 0.12(x + 2000) =$ _____ (3)

Step 5
$10x + 12(x + 2000) = 35000$
$10x + 12x + 24000 = 35000$
$22x =$ _____ ; $x = 500$
(4)

Step 6 $500 is invested at 10%; _____ is invested at 12%.
(5)

---

(1) $x + 2000$ (2) $0.12(x + 2000)$ (3) 350 (4) 11000 (5) $2500

---

7. Two investments produce an annual income of $1060. One investment earns 10% and the other earns 11%. How much is invested at each rate if the amount invested at 11% is $6200 more than the amount invested at 10%?

8. An amount of money is invested at 9% and $1200 more than that amount is invested at 10%. How much is invested at each rate if the total income is $1013?

9. An amount of $34,000 is invested, part at 8% and the remainder at 9%. Find the amount invested at each rate if the yearly income on each investment is the same.

10. An amount of $25,000 is invested, part at 10% and the remainder at 11%. Find the amount invested at each rate if the yearly income is $2590.

11. An amount of money is invested at 9% and three times that amount is invested at 12%. How much is invested at each rate if the total income is $675.

12. An amount of money is invested at 8% and twice that amount is invested at 10%. How much is invested at each rate if the total income is $1288?

Complete Example (see Examples 3–4)

How much coffee worth 50 cents a kilogram (kg) must a grocer blend with 15 kg of coffee worth 30 cents a kilogram to obtain a mixture worth 38 cents a kilogram?

Steps 1–2 Kilograms of 50-cent coffee: x

316

9.10 APPLICATIONS

Step 3

	Value of 1 kg in cents	Number of kilograms	Value of coffee in cents
0.50 Coffee	50	x	50x
0.30 Coffee	30	15	30(15)
0.38 Coffee	38	_____ (1)	38(x + 15)

Step 4

$$\begin{bmatrix} \text{value of} \\ x \text{ kilograms} \\ \text{of 50-cent} \\ \text{coffee} \end{bmatrix} + \begin{bmatrix} \text{value of} \\ 15 \text{ kilograms} \\ \text{of 30-cent} \\ \text{coffee} \end{bmatrix} = \begin{bmatrix} \text{value of} \\ (x + 15) \\ \text{kilograms of} \\ 38\text{-cent coffee} \end{bmatrix}$$

$$50x \quad + \quad \underline{}_{(2)} \quad = \quad 38(x + 15)$$

Step 5

$$50x + 450 = 38x + \underline{}_{(3)}$$

$$\underline{}_{(4)} x = 120$$

$$x = \underline{}_{(5)}$$

Step 6   The grocer should use 10 kg of the 50-cent coffee.

---

(1) x + 15   (2) 30(15) or 450   (3) 570   (4) 12   (5) 10

---

13. How many grams of metal worth 50¢ a gram should be mixed with 20 grams of metal worth 32¢ a gram to produce an alloy worth 40¢ a gram?

14. How many pounds of dog food worth 5¢ a pound should a pet store owner mix with 15 pounds of dog food worth 8¢ a pound to produce a mixture worth 6¢ a pound?

15. Fine powder is worth 30¢ a kilogram, and coarse powder is worth 12¢ a kilogram. How many kilograms of the fine power should be mixed with 50 kg of the coarse powder for the mixture to sell for 20¢ a kilogram?

16. A man uses 60 kg of fine powder worth 30¢ a kilogram and a coarse powder worth 25¢ a kilogram to make a mixture that he wishes to sell for 28¢ a kilogram. How many kilograms of the coarse powder does he use?

17. How many quarts of a 20% solution of acid should be added to 30 quarts of a 50% solution of acid to obtain a 40% solution of acid?

18. How many quarts of a 40% salt solution must be added to 40 quarts of a 10% salt solution to obtain a 20% salt solution?

19. How many ounces of an alloy containing 50% aluminum must be melted with an alloy containing 70% aluminum to obtain 40 ounces of an alloy containing 55% aluminum?

20. How many pounds of an alloy containing 60% copper must be melted with an alloy containing 20% copper to obtain 8 pounds of an alloy containing 30% copper?

317

PRODUCTS AND FACTORS

CHAPTER NINE REVIEW EXERCISES

Write as a polynomial.

1. a. $3x(x^2 + x)$ b. $2xy(y - x)$ c. $-(x^2 - y + 1)$
2. a. $a(2 - a)$ b. $-b(a - b)$ c. $3b(a + b + c)$

Factor.

3. a. $3a^2 - 6a^2b$ b. $2x^3 + 4x^2 + 6x$ c. $-y^2 - y^3$
4. a. $a^2 + a^2b$ b. $4b - 4$ c. $b - b^2 - b^3$

Write as a polynomial.

5. a. $(x - 2)(x + 3)$ b. $(2a - 3)(3a - 4)$ c. $(2a - 3)^2$
6. a. $(x + a)(x - 2a)$ b. $(2x - b)(x + b)$ c. $(2b + 1)^2$

Factor.

7. a. $x^2 - 4x - 21$ b. $10a^2 + 17a + 3$ c. $4x^2 - 9$
8. a. $a^2 - 10a + 21$ b. $3b^2 + 4b + 1$ c. $2b^2 + 3b - 2$
9. a. $2x^2 + 14x + 24$ b. $3y^2 + 24y - 60$ c. $4x^3 - 4x$
10. a. $x^2 - 3ax + 2a^2$ b. $x^2 - a^2$ c. $4b^2 + 6bc - 4c^2$

11. Solve.
    a. $3(x - 5) = 45$ b. $32 - 6b = 4(3 + b)$ c. $-(b - 2) = 26 + 3b$

12. Factor the right-hand member of the formula $A = 2krh + 2kr^2$ and evaluate for $k = 3.14$, $r = 7$, and $h = 10$.

13. The sum of two numbers is 24. If one of the numbers is represented by x, how can the second number be represented in terms of x?

14. How can the value (in cents) of x dimes be represented in terms of x?

15. How can the value (in cents) of $(x + 3)$ quarters be represented in terms of x?

16. If oranges cost $1.85 per dozen, how can the cost (in cents) of $(x + 4)$ dozen oranges be represented in terms of x?

17. Two investments produce an annual interest of $324. An amount of $500 more is invested at 12% than at 10%. How much is invested at each rate?

18. One number is six more than a second number. Ten times the smaller number minus four times the larger number equals six. Find the numbers.

19. One number is 10 more than a second number. Eight times the smaller number added to three times the larger number equals 129. Find the numbers.

20. A woman had $2.65 in change, consisting of eight more nickels than dimes. How many of each coin did she have?

10 PROPERTIES OF FRACTIONS

Fractions were introduced in Chapters 2 and 3. The treatment in this chapter will provide you with an opportunity to review this important topic and extend your skills as we apply the basic properties of fractions to algebraic expressions.

You may want to refer to the Summary of Part I to review some of the properties of fractions which we will expand upon in this chapter and Chapter 11.

10.1 FORMS OF FRACTIONS; GRAPHICAL REPRESENTATIONS

In earlier sections we have used the fraction $\frac{a}{b}$ to indicate the quotient $a \div b$. We called a the numerator (or dividend) and b the denominator (or divisor).

Example 1
a. $5 \div 8$
$= \frac{5}{8}$

b. $x \div 2y$
$= \frac{x}{2y}$

c. $(3x + 1) \div (y - 2)$
$= \frac{3x + 1}{y - 2}$

We defined a quotient $\frac{a}{b}$ to be a number q such that

$$b \cdot q = a$$

For example,

$$\frac{6}{2} = 3 \quad \text{because} \quad 2 \cdot 3 = 6, \quad \text{and} \quad \frac{12}{3} = 4 \quad \text{because} \quad 3 \cdot 4 = 12$$

We can also view a quotient in terms of a product. For example,

$$\frac{6}{2} = 6 \cdot \frac{1}{2}, \quad \frac{12}{3} = 12 \cdot \frac{1}{3}, \quad \text{and} \quad \frac{3}{4} = 3 \cdot \frac{1}{4}$$

In general,

$$\frac{a}{b} = a \cdot \frac{1}{b} \quad (b \neq 0)$$

PROPERTIES OF FRACTIONS

Example 2 a. $\dfrac{3}{7} = 3 \cdot \dfrac{1}{7}$ b. $\dfrac{x+2}{5} = (x+2) \cdot \dfrac{1}{5}$ c. $\dfrac{3}{y-1} = 3 \cdot \dfrac{1}{y-1}$

We can use this property to help us graph arithmetic fractions on a number line. For example, to graph the fraction $\dfrac{3}{4}$, which is equivalent to $3 \cdot \dfrac{1}{4}$, we divide a unit on the number line into four equal parts and then mark a point at the third quarter, as shown in Figure 10.1. In general, we can locate the graph of any fraction by dividing a unit on the number line into a number of equal parts corresponding to the denominator of the fraction, and then counting off the number of parts corresponding to the numerator.

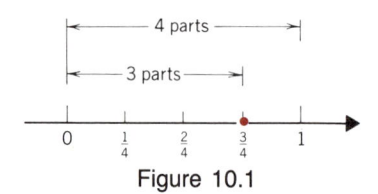

Figure 10.1

Example 3 Graph the fractions $\dfrac{-3}{2}$, $\dfrac{-1}{4}$, and $\dfrac{5}{4}$.

Solution

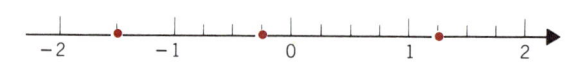

Fractions can involve algebraic expressions. In such cases, since division by 0 is undefined, we must restrict variables so that a divisor is never 0. In our work, *we will assume that no denominator is 0 unless otherwise specified.* For example,

For $\dfrac{3}{x}$, we assume that $x \neq 0$;

for $\dfrac{3x}{y+5}$, we assume that $y \neq -5$.

SIGNS OF FRACTIONS

There are three signs associated with a fraction: the sign of the numerator, the sign of the denominator, and the sign of the fraction itself.

$$+\dfrac{-6}{+3}$$

sign of fraction → + , sign of numerator, sign of denominator

Fractions that have different signs may have the same value. For example,

$$+\dfrac{+6}{+3} = +2 = 2, \qquad -\dfrac{-6}{+3} = -(-2) = 2$$

$$+\dfrac{-6}{-3} = +2 = 2, \qquad -\dfrac{+6}{-3} = -(-2) = 2$$

Each fraction above names the number 2.

$$+\dfrac{-6}{+3} = +(-2) = -2, \qquad -\dfrac{+6}{+3} = -(+2) = -2$$

10.1 FORMS OF FRACTIONS; GRAPHICAL REPRESENTATIONS

$$+\frac{+6}{-3} = +(-2) = -2, \quad -\frac{-6}{-3} = -(+2) = -2$$

Each fraction above names the number -2.

The above examples suggest the following rule.

> **Any two of the three signs of a fraction may be changed without changing the value of the fraction.**

For example,

$$\frac{8}{4} = \begin{cases} \text{Change sign of numerator and denominator.} \longrightarrow \frac{-8}{-4} \\ \text{Change sign of numerator and fraction.} \longrightarrow -\frac{-8}{4} \\ \text{Change sign of fraction and denominator.} \longrightarrow -\frac{8}{-4} \end{cases}$$

and

$$\frac{-8}{4} = \begin{cases} \text{Change sign of numerator and denominator.} \longrightarrow \frac{8}{-4} \\ \text{Change sign of numerator and fraction.} \longrightarrow -\frac{8}{4} \\ \text{Change sign of fraction and denominator.} \longrightarrow -\frac{-8}{-4} \end{cases}$$

In general,

$$\frac{a}{b} = \frac{-a}{-b} = -\frac{-a}{b} = -\frac{a}{-b}$$

and

$$\frac{-a}{b} = \frac{a}{-b} = -\frac{a}{b} = -\frac{-a}{-b}$$

We will refer to the two forms $\frac{a}{b}$ and $\frac{-a}{b}$ as standard forms. Note that the standard form has positive signs on the denominator and on the fraction itself. Although the form $-\frac{a}{b}$ is not in standard form, it is a form that is also commonly used.

Example 4 Write each fraction in standard form.

a. $-\frac{3}{x}$ b. $\frac{-x}{-2}$ c. $-\frac{y}{-x}$

Solutions a. $-\frac{3}{x} = \frac{-3}{x}$ b. $\frac{-x}{-2} = \frac{x}{2}$ c. $-\frac{y}{-x} = \frac{y}{x}$

If the numerator contains more than one term, there are alternative standard forms for a fraction.

PROPERTIES OF FRACTIONS

Example 5 Write $-\dfrac{x-3}{4}$ in standard form.

Solution Since

$$-\dfrac{x-3}{4} = \dfrac{-(x-3)}{4}$$

$$= \dfrac{-x+3}{4}$$

$$= \dfrac{3-x}{4}$$

any of the three forms on the right-hand side of the equals sign may be taken as standard form.

Common Errors When we write a standard form of a fraction such as $-\dfrac{x-3}{4}$, we must be careful how we change the signs in the numerator. The use of parentheses helps us avoid errors. Note that in the above example

$$-\dfrac{x-3}{4} = \dfrac{-(x-3)}{4} = \dfrac{-x+3}{4}$$

In particular note that

$$-\dfrac{x-3}{4} \neq \dfrac{-x-3}{4}$$

EXERCISES 10.1
Represent each quotient in fractional form. See Example 1.

Complete Examples
a. $9 \div 4 =$ _____ (1)

b. $(2x - y) \div 3y =$ _____ (2)

(1) $\dfrac{9}{4}$ (2) $\dfrac{2x-y}{3y}$

1. $4 \div 7$
2. $7 \div 2$
3. $3x \div y$
4. $x \div 3y$

5. $7 \div (x-y)$
6. $(2x+y) \div 3$
7. $(x-3) \div (4x+1)$
8. $(4y-2) \div (y+3)$

322

10.1 FORMS OF FRACTIONS; GRAPHICAL REPRESENTATIONS

Write each quotient in the form $a \cdot \dfrac{1}{b}$. See Example 2.

Complete Examples

a. $\dfrac{5}{9} = 5 \cdot \underline{}_{(1)}$

b. $\dfrac{x-3}{4} = (x-3) \cdot \underline{}_{(2)}$

c. $\dfrac{5}{y+1} = 5 \cdot \underline{}_{(3)}$

(1) $\dfrac{1}{9}$ (2) $\dfrac{1}{4}$ (3) $\dfrac{1}{y+1}$

9. $\dfrac{4}{7}$

10. $\dfrac{7}{9}$

11. $\dfrac{9}{5}$

12. $\dfrac{3}{7}$

13. $\dfrac{x-3}{4}$

14. $\dfrac{y+1}{6}$

15. $\dfrac{2}{x+3}$

16. $\dfrac{5}{y-2}$

Graph each set of numbers on a number line. Use a separate graph for each exercise. See Example 3.

Complete Examples

a. $\dfrac{2}{3}, \dfrac{-4}{3}, \dfrac{5}{3}$

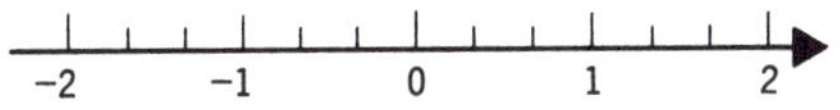

b. $\dfrac{-1}{4}, \dfrac{5}{4}, \dfrac{-7}{4}$

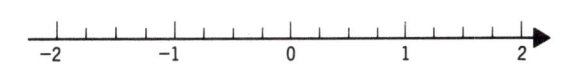

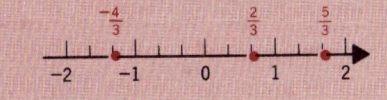

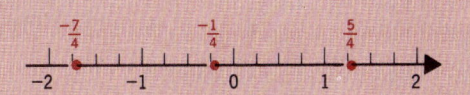

17. $\dfrac{1}{4}, \dfrac{3}{4}$

18. $\dfrac{1}{3}, \dfrac{2}{3}$

19. $\dfrac{1}{2}, \dfrac{5}{2}$

20. $\dfrac{-1}{4}, \dfrac{-3}{4}$

21. $\dfrac{-5}{6}, \dfrac{1}{6}$

22. $\dfrac{-1}{2}, \dfrac{1}{2}$

23. $\dfrac{-5}{2}, \dfrac{5}{4}$

24. $3, \dfrac{3}{4}, \dfrac{3}{2}$

25. $-3, \dfrac{-3}{4}, \dfrac{3}{2}$

26. $\dfrac{-2}{3}, \dfrac{1}{3}, 0$

27. $\dfrac{2}{5}, \dfrac{3}{5}, \dfrac{4}{5}$

28. $3, \dfrac{-5}{3}, 0$

323

PROPERTIES OF FRACTIONS

Rewrite each fraction in standard form. See Examples 4 and 5.

Complete Examples

a. $-\dfrac{4x}{y} = $ _____ (1)

b. $\dfrac{x-1}{-3} = \dfrac{-(x-1)}{\underline{}}$ (2)

$= \dfrac{-x+1}{3} = $ _____ (3)

---

(1) $\dfrac{-4x}{y}$ (2) 3 (3) $\dfrac{1-x}{3}$

---

29. $\dfrac{-3}{-5}$

30. $-\dfrac{-1}{2}$

31. $-\dfrac{2}{-7}$

32. $-\dfrac{-1}{-3}$

33. $\dfrac{2}{-5}$

34. $-\dfrac{-2}{5}$

35. $-\dfrac{-a}{-b}$

36. $-\dfrac{-a}{b}$

37. $\dfrac{a}{-b}$

38. $-\dfrac{a}{-b}$

39. $\dfrac{-x}{y}$

40. $-\dfrac{3y}{x}$

41. $-\dfrac{7x}{-8y}$

42. $-\dfrac{2c}{-1}$

43. $-\dfrac{-c}{-1}$

44. $\dfrac{c}{-1}$

45. $-\dfrac{x+2}{4}$

46. $\dfrac{x+3}{-3}$

47. $-\dfrac{x+5}{-4}$

48. $-\dfrac{x-3}{2}$

49. $-\dfrac{2x-1}{x+2}$

50. $-\dfrac{4x-3}{x-5}$

51. $-\dfrac{-x+3}{2}$

52. $-\dfrac{-x-4}{3}$

53. For what value of x is $\dfrac{5}{x}$ undefined?

54. For what value of y is $\dfrac{3}{y}$ undefined?

55. For what value of x is $\dfrac{7}{x-3}$ undefined?

56. For what value of y is $\dfrac{5}{y+2}$ undefined?

57. Use a numerical example to show that $-\dfrac{x-1}{2}$ is not equivalent to $\dfrac{-x-1}{2}$.

58. Use a numerical example to show that $-\dfrac{-x+2}{3}$ is not equivalent to $\dfrac{x+2}{3}$.

10.2 REDUCING FRACTIONS TO LOWEST TERMS

In algebraic expressions involving fractions, as well as arithmetic fractions, we use the fundamental principle of fractions to reduce a fraction to lowest terms.

10.2 REDUCING FRACTIONS TO LOWEST TERMS

> If both the numerator and the denominator of a given fraction are divided by the same nonzero number, the resulting fraction is equivalent to the given fraction.

Example 1 a. $\dfrac{\cancel{x} \cdot y}{\cancel{x} \cdot 3} = \dfrac{y}{3}$ b. $\dfrac{\cancel{(x+2)}(x-4)}{\cancel{(x+2)}(x+1)} = \dfrac{x-4}{x+1}$

As with arithmetic fractions, when we reduce fractions containing variables, it is easiest to first write the numerator and denominator in factored form and then divide each by their common factors.

Example 2 a. $\dfrac{10y^2}{4y}$ b. $\dfrac{xy^2}{-x^2 y}$ c. $\dfrac{-6a^3 b}{15ab^2}$

$= \dfrac{5 \cdot \cancel{2} \cancel{y} y}{2 \cdot \cancel{2} \cancel{y}}$ $= \dfrac{\cancel{x} \cancel{y} y}{-1 \cancel{x} x \cancel{y}}$ $= \dfrac{-2 \cdot \cancel{3} \cancel{a} aa \cancel{b}}{\cancel{3} \cdot 5 \cancel{a} \cancel{b} b}$

$= \dfrac{5y}{2}$ $= \dfrac{-y}{x}$ $= \dfrac{-2a^2}{5b}$

It is *critical* that when we reduce a fraction in which the numerator and/or the denominator contains more than one term, we must factor wherever possible before applying the fundamental theorem.

Example 3 Reduce

a. $\dfrac{6x - 4y}{12x - 8y}$ b. $\dfrac{x^2 + x - 12}{x^2 + 2x - 15}$

Solutions First, we factor the numerators and denominators and then divide out common factors.

a. $\dfrac{6x - 4y}{12x - 8y} = \dfrac{\cancel{2}\cancel{(3x-2y)}}{2 \cdot \cancel{2}\cancel{(3x-2y)}}$ b. $\dfrac{x^2 + x - 12}{x^2 + 2x - 15} = \dfrac{(x+4)\cancel{(x-3)}}{(x+5)\cancel{(x-3)}}$

$= \dfrac{1}{2}$ $= \dfrac{x+4}{x+5}$

AN ALTERNATIVE METHOD OF REDUCING FRACTIONS

We have been using the fundamental principle of fractions to reduce fractions. Sometimes, we can reduce fractions by a more direct method.

Consider the quotient

$$\dfrac{x^5}{x^2} = \dfrac{\cancel{x} \cdot \cancel{x} \cdot x \cdot x \cdot x}{\cancel{x} \cdot \cancel{x}} = x^3$$

Note that $x^3 = x^{5-2}$. Hence,

$$\dfrac{x^5}{x^2} = x^{5-2}$$

325

PROPERTIES OF FRACTIONS

In general, for $\dfrac{a^m}{a^n}$, where $n < m$, we have

$$\dfrac{a^m}{a^n} = \dfrac{\overbrace{aaa\cdots a}^{n \text{ factors}}\;\overbrace{aa\cdots a}^{(m-n)\text{ factors}}}{\underbrace{aaa\cdots a}_{n \text{ factors}}} = \overbrace{aaa\cdots a}^{(m-n)\text{ factors}}$$

which, in exponential form, is written as a^{m-n}. Thus,

$$\dfrac{a^m}{a^n} = a^{m-n} \quad (n < m) \tag{1}$$

Example 4

a. $\dfrac{x^5}{x^2} = x^{5-2}$
$= x^3$

b. $\dfrac{12x^3y^4}{4x^2y} = \left(\dfrac{12}{4}\right)\left(\dfrac{x^3}{x^2}\right)\left(\dfrac{y^4}{y}\right)$
$= 3x^{3-2}y^{4-1}$
$= 3xy^3$

If the greater exponent is in the denominator, that is, if n is greater than m, then

$$\dfrac{a^m}{a^n} = \dfrac{\overbrace{aaaaa\cdots a}^{m \text{ factors}}}{\underbrace{aaaaa\cdots a}_{m \text{ factors}}\;\underbrace{a\cdots a}_{(n-m)\text{ factors}}}$$

or

$$\dfrac{a^m}{a^n} = \dfrac{1}{a^{n-m}} \quad (n > m) \tag{2}$$

Thus, when we divide powers with the same base, either we can factor each power and divide out common factors or we can use Properties (1) and (2) above.

Example 5 Reduce $\dfrac{6x^3y}{2xy^4}$ by

a. Using the fundamental principle of fractions.
b. Using Properties (1) and (2).

Solutions

a. $\dfrac{6x^3y}{2xy^4} = \dfrac{\cancel{2}\cdot 3\,\cancel{x}xx\,\cancel{y}}{\cancel{2}\,\cancel{x}\,\cancel{y}yyy} = \dfrac{3x^2}{y^3}$

b. $\dfrac{6x^3y}{2xy^4} = \dfrac{3x^{3-1}}{y^{4-1}} = \dfrac{3x^2}{y^3}$

Common Errors In Chapter 3 we noted some common errors that are made in working with fractions. We again call your attention to the following kinds of errors so that you may avoid them. Note that

$$\dfrac{1+3}{4+3} \ne \dfrac{1+\cancel{3}}{4+\cancel{3}} = \dfrac{1}{4}$$

From the fundamental principle of fractions,

$$\dfrac{1\cdot \cancel{3}}{4\cdot \cancel{3}} = \dfrac{1}{4}$$

10.2 REDUCING FRACTIONS TO LOWEST TERMS

Thus, while we can divide out common factors, *we cannot divide out common terms*. As another example, note that

$$\frac{2+4}{2} \neq \frac{\cancel{2}+4}{\cancel{2}} = 4$$

From the fundamental principle,

$$\frac{\cancel{2} \cdot 4}{\cancel{2}} = 4$$

As another example of a common error, note that:

$$\frac{2x+4}{2} \neq \frac{\cancel{2}x+4}{\cancel{2}}$$
$$= x + 4$$

In the expression

$$\frac{2x+4}{2}$$

we cannot "divide out" the 2's until we write 2 as a factor of the entire numerator. Thus,

$$\frac{2x+4}{2} = \frac{\cancel{2}(x+2)}{\cancel{2}} = x + 2$$

EXERCISES 10.2

Reduce each fraction to an equivalent fraction in lowest terms by first completely factoring the numerator and the denominator and then dividing each by their common factors. Express your answer in standard form.

Complete Examples

a. $\dfrac{9x^2}{6x} = \dfrac{3 \cdot 3xx}{2 \cdot \underline{\qquad}}$
$\qquad\qquad\qquad (1)$

$= \dfrac{\underline{\qquad}}{(2)}$

b. $\dfrac{-9x^3y}{21xy^2} = \dfrac{-3 \cdot 3xxxy}{3 \cdot 7\underline{\qquad}}$
$\qquad\qquad\qquad\qquad (3)$

$= \dfrac{\underline{\qquad}}{(4)}$

---

(1) $3x$ (2) $\dfrac{3x}{2}$ (3) xyy (4) $\dfrac{-3x^2}{7y}$

---

1. $\dfrac{2x^3}{8x^4}$

2. $\dfrac{8x^2}{12x^5}$

3. $\dfrac{-6y}{9y^5}$

4. $\dfrac{-8y}{18y^3}$

5. $\dfrac{14}{-28x}$

6. $\dfrac{48}{-9x^3}$

7. $\dfrac{x^3y}{x}$

8. $\dfrac{xy^4}{y^2}$

PROPERTIES OF FRACTIONS

9. $\dfrac{-x^2}{x^3y^2}$
10. $\dfrac{-y}{xy^3}$
11. $\dfrac{4ax^2}{2a}$
12. $\dfrac{6bx}{3b}$

13. $\dfrac{12bx^4}{8bx^2}$
14. $\dfrac{12ax^5}{20ax}$
15. $\dfrac{5ab^2c^3}{4abc}$
16. $\dfrac{12a^2b^3c}{10ab^2c}$

17. $\dfrac{26a^3b^2c}{6ab^2c^3}$
18. $\dfrac{24abc}{6a^2b^2c^2}$

Complete Examples (see Example 3)

a. $\dfrac{4a + 4b}{6a + 6b} = \dfrac{2 \cdot 2(a + b)}{2 \cdot 3(\underline{})}$
    (1)

   $= \underline{}$
   (2)

b. $\dfrac{x^2 + 5x + 4}{x^2 - 16} = \dfrac{(x + 4)(x + 1)}{(x + 4)(\underline{})}$
   (3)

   $= \underline{}$
   (4)

(1) $a + b$ (2) $\dfrac{2}{3}$ (3) $x - 4$ (4) $\dfrac{x + 1}{x - 4}$

19. $\dfrac{3(a + b)}{4(a + b)}$
20. $\dfrac{4(a + 2b)}{6(a + 2b)}$
21. $\dfrac{12(x - y)}{-3}$
22. $\dfrac{15(a + b)}{-5}$

23. $\dfrac{(a - b)}{(a - b)}$
24. $\dfrac{(2x - y)}{(2x - y)}$
25. $\dfrac{2x + 2y}{-(x + y)}$
26. $\dfrac{3x - 3y}{-(3x + 3y)}$

27. $\dfrac{2x - 2a}{(x - a)^2}$
28. $\dfrac{3x - 3a}{2(x - a)^2}$
29. $\dfrac{-4x}{4x^2 + 16x}$
30. $\dfrac{-3x}{6x^2 + 9x}$

31. $\dfrac{x + 1}{x^2 + 2x + 1}$
32. $\dfrac{x - 4}{x^2 - 3x - 4}$
33. $\dfrac{a - b}{a^2 - 2ab + b^2}$
34. $\dfrac{a - b}{a^2 - b^2}$

35. $\dfrac{(a - b)^2}{a^2 - b^2}$
36. $\dfrac{(x - 2y)^2}{x^2 - 4y^2}$
37. $\dfrac{a^2 - 3a}{a^2 - 2a - 3}$
38. $\dfrac{a^2 - a}{a^2 + a - 2}$

39. $\dfrac{x^2 + x - 6}{x^2 - 9}$
40. $\dfrac{x^2 + 5x + 6}{x^2 - 4}$
41. $\dfrac{a^2 + 6a + 9}{a^2 + 2a - 3}$
42. $\dfrac{x^2 + 5x + 6}{x^2 + 6x + 9}$

10.2 REDUCING FRACTIONS TO LOWEST TERMS

Reduce each fraction: a. By using the fundamental principle of fractions. b. By using Properties (1) and (2) on page 326. See Examples 4 and 5.

Complete Example

$$\frac{3x^2y^4}{6x^3y}$$

a. $\frac{3x^2y^4}{6x^3y} = \frac{\cancel{3} \cdot x \cdot x \cdot \cancel{y} \cdot yyy}{2 \cdot \cancel{3} \cdot x \cdot x \cdot x \cdot \cancel{y}}$

= _____
(1)

b. $\frac{3x^2y^4}{6x^3y} = \left(\frac{3}{6}\right)\left(\frac{x^2}{x^3}\right)\left(\frac{y^4}{y}\right) = \frac{y^{4-1}}{2x^{3-2}}$

= _____
(2)

(1) $\frac{y^3}{2x}$ (2) $\frac{y^3}{2x}$

43. $\frac{12x^3y^2}{3xy}$

44. $\frac{18x^4y^2}{6xy}$

45. $\frac{15x^3y^3}{-3x^4y}$

46. $\frac{25x^2y^2}{-5x^5y}$

47. $\frac{-2x^3y}{-30x^2y^3}$

48. $\frac{-7x^3y^4}{-35x^5y}$

Reduce each fraction if possible. Select the correct response, a or b. See Common Errors on pages 326–327.

49. $\frac{x+2}{y+2}$;   a. $\frac{x}{y}$ b. Already in lowest terms

50. $\frac{2x+3}{2y}$;   a. $\frac{x+3}{y}$ b. Already in lowest terms

51. $\frac{2x+4}{4}$;   a. $\frac{x+2}{2}$ b. $2x$

52. $\frac{6+3y}{3y}$;   a. $\frac{2+y}{y}$ b. 6

53. $\frac{3(x-2y)}{3x}$;   a. $-2y$ b. $\frac{x-2y}{x}$

54. $\frac{3a+a^2}{3a}$;   a. a^2 b. $\frac{3+a}{3}$

329

PROPERTIES OF FRACTIONS

55. $\dfrac{y^2 - 1}{y - 1}$; a. $y + 1$ b. y

56. $\dfrac{a^3}{a^4 - a^3}$; a. $\dfrac{1}{a - 1}$ b. $\dfrac{1}{a^4}$

57. Use a numerical example to show that $\dfrac{x + y}{x}$ is not equivalent to y.

58. Use a numerical example to show that $\dfrac{2x + y}{2}$ is not equivalent to $x + y$.

10.3 QUOTIENTS OF POLYNOMIALS

MONOMIAL DIVISORS

In Section 10.2 we simplified a quotient by reducing the fraction. In this section, we will study two alternative methods of rewriting quotients in equivalent forms.

We use the first method when the divisor is a monomial. As noted in Section 10.1,

$$\frac{a + b + c}{d} = (a + b + c) \cdot \frac{1}{d}$$

and by the distributive property,

$$(a + b + c) \cdot \frac{1}{d} = a \cdot \frac{1}{d} + b \cdot \frac{1}{d} + c \cdot \frac{1}{d}$$

$$= \frac{a}{d} + \frac{b}{d} + \frac{c}{d}$$

Therefore,

$$\frac{a + b + c}{d} = \frac{a}{d} + \frac{b}{d} + \frac{c}{d}$$

Thus, a fraction whose numerator is a sum or difference of several terms can be expressed as the sum or difference of fractions whose numerators are the terms of the original numerator and whose denominators are the same as the original denominator.

Example 1 a. $\dfrac{2x^3 + 4x^2 + 2x}{2x} = \dfrac{2x^3}{2x} + \dfrac{4x^2}{2x} + \dfrac{2x}{2x}$

$= x^2 + 2x + 1$

10.3 QUOTIENTS OF POLYNOMIALS

Example 2   b. $\dfrac{3x^2 + 2x + 1}{x} = \dfrac{3\overset{x}{\cancel{x^2}}}{\cancel{x}} + \dfrac{2\cancel{x}}{\cancel{x}} + \dfrac{1}{x}$

$= 3x + 2 + \dfrac{1}{x}$

POLYNOMIAL DIVISORS

We use the next method when the divisor is a polynomial. In this case, we use a process similar to arithmetic long division, as the following examples illustrate.

$21\overline{)674}$ $x + 3\overline{)x^2 + x - 7}$

Divide 2 into 6; the quotient is 3.
3
$21\overline{)674}$

Divide x into x^2; the quotient is x.
x
$x + 3\overline{)x^2 + x - 7}$

Multiply 3 by 21: $3 \cdot 21 = 63$.
3
$21\overline{)674}$
63

Multiply x by $x + 3$: $x(x + 3) = x^2 + 3x$.
x
$x + 3\overline{)x^2 + x - 7}$
$x^2 + 3x$

Subtract. (Change the sign and add.)
3
$21\overline{)\ 674}$
-63
4

Subtract. (Change the signs and add.)
x
$x + 3\overline{)\ x^2 + x - 7}$
$-x^2 - 3x$
$-2x$

"Bring down" 4.
3
$21\overline{)\ 674}$
$63\downarrow$
44

"Bring down" -7.
x
$x + 3\overline{)\ x^2 + x - 7}$
$-x^2 - 3x\downarrow$
$-2x - 7$

Divide 2 into 4; the quotient is 2.
32
$21\overline{)\ 674}$
-63
44

Divide x into $-2x$; the quotient is -2.
$x - 2$
$x + 3\overline{)\ x^2 + x - 7}$
$-x^2 - 3x$
$-2x - 7$

Multiply 2 by 21: $2 \cdot 21 = 42$.
32
$21\overline{)\ 674}$
-63
44
42

Multiply -2 by $x + 3$:
$-2(x + 3) = -2x - 6$.
$x - 2$
$x + 3\overline{)\ x^2 + x - 7}$
$-x^2 - 3x$
$-2x - 7$
$-2x - 6$

331

PROPERTIES OF FRACTIONS

Subtract. (Change the signs and add.)

$$\begin{array}{r} 32 \\ 21\overline{\smash{)}674} \\ -63 \\ \hline 44 \\ 42 \\ \hline 2 \end{array}$$

Subtract. (Change the signs and add.)

$$\begin{array}{r} x - 2 \\ x+3\overline{\smash{)}x^2 + x - 7} \\ -x^2 - 3x \\ \hline -2x - 7 \\ +2x + 6 \\ \hline -1 \end{array}$$

The remainder is 2.
$$\begin{array}{r} 32\tfrac{2}{21} \\ 21\overline{\smash{)}674} \\ -63 \\ \hline 44 \\ -42 \\ \hline 2 \end{array}$$

Ans. $32\dfrac{2}{21}$

The remainder is -1.
$$\begin{array}{r} x - 2 + \tfrac{-1}{x+3} \\ x+3\overline{\smash{)}x^2 + x - 7} \\ -x^2 - 3x \\ \hline -2x - 7 \\ +2x + 6 \\ \hline -1 \end{array}$$

Ans. $x - 2 + \dfrac{-1}{x + 3}$

As always, the division is not valid if the divisor is 0. Thus, in the example where the divisor is $x + 3$, we must restrict x from having a value of -3.

When using the long division process, it is convenient to write the dividend in descending powers of the variable. Furthermore, it is helpful to insert a term with a zero coefficient for all powers of the variable that are missing between the highest-degree term and the lowest-degree term.

Example 3   Divide $3x - 1 + 4x^3$ by $2x - 1$.

Solution   We first rewrite $3x - 1 + 4x^3$ as

$$4x^3 + 0x^2 + 3x - 1 \quad \text{— Descending powers}$$

with 0 coefficient for the missing x^2 term.

Now, using the techniques of Example 2, we obtain

$$\begin{array}{r}
2x^2 + x + 2 \\
2x-1\overline{\smash{)}4x^3 + 0x^2 + 3x - 1} \\
4x^3 - 2x^2 \\
\hline
2x^2 + 3x \\
2x^2 - x \\
\hline
4x - 1 \\
4x - 2 \\
\hline
1
\end{array}$$

$2x^2(2x - 1)$ Subtract
$x(2x - 1)$ Subtract
$2(2x - 1)$ Subtract   Remainder

- $4x^3 \div 2x = 2x^2$
- $2x^2 \div 2x = x$
- $4x \div 2x = 2$

The quotient is $2x^2 + x + 2 + \dfrac{1}{2x - 1}$.

10.3 QUOTIENTS OF POLYNOMIALS

EXERCISES 10.3
Rewrite each quotient using the methods of this section. See Example 1.

Complete Examples

a. $\dfrac{6x - 8}{2}$

$= \dfrac{\cancel{6}^{3}x}{\cancel{2}} - \dfrac{\cancel{8}^{4}}{\cancel{2}}$

$= \underline{}$
(1)

b. $\dfrac{12x^3 - 8x^2 + 4x}{2x}$

$= \dfrac{\cancel{12}^{6x^2}\cancel{x^3}}{\cancel{2}\cancel{x}} - \dfrac{\cancel{8}^{4x}\cancel{x^2}}{\cancel{2}\cancel{x}} + \dfrac{\cancel{4}^{2}\cancel{x}}{\cancel{2}\cancel{x}}$

$= \underline{}$
(2)

c. $\dfrac{2x^3 - x^2 + 4}{-x}$

$= \dfrac{2\cancel{x^3}^{x^2}}{-\cancel{x}} + \dfrac{-\cancel{x^2}^{x}}{-\cancel{x}} + \dfrac{4}{-x}$

$= \underline{}$
(3)

(1) $3x - 4$ (2) $6x^2 - 4x + 2$ (3) $-2x^2 + x + \dfrac{-4}{x}$

1. $\dfrac{8x - 4}{4}$
2. $\dfrac{6x + 3}{3}$
3. $\dfrac{y^2 + 2y}{y}$
4. $\dfrac{y^2 - 4y}{y}$
5. $\dfrac{3x^2 + 9x}{3x}$
6. $\dfrac{4x^2 - 2x}{2x}$
7. $\dfrac{3y^3 - 2y^2 + y}{y}$
8. $\dfrac{y^4 + 2y^3 + y^2}{y^2}$
9. $\dfrac{6y^3 - 3y^2 + 9y}{3y}$
10. $\dfrac{12y^3 + 4y^2 - 8y}{4y}$
11. $\dfrac{4xy^2 - x^2y + xy}{xy}$
12. $\dfrac{9x^2y^2 + 3xy^2 - 3x^2y}{3xy}$
13. $\dfrac{9x^2 + 6x - 1}{3}$
14. $\dfrac{8x^2 + 4x - 1}{4}$
15. $\dfrac{y^2 + 2y - 1}{y}$
16. $\dfrac{6y^2 + 4y - 3}{2y}$
17. $\dfrac{9x^4 - 6x^2 - 2}{3x^2}$
18. $\dfrac{6x^4 - 6x^2 - 4}{6x^2}$
19. $\dfrac{y^3 - 3y^2 + 2y - 1}{y}$
20. $\dfrac{2y^3 + 8y^2 + 2y - 1}{2y}$
21. $\dfrac{xy^2 + xy + x}{xy}$
22. $\dfrac{x^3y^2 + x^2y^3 + xy}{xy^2}$
23. $\dfrac{2x^2y^2 - 4xy^2 + 6xy}{2xy^2}$
24. $\dfrac{8x^3y + 4x^2y - 4xy}{4x^2y}$

333

PROPERTIES OF FRACTIONS

Use long division to find each quotient.

Complete Example (see Example 2)

$(x^2 - 3x - 4) \div (x + 1)$

$x^2 \div x = x$
$-4x \div x = $ _____ (1)

$$\begin{array}{r} x - 4 \\ x + 1 \overline{)x^2 - 3x - 4} \end{array}$$

$x(x + 1)$ ⟶ $x^2 + x$
Subtract; change signs of $x^2 + x$ and add. ⟶ $-4x - 4$
$-4(x + 1)$ ⟶ $-4x - 4$
Subtract; change signs of $-4x - 4$ and add ⟶ _____ ← Remainder
$\qquad\qquad\qquad\qquad\qquad\qquad\qquad\qquad\qquad$ (2)

The quotient is _____ .
$\qquad\qquad\qquad\;\;$ (3)

---

(1) -4 (2) 0 (3) $x - 4$

---

25. $(x^2 + 5x - 6) \div (x - 1)$
26. $(x^2 + x - 6) \div (x - 2)$
27. $(x^2 + 6x + 5) \div (x + 5)$
28. $(x^2 - 4x + 4) \div (x - 2)$
29. $(2x^2 - 7x - 4) \div (x - 4)$
30. $(2x^2 - x - 3) \div (x + 1)$
31. $(2x^2 + 5x - 3) \div (2x - 1)$
32. $(2x^2 - 9x - 5) \div (2x + 1)$
33. $(4x^2 + 4x - 3) \div (2x - 1)$
34. $(4x^2 - 8x - 5) \div (2x + 1)$
35. $(2x^3 + 3x^2 - x + 2) \div (x + 2)$
36. $(3x^3 - x^2 - 4x + 2) \div (x - 1)$

Complete Example (see Example 2)

$(2x^2 + 3x - 3) \div (2x - 1)$

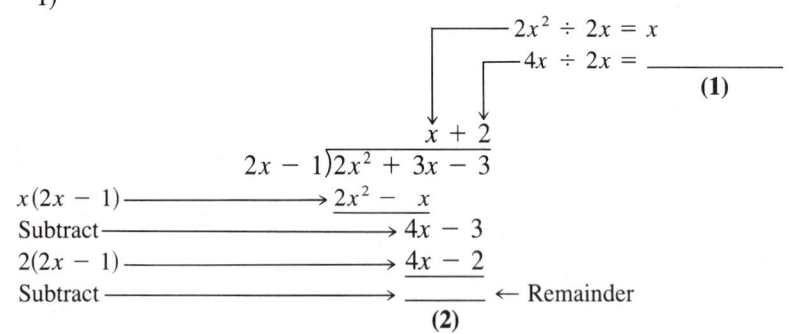

$2x^2 \div 2x = x$
$-4x \div 2x = $ _____ (1)

$$\begin{array}{r} x + 2 \\ 2x - 1 \overline{)2x^2 + 3x - 3} \end{array}$$

$x(2x - 1)$ ⟶ $2x^2 - x$
Subtract ⟶ $4x - 3$
$2(2x - 1)$ ⟶ $4x - 2$
Subtract ⟶ _____ ← Remainder
$\qquad\qquad\qquad\qquad\qquad\qquad$ (2)

The quotient is _____ .
$\qquad\qquad\qquad\qquad\;\;$ (3)

---

(1) 2 (2) -1 (3) $x + 2 + \dfrac{-1}{2x - 1}$

10.4 BUILDING FRACTIONS

37. $(x^2 + 3x + 1) \div (x + 2)$ **38.** $(x^2 - x + 3) \div (x + 1)$ **39.** $(x^2 + 3x - 9) \div (x + 5)$

40. $(x^2 - 2x - 2) \div (x - 3)$ **41.** $(2x^2 + x - 2) \div (x + 1)$ **42.** $(3x^2 - 8x - 1) \div (x - 3)$

43. $(4x^2 - 4x - 5) \div (2x + 1)$ **44.** $(6x^2 + x + 2) \div 3x + 2)$

Complete Example (see Examples 2 and 3)

$(-1 + x^2) \div (x + 1)$

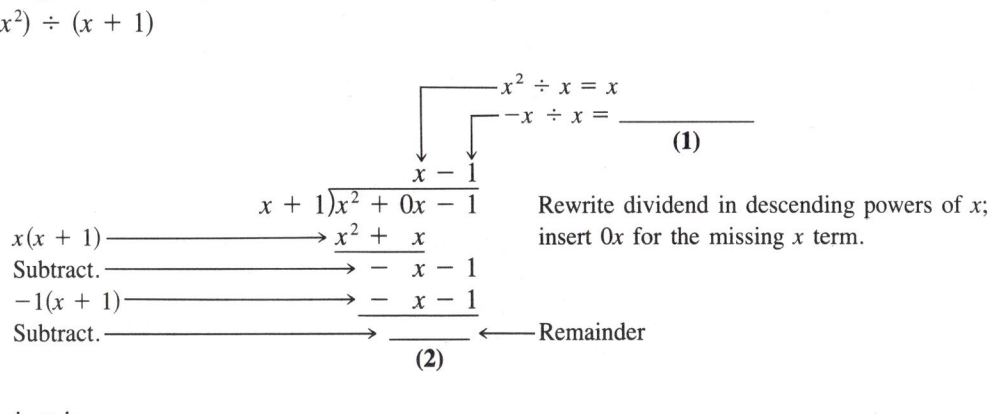

$$\begin{array}{r} x - 1 \\ x + 1 \overline{)x^2 + 0x - 1} \\ x^2 + x \\ \underline{} - x - 1 \\ - x - 1 \\ \underline{} \end{array}$$

$x^2 \div x = x$
$-x \div x = $ _____ (1)

$x(x + 1) \rightarrow x^2 + x$
Subtract.
$-1(x + 1) \rightarrow - x - 1$
Subtract. $\rightarrow$ _____ $\leftarrow$ Remainder
(2)

Rewrite dividend in descending powers of x; insert $0x$ for the missing x term.

The quotient is _____ .
(3)

---

(1) -1 (2) 0 (3) $x - 1$

---

45. $(x^2 - 49) \div (x - 7)$ **46.** $(x^2 - 4) \div (x + 2)$ **47.** $(-7 + x^2) \div (x + 6)$

48. $(-10 + x^2) \div (x - 7)$ **49.** $(1 + 2x^3 - x^2) \div (x - 1)$ **50.** $(2x + 4x^3 - 3) \div (x + 2)$

10.4 BUILDING FRACTIONS

In algebra, as in arithmetic, we use the fundamental principle to build fractions. We repeat this important principle here.

> **If both the numerator and denominator of a given fraction are multiplied by the same nonzero number, the resulting fraction is equivalent to the given fraction.**

In symbols:

$$\frac{a}{b} = \frac{a \cdot c}{b \cdot c}, \quad c \neq 0$$

335

PROPERTIES OF FRACTIONS

> To change $\dfrac{a}{b}$ to a fraction with a denominator bc:
>
> 1. Divide b, the denominator of the given fraction, into bc, the denominator to be obtained, to find the *building factor* c.
> 2. Multiply the numerator and denominator of the given fraction by the building factor c.

Example 1 Express the first fraction as a fraction with the indicated denominator.

$$\frac{3}{x^2y} = \frac{?}{x^3y^2}$$

Solution First, we obtain the building factor by dividing x^3y^2 by x^2y to get

$$(x^3y^2) \div (x^2y) = xy$$

Then, we can multiply the numerator and denominator of the first fraction by this building factor to obtain

$$\frac{3(xy)}{x^2y(xy)} = \frac{3xy}{x^3y^2}$$

If negative signs are attached to any part of the fraction, it is usually convenient to write the fraction in standard form before building it.

Example 2 Change $\dfrac{3}{-x^2y}$ to $\dfrac{?}{x^3y^2}$.

Solution First, we write $\dfrac{3}{-x^2y}$ in standard form $\quad \dfrac{3}{-x^2y} = \dfrac{-3}{x^2y}$

and then proceed as in Example 1.

It is helpful, when obtaining the building factor, to first write denominators in factored form.

Example 3 Change $\dfrac{3}{x-3}$ to $\dfrac{?}{x^2 - 7x + 12}$.

Solution First, we factor $x^2 - 7x + 12$ to get

$$\frac{3}{x-3} = \frac{?}{(x-3)(x-4)}$$

Next, we obtain the building factor

$$(x-3)(x-4) \div (x-3) = (x-4)$$

Now, we multiply the numerator and denominator of $\dfrac{3}{x-3}$ by the building factor $(x-4)$, to obtain

$$\frac{3(x-4)}{(x-3)(x-4)}$$

10.4 BUILDING FRACTIONS

EXERCISES 10.4
Express each fraction as a fraction with the indicated denominator.

Complete Example (see Examples 1 and 2)

$$\frac{2}{-5a} = \frac{?}{10a^2}$$

$$\frac{2}{-5a} = \underline{\qquad}_{(1)}$$ Write in standard form.

$$(10a^2 \div 5a = \underline{\qquad}_{(2)})$$ Obtain building factor.

$$\frac{-2(2a)}{5a(2a)} = \underline{\qquad}_{(3)}$$ Multiply numerator and denominator of given fraction by building factor $2a$.

(1) $\dfrac{-2}{5a}$ (2) $2a$ (3) $\dfrac{-4a}{10a^2}$

1. $\dfrac{5}{3x} = \dfrac{?}{6x}$
2. $\dfrac{6}{-7a} = \dfrac{?}{14a^2}$
3. $\dfrac{-a}{b} = \dfrac{?}{12b^3}$
4. $-\dfrac{3a}{5b} = \dfrac{?}{15ab}$
5. $\dfrac{-x^2}{y^2} = \dfrac{?}{3y^3}$
6. $\dfrac{-ax}{by} = \dfrac{?}{ab^2y}$

Complete Example

$$ab = \frac{?}{ab^2}$$

$$(ab^2 \div 1 = \underline{\qquad}_{(1)})$$ Obtain building factor.

$$\frac{ab(ab^2)}{1(ab^2)} = \underline{\qquad}_{(2)}$$ Multiply numerator and denominator of given fraction by building factor ab^2.

(1) ab^2 (2) $\dfrac{a^2b^3}{ab^2}$

7. $2 = \dfrac{?}{36}$
8. $-x = \dfrac{?}{y^2}$
9. $y = \dfrac{?}{xy}$
10. $3a = \dfrac{?}{9b^2}$
11. $x^2 = \dfrac{?}{3x^2y}$
12. $-2b^2 = \dfrac{?}{4a^2b}$

PROPERTIES OF FRACTIONS

Complete Example

$\dfrac{1}{3} = \dfrac{?}{3(x - a)}$

$[3(x - a) \div 3 = \underline{\quad\quad}_{(1)}]$

$\dfrac{1\,(x - a)}{3\,(x - a)} = \underline{\quad\quad}_{(2)}$

Obtain building factor.

Multiply numerator and denominator of given fraction by building factor $(x - a)$.

---

(1) $(x - a)$ (2) $\dfrac{x - a}{3(x - a)}$

---

13. $\dfrac{1}{2} = \dfrac{?}{2(x + y)}$

14. $\dfrac{2}{3} = \dfrac{?}{6(x - y)}$

15. $\dfrac{-2a}{5} = \dfrac{?}{5(a + 4)}$

16. $\dfrac{3b}{-4} = \dfrac{?}{4(a + b)^2}$

17. $2a = \dfrac{?}{a + 3}$

18. $3x = \dfrac{?}{6(x - 2)}$

19. $\dfrac{3}{x - y} = \dfrac{?}{(x - y)(x + y)}$

20. $\dfrac{2x}{-(x - y)} = \dfrac{?}{(x - y)(x - y)}$

21. $-\dfrac{3}{2x - 1} = \dfrac{?}{(x + 1)(2x - 1)}$

22. $\dfrac{-1}{a + b} = \dfrac{?}{(2a - b)(a + b)}$

23. $\dfrac{7a}{b + 2} = \dfrac{?}{(b - 3)(b + 2)}$

24. $\dfrac{6x^2}{3x - 4} = \dfrac{?}{(3x - 4)(2x + 5)}$

Complete Example (see Example 3)

$\dfrac{-2}{x + 2} = \dfrac{?}{x^2 - 4}$

$\dfrac{-2}{x + 2} = \dfrac{?}{(x + 2)(\underline{\quad})_{(1)}}$

$[(x + 2)(x - 2) \div (x + 2) = (\underline{\quad})_{(2)}]$

$= \dfrac{-2(x - 2)}{(x + 2)(\underline{\quad})_{(3)}}$

Factor denominator $x^2 - 4$.

Obtain building factor.

Multiply numerator and denominator of given fraction by building factor $(x - 2)$.

---

(1) $x - 2$ (2) $x - 2$ (3) $x - 2$

---

25. $\dfrac{a}{a - 3} = \dfrac{?}{a^2 - 3a}$

26. $\dfrac{2}{b} = \dfrac{?}{b + b^2}$

27. $\dfrac{-3}{x + y} = \dfrac{?}{x^2 - y^2}$

28. $\dfrac{-2}{a - b} = \dfrac{?}{a^2 - b^2}$

29. $\dfrac{y}{y - 1} = \dfrac{?}{y^2 + y - 2}$

30. $\dfrac{-1}{x - 1} = \dfrac{?}{2x^2 - 4x + 2}$

31. $\dfrac{x + 1}{x^2 - x} = \dfrac{?}{x^3 - 2x^2 + x}$

32. $\dfrac{y + 1}{y^2 - 1} = \dfrac{?}{(y - 1)(y^2 + 2y + 1)}$

10.5 INTEGER EXPONENTS; SCIENTIFIC NOTATION

In this section, we will introduce another symbol for a fraction of the form $\frac{1}{a^n}$ and then we will use this symbol to write certain numbers in simpler form.

INTEGER EXPONENTS

Recall that we have defined a power a^n (where n is a natural number) as follows:

$$a^n = a \cdot a \cdot a \cdots \cdot a \quad (n \text{ factors})$$

We will now give meaning to powers in which the exponent is 0 or a negative integer. First, let us consider the quotient a^4/a^4. Using the property of quotients of powers on page 326, we have

$$\frac{a^4}{a^4} = a^{4-4} = a^0$$

Note that for any a not equal to zero, the left-hand member equals 1 and the right-hand member equals a^0. In general, we define:

$$a^0 = 1$$

for any number a except 0.

Example 1 a. $3^0 = 1$ b. $425^0 = 1$ c. $(x^2 y)^0 = 1$

Now consider the quotient a^4/a^7. Using the two quotient laws for powers on page 326, we have

$$\frac{a^4}{a^7} = a^{4-7} = a^{-3}$$

and

$$\frac{a^4}{a^7} = \frac{1}{a^{7-4}} = \frac{1}{a^3}$$

Thus, for any a not equal to 0, we can view a^{-3} as equivalent to $\frac{1}{a^3}$. In general, we define

$$a^{-n} = \frac{1}{a^n}$$

for any number a except zero.

Example 2 a. $3^{-2} = \frac{1}{3^2}$ b. $10^{-4} = \frac{1}{10^4}$ c. $y^{-5} = \frac{1}{y^5}$

SCIENTIFIC NOTATION

Very large numbers such as

$$5{,}980{,}000{,}000{,}000{,}\,000{,}000{,}000{,}000{,}000$$

and very small numbers such as

$$0.000\ 000\ 000\ 000\ 000\ 000\ 000\ 001\ 67$$

PROPERTIES OF FRACTIONS

occur in many scientific areas. Large numbers can be rewritten in a more compact and useful form by using powers with positive exponents. We can also rewrite small numbers by using powers with negative exponents that have been introduced in this section.

First, let us consider some factored forms of 38,400 in which one of the factors is a power of 10.

$$38,400 = 3840 \times 10$$
$$= 384 \times 10^2$$
$$= 38.4 \times 10^3$$
$$= 3.84 \times 10^4$$

Although any one of such factored forms may be more useful than the original form of the number, a special name is given to the last form. A number expressed as the product of a number between 1 and 10 (including 1) and a power of 10 is said to be in **scientific form** or **scientific notation.** For example,

$$4.18 \times 10^4, \quad 9.6 \times 10^2, \quad \text{and} \quad 4 \times 10^5$$

are in scientific form.

Now, let us consider some factored forms of 0.0057 in which one of the factors is a power of 10.

$$0.0057 = \frac{0.057}{10} = 0.057 \times \frac{1}{10} = 0.057 \times 10^{-1}$$
$$= \frac{0.57}{100} = 0.57 \times \frac{1}{10^2} = 0.57 \times 10^{-2}$$
$$= \frac{5.7}{1000} = 5.7 \times \frac{1}{10^3} = 5.7 \times 10^{-3}$$

In this case, 5.7×10^{-3} is the scientific form for 0.0057.

To write a number in scientific form:

1. Move the decimal point so that there is one nonzero digit to the left of the decimal point.
2. Multiply the result by a power of ten with an exponent equal to the number of places the decimal point was moved. The exponent is positive if the decimal point has been moved to the left and it is negative if the decimal point has been moved to the right.

Example 3
   a. $248. = 2.48 \times 10^2$ b. $38.05 = 3.805 \times 10^1$

   c. $0.044 = 4.4 \times 10^{-2}$ d. $0.00241 = 2.41 \times 10^{-3}$

If a number is written in scientific form and we want to rewrite it in standard form, we simply reverse the above procedure.

Example 4
   a. 3.84×10^4 b. 5.6×10^{-2}
       $= 3.8400 \times 10^4$ $= 05.6 \times 10^{-2}$
       $= 38,400$ $= 0.056$

10.5 INTEGER EXPONENTS; SCIENTIFIC NOTATION

Common Error   Note that

$$3x^{-2} \ne \frac{1}{3x^2}$$

The exponent only applies to the x, not the 3. Thus,

$$3x^{-2} = 3 \cdot \frac{1}{x^2} = \frac{3}{x^2}$$

EXERCISES 10.5

Write each expression as a basic numeral or basic fraction without using negative or zero exponents. See Examples 1 and 2.

Complete Examples

a. $6^{-2} = \dfrac{1}{6^2}$
 = _____
   (1)

b. $3 \cdot 2^{-1} = 3 \cdot \dfrac{1}{2^1}$
 = _____
   (2)

c. $4 \cdot x^{-3} = 4 \cdot \dfrac{1}{x^3}$
 = _____
   (3)

(1) $\dfrac{1}{36}$ (2) $\dfrac{3}{2}$ (3) $\dfrac{4}{x^3}$

1. 5^{-2}
2. 4^{-3}
3. x^{-6}
4. y^{-2}
5. $(8x)^0$
6. $(3y)^0$
7. $3 \cdot 4^{-3}$
8. $4 \cdot 7^{-2}$
9. $3 \cdot 10^{-3}$
10. $2 \cdot 10^{-2}$
11. $4x^{-2}$
12. $7x^{-4}$

Write each expression in a nonfraction form using negative exponents. See Example 2.

Complete Examples

a. $\dfrac{1}{3^2} =$ _____
   (1)

b. $\dfrac{5}{10^2} = 5 \cdot \dfrac{1}{10^2}$
 $= 5 \cdot$ _____
   (2)

c. $\dfrac{x}{8} = x \cdot \dfrac{1}{2^3}$
 $= x \cdot$ _____
   (3)

(1) 3^{-2} (2) 10^{-2} (3) 2^{-3}

13. $\dfrac{1}{2^3}$
14. $\dfrac{1}{4^3}$
15. $\dfrac{1}{5^2}$
16. $\dfrac{1}{3^4}$
17. $\dfrac{1}{4}$
18. $\dfrac{1}{9}$
19. $\dfrac{x}{25}$
20. $\dfrac{y}{36}$
21. $\dfrac{2}{10^2}$
22. $\dfrac{3}{10^4}$
23. $\dfrac{x}{10^3}$
24. $\dfrac{y}{10^5}$

PROPERTIES OF FRACTIONS

Complete each factored form.

25. $234 = 23.4 \times 10 = 2.34 \times$ _____

26. $4800 = 480 \times 10 = 48 \times 10^2 = 4.8 \times$ _____

27. $0.074 = 0.74 \times 10^{-1} = 7.4 \times$ _____

28. $0.006 = 0.06 \times 10^{-1} = 0.6 \times 10^{-2} = 6 \times$ _____

Express each of the following in scientific form. See Example 3.

Complete Examples

a. 62,000,000

  $62{,}000{,}000 \times 10^7$

  = _____
     (1)

b. 0.000431

  0.000431×10^{-4}

  = _____
     (2)

---

(1) 6.2×10^7 (2) 4.31×10^{-4}

---

29. 483 **30.** 5420 **31.** 0.072 **32.** 0.0063

33. 4000 **34.** 230,000 **35.** 0.00063 **36.** 0.000007

Express each of the following in standard decimal form. See Example 4.

Complete Examples

a. 1.47×10^5

  1.47000

  = _____
     (1)

b. 4.2×10^{-3}

  004.2

  = _____
     (2)

---

(1) 147,000   (2) 0.0042

---

37. 4.3×10^4 **38.** 6.1×10^3 **39.** 5.7×10^{-4} **40.** 6.8×10^{-2}

41. 8.234×10^7 **42.** 1.413×10^4 **43.** 8×10^{-6} **44.** 2×10^{-5}

45. The mass of the earth is approximately

  5,980,000,000,000,000,000,000,000,000 grams

  Write this number in scientific form.

46. The mass of the hydrogen atom is approximately

  0.000 000 000 000 000 000 000 001 67 gram

  Write this number in scientific form.

342

47. Light travels at a speed of 300,000,000 meters per second. Write this number in scientific form.

48. Visible blue light has a wavelength of 0,000 000 45 meters. Write this number in scientific form.

49. The average body cell of an animal has a diameter of 0.000 015 meters. Write this number in scientific form.

50. The diameter of the earth is approximately 6,450,000 meters. Write this number in scientific form.

CHAPTER TEN REVIEW EXERCISES

1. Graph the following numbers. $\quad \frac{-27}{4}, \frac{-5}{2}, 2, \frac{11}{2}, \frac{37}{4}$

2. Rewrite in the form $a \cdot \frac{1}{b}$.

   a. $\frac{4}{9}$
   b. $\frac{x+6}{3}$
   c. $\frac{2y}{x+y^2}$

3. Rewrite in the form $\frac{ac}{b}$.

   a. $\frac{2}{3}(x-3)$
   b. $-\frac{1}{3}(x^2+1)$
   c. $-\frac{3}{4}(2x+y)$

4. Change to equivalent fractions in standard form.

   a. $-\frac{3}{x+y}$
   b. $-\frac{-a}{x}$
   c. $-\frac{b-2}{4}$

5. Reduce to lowest terms.

   a. $\frac{x^2 y^2 z^2}{xy^3}$
   b. $\frac{b-3}{b^2-2b-3}$
   c. $\frac{a^2+a}{a^3-a}$

6. Simplify the quotient.

   a. $\frac{3x^3 + 6x^2 - 9x}{3x}$
   b. $\frac{8x^4 - 4x^2 + 3}{2x^2}$
   c. $\frac{6x^3 - 4x - 1}{2x}$

7. Simplify the quotient by using long division.

   a. $\frac{2x^2 - 5x + 1}{x - 1}$
   b. $\frac{2x^3 + x - 3}{x + 2}$
   c. $\frac{2x^2 + 5x + 3}{x + 1}$

8. Build each fraction to an equivalent fraction with denominator shown.

   a. $\frac{3}{x-y} = \frac{?}{2(x-y)}$
   b. $\frac{x}{x-2} = \frac{?}{x^2 - 3x + 2}$
   c. $\frac{2x}{x+3} = \frac{?}{x^2 - 9}$

9. Rewrite each expression without using zero or negative exponents.

   a. $2 \cdot 5^{-2}$
   b. $3^0 \cdot x^{-2}$
   c. $4y^{-3}$

343

PROPERTIES OF FRACTIONS

10. Express in scientific form.

    a. 34,700,000   b. 0.000873   c. 0.000004

Express in standard decimal form.

    d. 4.83×10^4 e. 3.81×10^{-4} f. 4.03×10^{-7}

11 OPERATIONS WITH FRACTIONS

The properties for the basic operations on arithmetic fractions that were introduced in Chapter 3 are also applicable for fractions that include variables. You may want to review these properties in Chapter 3 or in the Summary of Part 1 before we expand upon them in this chapter.

11.1 PRODUCTS OF FRACTIONS

In Section 3.6 we defined the product of two fractions as follows.

> **The product of two fractions is a fraction whose numerator is the product of the numerators and whose denominator is the product of the denominators of the given fractions.**

In symbols,

$$\frac{a}{b} \cdot \frac{c}{d} = \frac{ac}{bd}$$

Any common factor occurring in both a numerator and a denominator of either fraction can be divided out either before or after multiplying.

Example 1   Find the product of $\dfrac{12x^2}{5y} \cdot \dfrac{10y^3}{3x^2y}$

Solution   First, we divide the numerator and denominator by the common factors to get

$$\frac{\cancel{12}^{4}\cancel{x^2}}{\cancel{5}\cancel{y}} \cdot \frac{\cancel{10}^{2y}\cancel{y^3}}{\cancel{3}\cancel{x^2}\cancel{y}}$$

345

OPERATIONS WITH FRACTIONS

Now, multiplying the remaining factors of the numerators and denominators yields

$$\frac{4 \cdot 2 \cdot y}{1} = 8y$$

If a negative sign is attached to any of the factors, it is advisable to proceed as if all the factors were positive and then attach the appropriate sign to the result. A positive sign is attached if there are no negative signs or an even number of negative signs on the factors; a negative sign is attached if there is an odd number of negative signs on the factors.

Example 2

a. $\dfrac{-3x}{4} \cdot \dfrac{2}{3y} = -\dfrac{\cancel{3}x}{\cancel{4}} \cdot \dfrac{\cancel{2}}{\cancel{3}y} = -\dfrac{x}{2y}$

b. $\dfrac{-4x}{3} \cdot \dfrac{-6}{y} = \dfrac{4x}{\cancel{3}} \cdot \dfrac{\cancel{6}^{2}}{y} = \dfrac{8x}{y}$

When the fractions contain algebraic expressions, it is necessary to factor wherever possible and divide out common factors before multiplying.

Example 3 Find the product of $\dfrac{x^2 - x}{2x + 6} \cdot \dfrac{x^2 + 5x + 6}{x^2 - 1}$.

Solution First, we must factor the numerators and denominators to get

$$\frac{x(x - 1)}{2(x + 3)} \cdot \frac{(x + 3)(x + 2)}{(x + 1)(x - 1)}$$

Now, dividing out common factors yields

$$\frac{x\cancel{(x-1)}}{2\cancel{(x+3)}} \cdot \frac{\cancel{(x+3)}(x + 2)}{\cancel{(x-1)}(x + 1)}$$

We now multiply the remaining factors of the numerators and denominators to obtain

$$\frac{x(x + 2)}{2(x + 1)} = \frac{x^2 + 2x}{2(x + 1)}$$

Note that when writing fractional answers, we will multiply out the numerator and leave the denominator in factored form. Very often, fractions are more useful in this form.

In algebra, we often rewrite an expression such as $a\left(\dfrac{b}{c}\right)$ as an equivalent expression $\dfrac{ab}{c}$. Use whichever form is most convenient for a particular problem.

Example 4

a. $3\left(\dfrac{x}{4}\right) = \dfrac{3x}{4}$ b. $4\left(\dfrac{1}{y}\right) = \dfrac{4}{y}$ c. $-2\left(\dfrac{x}{x + y}\right) = \dfrac{-2x}{x + y}$

Common Errors Remember that we can only divide out *common factors*, never common terms! For example,

$$\frac{1}{3 + x} \cdot \frac{4 + x}{y} \neq \frac{1}{3 + \cancel{x}} \cdot \frac{4 + \cancel{x}}{y} = \frac{4}{3y}$$

11.1 PRODUCTS OF FRACTIONS

because x is a term and *cannot* be divided out. Similarly,

$$\frac{1}{3x} \cdot \frac{3y + 2}{5} \neq \frac{1}{\cancel{3}x} \cdot \frac{\cancel{3}y + 2}{5x} = \frac{y + 2}{5x}$$

because 3 is not a factor of the entire numerator $3y + 2$.

EXERCISES 11.1
Write each product as a single term.

Complete Example (see Examples 1 and 2)

$$\frac{-5xy}{3} \cdot \frac{9y}{10x^2 y}$$

$$= \frac{-\cancel{5}\,\cancel{x}y}{\cancel{3}} \cdot \frac{\overset{3}{\cancel{9}}\,\cancel{y}}{10\cancel{x}^2\,\cancel{y}}$$

$$= \frac{}{\underset{(3)}{}} \quad \underset{(1)\ (2)}{}$$

Divide numerator and denominator by common factors.

Multiply remaining factors of numerators and remaining factors of denominators. Prefix appropriate sign to the product.

(1) 2   (2) x (3) $\dfrac{-3y}{2x}$

1. $\dfrac{1}{3} \cdot \dfrac{3y}{1}$

2. $\dfrac{2}{3} \cdot \dfrac{9x^2}{4}$

3. $\dfrac{6x^3}{5} \cdot \dfrac{2}{3x}$

4. $\dfrac{7a}{3} \cdot \dfrac{1}{a^3}$

5. $6x^2 y \cdot \dfrac{2}{3x^2}$

6. $5x^2 y^2 \cdot \dfrac{1}{x^3 y^3}$

7. $\dfrac{-6xy}{3} \cdot \dfrac{4x}{8xy^2}$

8. $\dfrac{-24ab^2}{8a} \cdot \dfrac{21a^2 b}{14b}$

9. $\dfrac{-21r^2 s}{8t} \cdot \dfrac{-14t^2}{3rs}$

10. $\dfrac{-12a^2 b}{5c} \cdot \dfrac{10bc^2}{24a^3 b}$

11. $\dfrac{-6xyz}{4a^2 b} \cdot \dfrac{10ab^2}{15xyz^2}$

12. $\dfrac{-56x^3 yz^2}{24xy^2} \cdot \dfrac{-48z}{28x^2 z^3}$

Complete Example (see Example 3)

$$\frac{x^2 + x - 12}{x^2 - 16} \cdot \frac{x^2 - 2x - 8}{x^2 - 4}$$

$$= \frac{(x + 4)(x - 3)}{(x + 4)(\underset{(1)}{})} \cdot \frac{(x - 4)(x + 2)}{(x + 2)(\underset{(2)}{})}$$

$$= \frac{}{\underset{(3)}{}}$$

Factor numerators and denominators.

Divide numerators and denominators by common factors.

Multiply remaining factors of numerators and remaining factors of denominators.

(1) $x - 4$ (2) $x - 2$ (3) $\dfrac{x - 3}{x - 2}$

347

OPERATIONS WITH FRACTIONS

13. $\dfrac{3x-9}{5x-15} \cdot \dfrac{10x-5}{8x-4}$

14. $\dfrac{2x+4}{3x-9} \cdot \dfrac{x-3}{x+2}$

15. $\dfrac{5a+25}{2a} \cdot \dfrac{4a}{2a+10}$

16. $\dfrac{2a-4b}{8a+24b} \cdot \dfrac{2a+6b}{4a-8b}$

17. $\dfrac{2x+3y}{x-2y} \cdot \dfrac{3x-6y}{x-2y} \cdot \dfrac{x-2y}{6x+9y}$

18. $\dfrac{7x+14}{14x-28} \cdot \dfrac{2x-4}{x+2} \cdot \dfrac{x-3}{x+1}$

19. $\dfrac{x^2-3x-10}{x^2+2x-35} \cdot \dfrac{x^2+4x-21}{x^2+9x+14}$

20. $\dfrac{4y^2-1}{y^2-16} \cdot \dfrac{y^2-4y}{2y+1}$

21. $\dfrac{6x^2-x-2}{12x^2+5x-2} \cdot \dfrac{8x^2-6x+1}{4x^2-1}$

22. $\dfrac{y^2-y-20}{y^2+7y+12} \cdot \dfrac{y^2+9y+18}{y^2-7y+10}$

23. $\dfrac{x^2+x-2}{x^2-3x+2} \cdot \dfrac{x^2-x-2}{x^2+5x+6}$

24. $\dfrac{x^2+x-6}{2x^2+6x} \cdot \dfrac{8x^2}{x^2-5x+6}$

25. $\dfrac{x^2-4}{x^2-1} \cdot \dfrac{x-1}{2x^2+4x}$

26. $\dfrac{a^2+a}{2a+1} \cdot \dfrac{10a+5}{3a+3}$

27. $\dfrac{x^2-4}{x^2-5x+6} \cdot \dfrac{x^2-2x-3}{x^2+3x+2}$

28. $\dfrac{x^2+3x}{x^2-3x-4} \cdot \dfrac{x^2-5x+4}{x^2+2x-3}$

29. $\dfrac{y^2-y-20}{y^2-6y+5} \cdot \dfrac{y^2+5y-6}{y^2+7y+12} \cdot \dfrac{y^2-9}{y^2-36}$

30. $\dfrac{x^2-xy}{xy+y^2} \cdot \dfrac{x^2-4y^2}{x^2-y^2} \cdot \dfrac{x^2-2xy-3y^2}{x^2-5xy+6y^2}$

Change each of the following to the form $\dfrac{ab}{c}$. See Example 5.

Complete Examples

a. $\dfrac{3}{4}a = \dfrac{3a}{(1)}$

b. $\dfrac{5}{b}a = \dfrac{5a}{(2)}$

c. $\dfrac{2}{3}(x-y) = \dfrac{2(x-y)}{(3)}$

d. $-\dfrac{1}{3}x = \dfrac{-x}{(4)}$

(1) 4   (2) b   (3) 3   (4) 3

31. $\dfrac{2}{3}x$

32. $\dfrac{3}{4}y$

33. $-\dfrac{2}{5}a$

34. $-\dfrac{4}{7}b$

35. $\dfrac{3}{4}(a-b)$

36. $\dfrac{2}{3}(b-c)$

37. $-\dfrac{3}{5}(2x-y)$

38. $-\dfrac{4}{7}(x-2y)$

Change each of the following to the form $\frac{a}{b}c$. See Example 5.

Complete Examples

a. $\dfrac{2a}{5} = \dfrac{2}{5}$ ____(1)____   b. $\dfrac{b}{3} = \dfrac{1}{3}$ ____(2)____   c. $\dfrac{2(x+2y)}{5} =$ ____(3)____ $(x+2y)$

(1) a (2) b (3) $\dfrac{2}{5}$

39. $\dfrac{3x}{7}$ 40. $\dfrac{4y}{3}$ 41. $\dfrac{-5a}{7}$ 42. $\dfrac{-b}{5}$

43. $\dfrac{5(a-b)}{2}$ 44. $\dfrac{-3(a+2b)}{4}$ 45. $\dfrac{x+y}{7}$ 46. $\dfrac{-(x+y)}{5}$

11.2 QUOTIENTS OF FRACTIONS

The quotient of two fractions equals the product of the dividend and the reciprocal of the divisor.

That is, to divide one fraction by another, we invert the divisor and multiply. In symbols,

$$\frac{a}{b} \div \frac{c}{d} = \frac{a}{b} \cdot \frac{d}{c}$$

Example 1

a. $x \div \dfrac{y}{4} = \dfrac{x}{1} \cdot \dfrac{4}{y}$ ——— Change to multiplication.

    ——— Divisor inverted.

$= \dfrac{4x}{y}$

b. $\dfrac{2}{x} \div \dfrac{3}{y} = \dfrac{2}{x} \cdot \dfrac{y}{3}$ ——— Change to multiplication.

    ——— Divisor inverted.

$= \dfrac{2y}{3x}$

As in multiplication, when fractions in a quotient have signs attached, it is advisable to proceed with the problem as if all the factors were positive and then attach the appropriate sign to the solution.

OPERATIONS WITH FRACTIONS

Example 2
$$\frac{-3x}{2y} \div \frac{9x^2}{8y^2} = -\frac{3x}{2y} \cdot \frac{8y^2}{9x^2}$$

— Change to multiplication.
— Divisor inverted.

$$= -\frac{\cancel{3}\cancel{x}}{\cancel{2}\cancel{y}} \cdot \frac{\cancel{8}y^2}{\cancel{9}\cancel{x}^2} = \frac{-4y}{3x}$$

Some quotients occur so frequently that it is helpful to recognize equivalent forms directly. One case is

$$1 \div \frac{a}{b} = \frac{1}{1} \cdot \frac{b}{a} = \frac{b}{a}$$

In general,

$$a \div \frac{b}{c} = \frac{a}{1} \cdot \frac{c}{b} = \frac{ac}{b}$$

Example 3 a. $1 \div \frac{x}{2y} = \frac{2y}{x}$ b. $1 \div \frac{3}{x+y} = \frac{x+y}{3}$ c. $x \div \frac{3}{y} = \frac{xy}{3}$

When the fractions in a quotient involve algebraic expressions, it is necessary to factor wherever possible and divide out common factors before multiplying.

Example 4

a. $\frac{x^2 - 4}{3x} \div \frac{x+2}{x} = \frac{(x+2)(x-2)}{3\cancel{x}} \cdot \frac{\cancel{x}}{(x+2)}$

— Change to multiplication.
— Divisor inverted.

$$= \frac{x-2}{3}$$

b. $\frac{a^2 - 3a + 2}{a^2 - 5a + 6} \div \frac{a^2 - a}{2a} = \frac{\cancel{(a-2)}\cancel{(a-1)}}{\cancel{(a-2)}(a-3)} \cdot \frac{2\cancel{a}}{\cancel{a}\cancel{(a-1)}}$

— Change to multiplication.
— Divisor inverted.

$$= \frac{2}{a-3}$$

EXERCISES 11.2
Write each quotient as a single term.

Complete Example (see Examples 1, 2, and 3)

$$\frac{-3x^2y}{b^2} \div \frac{9x^2y}{ab}$$

Invert divisor and multiply.

$$= \frac{-3x^2y}{b^2} \cdot \frac{ab}{\underline{}} = \underline{}$$
  (1)   (2)

(1) $9x^2y$ (2) $\frac{-a}{3b}$

350

11.2 QUOTIENTS OF FRACTIONS

1. $\dfrac{2c}{3d} \div \dfrac{4c}{6d}$
2. $\dfrac{c^2}{d} \div \dfrac{c^4}{d^2}$
3. $\dfrac{15}{27ab} \div \dfrac{16b}{9a}$
4. $\dfrac{a}{b^2} \div \dfrac{ab^2}{b^3}$

5. $\dfrac{-x^2y^2}{u^2v^2} \div \dfrac{xy^2}{u^2v}$
6. $\dfrac{14a^2b^3}{15x^2y} \div \dfrac{-21a^2b^2}{35xy}$
7. $\dfrac{3xy}{4} \div (-12y^2)$
8. $\dfrac{36x^3}{7y} \div (-3x^2)$

9. $\dfrac{2ab^2}{3} \div (4a^2b)$
10. $\dfrac{3a^2b^4}{7} \div (6ab^3)$
11. $y \div \dfrac{3x}{y^3}$
12. $x \div \dfrac{-2y}{x^2}$

13. $16y^2 \div \dfrac{4y}{3}$
14. $ax^2 \div \dfrac{x^2}{b}$
15. $1 \div \dfrac{x}{2y}$
16. $1 \div \dfrac{x^2}{3y}$

17. $1 \div \dfrac{2a}{a+b}$
18. $1 \div \dfrac{a^2}{a-b}$

Complete Example (see Example 4)

$\dfrac{x^2 - 3x - 4}{x^2 - 2x - 8} \div \dfrac{2x^2 - 7x - 15}{x^2 - 3x - 10}$

Invert divisor and multiply.

$= \dfrac{x^2 - 3x - 4}{x^2 - 2x - 8} \cdot \dfrac{x^2 - 3x - 10}{\underline{\qquad(1)\qquad}}$

Factor where possible and simplify.

$= \dfrac{(x-4)(x+1)}{(x-4)(\underline{\qquad(2)\qquad})} \cdot \dfrac{(x-5)(x+2)}{(2x+3)(\underline{\qquad(3)\qquad})}$

$= \dfrac{\underline{\qquad\qquad}}{(4)}$

(1) $2x^2 - 7x - 15$ (2) $x + 2$ (3) $x - 5$ (4) $\dfrac{x+1}{2x+3}$

19. $\dfrac{a^2 - ab}{ab} \div \dfrac{2a - 2b}{ab}$
20. $\dfrac{2x - 2y}{xy} \div \dfrac{4x - 4y}{xy}$
21. $\dfrac{6a - 12}{3a + 9} \div \dfrac{4a - 8}{5a + 15}$

22. $\dfrac{x^2 + xy}{x^2 - xy} \div \dfrac{x + y}{4x - 4y}$
23. $\dfrac{10x^2 - 5x}{12x^3 + 24x^2} \div \dfrac{2x^2 - x}{2x^2 + 4x}$
24. $\dfrac{ax - ay}{bx + by} \div \dfrac{cx - cy}{dx + dy}$

25. $\dfrac{4x^2 - y^2}{x^2 - 4y^2} \div \dfrac{2x - y}{x - 2y}$
26. $\dfrac{x^2 - 9y^2}{16x^2 - y^2} \div \dfrac{x + 3y}{4x - y}$

27. $\dfrac{y^2 - 6y + 5}{y^2 + 8y + 7} \div \dfrac{y^2 - 3y - 10}{y^2 + 3y + 2}$
28. $\dfrac{x^2 - 8x + 15}{x^2 + 9x + 14} \div \dfrac{x^2 + 4x - 21}{x^2 - 6x - 16}$

29. $\dfrac{x^2 - x - 6}{x^2 + 2x - 15} \div \dfrac{x^2 - 4}{x^2 - 25}$
30. $\dfrac{6x^2 - x - 2}{12x^2 + 5x - 2} \div \dfrac{4x^2 - 1}{8x^2 - 6x + 1}$

351

OPERATIONS WITH FRACTIONS

31. $\dfrac{y^2 - y - 20}{y^2 - 6y + 5} \div \dfrac{y^2 - 7y + 10}{y^2 + y - 2}$

32. $\dfrac{x^2 - x - 2}{x^2 + 5x - 6} \div \dfrac{x^2 - 3x - 4}{x^2 - x - 12}$

33. $\dfrac{2x^2 - x - 28}{3x^2 - x - 2} \div \dfrac{4x^2 + 16x + 7}{3x^2 + 11x + 6}$

34. $\dfrac{y^2 + 7y + 10}{y^2 + 7y + 12} \div \dfrac{y^2 + 6y + 5}{y^2 + 8y + 16}$

35. $\dfrac{3y + 2}{5y^2 - y} \cdot \dfrac{2y^2 - y}{2y^2 - y - 1} \div \dfrac{6y^2 + y - 2}{10y^2 + 3y - 1}$

36. $\dfrac{x^2 + 4x + 3}{x^2 - 8x + 7} \cdot \dfrac{x^2 - 2x - 35}{x^2 - 7x - 8} \div \dfrac{x^2 + 8x + 15}{x^2 - 9x + 8}$

11.3 SUMS AND DIFFERENCES OF FRACTIONS WITH LIKE DENOMINATORS

In Section 3.4 we defined the sum of two or more fractions as follows:

> **The sum of two or more fractions with common denominators is a fraction with the same denominator and a numerator equal to the sum of the numerators of the original fractions.**

In general,

$$\frac{a}{c} + \frac{b}{c} = \frac{a + b}{c}$$

Example 1 a. $\dfrac{2}{x} + \dfrac{5}{x} = \dfrac{2 + 5}{x} = \dfrac{7}{x}$ b. $\dfrac{3}{x + 1} + \dfrac{x}{x + 1} = \dfrac{3 + x}{x + 1}$

When subtraction is involved, it is helpful to change to standard form before adding.

Example 2 a. $\dfrac{3}{x} - \dfrac{5}{x} = \dfrac{3}{x} + \dfrac{-5}{x}$ b. $\dfrac{4x}{y} - \dfrac{x}{y} = \dfrac{4x}{y} + \dfrac{-x}{y}$

$\qquad\qquad\quad = \dfrac{3 + (-5)}{x} \qquad\qquad\qquad\quad = \dfrac{4x + (-x)}{y}$

$\qquad\qquad\quad = \dfrac{-2}{x} \qquad\qquad\qquad\qquad\quad = \dfrac{3x}{y}$

We must be especially careful with binomial numerators. For example, we should rewrite

$$\frac{5}{x + 1} - \frac{x - 2}{x + 1}$$

as

$$\frac{5}{x + 1} + \frac{-(x - 2)}{x + 1}$$

where *the entire numerator is enclosed within parentheses*.

11.3 SUMS AND DIFFERENCES OF FRACTIONS WITH LIKE DENOMINATORS

Example 3

$$\frac{5}{x+1} - \frac{x-2}{x+1} = \frac{5}{x+1} + \frac{-(x-2)}{x+1}$$

The sign of *each term of* the binomial is changed.

$$= \frac{5}{x+1} + \frac{-x+2}{x+1}$$

$$= \frac{5 + (-x) + 2}{x+1} = \frac{7-x}{x+1}$$

EXERCISES 11.3

Write each sum or difference as a single term.

Complete Examples (see Examples 1 and 2)

a. $\dfrac{x}{5} + \dfrac{y}{5}$

$= \underline{}$
  (1)

b. $\dfrac{2x}{7} - \dfrac{3y}{7}$

$= \dfrac{2x}{7} + \dfrac{-3y}{\underline{}}$
               (2)

$= \underline{}$
   (3)

Write in standard form.

Add numerators.

---

(1) $\dfrac{x+y}{5}$ (2) 7 (3) $\dfrac{2x-3y}{7}$

---

1. $\dfrac{2x}{9} + \dfrac{5y}{9}$

2. $\dfrac{4x}{7} + \dfrac{2y}{7}$

3. $\dfrac{5x}{11} - \dfrac{3}{11}$

4. $\dfrac{2}{13} - \dfrac{3y}{13}$

5. $\dfrac{2}{5} + \dfrac{1}{5} - \dfrac{x}{5}$

6. $\dfrac{4}{7} - \dfrac{x}{7} + \dfrac{y}{7}$

Complete Examples (see Examples 1 and 2)

a. $\dfrac{3}{2a} - \dfrac{1}{2a}$

$= \dfrac{3}{2a} + \dfrac{-1}{\underline{}}$
                (1)

$= \dfrac{\cancel{2}}{\cancel{2}a}$

$= \underline{}$
    (2)

b. $\dfrac{3x}{4y} - \dfrac{x}{4y}$

$= \dfrac{3x}{4y} + \dfrac{-x}{\underline{}}$
                (3)

$= \dfrac{\cancel{2}x}{\cancel{4}y}$
  $ \; 2$

$= \underline{}$
    (4)

Write in standard form.

Add numerators and simplify.

---

(1) $2a$ (2) $\dfrac{1}{a}$ (3) $4y$ (4) $\dfrac{x}{2y}$

353

OPERATIONS WITH FRACTIONS

7. $\dfrac{5}{2a} + \dfrac{3}{2a}$
8. $\dfrac{6}{5a} + \dfrac{4}{5a}$
9. $\dfrac{7}{2b} - \dfrac{5}{2b}$
10. $\dfrac{5}{3a} - \dfrac{8}{3a}$

11. $\dfrac{7}{3x} + \dfrac{7}{3x} - \dfrac{2}{3x}$
12. $\dfrac{2}{7x} - \dfrac{10}{7x} - \dfrac{13}{7x}$
13. $\dfrac{4x}{5y} - \dfrac{x}{5y} + \dfrac{2x}{5y}$
14. $\dfrac{7x}{3y} - \dfrac{5x}{3y} - \dfrac{14x}{3y}$

Complete Examples (see Examples 1 and 2)

a. $\dfrac{x-y}{a} + \dfrac{y}{a}$

$= \dfrac{x-y+y}{\underline{\qquad(1)\qquad}}$

$= \underline{\qquad(2)\qquad}$

b. $\dfrac{x+y}{x} + \dfrac{x-y}{x}$ Add numerators.

$= \dfrac{x+y+x-y}{\underline{\qquad(3)\qquad}}$ Simplify.

$= \dfrac{2\cancel{x}}{\cancel{x}} = \underline{\qquad(4)\qquad}$

---

(1) a (2) $\dfrac{x}{a}$ (3) x (4) 2

---

15. $\dfrac{x+1}{2} - \dfrac{3}{2}$
16. $\dfrac{x-2}{5} - \dfrac{3}{5}$
17. $\dfrac{x-2y}{3x} + \dfrac{x+3y}{3x}$
18. $\dfrac{3-x}{2y} + \dfrac{4-x}{2y}$

19. $\dfrac{x+1}{2a} + \dfrac{x-1}{2a}$
20. $\dfrac{2x-y}{3y} + \dfrac{2x+2y}{3y}$
21. $\dfrac{x^2-x}{2} - \dfrac{x^2}{2} + \dfrac{3x}{2}$

22. $\dfrac{2x^2-x}{3} - \dfrac{x^2}{3} + \dfrac{5x}{3}$
23. $\dfrac{2x+y}{y} + \dfrac{x-2y}{y} + \dfrac{x+y}{y}$
24. $\dfrac{x-2y}{2x} + \dfrac{x+y}{2x} + \dfrac{2x+y}{2x}$

Complete Example (see Example 3)

$\dfrac{x-2}{a+b} - \dfrac{2x+1}{a+b}$

$= \dfrac{(x-2)}{a+b} + \dfrac{-(2x+1)}{\underline{\qquad(1)\qquad}}$ Insert parentheses and write in standard form.

$= \dfrac{(x-2)-(2x+1)}{a+b}$ Add numerators.

$= \dfrac{x-2-2x-1}{a+b}$ Remove parentheses. Simplify.

$= \underline{\qquad(2)\qquad}$

---

(1) $a+b$ (2) $\dfrac{-x-3}{a+b}$

354

11.3 SUMS AND DIFFERENCES OF FRACTIONS WITH LIKE DENOMINATORS

25. $\dfrac{2x+3}{2} - \dfrac{x-3}{2}$

26. $\dfrac{2x-y}{3} - \dfrac{3x-y}{3}$

27. $\dfrac{2a+b}{a-b} - \dfrac{a-2b}{a-b}$

28. $\dfrac{2a-b}{b} - \dfrac{a-2b}{b}$

29. $\dfrac{3}{a+b} - \dfrac{a+3}{a+b}$

30. $\dfrac{b-1}{a} - \dfrac{b+1}{a}$

31. $\dfrac{2x-y}{x+y} - \dfrac{x-3y}{x+y} + \dfrac{2x}{x+y}$

32. $\dfrac{2u-3v}{u+2} - \dfrac{u+2v}{u+2} + \dfrac{u}{u+2}$

33. $\dfrac{3}{x+2y} - \dfrac{x+3}{x+2y} + \dfrac{x+1}{x+2y}$

34. $\dfrac{2x-y}{x-y} + \dfrac{x-2y}{x-y} - \dfrac{3x-3y}{x-y}$

35. $\dfrac{2a+b}{3} + \dfrac{4a-2b}{3}$

36. $\dfrac{6x-6y}{5} + \dfrac{4x-4y}{5}$

37. $\dfrac{2x+y}{2} + \dfrac{4x+y}{2}$

38. $\dfrac{x-y}{4} + \dfrac{3x-7y}{4}$

39. $\dfrac{3u+2v}{4u-2v} - \dfrac{u+2v}{4u-2v}$

40. $\dfrac{u+7}{2u-4v} - \dfrac{5-u}{2u-4v}$

41. $\dfrac{x}{2x+4} - \dfrac{2x-3}{2x+4} + \dfrac{3x-5}{2x+4}$

42. $\dfrac{x+y}{2(x-y)} - \dfrac{2y-2x}{2(x-y)} + \dfrac{x-3y}{2(x-y)}$

Complete Example

$\dfrac{3}{x^2+2x+1} - \dfrac{2-x}{x^2+2x+1}$

$= \dfrac{3}{x^2+2x+1} + \dfrac{-(2-x)}{(1)}$ Insert parentheses and write in standard form.

$= \dfrac{3-(2-x)}{x^2+2x+1}$ Add numerators and simplify.

$= \dfrac{x+1}{x^2+2x+1}$

$= \dfrac{(x+1)}{(x+1)(x+1)} =$ _____   Factor denominator and reduce to lowest terms.
(2)

(1) $x^2 + 2x + 1$ (2) $\dfrac{1}{x+1}$

43. $\dfrac{x+1}{x^2-2x+1} - \dfrac{5-3x}{x^2-2x+1}$

44. $\dfrac{2x+1}{x^2-x-6} - \dfrac{x-1}{x^2-x-6}$

45. $\dfrac{2x-3}{x^2+3x-4} + \dfrac{7-x}{x^2+3x-4}$

46. $\dfrac{u+3v}{u^2-v^2} + \dfrac{3u+v}{u^2-v^2}$

47. $\dfrac{x^2-2}{x^2-x} - \dfrac{2-4x}{x^2-x} - \dfrac{1}{x^2-x}$

48. $\dfrac{3x^2-4}{x^2-4} - \dfrac{x^2}{x^2-4} - \dfrac{4}{x^2-4}$

355

OPERATIONS WITH FRACTIONS

11.4 SUMS OF FRACTIONS WITH UNLIKE DENOMINATORS

In Section 11.3, we added algebraic fractions with like denominators. In this section, we will add algebraic fractions with unlike denominators. The procedure is similar to the one that we use in Section 3.4 to add arithmetic fractions with unlike denominators.

The LCD of a set of algebraic fractions is the simplest algebraic expression that is a multiple of each of the denominators in the set. Thus, the LCD of the fractions

$$\frac{3}{x}, \quad \frac{2}{x^2 - 1}, \quad \text{and} \quad \frac{1}{x^2(x - 1)}$$

is

$$x^2(x + 1)(x - 1)$$

because this is the simplest expression that is a multiple of each of the denominators.

> To find the LCD:
> 1. Completely factor each denominator, aligning common factors when possible.
> 2. Include in the LCD each of these factors the greatest number of times it occurs in any single denominator.

Example 1 Find the LCD of the fractions

$$\frac{3}{x}, \quad \frac{2}{x^2 - 1}, \quad \text{and} \quad \frac{1}{x^2(x - 1)}$$

Solution Following the method of Example 1, we get

$$x = x$$
$$x^2 - 1 = (x + 1)(x - 1)$$
$$x^2(x - 1) = x \qquad (x - 1)x$$

The LCD contains
the factors $\quad x \, (x + 1)(x - 1)x$
whose product is $\quad x^2(x + 1)(x - 1)$

Thus, the LCD is $x^2(x + 1)(x - 1)$.

We can add fractions with unlike denominators by first building the fractions to *equivalent fractions with like denominators and then adding*.

Example 2 Write the sum of $\frac{a}{2} + \frac{b}{5}$ as a single term.

Solution The LCD is 10. We build each fraction to a fraction with 10 as the denominator. Thus,

$$\frac{(5)a}{(5)2} + \frac{(2)b}{(2)5}$$

11.3 SUMS OF FRACTIONS WITH UNLIKE DENOMINATORS

is equivalent to

$$\frac{5a}{10} + \frac{2b}{10}$$

from which we obtain

$$\frac{5a + 2b}{10}$$

Sometimes, the fractions have denominators that are binomials.

Example 3 Write the sum of $\dfrac{x}{x + 2} + \dfrac{x + 1}{x - 1}$ as a single term.

Solution The LCD is $(x + 2)(x - 1)$. We build each fraction to a fraction with denominator $(x + 2)(x - 1)$, inserting parentheses as needed, and get

$$\frac{x}{x + 2} + \frac{x + 1}{x - 1} = \frac{(x - 1)x}{(x - 1)(x + 2)} + \frac{(x + 1)(x + 2)}{(x - 1)(x + 2)}$$

Now that we have like denominators, we can add the numerators, simplify, and obtain

$$\frac{x(x - 1) + (x + 1)(x + 2)}{(x - 1)(x + 2)} = \frac{x^2 - x + x^2 + 3x + 2}{(x - 1)(x + 2)}$$

$$= \frac{2x^2 + 2x + 2}{(x - 1)(x + 2)}$$

Example 4 Write the sum of $\dfrac{x}{x^2 - 4} + \dfrac{1}{x^2 - x - 2}$ as a single term.

Solution First we factor the denominators in order to obtain the LCD.

$$x^2 - 4 = (x + 2)(x - 2)$$
$$x^2 - x - 2 = \quad\quad (x - 2)(x + 1)$$

The LCD is $(x + 2)(x - 2)(x + 1)$.

We now build each fraction to fractions with this denominator and get

$$\frac{x}{x^2 - 4} + \frac{1}{x^2 - x - 2} = \frac{x}{(x - 2)(x + 2)} + \frac{1}{(x - 2)(x + 1)}$$

$$= \frac{(x + 1)x}{(x + 1)(x - 2)(x + 2)} + \frac{1(x + 2)}{(x - 2)(x + 1)(x + 2)}$$

We can now add the numerators, simplify, and obtain

$$\frac{x(x + 1) + (x + 2)}{(x + 1)(x - 2)(x + 2)} = \frac{x^2 + x + x + 2}{(x + 1)(x - 2)(x + 2)}$$

$$= \frac{x^2 + 2x + 2}{(x + 1)(x - 2)(x + 2)}$$

Common Errors Note that we can add only those fractions with like denominators. Thus,

$$\frac{2}{x} + \frac{3}{y} \neq \frac{5}{x + y}$$

OPERATIONS WITH FRACTIONS

Also, we add only the numerators of fractions with like denominators. Thus,

$$\frac{3}{2x} + \frac{4}{2x} \neq \frac{7}{4x}$$

EXERCISES 11.4
Find the lowest common denominator for each set of fractions.

Complete Example (see Example 1)

$$\frac{1}{3x}, \frac{1}{6x^2}, \frac{3}{4x}$$

Completely factor each denominator and align common factors.

$3x = 3 \cdot \quad x$

$6x^2 = 3 \cdot \quad x \quad \cdot \quad x \quad \cdot \quad \underline{\quad}$
                                                      (1)

Include in the LCD each of these factors the greatest number of times it occurs in any single denominator.

$4x = \quad \underline{\quad} \quad \cdot \quad 2 \quad \cdot \quad \underline{\quad}$
         (2)   (3)

LCD: $3 \cdot \underline{\quad} \cdot x \cdot 2 \cdot \underline{\quad} = \underline{\quad}$
            (4)   (5)   (6)

---

(1) 2 (2) x (3) 2 (4) x (5) 2 (6) $12x^2$

---

1. $\dfrac{2}{x}, \dfrac{3}{x^2}, \dfrac{1}{y}$

2. $\dfrac{a}{x^2}, \dfrac{a}{x^2y}, \dfrac{1}{z}$

3. $\dfrac{1}{xy}, \dfrac{2}{yz}, \dfrac{3}{xz}$

4. $\dfrac{a}{x^2y}, \dfrac{2}{xyz}, \dfrac{3}{yz^2}$

5. $\dfrac{1}{8xy}, \dfrac{2}{3x^2}, \dfrac{2}{4xy}$

6. $\dfrac{2}{4xy}, \dfrac{3}{6yz^2}, \dfrac{1}{3xy^2z}$

Complete Example (see Example 1)

$$\frac{2}{x-2}, \frac{3x}{x^2+x-6}, \frac{3}{x+3}$$

Completely factor each denominator and align common factors.

$x - 2 = (x - 2)$

$x^2 + x - 6 = (x - 2)(\underline{\quad})$
                              (1)

Include in the LCD each of these factors the greatest number of times it occurs in any single denominator.

$x + 3 = \quad (x + 3)$

LCD: $(\underline{\quad})(\underline{\quad})$
       (2)   (3)

---

(1) $x + 3$ (2) $x - 2$ (3) $x + 3$

358

11.4 SUMS OF FRACTIONS WITH UNLIKE DENOMINATORS

7. $\dfrac{2}{x^2 - y^2}, \dfrac{1}{x - y}$
8. $\dfrac{3}{x^2 + 2x}, \dfrac{4}{x + 2}$
9. $\dfrac{3}{x^2}, \dfrac{4}{x^2 + 2x}$
10. $\dfrac{5}{x^2 + 2x + 1}, \dfrac{3}{x^2 + 4x + 3}$

11. $\dfrac{2}{x^2 + 3x - 4}, \dfrac{3}{(x - 1)^2}, \dfrac{2}{x + 4}$
12. $\dfrac{3}{a^2 - a - 6}, \dfrac{a + 2}{a^2 + 7a + 10}, \dfrac{3}{(a - 3)^2}$

Rewrite each sum as a single term.

Complete Example (see Example 2)

$\dfrac{3}{5x} + \dfrac{7}{10}$

$= \dfrac{(2)3}{(\underline{\quad})5x} + \dfrac{7(x)}{10(\underline{\quad})}$
$\qquad \text{(1)} \qquad \qquad \text{(2)}$

Find LCD $10x$ and build each fraction to a fraction with denominator $10x$.

$= \dfrac{6 + 7x}{\underline{\quad}}$
$\qquad \text{(3)}$

Add numerators.

---

(1) 2   (2) x (3) $10x$

---

13. $\dfrac{x}{8} + \dfrac{5x}{2}$
14. $\dfrac{2y}{3} + \dfrac{5y}{12}$
15. $\dfrac{5}{ax} + \dfrac{3}{x}$
16. $\dfrac{2}{ax} + \dfrac{3}{a}$

17. $\dfrac{2}{x} + \dfrac{4}{3y}$
18. $\dfrac{2}{3x} + \dfrac{1}{2y}$
19. $\dfrac{1}{x} + \dfrac{1}{y} + \dfrac{1}{z}$
20. $\dfrac{1}{x} + \dfrac{2}{y} + \dfrac{3}{z}$

Complete Example (see Example 2)

$\dfrac{x + 1}{4} + \dfrac{x - 1}{6}$

$= \dfrac{(3)(x + 1)}{(\underline{\quad})4} + \dfrac{(x - 1)(2)}{6(\underline{\quad})}$
$\qquad \text{(1)} \qquad \qquad \text{(2)}$

Find LCD 12 and build each fraction to a fraction with denominator 12. Insert parentheses as needed.

$= \dfrac{3(x + 1) + 2(x - 1)}{\underline{\quad}}$
$\qquad \text{(3)}$

$= \dfrac{3x + 3 + 2x - 2}{12} = \underline{\quad}$
$\qquad \qquad \qquad \qquad \text{(4)}$

Remove parentheses and simplify.

---

(1) 3   (2) 2   (3) 12   (4) $\dfrac{5x + 1}{12}$

359

OPERATIONS WITH FRACTIONS

21. $\dfrac{3y-2}{3} + \dfrac{2y-1}{6}$

22. $\dfrac{3x+4}{2} + \dfrac{4x-1}{4}$

23. $\dfrac{x-2}{6} + \dfrac{x+3}{2}$

24. $\dfrac{5x+1}{6x} + \dfrac{3x-2}{2x}$

25. $\dfrac{x-y}{2x} + \dfrac{x+y}{3x}$

26. $\dfrac{2x-y}{4y} + \dfrac{x-2y}{3y}$

Complete Example (see Example 3)

$\dfrac{x}{x+1} + \dfrac{x-2}{x+2}$

$= \dfrac{(x+2)x}{(\underline{})(x+1)} + \dfrac{(x-2)(x+1)}{(x+2)(\underline{})}$
$(1)(2)$

$= \dfrac{x(x+2) + (x-2)(x+1)}{(x+2)(\underline{})}$
(3)

$= \dfrac{x^2 + 2x + x^2 + x - 2x - 2}{(x+2)(x+1)}$

$= \underline{}$
(4)

Find LCD $(x+1)(x+2)$ and build each fraction to a fraction with denominator $(x+1)(x+2)$. Insert parentheses as needed.

Add numerators.

Remove parentheses in numerator.

Simplify.

(1) $x+2$ (2) $x+1$ (3) $x+1$ (4) $\dfrac{2x^2 + x - 2}{(x+2)(x+1)}$

27. $\dfrac{3}{x+3} + \dfrac{4}{x-3}$

28. $\dfrac{2}{x+2} + \dfrac{3}{x-2}$

29. $\dfrac{x}{x-y} + \dfrac{2y}{x+y}$

30. $\dfrac{y}{x-y} + \dfrac{x}{x+y}$

31. $\dfrac{x}{x-2} + \dfrac{x+1}{x+2}$

32. $\dfrac{x-3}{x+1} + \dfrac{2x}{x-1}$

33. $\dfrac{y+1}{y-3} + \dfrac{y-4}{y+2}$

34. $\dfrac{y-2}{y+1} + \dfrac{y+3}{y+2}$

Complete Example (see Example 4)

$\dfrac{x}{x^2-9} + \dfrac{1}{x^2-x-6}$

$= \dfrac{x}{(x+3)(\underline{})} + \dfrac{1}{(x+2)(\underline{})}$
$(1)(2)$

$= \dfrac{(x+2)x}{(\underline{})(x+3)(x-3)} + \dfrac{1(x+3)}{(x+2)(x-3)(x+3)}$
(4)

$= \dfrac{x(x+2) + (x+3)}{(x+3)(x-3)(x+2)}$

Factor denominators.

Find LCD $(x+3)(x-3)(\underline{})$ and build
(3)
each fraction to a fraction with this denominator.

Add numerators.

360

$$= \frac{x^2 + 2x + x + 3}{(x+3)(x-3)(x+2)}$$ Remove parentheses in numerator.

$$= \frac{\underline{}}{(5)}$$ Simplify.

---

(1) $x-3$ (2) $x-3$ (3) $x+2$ (4) $x+2$ (5) $\dfrac{x^2+3x+3}{(x+3)(x-3)(x+2)}$

---

35. $\dfrac{y}{y^2-1} + \dfrac{3}{y^2+2y+1}$ 36. $\dfrac{4}{y^2-5y+4} + \dfrac{y}{y^2-16}$ 37. $\dfrac{2x}{x^2+3x+2} + \dfrac{x}{x^2-4}$

38. $\dfrac{x}{x^2-x-6} + \dfrac{3x}{x^2+3x+2}$ 39. $\dfrac{x-1}{x^2+3x} + \dfrac{x}{x^2+6x+9}$ 40. $\dfrac{x}{x^2+5x+4} + \dfrac{x-2}{x^2+x}$

11.5 DIFFERENCES OF FRACTIONS WITH UNLIKE DENOMINATORS

We subtract fractions with unlike denominators in a similar way that we add such fractions. However, we first write each fraction in standard form. Thus, any fraction in the form

$$\frac{a}{b} - \frac{c}{d}$$

is first written as

$$\frac{a}{b} + \frac{-c}{d}$$

We can now add the fractions as we did in Section 11.4.

Example 1 Write the difference $\dfrac{2}{3x} - \dfrac{3}{4x}$ as a single term.

Solution We begin by writing $-\dfrac{3}{4x}$ in standard form as $\dfrac{-3}{4x}$. The LCD is $12x$. We build each fraction to an equivalent fraction with this denominator to get

$$\frac{(4)2}{(4)3x} + \frac{-3(3)}{4x(3)} = \frac{8}{12x} + \frac{-9}{12x}$$

Now, adding numerators yields

$$\frac{8-9}{12x} = \frac{-1}{12x}$$

Again, special care must be taken with binomial numerators.

361

OPERATIONS WITH FRACTIONS

Example 2 Write the difference of $\dfrac{1}{2} - \dfrac{x-1}{3}$ as a single term.

Solution $\dfrac{1}{2} - \dfrac{x-1}{3}$ should first be written as

$$\dfrac{1}{2} + \dfrac{-(x-1)}{3}$$

where *the entire numerator is enclosed within parentheses*. Then, we obtain the LCD 6 and build each fraction to fractions with denominator 6, add numerators, and simplify.

$$\dfrac{(3)1}{(3)2} + \dfrac{-(x-1)(2)}{3(2)} = \dfrac{3}{6} + \dfrac{-2(x-1)}{6}$$

$$= \dfrac{3 - 2x + 2}{6}$$

$$= \dfrac{-2x + 5}{6}$$

The next examples involve binomial denominators.

Example 3 Write the difference of $\dfrac{x}{x+2} - \dfrac{1}{x-1}$ as a single term.

Solution We begin by writing $-\dfrac{1}{x-1}$ in standard form as $\dfrac{-1}{x-1}$. The LCD is $(x-1)(x+2)$ and we build each fraction to an equivalent fraction with this denominator to get

$$\dfrac{x}{x+2} + \dfrac{-1}{x-1} = \dfrac{(x-1)x}{(x-1)(x+2)} + \dfrac{-1(x+2)}{(x-1)(x+2)}$$

Now adding numerators and simplifying yields

$$\dfrac{x(x-1) - 1(x+2)}{(x-1)(x+2)} = \dfrac{x^2 - x - x - 2}{(x-1)(x+2)}$$

$$= \dfrac{x^2 - 2x - 2}{(x-1)(x+2)}$$

Example 4 Write the difference of

$$\dfrac{x}{x^2 - 9} - \dfrac{1}{x^2 + 4x - 21}$$

as a single term.

Solution We first factor the denominators and write the fractions in standard form to get

$$\dfrac{x}{x^2 - 9} - \dfrac{1}{x^2 + 4x - 21} = \dfrac{x}{(x-3)(x+3)} + \dfrac{-1}{(x+7)(x-3)}$$

We find the LCD $(x+7)(x-3)(x+3)$ and build each fraction to an equivalent fraction with this denominator to get

11.5 DIFFERENCES OF FRACTIONS WITH UNLIKE DENOMINATORS

$$\frac{(x+7)x}{(x+7)(x-3)(x+3)} + \frac{-1(x+3)}{(x+7)(x-3)(x+3)}$$

Now, adding numerators and simplifying yield

$$\frac{x(x+7) - (x+3)}{(x+7)(x-3)(x+3)} = \frac{x^2 + 7x - x - 3}{(x+7)(x-3)(x+3)}$$

$$= \frac{x^2 + 6x - 3}{(x+7)(x-3)(x+3)}$$

EXERCISES 11.5
Rewrite each difference as a single term.

Complete Example (see Example 1)

$$\frac{3}{5x} - \frac{2}{3x}$$

Write in standard form; find LCD $15x$ and build each fraction to a fraction with denominator ____(1)____ .

$$= \frac{(3)3}{(\underline{})5x} + \frac{-2(5)}{3x(\underline{})}$$

$$= \frac{9}{15x} + \frac{-10}{15x}$$

Add numerators.

$$= \underline{}$$

---

(1) $15x$ (2) 3 (3) 5 (4) $\dfrac{-1}{15x}$

---

1. $\dfrac{x}{5} - \dfrac{y}{2}$
2. $\dfrac{x}{8} - \dfrac{y}{3}$
3. $\dfrac{y}{4} - \dfrac{x}{6}$
4. $\dfrac{y}{6} - \dfrac{x}{9}$

5. $\dfrac{3}{x} - \dfrac{1}{2x}$
6. $\dfrac{2}{y} - \dfrac{1}{3y}$
7. $\dfrac{1}{2x} - \dfrac{1}{6x}$
8. $\dfrac{2}{3y} - \dfrac{4}{9y}$

9. $\dfrac{2}{x} - \dfrac{3}{y}$
10. $\dfrac{4}{y} - \dfrac{1}{x}$
11. $\dfrac{3}{2x} - \dfrac{5}{6y}$
12. $\dfrac{1}{3y} - \dfrac{3}{4x}$

Complete Example (see Example 2)

$$\frac{x+2}{6} - \frac{2x-1}{3}$$

Insert parentheses and write in standard form; find LCD 6 and build each fraction to a fraction with denominator ____(1)____ .

$$= \frac{(x+2)}{6} + \frac{-(2x-1)(2)}{3(\underline{})}$$

363

OPERATIONS WITH FRACTIONS

$$= \frac{(x+2)-2(2x-1)}{(3)}$$ Add numerators.

$$= \frac{x+2-4x+2}{6}$$ Remove parentheses and simplify.

$$= \frac{}{(4)}$$

---

(1) 6   (2) 2   (3) 6   (4) $\dfrac{-3x+4}{6}$

---

13. $\dfrac{x-2}{6} - \dfrac{x+1}{3}$ 14. $\dfrac{2x+1}{3} - \dfrac{x-1}{9}$ 15. $\dfrac{3y-2}{3} - \dfrac{2y-1}{6}$ 16. $\dfrac{3x+4}{2} - \dfrac{4x-1}{4}$

17. $\dfrac{2-x}{6} - \dfrac{3+x}{2}$ 18. $\dfrac{y+2}{3} - \dfrac{y-4}{6}$ 19. $\dfrac{5x+1}{6x} - \dfrac{3x-2}{2x}$ 20. $\dfrac{2b-c}{2c} - \dfrac{c+a}{c}$

21. $\dfrac{x-y}{2x} - \dfrac{x+y}{3x}$ 22. $\dfrac{4y-9}{3y} - \dfrac{3y-8}{4y}$ 23. $\dfrac{2a-b}{4b} - \dfrac{a-3b}{6a}$ 24. $\dfrac{a-b}{ab} - \dfrac{b-c}{bc}$

Complete Example (see Example 3)

$$\frac{2}{x+3} - \frac{3}{2x+6}$$

$$= \frac{2}{x+3} + \frac{-3}{2(\underline{\quad(1)\quad})}$$ Factor denominators where possible. Write in standard form.

$$= \frac{(2)2}{(\underline{\quad(2)\quad})(x+3)} + \frac{-3}{2(x+3)}$$ Find LCD $2(x+3)$ and build each fraction to a fraction with denominator $2(x+3)$.

$$= \frac{4}{2(x+3)} + \frac{-3}{2(x+3)}$$

$$= \frac{}{(3)}$$ Add numerators.

---

(1) $x+3$ (2) 2   (3) $\dfrac{1}{2(x+3)}$

---

25. $\dfrac{2}{x+y} - \dfrac{1}{2x+2y}$ 26. $\dfrac{2}{x+1} - \dfrac{3}{2x+2}$ 27. $\dfrac{5}{6x+6} - \dfrac{3}{2x+2}$ 28. $\dfrac{7}{5y-10} - \dfrac{5}{3y-6}$

29. $\dfrac{3}{2a+b} - \dfrac{2}{4a+2b} + \dfrac{1}{8a+4b}$ 30. $\dfrac{3}{2x+3y} - \dfrac{5}{4x+6y} + \dfrac{1}{8x+12y}$

364

11.5 DIFFERENCES OF FRACTIONS WITH UNLIKE DENOMINATORS

Complete Example (see Example 3)

$$\frac{x}{x+3} - \frac{1}{x-2}$$

$$= \frac{(x-2)x}{(\underline{})(x+3)} + \frac{-1(x+3)}{(x-2)(\underline{})}$$
$$(1)(2)$$

$$= \frac{x(x-2) - (x+3)}{(x-2)(x+3)}$$

$$= \frac{x^2 - 2x - x - 3}{(x-2)(x+3)}$$

$$= \underline{}$$
$$(3)$$

Write in standard form.
Find LCD $(x-2)(x+3)$ and build each fraction to a fraction with denominator $(x-2)(x+3)$.

Add numerators.

Remove parentheses in numerator and simplify.

---

(1) $x - 2$ (2) $x + 3$ (3) $\dfrac{x^2 - 3x - 3}{(x-2)(x+3)}$

---

31. $\dfrac{x}{x+3} - \dfrac{x}{x-3}$ **32.** $\dfrac{2}{x+2} - \dfrac{3}{x+3}$ **33.** $\dfrac{3}{3x-4} - \dfrac{5}{5x+6}$

34. $\dfrac{1}{x+y} - \dfrac{1}{x-y}$ **35.** $\dfrac{1}{2a+1} - \dfrac{3}{a-2} + \dfrac{2}{2a+1}$ **36.** $\dfrac{x}{2x-y} - \dfrac{y}{x-2y} + \dfrac{y}{2x-y}$

Complete Example (see Example 3)

$$\frac{x-2}{x+1} - \frac{x+2}{x-1}$$

$$= \frac{(x-2)}{(x+1)} + \frac{-(x+2)}{(x-1)}$$

$$= \frac{(x-1)(x-2)}{(\underline{})(x+1)} + \frac{-(x+2)(x+1)}{(x-1)(\underline{})}$$
$$(1)(2)$$

$$= \frac{(x-1)(x-2) - (x+2)(x+1)}{(x-1)(\underline{})}$$
$$(3)$$

$$= \frac{(x^2 - 3x + 2) - (x^2 + 3x + 2)}{(x-1)(x+1)}$$

$$= \frac{x^2 - 3x + 2 - x^2 - 3x - 2}{(x-1)(x+1)}$$

$$= \underline{}$$
$$(4)$$

Enclose numerators and denominators in parentheses and write in standard form.

Find LCD $(x-1)(x+1)$ and build each fraction to a fraction with denominator $(x-1)(x+1)$.

Add numerators.

Perform indicated multiplication. Write products in parentheses.

Remove parentheses in numerator and simplify.

---

(1) $x - 1$ (2) $x + 1$ (3) $x + 1$ (4) $\dfrac{-6x}{(x-1)(x+1)}$

365

OPERATIONS WITH FRACTIONS

37. $\dfrac{x-2}{x+2} - \dfrac{x+2}{x-2}$

38. $\dfrac{y-4}{y-2} - \dfrac{y-7}{y-5}$

39. $\dfrac{x+1}{x+2} - \dfrac{x+2}{x+3}$

40. $\dfrac{x+y}{x-y} - \dfrac{x-y}{x+y}$

41. $\dfrac{2a-3b}{a+b} - \dfrac{a+b}{a-b}$

42. $\dfrac{a+2b}{2a-b} - \dfrac{2a+b}{a-2b}$

Complete Example (see Example 4)

$\dfrac{x}{x^2+5x+6} - \dfrac{1}{x^2+7x+12}$

$= \dfrac{x}{(x+3)(\underline{\qquad})} + \dfrac{-1}{(x+4)(\underline{\qquad})}$
$\quad\quad\quad\quad\quad (1) \quad\quad\quad\quad\quad\quad (2)$

Factor denominators and write in standard form.

$= \dfrac{x(x+4)}{(x+3)(x+2)(\underline{\qquad})} + \dfrac{-1(x+2)}{(x+4)(x+3)(\underline{\qquad})}$
$\quad\quad\quad\quad\quad\quad\quad (3) \quad\quad\quad\quad\quad\quad\quad\quad\quad (4)$

Find LCD $(x+3)(x+2)(x+4)$ and build each fraction to a fraction with this denominator.

$= \dfrac{x(x+4) - (x+2)}{(x+3)(x+2)(x+4)}$

Add numerators.

$= \dfrac{x^2 + 4x - x - 2}{(x+3)(x+2)(x+4)}$

Remove parentheses in numerator and simplify.

$= \underline{\qquad\qquad\qquad}$
$\quad\quad (5)$

(1) $x+2$ (2) $x+3$ (3) $x+4$ (4) $x+2$ (5) $\dfrac{x^2+3x-2}{(x+3)(x+2)(x+4)}$

43. $\dfrac{1}{x^2-x-2} - \dfrac{1}{x^2+2x+1}$

44. $\dfrac{2}{x^2-5x+6} - \dfrac{5}{x^2+2x-15}$

45. $\dfrac{3x}{x^2+3x-10} - \dfrac{2x}{x^2+x-6}$

46. $\dfrac{2}{x^2-x-6} - \dfrac{3}{x^2-9}$

47. $\dfrac{5x}{x^2+3x+2} - \dfrac{3x-6}{x^2+4x+4}$

48. $\dfrac{8}{x^2-4y^2} - \dfrac{2}{x^2-5xy+6y^2}$

11.6 COMPLEX FRACTIONS

A fraction that contains one or more fractions in either its numerator or denominator or both is called a **complex fraction**. For example,

$$\dfrac{\frac{1}{2}}{\frac{3}{4}} \quad \text{and} \quad \dfrac{a + \frac{1}{3}}{a - \frac{1}{2}}$$

11.6 COMPLEX FRACTIONS

are complex fractions. Like simple fractions, complex fractions represent quotients. For example,

$$\frac{\frac{1}{2}}{\frac{3}{4}} = \frac{1}{2} \div \frac{3}{4} \qquad (1)$$

and

$$\frac{a + \frac{1}{3}}{a - \frac{1}{2}} = \left(a + \frac{1}{3}\right) \div \left(a - \frac{1}{2}\right) \qquad (2)$$

In cases like Equation (1), in which the numerator and denominator of the complex fraction do not contain sums or differences, we can simply invert the divisor and multiply. In fact we can also use the fundamental principle of fractions to simplify complex fractions of the Form (1) above.

$$\frac{\frac{1}{2}}{\frac{3}{4}} = \frac{1}{2} \div \frac{3}{4} = \frac{1}{2} \cdot \frac{4}{3}$$

— Divisor inverted.
— Multiply.

$$= \frac{1}{\cancel{2}} \cdot \frac{\cancel{4}^{2}}{3} = \frac{2}{3}$$

In cases like Equation (2), in which the numerator or denominator of the complex fraction contains sums or differences, we *cannot* simply invert the divisor and multiply. However, we can use the fundamental principle of fractions to simplify complex fractions. In fact, we can also use the fundamental principle to simplify complex fractions of Form (1) above.

Example 1 Simplify $\dfrac{\frac{1}{2}}{\frac{3}{4}}$ by using the fundamental principle of fractions.

Solution We multiply the numerator and denominator by the LCD of all fractions in the numerator and denominator; in this case, the LCD is . The result is a simple fraction equivalent to the given complex fraction.

$$\frac{\frac{1}{2}}{\frac{3}{4}} = \frac{4\left(\frac{1}{2}\right)}{4\left(\frac{3}{4}\right)} = \frac{\cancel{4}^{2}\left(\frac{1}{\cancel{2}}\right)}{\cancel{4}\left(\frac{3}{\cancel{4}}\right)} = \frac{2}{3}$$

Multiply numerator and denominator by 4, the LCD of 2 and 4.

—Result is a simple fraction.

The simplification of Equation (2) above appears in the next example.

367

OPERATIONS WITH FRACTIONS

Example 2 Simplify $\dfrac{a + \frac{1}{3}}{a - \frac{1}{2}}$.

Solution We multiply the numerator and denominator by the LCD of all fractions in the numerator and denominator; in this case, the LCD is 6. We obtain

$$\frac{a + \frac{1}{3}}{a - \frac{1}{2}} = \frac{6\left(a + \frac{1}{3}\right)}{6\left(a - \frac{1}{2}\right)} = \frac{6a + 6\left(\frac{1}{3}\right)}{6a - 6\left(\frac{1}{2}\right)} = \frac{6a + 2}{6a - 3}$$

Multiply numerator and denominator by 6, the LCD of 3 and 2.

In Exercises 11.6, we will simplify all complex fractions using the fundamental principle of fractions.

EXERCISES 11.6
Simplify each complex fraction by using the fundamental principle of fractions.

Complete Example (see Example 1)

$$\frac{\frac{3c^2}{4d^4}}{\frac{5c^3}{12d^4}} = \frac{(12d^4)\frac{3c^2}{4d^4}}{(12d^4)\frac{5c^3}{12d^4}}$$

Find the LCD _____(1)_____ for the fractions in the numerator and denominator; multiply numerator and denominator by the LCD.

$$= \frac{9c^2}{\underline{\qquad(2)\qquad}}$$

Divide numerator and denominator by common factors.

$$= \frac{}{\underline{\qquad(3)\qquad}}$$

(1) $12d^4$ (2) $5c^3$ (3) $\dfrac{9}{5c}$

1. $\dfrac{\frac{3}{4}}{\frac{1}{2}}$

2. $\dfrac{\frac{2}{3}}{\frac{4}{5}}$

3. $\dfrac{\frac{5}{6}}{\frac{2}{3}}$

4. $\dfrac{\frac{4}{9}}{\frac{2}{3}}$

5. $\dfrac{\frac{4}{5}}{\frac{7}{10}}$

6. $\dfrac{\frac{8}{5}}{\frac{16}{7}}$

7. $\dfrac{\frac{3x}{y}}{\frac{x}{2y}}$

8. $\dfrac{\frac{2x}{y}}{\frac{x}{3y}}$

11.6 COMPLEX FRACTIONS

9. $\dfrac{\dfrac{3}{ax}}{\dfrac{9}{bx}}$

10. $\dfrac{\dfrac{2a}{bx}}{\dfrac{3a}{cx}}$

11. $\dfrac{\dfrac{2x}{y^2}}{\dfrac{6x^2}{5y}}$

12. $\dfrac{\dfrac{4x^2}{3y}}{\dfrac{2x}{9y^2}}$

13. $\dfrac{\dfrac{1}{3}}{\dfrac{5}{}}$ (i.e. $\dfrac{1/3}{5}$)

14. $\dfrac{\dfrac{1}{2}}{7}$

15. $\dfrac{\dfrac{1}{2x}}{5}$

16. $\dfrac{\dfrac{1}{3y}}{2}$

17. $\dfrac{5}{\dfrac{10}{3}}$

18. $\dfrac{4}{\dfrac{12}{7}}$

19. $\dfrac{3x}{\dfrac{x}{2}}$

20. $\dfrac{\dfrac{4}{y}}{\dfrac{y}{6}}$

Complete Example (see Example 2)

$\dfrac{1 - \dfrac{1}{3}}{2 + \dfrac{5}{6}}$

Find the LCD for all fractions in numerator and denominator, 6. Multiply each term in numerator and each term in denominator by 6.

$= \dfrac{(6)1 - (\overset{2}{\cancel{6}})\dfrac{1}{\cancel{3}}}{(\underline{\quad}_{(1)})2 + (\underline{\quad}_{(2)})\dfrac{5}{6}}$

Simplify.

$= \dfrac{6-2}{12 + \underline{\quad}_{(3)}} = \underline{\quad}_{(4)}$

(1) 6   (2) 6   (3) 5   (4) $\dfrac{4}{17}$

21. $\dfrac{\dfrac{2}{3}}{3 - \dfrac{1}{3}}$

22. $\dfrac{1 + \dfrac{1}{5}}{\dfrac{2}{5}}$

23. $\dfrac{1 - \dfrac{1}{3}}{2 + \dfrac{2}{3}}$

24. $\dfrac{3 + \dfrac{1}{10}}{2 + \dfrac{3}{5}}$

25. $\dfrac{\dfrac{1}{2} - \dfrac{3}{8}}{\dfrac{5}{4} + \dfrac{1}{2}}$

26. $\dfrac{\dfrac{2}{3} - \dfrac{1}{6}}{\dfrac{1}{3} + \dfrac{5}{6}}$

27. $\dfrac{\dfrac{1}{2} + \dfrac{1}{3}}{\dfrac{1}{3} - \dfrac{1}{6}}$

28. $\dfrac{\dfrac{3}{4} - \dfrac{1}{2}}{\dfrac{1}{6} + \dfrac{1}{3}}$

OPERATIONS WITH FRACTIONS

Complete Example (see Example 2)

$$\frac{x + \frac{2}{y}}{x - \frac{y}{x}}$$

$$= \frac{(xy)x + (x\cancel{y})\frac{2}{\cancel{y}}}{(\underline{}_{(2)})x - (\underline{}_{(3)})\frac{y}{x}}$$

$$= \frac{x^2y + 2x}{x^2y - \underline{}_{(4)}}.$$

Find the LCD for all fractions in the numerator and denominator, xy. Multiply each term in numerator and each term in denominator by $\underline{}_{(1)}$.

Simplify.

---

(1) xy (2) xy (3) xy (4) y^2

---

29. $\dfrac{2 - \frac{a}{b}}{2 - \frac{b}{a}}$ 30. $\dfrac{4 - \frac{1}{x^2}}{2 - \frac{1}{x}}$ 31. $\dfrac{y - \frac{1}{y}}{y + \frac{1}{y}}$ 32. $\dfrac{a + \frac{a}{b}}{1 + \frac{1}{b}}$

33. $\dfrac{\frac{x}{3y} - \frac{1}{2}}{\frac{4}{3y} - \frac{2}{x}}$ 34. $\dfrac{\frac{2}{y} + \frac{1}{2y}}{y + \frac{y}{2}}$ 35. $\dfrac{\frac{3}{2b} - \frac{1}{b}}{\frac{4}{a} + \frac{3}{2a}}$ 36. $\dfrac{\frac{1}{ab} - \frac{1}{b}}{\frac{1}{a} - \frac{1}{ab}}$

37. $\left(a - \dfrac{a}{b}\right) \div \left(2 + \dfrac{3}{b}\right)$ 38. $\left(y - \dfrac{2}{y}\right) \div \left(y + \dfrac{2}{y}\right)$

39. $\left(4 - \dfrac{1}{x^2}\right) \div \left(2 + \dfrac{1}{x}\right)$ 40. $\left(9 - \dfrac{1}{x^2}\right) \div \left(3 + \dfrac{1}{x}\right)$

11.7 FRACTIONAL EQUATIONS

To solve an equation containing fractions, it is usually easiest to first find an equivalent equation that is free of fractions. We do this by multiplying each member of an equation by the lowest common denominator of the fractions.

Although we can apply the algebraic properties we have studied in any order, the following steps show the order most helpful in solving an equation when the solution is not obvious. Of course, not all the steps are always necessary.

11.7 FRACTIONAL EQUATIONS

> To solve an equation:
> 1. "Clear fractions," if there are any, by multiplying each member of the equation by the LCD.
> 2. Write any expression that contains parentheses as an expression without parentheses.
> 3. Combine any like terms in either member.
> 4. Obtain all terms containing the variable in one member and all terms not containing the variable in the other member.
> 5. Divide each member by the coefficient of the variable if it is different from 1.
> 6. Check the answer if each member of the equation has been multiplied by an expression containing a variable.

Example 1 Solve $\frac{x}{3} - 2 = \frac{4}{5}$.

Solution We multiply each member by the LCD 15 to obtain an equivalent equation that does not contain a fraction.

$$(\overset{5}{\cancel{15}})\left(\frac{x}{3}\right) - 15(2) = \overset{3}{\cancel{15}}\left(\frac{4}{5}\right)$$

$$5x - 30 = 12$$

$$5x = 42$$

$$x = \frac{42}{5}$$

The multiplication property of equality (Section 8.1) allows us to multiply each member of an equation by a *nonzero* value in order to obtain an equivalent equation. Thus, to solve the equation

$$\frac{3}{4} = 8 - \frac{2x + 11}{x - 5}$$

we would multiply each member by the LCD $4(x - 5)$. We note that x cannot equal 5 since $4(x - 5)$ equals 0 if $x = 5$. The entire solution is shown in the next example.

Example 2 Solve $\frac{3}{4} = 8 - \frac{2x + 11}{x - 5}$.

Solution We multiply each member by the LCD $4(x - 5)$ to get

$$\cancel{4}(x - 5) \cdot \frac{3}{\cancel{4}} = 4(x - 5)8 - 4(x \cancel{- 5}) \cdot \frac{(2x + 11)}{(x \cancel{- 5})}$$

$$3(x - 5) = 32(x - 5) - 4(2x + 11)$$

Applying the distributive property, we obtain

$$3x - 15 = 32x - 160 - 8x - 44$$

$$3x - 15 = 24x - 204$$

OPERATIONS WITH FRACTIONS

Solving for x yields
$$-21x = -189; \quad x = 9$$

Note that $4(x - 5)$ is *not equal* to zero for $x = 9$. Thus, $x = 9$ is a valid solution for the equation.

When equations contain more than one variable, it is sometimes desirable to solve for one variable in terms of the other variable(s).

Example 3 Solve $\dfrac{4 + b}{x} = \dfrac{2a}{3c}$ for x in terms of a, b, and c.

Solution We multiply each member by the LDC $3xc$ to get

$$3xc \cdot \frac{(4 + b)}{x} = 3xc \cdot \frac{(2a)}{3c}$$

$$3c(4 + b) = x(2a)$$

$$12c + 3bc = 2ax$$

Now, dividing each member by $2a$, we obtain

$$\frac{12c + 3bc}{2a} = \frac{2ax}{2a}$$

$$x = \frac{12c + 3bc}{2a}$$

PROPORTIONS

Proportions are special types of fractional equations. We can use the following property that we introduced in Section 2.8 to simplify work involving proportions.

> **In any proportion, the product of the extremes is equal to the product of the means.**

In symbols,

$$\text{If } \frac{a}{b} = \frac{c}{d}, \text{ then } a \cdot d = b \cdot c$$

Example 4 Solve the proportion $\dfrac{x - 2}{3} = \dfrac{x}{4}$.

Solution By setting the product of the extremes equal to the product of the means, we obtain

$$4(x - 2) = 3x$$
$$4x - 8 = 3x$$
$$x = 8$$

11.7 FRACTIONAL EQUATIONS

EXERCISES 11.7
Solve.

Complete Example (see Example 1)

$$2 + \frac{3}{a} = \frac{1}{2}$$

$$(2a)2 + (\underset{(2)}{\underline{}})\frac{3}{a} = (\underset{(3)}{\underline{}})\frac{1}{2}$$

$$4a + 6 = \underset{(4)}{\underline{}}$$

$$3a = -6$$

$$a = \underset{(5)}{\underline{}}$$

Multiply each member by LCD $\underset{(1)}{\underline{}}$.

Complete the solution.

---

(1) $2a$ (2) $2a$ (3) $2a$ (4) a (5) -2

---

1. $\dfrac{5x}{2} - 1 = x + \dfrac{1}{2}$

2. $y - \dfrac{3}{10} = \dfrac{1}{2} + \dfrac{3y}{5}$

3. $\dfrac{5y}{6} - \dfrac{1}{6} = \dfrac{2y}{3} + \dfrac{5}{6}$

4. $\dfrac{2t}{3} - \dfrac{1}{4} = \dfrac{25}{12} + \dfrac{t}{3}$

5. $\dfrac{x}{6} - \dfrac{7}{3} = \dfrac{2x}{9} - \dfrac{x}{4}$

6. $\dfrac{8x}{3} - 3 = \dfrac{2x}{3} - 6$

7. $4 + \dfrac{4}{y} = \dfrac{12}{y}$

8. $1 + \dfrac{3}{x} = \dfrac{12}{x}$

9. $2 + \dfrac{5}{z} = \dfrac{11}{z}$

10. $2 + \dfrac{5}{2x} = \dfrac{3}{x} + \dfrac{3}{2}$

11. $3 - \dfrac{1}{x} = \dfrac{7}{5x} - \dfrac{9}{5}$

12. $\dfrac{1}{3} - \dfrac{1}{4} = \dfrac{10}{3x} - \dfrac{1}{18}$

Complete Example (see Example 1)

$$\frac{x+5}{10} - \frac{5-x}{5} = 1$$

$$\frac{(x+5)}{10} + \frac{-(5-x)}{5} = 1$$

$$\frac{(\cancel{10})(x+5)}{\cancel{10}} + \frac{-(\cancel{10})(5-x)}{\cancel{5}}^{2} = (\underset{(2)}{\underline{}})1$$

$$(x+5) - 2(\underset{(3)}{\underline{}}) = 10$$

$$x + 5 - 10 + 2x = 10$$

$$3x - 5 = 10$$

$$3x = 15$$

$$x = \underset{(4)}{\underline{}}$$

Enclose numerators in parentheses and write in standard form.

Multiply each member by LCD $\underset{(1)}{\underline{}}$.

Complete the solution.

---

(1) 10 (2) 10 (3) $5 - x$ (4) 5

373

OPERATIONS WITH FRACTIONS

13. $\dfrac{y+12}{9} = \dfrac{y-9}{2}$

14. $\dfrac{y+1}{4} - \dfrac{3}{2} = \dfrac{2y-9}{10}$

15. $\dfrac{x-1}{10} + \dfrac{19}{15} = \dfrac{x}{3}$

16. $\dfrac{2x}{3} - \dfrac{2x+5}{6} = \dfrac{1}{2}$

17. $\dfrac{x+6}{2} - 1 = 5$

18. $\dfrac{2x-2}{2} + 2 = \dfrac{1}{2}$

Complete Example (see Example 2)

$$\dfrac{5}{2} = \dfrac{1}{2} + \dfrac{3x-2}{x+1}$$

$$2(x+1)\dfrac{5}{2} = 2(x+1)\dfrac{1}{2} + \underline{}_{(1)} \cdot \dfrac{3x-2}{x+1}$$

$$5(x+1) = \underline{}_{(2)} + 2(3x-2)$$

$$5x + 5 = x + 1 + 6x - 4$$

$$5x + 5 = \underline{}_{(3)}$$

$$-2x = -8$$

$$x = \underline{}_{(4)}$$

Multiply each member of the equation by the LCD $2(x+1)$. Note: x cannot be -1.

Complete the solution.

(1) $2(x+1)$ (2) $x+1$ (3) $7x-3$ (4) 4

19. $\dfrac{x-2}{x} = \dfrac{14}{3x} - \dfrac{1}{3}$

20. $\dfrac{3}{2x} - \dfrac{x-3}{2x} = \dfrac{5}{2x} - 1$

21. $\dfrac{2-y}{5y} = \dfrac{4}{15y} - \dfrac{1}{6}$

22. $\dfrac{2z-5}{z} - \dfrac{3}{z} = -\dfrac{2}{3}$

23. $\dfrac{x-3}{2x} + \dfrac{3x-7}{2x} = \dfrac{1}{3}$

24. $\dfrac{4}{x} + \dfrac{5}{2} = \dfrac{4x+5}{2x} - \dfrac{2x-3}{5x}$

25. $\dfrac{x}{x-3} = \dfrac{3}{x-3} + 2$

26. $\dfrac{x}{x-2} - 7 = \dfrac{2}{x-2}$

Solve for x in terms of the other variables. Assume that no denominator equals zero. See Example 3.

Complete Example

$$\dfrac{a+b}{x} = \dfrac{3}{c}$$

$$(xc)\dfrac{(a+b)}{x} = (\underline{}_{(1)})\dfrac{3}{c}$$

Multiply each member of the equation by LCD xc.

374

11.7 FRACTIONAL EQUATIONS

Simplify and complete solution.

$$c(a+b) = \underline{\quad(2)\quad}$$
$$ca + \underline{\quad(3)\quad} = 3x$$
$$x = \frac{ca+cb}{\underline{\quad(4)\quad}}$$

---

(1) xc (2) $3x$ (3) cb (4) 3

---

27. $\dfrac{a}{x} + \dfrac{b}{x} = 2$

28. $\dfrac{a+c}{x} = \dfrac{2}{b}$

29. $\dfrac{a}{x+1} = \dfrac{2a}{x-2}$

30. $\dfrac{2x+a}{x-b} = \dfrac{3}{2}$

31. $\dfrac{3}{4-x} = \dfrac{a}{b}$

32. $\dfrac{3}{x-a} - \dfrac{2}{x+b} = 0$

33. $\dfrac{a}{x+6} - \dfrac{a}{x} = \dfrac{3}{x}$

34. $\dfrac{a+b}{x+4} - \dfrac{a}{x} = \dfrac{-b}{x}$

35. $\dfrac{x+2b}{x+b} = \dfrac{b}{x+b} + 2$

36. $\dfrac{a}{x-a} = 2 - \dfrac{x+2a}{x-a}$

Complete Example (see Example 4)

$$\frac{3}{x+2} = \frac{2}{x+1}$$

$$3(\underline{\quad(1)\quad}) = 2(\underline{\quad(2)\quad})$$
$$3x + 3 = 2x + 4$$
$$x = \underline{\quad(3)\quad}$$

Set the product of the extremes equal to the product of the means.
Note: x cannot be -2 or -1.

Complete the solution.

---

(1) $x+1$ (2) $x+2$ (3) 1

---

37. $\dfrac{3}{5} = \dfrac{x}{x+2}$

38. $\dfrac{2}{x+4} = \dfrac{2}{3x}$

39. $\dfrac{3}{2y-1} = \dfrac{7}{3y+1}$

40. $\dfrac{7}{4-x} = \dfrac{4}{7+x}$

41. $\dfrac{2}{x-9} = \dfrac{9}{x+12}$

42. $\dfrac{4}{y-5} = \dfrac{-5}{y+4}$

375

OPERATIONS WITH FRACTIONS

11.8 APPLICATIONS

The word problems in the following exercises lead to equations involving fractions. At this time, you may want to review the steps suggested on page 264 to solve word problems and the steps suggested on page 371 to solve equations that contain fractions.

Example 1 If $\frac{2}{3}$ of a certain number is added to $\frac{1}{4}$ of the number, the result is 11. Find the number.

Solution *Steps 1–2* We first write what we want to find (the number) as a word phrase. Then, we represent the number in terms of a variable.

$$\text{The number: } x$$

Step 3 A sketch is not applicable.

Step 4 Now we can write an equation. Remember that "of" indicates multiplication.

$$\frac{2}{3} \cdot x + \frac{1}{4} \cdot x = 11$$

Step 5 Solving the equation yields

$$(12)\frac{2}{3}x + (12)\frac{1}{4}x = (12)11$$

$$8x + 3x = 132$$

$$11x = 132$$

$$x = 12$$

Step 6 The number is 12.

Equations for word problems concerned with motion sometimes include fractions. The basic idea of motion problems is that the distance traveled d equals the product of the rate of travel r and the time of travel t. Thus, $d = rt$. We can solve this formula for r or t to obtain:

$$r = \frac{d}{t} \quad \text{and} \quad t = \frac{d}{r}$$

A table like the one shown in the next example is helpful in solving motion problems.

11.8 APPLICATIONS

Example 2 An express train travels 180 miles in the same time that a freight train travels 120 miles. If the express goes 20 miles per hour faster than the freight, find the rate of each.

Solution *Steps 1–2* We represent the two unknown quantities that we want to find as word phrases. Then, we represent the word phrases in terms of one variable.

$$\text{Rate of freight train:} \quad r$$
$$\text{Rate of express train:} \quad r + 20$$

Step 3 Next, we make a table showing the distances, rates, and times.

	Distance	Rate	Time: d/r
Freight train	120	r	$120/r$
Express train	180	$r + 20$	$180/(r + 20)$

Step 4 Because the times of both trains are the same, we can equate the expressions for time to obtain

$$\frac{120}{r} = \frac{180}{r + 120}$$

Step 5 We can now solve for r by first multiplying each member by the LCD $r(r + 120)$ and we get

$$\cancel{r}(r + 20)\frac{120}{\cancel{r}} = r(r + 20)\frac{180}{\cancel{r + 20}}$$

$$120(r + 20) = 180r$$
$$120r + 2400 = 180r$$
$$2400 = 60r$$
$$r = 40$$

Step 6 The freight train's speed is 40 mph and the express train's speed is $40 + 20$, or 60 mph.

EXERCISES 11.8
Solve. Follow the six steps outlined on page 264.

Complete Example (see Example 1)
If two-thirds of a certain number is added to three-fourths of the number, the result is 17. Find the number.

Steps 1–2   The number: x

Step 3   A sketch is not applicable.

OPERATIONS WITH FRACTIONS

Step 4
$$\frac{2}{3}x + \frac{3}{4}x = \underline{}_{(1)}$$

Step 5
$$(12)\frac{2}{\cancel{3}}^4 x + (12)\frac{3}{\cancel{4}}^3 x = (\underline{}_{(2)})17$$

$$8x + \underline{}_{(3)} = 204$$

$$17x = 204$$

$$x = \underline{}_{(4)}$$

Step 6   The number is 12.

---

(1) 17   (2) 12   (3) $9x$ (4) 12

---

1. If one-half of a certain number is added to three times the number, the result is $\frac{35}{2}$. Find the number.

2. If two-thirds of a certain number is subtracted from twice the number, the result is 20. Find the number.

3. Find two consecutive integers such that the sum of one-half the first and two-thirds of the next is 17.

4. Find two consecutive integers such that twice the second less one-half of the first is 14.

5. The denominator of a certain fraction is 6 more than the numerator, and the fraction is equivalent to $\frac{3}{4}$. Find the numerator.

6. The denominator of a certain fraction is eight more than the numerator, and the fraction is equivalent to $\frac{3}{5}$. Find the denominator.

7. A partner receives two-thirds of the profits of the partnership. How much must the business make in profits if this partner is to receive $160 per week from the partnership.

8. A man owns a three-eighths interest in a lot that was purchased for $4000. What should the lot sell for if the man is to obtain a $600 profit on his investment?

9. Fahrenheit (F) and Celsius (C) temperatures are related by the equation $F = \frac{9}{5}C + 32$. What is the Celsius temperature in a room in which the Fahrenheit temperature is 68?

10. Using the equation in Exercise 9, find the Celsius temperature of boiling water at sea level (212°F).

11. A woman has 75 kg of ore of which $\frac{1}{15}$ is copper. How many more kilograms of the same ore does she need to have 15 kg of copper?

12. A man saves one-sixth of his weekly wages. How much more than $100 per week must he make if he is to save $25 per week?

13. A student received grades of 72, 78, 84, and 94 on the first four tests. What grade would she need on the next test to have an average of 80?

14. A boy got 17 hits on his first 60 times at bat. How many hits does he need in the next 20 times at bat to have an average of 0.300?

11.8 APPLICATIONS

Complete Example (see Example 2)

A train travels 60 miles in the same time that a plane travels 360 miles. The plane goes 100 miles per hour faster than the train. Find the rate of each.

Steps 1–2 Rate of train: r; Rate of plane: _____ .
                                              (1)

Step 3 Make a table listing distance, rate, and time for the train and plane.

	Distance	Rate	Time: d/r
Train	____ (2)	r	$\dfrac{60}{r}$
Plane	360	$r + 100$	$\dfrac{360}{r + 100}$

Step 4 Because the times of the plane and train are equal, equate the expressions for time.

$$\underline{}_{(3)} = \frac{360}{r + 100}$$

Step 5 Clear fractions by multiplying each term by the LCD of all the fractions, $r(r + 100)$. Solve for r.

$$\cancel{r}(r + 100)\frac{60}{\cancel{r}} = r\cancel{(r + 100)}\frac{360}{\cancel{(r + 100)}}$$

$$60(r + 100) = 360r$$

$$60r + 6000 = 360r$$

$$6000 = 300r$$

$$r = \underline{}_{(4)}$$

Step 6 The train's speed is 20 mph and the plane's speed is 20 + 100 or 120 mph.

(1) $r + 100$ (2) 60   (3) $\dfrac{60}{r}$ (4) 20

---

15. A woman drives 120 miles in the same time that a man drives 80 miles. If the speed of the woman is 20 miles per hour greater than the speed of the man, find the speed of each.

16. An airplane travels 630 miles in the same time that an automobile covers 210 miles. If the speed of the airplane is 120 miles per hour greater than the speed of the automobile, find the speed of each.

17. A man rides 15 miles on his bicycle in the same time it takes him to walk 7 miles. If his rate riding is 2 miles per hour more than his rate walking, how fast does he walk?

18. A woman rides 10 miles in a car and then walks 4 miles on foot. If her rate driving is 20 times her rate walking, and if the whole trip takes her $2\frac{1}{4}$ hours, how fast does she walk?

379

OPERATIONS WITH FRACTIONS

19. Two men drive from town A to town B, a distance of 300 miles. If one man drives twice as fast as the other and arrives at town B 5 hours ahead of the other, how fast was each driving?

20. Two trains traveled from town A to town B, a distance of 400 miles. If one train traveled twice as fast as the other and arrived at town B 4 hours ahead of the other, how fast was each traveling?

21. A rocket travels 12,600 miles in the same time that a supersonic plane travels 3600 miles. If the rate of the rocket is 2700 miles per hour greater than the rate of the plane, find the rate of each.

22. Two rockets are propelled over a 5600-mile test range. One rocket travels twice as fast as the other. The faster rocket covers the entire distance in two hours less time than the slower. How fast in miles per hour are the rockets traveling?

23. Two cars start together and travel in the same direction, one going three times as fast as the other. At the end of 2 hours, they are 100 miles apart. How fast is each traveling?

24. Two trains start together and travel in opposite directions, one going twice as fast as the other. At the end of four hours, they are 216 miles apart. How fast is each train traveling?

CHAPTER ELEVEN REVIEW EXERCISES

Simplify.

1. a. $\dfrac{2xy^2}{3} \cdot \dfrac{x}{4y^2}$ b. $\dfrac{x^2 - 2x}{5} \cdot \dfrac{25}{x^2}$ c. $\dfrac{x^2 - 7x + 6}{x^2 - 1} \cdot \dfrac{x + 1}{x - 6}$

2. a. $\dfrac{2r}{3s} \div \dfrac{2r^2}{21s^2}$ b. $\dfrac{a^2 - b^2}{4} \div \dfrac{a^2 + ab}{4a - 4}$ c. $\dfrac{2x^2 - 5x - 3}{2x^2 + x} \div \dfrac{x - 3}{x^4}$

3. a. $\dfrac{x}{5} - \dfrac{y}{5} + \dfrac{3}{5}$ b. $\dfrac{x + 3}{y} - \dfrac{3}{y}$ c. $\dfrac{a - 2}{3} - \dfrac{a + 3}{3}$

4. a. $\dfrac{x}{3} + \dfrac{y}{6}$ b. $\dfrac{7}{x} + \dfrac{1}{3x}$ c. $\dfrac{5}{2y} - \dfrac{1}{y}$

5. a. $\dfrac{3}{x} - \dfrac{2}{3x}$ b. $\dfrac{3}{r} + \dfrac{5}{2s}$ c. $\dfrac{2}{ab^2} - \dfrac{3}{a^2b}$

6. a. $\dfrac{3}{a - b} + \dfrac{1}{a + b}$ b. $\dfrac{a}{a^2 - 1} - \dfrac{1}{a^2 + a}$ c. $\dfrac{1}{x^2 - 25} + \dfrac{5}{x^2 - 4x - 5}$

7. a. $\dfrac{\tfrac{3}{6}}{\tfrac{2}{9}}$ b. $\dfrac{\tfrac{2}{3} + \tfrac{1}{6}}{\tfrac{1}{3} + \tfrac{5}{6}}$ c. $\dfrac{1 + \tfrac{1}{2}}{3 - \tfrac{1}{4}}$

8. a. $\dfrac{1 - \tfrac{a}{b}}{1 + \tfrac{2}{b}}$ b. $\dfrac{x - \tfrac{x}{y}}{y - \tfrac{y}{x}}$ c. $\dfrac{\tfrac{1}{y} + 3}{2 - \tfrac{3}{y}}$

9. a. $\dfrac{x}{2} = -1 + \dfrac{2x}{3}$ b. $\dfrac{x}{3} + \dfrac{7}{9} = \dfrac{1}{3}$ c. $\dfrac{x + 1}{2} = \dfrac{2x - 9}{5} + 3$

380

10. a. $\dfrac{10}{x+4} - \dfrac{6}{x} = \dfrac{-4}{x}$ b. $\dfrac{14}{x-1} + \dfrac{1}{x} = \dfrac{8}{x}$ c. $1 - \dfrac{3+y}{2y} = \dfrac{3-y}{y}$

11. a. $\dfrac{6}{x} = \dfrac{16}{x+5}$ b. $\dfrac{2+y}{y} = \dfrac{3}{2}$ c. $\dfrac{y-2}{2y} = \dfrac{5}{2}$

Solve for x.

12. a. $\dfrac{b}{3} = \dfrac{2ax}{4}$ b. $\dfrac{b-x}{4} - \dfrac{b}{3} = \dfrac{x}{2}$ c. $\dfrac{a}{x-1} = \dfrac{2a}{x}$

13. If three times a certain number is divided by 10 more than that number, the result is $\frac{1}{2}$. What is the number?

14. A sample of 92 parts in a manufacturing plant proved to contain 3 defective parts. If the sample was a valid sample, how many defective parts would you expect to find in a run of 276 parts?

15. One car travels 90 miles in the same time that another car travels 60 miles. If the slower car is traveling 10 mph slower than the other car, find the rate of each.

12 FIRST-DEGREE EQUATIONS AND INEQUALITIES IN TWO VARIABLES

The language of mathematics is particularly effective in representing relationships between two or more variables. As an example, let us consider the distance traveled in a certain length of time by a car moving at a constant speed of 40 miles per hour. We can represent this relationship by

1. A word sentence: The distance traveled in miles is equal to forty times the number of hours traveled.
2. An equation: $d = 40t$.
3. A tabulation of values.
4. A graph showing the relationship between time and distance.

We have already used word sentences and equations to describe such relationships; in this chapter, we will deal with tabular and graphical representations.

12.1 SOLVING EQUATIONS IN TWO VARIABLES

ORDERED PAIRS

The equation $d = 40t$ pairs a distance d for each time t. For example,

if $t = 1$, then $d = 40$

if $t = 2$, then $d = 80$

if $t = 3$, then $d = 120$

and so on.

The pair of numbers 1 and 40, considered together, is called a **solution** of the equation $d = 40t$ because when we substitute 1 for t and 40 for d in the equation, we get a true statement. If we agree to refer to the paired numbers in a specified order in which the first number refers to time and the second number refers to distance, we can abbreviate the above solutions as (1, 40), (2, 80), (3, 120), and

FIRST-DEGREE EQUATIONS AND INEQUALITIES IN TWO VARIABLES

so on. We call such pairs of numbers **ordered pairs**, and we refer to the first and second numbers in the pairs as **components**. With this agreement, solutions of the equation $d = 40t$ are ordered pairs (t, d) whose components satisfy the equation. Some ordered pairs for t equal to 0, 1, 2, 3, 4, and 5 are

$$(0, 0), \quad (1, 40), \quad (2, 80), \quad (3, 120), \quad (4, 160), \quad \text{and} \quad (5, 200)$$

Such pairings are sometimes shown in one of the following tabular forms.

t	0	1	2	3	4	5
d	0	40	80	120	160	200

t	d
0	0
1	40
2	80
3	120
4	160
5	200

In any particular equation involving two variables, when we assign a value to one of the variables, the value for the other variable is determined and therefore dependent on the first. It is convenient to speak of the variable associated with the *first component* of an ordered pair as the **independent variable** and the variable associated with the *second component* of an ordered pair as the **dependent variable.** If the variables x and y are used in an equation, it is understood that replacements for x are first components and hence x is the independent variable and replacements for y are second components and hence y is the dependent variable.

For example, we can obtain pairings for equation

$$\underset{\text{independent variable}}{2x} + \underset{\text{dependent variable}}{y} = 4 \qquad (1)$$

by substituting a particular value of one variable into Equation (1) and solving for the other variable.

Example 1   Find the missing component so that the ordered pair is a solution to

$$2x + y = 4$$

a. (0, ?)   b. (1, ?)   c. (2, ?)

Solution

if $x = 0$, then $2(0) + y = 4$
$$y = 4$$

if $x = 1$, then $2(1) + y = 4$
$$y = 2$$

if $x = 2$, then $2(2) + y = 4$
$$y = 0$$

The three pairings can now be displayed as the three ordered pairs

$$(0, 4), \quad (1, 2), \quad \text{and} \quad (2, 0)$$

12.1 SOLVING EQUATIONS IN TWO VARIABLES

or in the tabular forms

x	0	1	2
y	4	2	0

or

x	y
0	4
1	2
2	0

EXPRESSING A VARIABLE EXPLICITLY

We can add $-2x$ to both members of $2x + y = 4$ to get

$$-2x + 2x + y = -2x + 4$$
$$y = -2x + 4 \qquad (2)$$

In Equation (2), where y is by itself, we say that y is expressed *explicitly* in terms of x. It is often easier to obtain solutions if equations are first expressed in such form because the dependent variable is expressed explicitly in terms of the independent variable.

For example, in Equation (2) above,

if $x = 0$, then $y = -2(0) + 4 = 4$

if $x = 1$, then $y = -2(1) + 4 = 2$

if $x = 2$, then $y = -2(2) + 4 = 0$

We get the same pairings that we obtained using Equation (1)

$(0, 4), \quad (1, 2), \quad \text{and} \quad (2, 0)$

We obtained Equation (2) by adding the same quantity, $-2x$, to each member of Equation (1), in that way getting y by itself. In general, we can write equivalent equations in two variables by using the properties we introduced in Chapter 3, where we solved first-degree equations in one variable.

Equations are equivalent if:
1. The same quantity is added to or subtracted from equal quantities.
2. Equal quantities are multiplied or divided by the same nonzero quantity.

Example 2   Solve $2y - 3x = 4$ explicitly for y in terms of x and obtain solutions for $x = 0$, $x = 1$, and $x = 2$.

Solution   First, adding $3x$ to each member we get

$$2y - 3x + 3x = 4 + 3x$$
$$2y = 4 + 3x$$

Now, dividing each member by 2, we obtain

$$\frac{2y}{2} = \frac{4 + 3x}{2}$$

$$y = \frac{4 + 3x}{2}$$

FIRST-DEGREE EQUATIONS AND INEQUALITIES IN TWO VARIABLES

In this form, we obtain values of y for given values of x as follows:

$$\text{if } x = 0, \quad y = \frac{4 + 3(0)}{2} = \frac{4 + 0}{2} = 2$$

$$\text{if } x = 1, \quad y = \frac{4 + 3(1)}{2} = \frac{4 + 3}{2} = \frac{7}{2}$$

$$\text{if } x = 2, \quad y = \frac{4 + 3(2)}{2} = \frac{4 + 6}{2} = 5$$

In this case, three solutions are (0, 2), (1, 7/2), and (2, 5).

FUNCTION NOTATION

Sometimes, we use a special notation to name the second component of an ordered pair that is paired with a specified first component. The symbol $f(x)$, which is often used to name an algebraic expression in the variable x, can also be used to denote the value of the expression for specific values of x. For example, if

$$f(x) = -2x + 4 \qquad (3)$$

where $f(x)$ is playing the same role as y in Equation (2) on page 385, then $f(\ )$ represents the value of the expression $-2x + 4$ when x is replaced by

$$f(1) = -2(1) + 4 = 2$$

Similarly,

$$f(0) = -2(0) + 4 = 4$$

and

$$f(2) = -2(2) + 4 = 0$$

The symbol $f(x)$ is commonly referred to as **function notation.**

Example 3 If $f(x) = -3x + 2$, find $f(-2)$ and $f(2)$.

Solution Replace x with -2 to obtain

$$f(-2) = -3(-2) + 2 = 8$$

Replace x with 2 to obtain

$$f(2) = -3(2) + 2 = -4$$

EXERCISES 12.1

1. The equation $d = 4t$ relates the distance d traveled by a person walking 4 miles per hour to the length of time t he or she walks.
   a. Which symbols are variables?
   b. What is the effect on d of increasing t?
   c. Which symbol is the independent variable if a solution is given by (t, d)?
   d. Which symbol is the dependent variable?
   e. If t is assigned the value 3, what is the value of d?

12.1 SOLVING EQUATIONS IN TWO VARIABLES

2. The equation $F = 32m$ relates the force F of an object on earth to the mass m of the object.
   a. Which symbols are variables?
   b. What is the effect on F of increasing m?
   c. Which symbol is the independent variable if a solution is given by (m, F)?
   d. Which symbol is the dependent variable?
   e. If m is assigned the value 4, what is the value of F?

Find the missing component so that the ordered pair is a solution to the given equation. See Example 1.

Complete Example

$$-2x + 3y = 6$$

a. $(3, ?)$
b. $(-2, ?)$

Replace x with 3.

$$-2(\underline{\quad}_{(1)}) + 3y = 6$$

$$3y = \underline{\quad}_{(2)}$$

$$y = 4$$

Ans. $(3, 4)$

Replace x with -2.

$$-2(\underline{\quad}_{(3)}) + 3y = 6$$

$$3y = \underline{\quad}_{(4)}$$

$$y = \frac{2}{3}$$

Ans. $(-2, \underline{\quad}_{(5)})$

(1) 3   (2) 12   (3) -2 (4) 2   (5) $\frac{2}{3}$

3. $x + y = 3$
   a. $(2, ?)$ b. $(-1, ?)$

4. $3x + y = 10$
   a. $(-3, ?)$ b. $(3, ?)$

5. $4x + 2y = -2$
   a. $(2, ?)$ b. $(-2, ?)$

6. $x + 2y = -6$
   a. $(2, ?)$ b. $(4, ?)$

7. $3x + 2y = 7$
   a. $(-3, ?)$ b. $(3, ?)$

8. $2x - 2y = -4$
   a. $(-2, ?)$ b. $(2, ?)$

9. $2y - x = 1$
   a. $(-4, ?)$ b. $(4, ?)$

10. $3y - x = -3$
    a. $(-5, ?)$ b. $(5, ?)$

387

FIRST-DEGREE EQUATIONS AND INEQUALITIES IN TWO VARIABLES

Complete Examples

$y = 2x + 3$

a. $(-4, ?)$　　　　　　　　　　　　　　　　　b. $(0, ?)$

Replace x with -4.　　　　　　　　　　　　　Replace x with 0.

$y = 2(\underline{\quad}_{(1)}) + 3$　　　　　　　　　　　$y = 2(\underline{\quad}_{(3)}) + 3$

$= \underline{\quad}_{(2)}$　　　　　　　　　　　　　　　$= \underline{\quad}_{(4)}$

Ans. $(-4, -5)$　　　　　　　　　　　　　　　Ans. $\underline{\quad}_{(5)}$

---

(1) -4　　(2) -5　　(3) 0　　(4) 3　　(5) $(0, 3)$

---

11. $y = 2x + 3$
    a. $(2, ?)$　b. $(-3, ?)$

12. $y = 3x - 2$
    a. $(4, ?)$　b. $(-1, ?)$

13. $y = 4x - 1$
    a. $(3, ?)$　b. $(1, ?)$

14. $y = 4 - 2x$
    a. $(-1, ?)$　b. $(-2, ?)$

15. $y = 3x$
    a. $(0, ?)$　b. $(-3, ?)$

16. $y = -x$
    a. $(0, ?)$　b. $(2, ?)$

In Exercises 17–30, express y explicitly in terms of x. Then find the missing component so that the ordered pair is a solution to the given equation. See Example 2.

Complete Example

$y - 2x = 5;\ (2, ?)$

$y = \underline{\quad}_{(1)} + 5$　　　　　　Solve $y - 2x = 5$ for y in terms of x.

$y = 2(\underline{\quad}_{(2)}) + 5$　　　　　Replace x with 2.

$= \underline{\quad}_{(3)}$

Ans. $(\underline{\quad}_{(4)})$

---

(1) $2x$　　(2) 2　　(3) 9　　(4) $(2, 9)$

---

17. $y - 2x = 5$
    a. $(4, ?)$　b. $(1, ?)$

18. $y + 3x - 5 = 0$
    a. $(2, ?)$　b. $(1, ?)$

19. $y + 2x + 4 = 0$
    a. $(5, ?)$　b. $(3, ?)$

20. $y - 5x + 6 = 0$
    a. $(6, ?)$　b. $(4, ?)$

21. $y - 4x + 2 = 0$
    a. $(-2, ?)$　b. $(0, ?)$

22. $y - 2x - 4 = 0$
    a. $(-1, ?)$　b. $(0, ?)$

12.1 SOLVING EQUATIONS IN TWO VARIABLES

Complete Example

$2x - 3y = 6$; (2, ?)   Solve $2x - 3y = 6$ for y in terms of x.

$2x - 6 = 3y$

$3y = 2x - 6$

$y = \dfrac{2x - 6}{\underline{}}$
  (1)

Replace x with 2

$y = \dfrac{2(2) - 6}{3} = \underline{}$
  (2)

Ans. $\left(\underline{} \right)$
  (3)

---

(1) 3   (2) $\dfrac{-2}{3}$ (3) $\left(2, \dfrac{-2}{3}\right)$

---

23. $3x + 4y = 8$
    a. $(-2, ?)$ b. $(5, ?)$

24. $4x - 3y = 9$
    a. $(1, ?)$ b. $(-1, ?)$

25. $2x - 3y = 6$
    a. $(10, ?)$ b. $(5, ?)$

26. $3x + 2y = 4$
    a. $(2, ?)$ b. $(1, ?)$

27. $4x + 2y = 7$
    a. $(7, ?)$ b. $(5, ?)$

28. $3x - 4y = 5$
    a. $(-3, ?)$ b. $(5, ?)$

29. $4x - 5y = 3$
    a. $(-3, ?)$ b. $(0, ?)$

30. $3x + 5y = -2$
    a. $(0, ?)$ b. $(-2, ?)$

Find values for each expression for specified values of the variable. See Example 3.

Complete Example

If $f(x) = 2x - 4$, find

a. $f(-3)$

$f(-3) = 2(\underline{}) - 4$
             (1)

$= \underline{}$
  (2)

b. $f(0)$

$f(0) = 2(\underline{}) - 4$
           (3)

$= \underline{}$
  (4)

---

(1) -3 (2) -10 (3) 0   (4) -4

---

31. If $f(x) = x + 5$, find $f(-2)$ and $f(2)$.
32. If $f(x) = x - 3$, find $f(-3)$ and $f(3)$.
33. If $f(x) = 2x - 7$, find $f(-5)$ and $f(-1)$.
34. If $f(x) = 3x + 2$, find $f(-4)$ and $f(2)$.
35. If $f(x) = 4 - 3x$, find $f(-1)$ and $f(0)$.
36. If $f(x) = 3 - 2x$, find $f(-4)$ and $f(-2)$.

FIRST-DEGREE EQUATIONS AND INEQUALITIES IN TWO VARIABLES

37. If $f(x) = \dfrac{3x - 4}{2}$, find $f(2)$ and $f(6)$.

38. If $f(x) = \dfrac{5x + 2}{2}$, find $f(-6)$ and $f(-2)$.

39. If $f(x) = \dfrac{3x - 5}{5}$, find $f(-5)$ and $f(0)$.

40. If $f(x) = \dfrac{3x + 4}{4}$, find $f(4)$ and $f(0)$.

41. If $f(x) = \dfrac{2x - 3}{5}$, find $f(-1)$ and $f\left(\dfrac{5}{2}\right)$.

42. If $f(x) = \dfrac{3x - 1}{4}$, find $f(2)$ and $f\left(\dfrac{4}{3}\right)$.

12.2 GRAPHS OF ORDERED PAIRS

In Section 6.1, we saw that every number corresponds to a point in a line. Similarly, every ordered pair of numbers (x, y) corresponds to a point in a plane. To graph an ordered pair of numbers, we begin by constructing a pair of perpendicular number lines, called **axes**. The horizontal axis is called the **x axis**, the vertical axis is called the **y axis**, and their point of intersection is called the **origin**. These axes divide the plane into four **quadrants**, as shown in Figure 12.1.

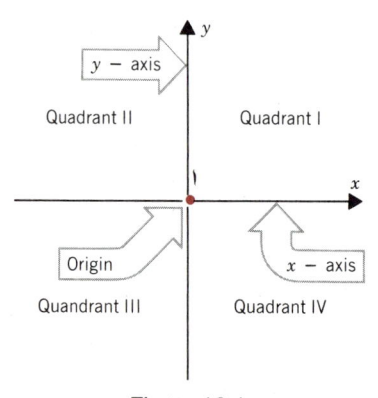

Figure 12.1

Now we can assign an ordered pair of numbers to a point in the plane by referring to the perpendicular distance of the point from each of the axes. If the first component is positive, the point lies to the right of the vertical axis; if negative, it lies to the left. If the second component is positive, the point lies above the horizontal axis; if negative, it lies below.

Example 1   Graph $(3, 2)$, $(-3, 2)$, $(-3, -2)$, and $(3, -2)$ on a rectangular coordinate system.

Solution   The graph of $(3, 2)$ lies 3 units *to the right* of the y axis and 2 units *above* the x axis;

the graph of $(-3, 2)$ lies 3 units *to the left* of the y axis and 2 units *above* the x axis;

the graph of $(-3, -2)$ lies 3 units *to the left* of the y axis and 2 units *below* the x axis;

the graph of $(3, -2)$ lies 3 units *to the right* of the y axis and 2 units *below* the x axis.

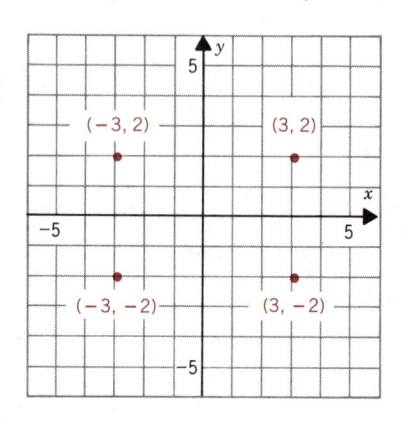

12.2 GRAPHS OF ORDERED PAIRS

The distance y that the point is located from the x axis is called the **ordinate** of the point, and the distance x that the point is located from the y axis is called the **abscissa** of the point. The abscissa and ordinate together are called the **rectangular** or **Cartesian coordinates** of the point (see Figure 12.2).

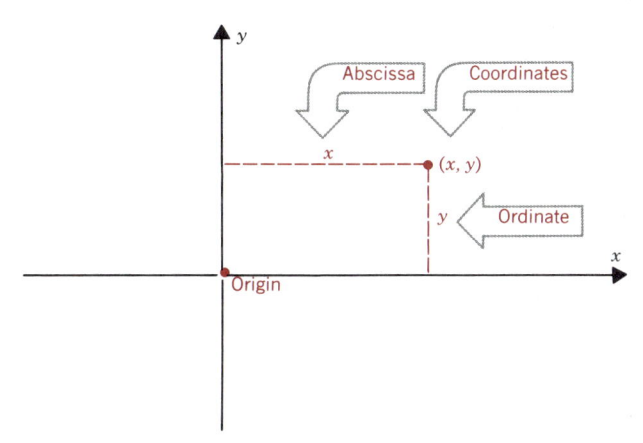

Figure 12.2

EXERCISES 12.2

In Exercises 1–8, graph each set of ordered pairs on a rectangular coordinate system. See Example 1.

Complete Example

a. $(-3, -1)$ d. $(4, -3)$
b. $(0, 2)$ e. $(-4, 3)$
c. $(2, 4)$ f. $(3, 0)$

(1)

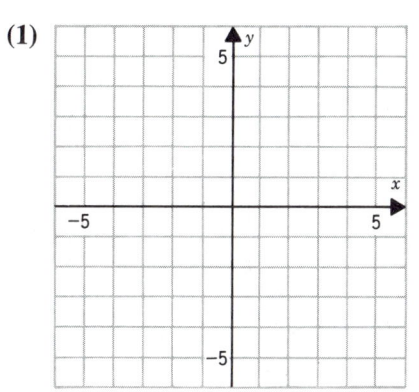

(1) Points plotted: $(-4, 3)$, $(2, 4)$, $(0, 2)$, $(3, 0)$, $(-3, -1)$, $(4, -3)$

391

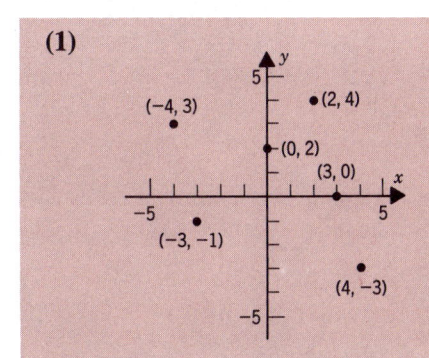

FIRST-DEGREE EQUATIONS AND INEQUALITIES IN TWO VARIABLES

1. a. (1, 2)   d. (−4, 5)
   b. (−2, 3)   e. (4, 4)
   c. (3, −1)   f. (0, 5)

2. a. (3, 4)   d. (0, 5)
   b. (−2, 0)   e. (4, −1)
   c. (2, −1)   f. (−5, 1)

3. a. (0, 0)   d. (0, −2)
   b. (0, 2)   e. (−5, 0)
   c. (0, 5)   f. (5, 0)

In Exercises 4 and 5, let the distance between successive marks on your axes represent 5 units.

4. a. (10, 5)   d. (0, 20)
   b. (−10, 5)   e. (−20, −20)
   c. (25, −5)   f. (25, 15)

5. a. (0, 30)   d. (30, −25)
   b. (−25, 0)   e. (−5, 0)
   c. (0, −20)   f. (5, −30)

In Exercises 6, 7, and 8, let the distance between successive marks on the x axis represent 1 unit and on the y axis 5 units.

6. a. (3, 25)   d. (−2, −20)
   b. (−1, 20)   e. (0, −10)
   c. (5, 0)   f. (−5, −30)

7. a. (2, −10)   d. (−1, 5)
   b. (1, −5)   e. (−2, 10)
   c. (0, 0)   f. (−3, 15)

8. a. (3, −24)   d. (0, 0)
   b. (2, −16)   e. (−1, 8)
   c. (1, −8)   f. (−2, 16)

9. Connect the points plotted in Exercise 7. What do you observe?

10. Connect the points plotted in Exercise 8. What do you observe?

11. In the rectangular coordinate system, name the (perpendicular) distance to a given point from:
    a. The x axis.   b. The y axis

12. Which component of the ordered pair (x, y) represents the (perpendicular) distance of a point from the x axis? The y axis?

13. Name the point corresponding to (0, 0).

14. Describe the location of the graphs of all ordered pairs of the form $(0, y)$ and $(x, 0)$.

15. Describe the location of the graphs of all ordered pairs (a, a), that is, all points whose first and second components are equal.

16. Graph (1, 2) and (3, 6) and draw a line connecting the points.
    a. Does the graph of (2, 4) lie on the line?
    b. If you extended the graph in both directions, would the graphs of (4, 8) and (−1, −2) lie on the line?
    c. Is the graph of (0, 0) on this line?

17. Locate the graphs of (2, 3) and (2, 5) and draw a line connecting them.
    a. Does the graph of (2, 4) lie on this line?
    b. Does the graph of (1, 3) lie on this line?
    c. If you extended the graph in both directions, would the graphs of (2, 6) and (2, −1) lie on the line?
    d. Is the graph of (0, 0) on the line?

18. How many points lie on any line in the plane?

19. What would be the least number of points necessary to determine a line?

20. Which axis, the horizontal or the vertical, is usually used to represent: a. The independent variable? b. The dependent variable?

12.3 GRAPHING FIRST-DEGREE EQUATIONS

In Section 12.1, we saw that a solution of an equation in two variables is an ordered pair. In Section 12.2, we saw that the components of an ordered pair are the coordinates of a point in a plane. Thus, to graph an equation in two variables, we graph the set of ordered pairs that are solutions to the equation. For example, we can find some solutions to the first-degree equation

$$y = x + 2$$

by letting x equal 0, -3, -2, and 3. Then,

for $x = 0$, $y = 0 + 2 = 2$
for $x = -3$, $y = -3 + 2 = -1$
for $x = -2$, $y = -2 + 2 = 0$
for $x = 3$, $y = 3 + 2 = 5$

and we obtain the solutions

$(0, 2)$, $(-3, -1)$, $(-2, 0)$, and $(3, 5)$

which can be displayed in a tabular form as shown below.

x	y
0	2
-3	-1
-2	0
3	5

x	0	-3	-2	3
y	2	-1	0	5

If we graph the points determined by these ordered pairs and pass a straight line through them, we obtain the graph of all solutions of $y = x + 2$, as shown in Figure 12.3. That is, *every* solution of $y = x + 2$ lies on the line, and *every* point on the line is a solution of $y = x + 2$.

The graphs of first-degree equations in two variables are always straight lines; therefore, such equations are also referred to as **linear equations.**

In the above example, the values we used for x were chosen at random; we could have used any values of x to find solutions to the equation.

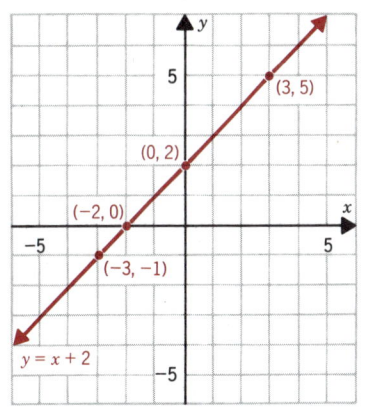

Figure 12.3

The graphs of any other ordered pairs that are solutions of the equation would also be on the line shown in Figure 12.3. In fact, each linear equation in two variables has an infinite number of solutions whose graph lies on a line. However, we only need to find two solutions because only two points are necessary to determine a straight line. A third point can be obtained as a check.

FIRST-DEGREE EQUATIONS AND INEQUALITIES IN TWO VARIABLES

> To graph a first-degree equation:
> 1. Construct a set of rectangular axes showing the scale and the variable represented by each axis.
> 2. Find two ordered pairs that are solutions of the equation to be graphed by assigning any convenient value to one variable and determining the corresponding value of the other variable.
> 3. Graph these ordered pairs.
> 4. Draw a straight line through the points.
> 5. Check by graphing a third ordered pair that is a solution of the equation and verify that it lies on the line.

Example 1   Graph the equation $y = 2x - 6$.

Solution   We first select *any two values* of x to find the associated values of y. We will use 1 and 4 for x.

$$\text{If } x = 1, \quad y = 2(1) - 6 = -4$$
$$\text{if } x = 4, \quad y = 2(4) - 6 = 2$$

Thus, two solutions of the equation are $(1, -4)$ and $(4, 2)$.

Next, we graph these ordered pairs and draw a straight line through the points as shown in the figure. We use arrowheads to show that the line extends infinitely far in both directions.

Any third ordered pair that satisfies the equation can be used as a check:

$$\text{if } x = 5, \quad y = 2(5) - 6 = 4$$

We then note that the graph of $(5, 4)$ also lies on the line.

To find solutions to an equation, as we have noted it is often easiest to first solve explicitly for y in terms of x.

Example 2   Graph $x + 2y = 4$.

Solution   We first solve for y in terms of x to get

$$2y = -x + 4$$
$$y = \frac{-x + 4}{2}$$

We now select *any two* values of x to find the associated values of y. We will use 2 and 0 for x.

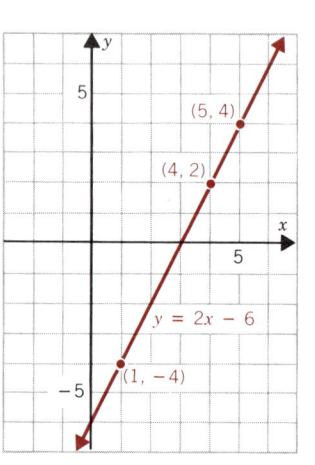

12.3 GRAPHING FIRST-DEGREE EQUATIONS

For $x = 2$, $y = \dfrac{-(2) + 4}{2} = 1$

for $x = 0$, $y = \dfrac{-(0) + 4}{2} = 2$

Thus, two solutions of the equation are $(2, 1)$ and $(0, 2)$.

Next, we graph these ordered pairs and pass a straight line through the points, as shown in the figure.

Any third ordered pair that satisfies the equation can be used as a check:

if $x = -2$, $y = \dfrac{-(-2) + 4}{2} = 3$

We then note that the graph of $(-2, 3)$ also lies on the line.

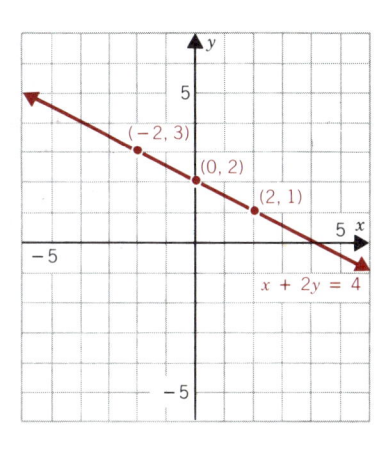

SPECIAL CASES OF LINEAR EQUATIONS

The equation $y = 2$ can be written as

$$0x + y = 2 \qquad (1)$$

and can be considered a linear equation in two variables where the coefficient of x is 0. Some solutions of $0x + y = 2$ are

$$(-1, 2), \quad (1, 2), \quad \text{and} \quad (4, 2)$$

In fact, any ordered pair of the form $(x, 2)$ is a solution of (1). Graphing the solutions yields a horizontal line as shown in Figure 12.4.

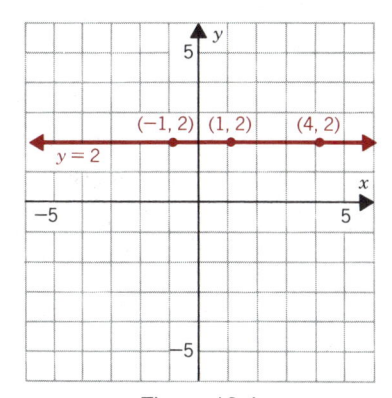

Figure 12.4

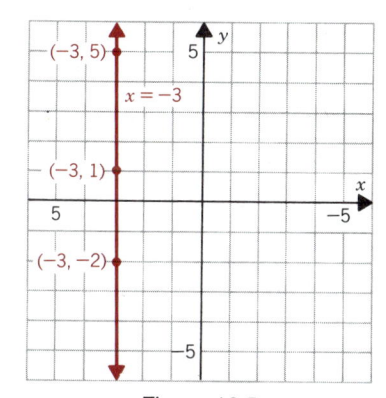

Figure 12.5

Similarly, an equation such as $x = -3$ can be written as

$$x + 0y = -3 \qquad (2)$$

and can be considered a linear equation in two variables where the coefficient of y is 0.

Some solutions of $x + 0y = -3$ are $(-3, 5)$, $(-3, 1)$, and $(-3, -2)$. In fact, any ordered pair of the form $(-3, y)$ is a solution of (2). Graphing the solutions yields a vertical line as shown in Figure 12.5.

395

FIRST-DEGREE EQUATIONS AND INEQUALITIES IN TWO VARIABLES

Example 3 Graph

a. $y = 3$
b. $x = 2$

Solutions

a. We may write $y = 3$ as $0x + y = 3$. Some solutions are $(1, 3)$, $(2, 3)$, and $(5, 3)$.

b. We may write $x = 2$ as $x + 0y = 2$. Some solutions are $(2, 4)$, $(2, 1)$, and $(2, -2)$.

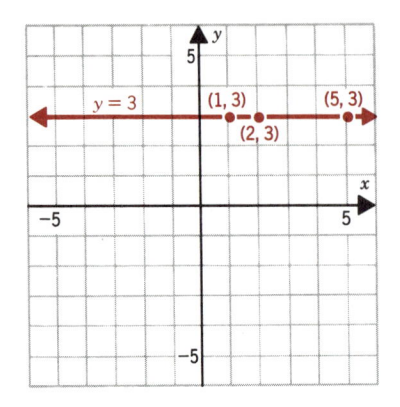

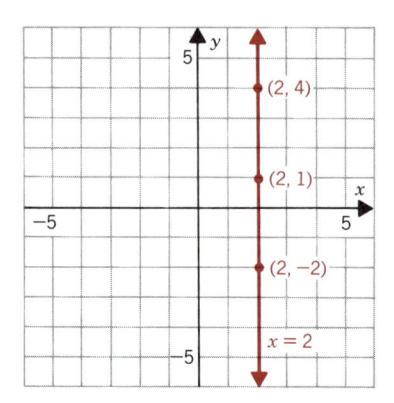

EXERCISES 12.3

1. Given $d = 4t$ (see Section 12.1, Exercise 1), find the value of d corresponding to each value of t and express your answer in the form of an ordered pair (t, d). Then graph each of the ordered pairs and connect them with a straight line.
   a. $(0, ?)$
   b. $(2, ?)$
   c. $(4, ?)$
   d. Where are all points located whose coordinates satisfy $d = 4t$?
   e. Check by obtaining additional solutions $(1, ?)$ and $(3, ?)$ and graphing these ordered pairs.
   f. What is the least number of points necessary to determine the line representing $d = 4t$?

2. Given $d = 2t$, find the value of d corresponding to each value of t and express your answer in the form of an ordered pair (t, d). Then graph each of the ordered pairs and connect them with a straight line.
   a. $(0, ?)$
   b. $(2, ?)$
   c. $(4, ?)$
   d. Where are all points located whose coordinates satisfy $d = 2t$?
   e. Check by obtaining additional solutions $(1, ?)$ and $(3, ?)$ and graphing these ordered pairs.
   f. What is the least number of points necessary to determine the line representing $d = 2t$?

In Exercises 3–14,

1. *Find any two ordered pairs that are solutions of the given equation.*
2. *Graph these ordered pairs and draw a straight line through them.*

396

12.3 GRAPHING FIRST-DEGREE EQUATIONS

3. Check your result by finding a third solution of the equation and verifying that its graph is a point on the line. Graph each equation on a separate set of axes.

Complete Example (see Examples 1 and 2)

$2y - x = -6$

$y = $ _____ (1)

$y = \dfrac{(-2) - 6}{2} = $ _____ (2)

$y = \dfrac{(2) - 6}{2} = $ _____ (3)

Solve for y in terms of x.

Take any two numbers for the first component, say -2 and 2. Find the second component of $(-2, ?)$ and $(2, ?)$ so that the ordered pairs are solutions of the equation.

Graph the ordered pairs $(-2, -4)$ and $(2, -2)$ and draw a straight line through them.

Check by noting that the graph of a third pair, say $(0, \underline{\quad}_{(4)})$, which satisfies the equation, also lies on the line.

(5)

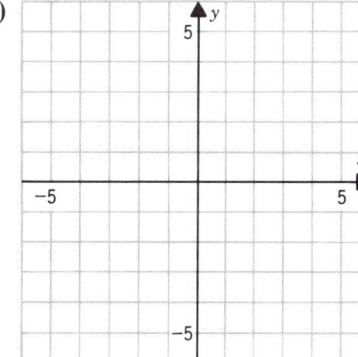

(1) $\dfrac{x-6}{2}$ (2) -4 (3) -2 (4) -3 (5)

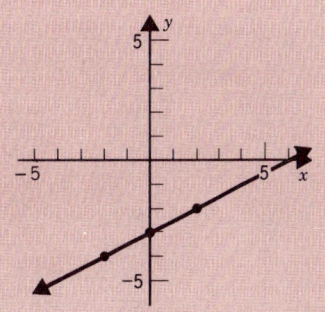

3. $y = x + 2$
4. $y = x - 2$
5. $y = 2x + 1$
6. $y = 3x - 1$

7. $y = 2x - 1$
8. $y + x = 4$
9. $y - 3x = 0$
10. $2y = 3x + 4$

11. $3y = 4 - x$
12. $3y + 2x = 12$
13. $2y + x - 6 = 0$
14. $y - 2x - 6 = 0$

397

FIRST-DEGREE EQUATIONS AND INEQUALITIES IN TWO VARIABLES

15. Consider the two ordered pairs (5, 3) and (2, 3).
   a. Graph these ordered pairs and draw a straight line through their graphs.
   b. Are the graphs of (1, 3), (2, 3), (5, 3), (−2, 3), and (−1, 3) on the line?
   c. Would the graph of (x, 3) lie on the line for any (all) x?
   d. Does the value of x have anything to do with the fact that a point lies on this line?
   e. Is $y = 0x + 3$ an equation for the line?
   f. Does $y = 3$ give a complete description of the line?

16. Consider the two ordered pairs (3, 5) and (3, 2).
   a. Graph these ordered pairs and draw a straight line through their graphs.
   b. Are the graphs of (3, −1), (3, 4), (3, 6) on the line?
   c. Would the graphs of (3, y) lie on the line for any (all) y?
   d. Does the value of y have anything to do with the fact that a point lies on this line?
   e. Is $x = 0y + 3$ an equation for the line?
   f. Does $x = 3$ give a complete description of the line?

Graph each of the following equations. See Example 3.

Complete Example

$y = -2$

Write $y = -2$ as

$\underline{}_{(1)} + y = -2$

If $x = -2$, $y = \underline{}_{(2)}$

if $x = 1$, $y = \underline{}_{(3)}$

if $x = 3$, $y = \underline{}_{(4)}$

(5)

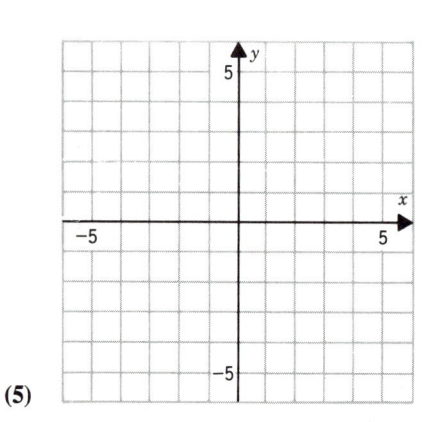

(1) $0x$ (2) -2 (3) -2 (4) -2 (5)

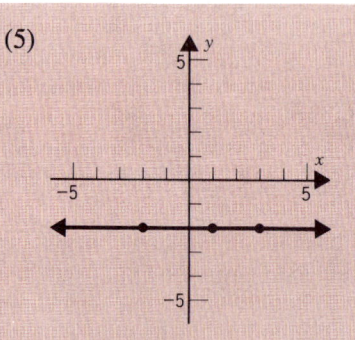

17. $x + 0y = 4$
18. $0x + y = -1$
19. $0x - 2y = 8$
20. $3x + 0y = 6$

21. $x = -2$
22. $y = -5$
23. $x = 0$
24. $y = 0$

25. $3y = 12$
26. $2x = 6$
27. $4x = -20$
28. $4y = -12$

12.4 INTERCEPT METHOD OF GRAPHING

In Section 12.3, we assigned values to *x* in equations in two variables to find the corresponding values of *y*. The solutions of an equation in two variables that are generally easiest to find are those in which either the first or second component is 0. For example, if we substitute 0 for *x* in the equation

$$3x + 4y = 12 \qquad (1)$$

we have

$$3(0) + 4y = 12$$
$$y = 3$$

Thus, a solution of Equation (1) is (0, 3). We can also find ordered pairs that are solutions of equations in two variables by assigning values to *y* and determining the corresponding values of *x*. In particular, if we substitute 0 for *y* in Equation (1), we get

$$3x + 4(0) = 12$$
$$x = 4$$

and a second solution of the equation is (4, 0). We can now use the ordered pairs (0, 3) and (4, 0) to graph Equation (1). The graph is shown in Figure 12.6. Notice that the line crosses the *x* axis at 4 and the *y* axis at 3. For this reason, the number 4 is called the **x intercept** of the graph, and the number 3 is called the **y intercept.**

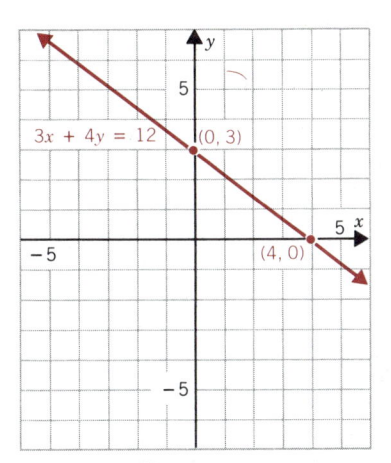

Figure 12.6

This method of drawing the graph of a linear equation is called the **intercept method of graphing.** Note that when we use this method of graphing a linear equation, there is no advantage in first expressing *y* explicitly in terms of *x*.

FIRST-DEGREE EQUATIONS AND INEQUALITIES IN TWO VARIABLES

Example 1 Graph $2x - y = 6$ by the intercept method.

Solution We find the x intercept by substituting 0 for y in the equation to obtain

$$2x - (0) = 6$$
$$x = 3$$

Now, we find the y intercept by substituting 0 for x in the equation to get

$$2(0) - y = 6$$
$$y = -6$$

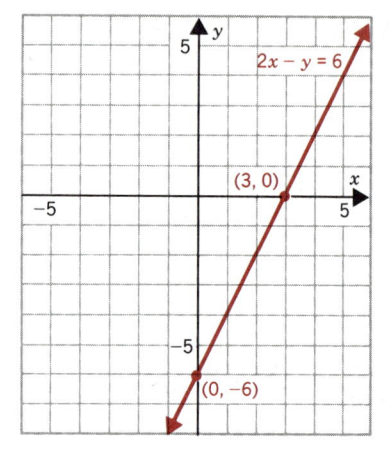

The ordered pairs $(3, 0)$ and $(0, -6)$ are solutions of $2x - y = 6$. Graphing these points and connecting them with a straight line give us the graph of $2x - y = 6$.

If the graph intersects the axes at or near the origin, the intercept method *is not* satisfactory. We must then graph an ordered pair that is a solution of the equation and whose graph is not the origin or is not too close to the origin.

Example 2 Graph $y = 3x$.

Solution We can substitute 0 for x and find

$$y = 3(0) = 0$$

Similarly, substituting 0 for y, we get

$$0 = 3x, \quad x = 0$$

Thus, 0 is both the x intercept and the y intercept. Since one point is not sufficient to graph $y = 3x$, we resort to the methods outlined in Section 12.3. Choosing any other value for x, say 2, we get

$$y = 3(2) = 6$$

Thus, $(0, 0)$ and $(2, 6)$ are solutions to the equation. The graph of $y = 3x$ is shown at the right.

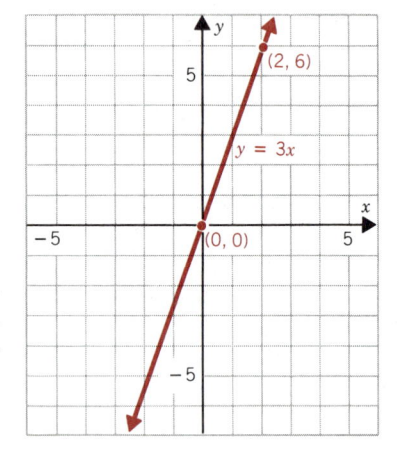

EXERCISES 12.4
Graph each equation by the intercept method. See Examples 1 and 2.

Complete Example

$$-2x + y = 4$$
$$-2(\underline{}_{(1)}) + y = 4$$
$$y = 4$$

Substitute 0 for x and solve for y.

$$-2x + (0) = 4$$
$$-2x = 4$$
$$x = \underline{}_{(2)}$$

Substitute 0 for y and solve for x.

400

12.5 SLOPE OF A LINE

The ordered pairs (0, ___(3)___) and (___(4)___ , 0) are solutions to the equation. Graph these points and draw a straight line through them. (5)

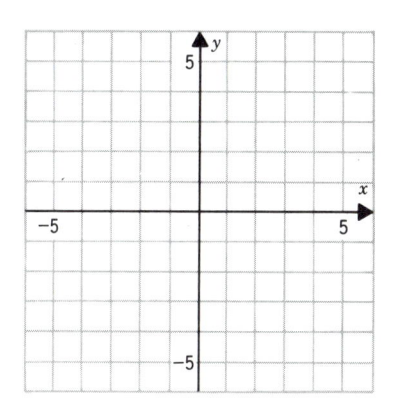

---

(1) 0   (2) −2   (3) 4   (4) −2   (5) [graph]

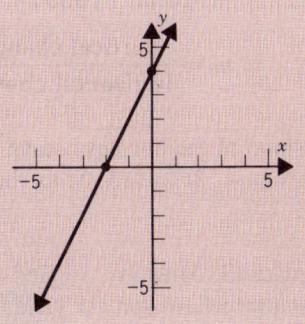

1. $x + y = 5$
2. $x - y = 4$
3. $2x + y = 8$
4. $x + 2y = 6$
5. $3x - y = 6$
6. $x - 2y = 4$
7. $2x + 3y = 12$
8. $3x + 5y = 15$
9. $3x - 4y = 12$
10. $4x - 5y = 20$
11. $y = x + 6$
12. $y = x + 4$
13. $y = 2x - 4$
14. $y = 3x + 9$
15. $y = 2x + 5$
16. $y = 3x - 7$
17. $x = 4 + y$
18. $x = 6 - 2y$
19. $x = 3y - 10$
20. $x = 5y + 5$
21. $2x + 3y = 1$
22. $4x - 3y = 1$
23. $4x + 5y = 1$
24. $5x - 4y = 1$
25. $2x - y = 0$
26. $x - 3y = 0$
27. $x - 2y = 0$
28. $x - 4y = 0$
29. $2x - 3y = 0$
30. $3x - 2y = 0$

12.5 SLOPE OF A LINE

SLOPE FORMULA

In this section, we will study an important property of a line. We will assign a number to a line, which we call *slope*, that will give us a measure of the "steepness" or "direction" of the line.

It is often convenient to use a special notation to distinguish between the rectangular coordinates of two different points. We can designate one pair of coordinates by (x_1, y_1) (read "x sub one, y sub one"), associated with a point P_1, and a second pair of coordinates by (x_2, y_2), associated with a second point P_2, as illustrated in

FIRST-DEGREE EQUATIONS AND INEQUALITIES IN TWO VARIABLES

Figure 12.7. Note that when going from P_1 to P_2, the vertical change (or vertical distance) between the two points is $y_2 - y_1$ and the horizontal change (or horizontal distance) is $x_2 - x_1$.

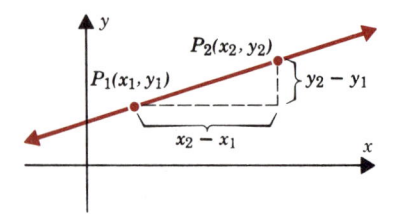

Figure 12.7

The ratio of the vertical change to the horizontal change is called the **slope** of the line containing the points P_1 and P_2. This ratio is usually designated by m. Thus,

$$m = \frac{\text{vertical change}}{\text{horizontal change}} = \frac{y_2 - y_1}{x_2 - x_1} \quad (x_2 \neq x_1) \qquad (1)$$

Example 1 Find the slope of the line containing the two points with coordinates $(-4, 2)$ and $(3, 5)$ as shown in the figure at the right.

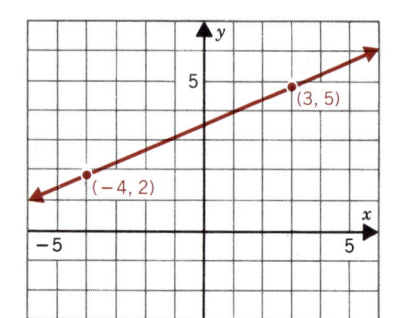

Solution We designate $(3, 5)$ as (x_2, y_2) and $(-4, 2)$ as (x_1, y_1). Substituting into Equation (1) yields

$$m = \frac{5 - 2}{3 - (-4)} = \frac{3}{7}$$

Note that we get the same result if we subsitute -4 and 2 for x_2 and y_2 and 3 and 5 for x_1 and y_1

$$m = \frac{2 - 5}{-4 - 3} = \frac{-3}{-7} = \frac{3}{7}$$

Lines with various slopes are shown in Figure 12.8 below. The slopes of lines that go up to the right are positive (Figure 12.8*a*) and the slopes of lines that go down to the right are negative (Figure 12.8*b*). And note (Figure 12.8*c*) that because all points on a horizontal line have the same y value, $y_2 - y_1$ equals zero for any two points and the slope of the line is simply

$$m = \frac{y_2 - y_1}{x_2 - x_1} = \frac{0}{x_2 - x_1} = 0$$

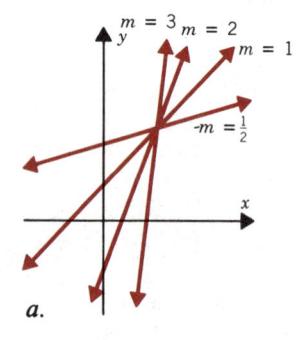

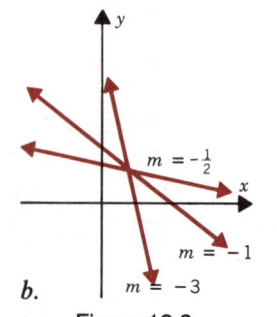

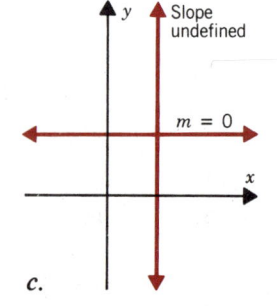

Figure 12.8

402

12.5 SLOPE OF A LINE

Also note (Figure 12.8c) that since all points on a vertical have the same x value, $x_2 - x_1$ equals zero for any two points. However,

$$m = \frac{y_2 - y_1}{x_2 - x_1} = \frac{y_2 - y_1}{0}$$

is undefined, so that a *vertical line does not have a slope*.

PARALLEL AND PERPENDICULAR LINES

Consider the lines shown in Figure 12.9. Line l_1 has slope $m_1 = 3$, and line l_2 has slope $m_2 = 3$. In this case,

$$m_1 = m_2 = 3$$

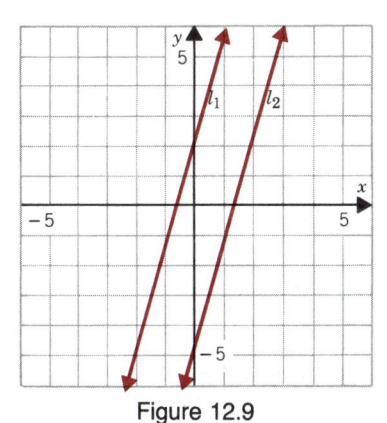

Figure 12.9

These lines will never intersect and are called **parallel lines**. Now consider the lines shown in Figure 12.10. Line l_1 has slope $m_1 = 1/2$ and line l_2 has slope $m_2 = -2$. In this case,

$$m_1 \cdot m_2 = \frac{1}{2} \cdot (-2) = -1$$

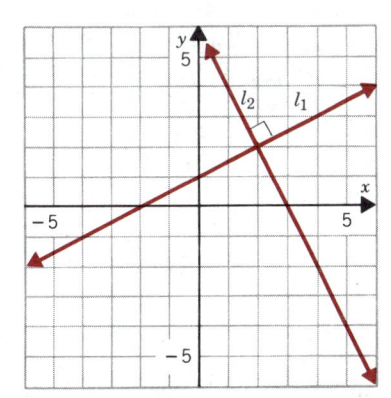

Figure 12.10

These lines meet to form a right angle and are called **perpendicular lines**.

FIRST-DEGREE EQUATIONS AND INEQUALITIES IN TWO VARIABLES

In general, if two lines have slopes m_1 and m_2:

> a. The lines are parallel if they have the same slope, that is, if $m_1 = m_2$.
> b. The lines are perpendicular if the product of their slopes is -1, that is, if $m_1 \cdot m_2 = -1$.

EXERCISES 12.5

In Exercises 1–12, find the slope of the line segment joining the given pairs of points. See Example 1.

Complete Example

$(2, -5), (4, 3)$ — Consider $(2, -5)$ as P_1, and $(4, 3)$ as P_2 and use the formula

$$m = \frac{3 - (-5)}{\underline{}_{(1)} - \underline{}_{(2)}} \qquad m = \frac{y_2 - y_1}{x_2 - x_1}$$

$$= \frac{3 + 5}{2} = \underline{}_{(3)}$$

(1) 4   (2) 2   (3) 4

1. (5, 2), (8, 7)
2. (4, 1), (6, 3)
3. (2, 4), (6, 1)
4. (4, 1), (2, 5)
5. (3, −2), (0, 1)
6. (−4, 0), (3, 5)
7. (−3, −4), (−7, 1)
8. (6, −2), (−3, −3)
9. (2, 5), (2, −3)
10. (3, −1), (3, 5)
11. (3, 4), (−2, 4)
12. (−2, 1), (3, 1)

13. a. Graph the points $A(-2, 5)$, $B(1, -1)$, and $C(3, -5)$.
     b. Find the slope of the line passing through points A and B.
     c. Find the slope of the line passing through points B and C.
     d. Do points A, B, and C lie on the same line?

14. a. Graph the points $A(-1, 5)$, $B(1, -1)$ and $C(2, -4)$.
     b. Find the slope of the line passing through points A and B.
     c. Find the slope of the line passing through points B and C.
     d. Do points A, B, and C lie on the same line?

15. a. Graph the points $A(2, 2)$, $B(3, 5)$ and $C(4, 7)$.
     b. Find the slope of the line passing through points A and B.
     c. Find the slope of the line passing through points B and C.
     d. Do points A, B, and C lie on the same line?

16. a. Graph the points $A(-1, 3)$, $B(0, 5)$ and $C(2, 6)$.
     b. Find the slope of the line passing through points A and B.
     c. Find the slope of the line passing through points B and C.
     d. Do points A, B, and C lie on the same line?

17. a. Find the slope of the line passing through the points $(5, 4)$ and $(3, 0)$.
     b. Find the slope of the line passing through the points $(-1, 8)$ and $(-4, 2)$.
     c. Graph the two lines.
     d. Are the lines parallel?

18. a. Find the slope of the line passing through the points $(-4, 2)$ and $(2, -2)$.
    b. Find the slope of the line passing through the points $(3, 0)$ and $(-3, 4)$.
    c. Graph the two lines.
    d. Are the lines parallel?

19. a. Find the slope of the line passing through the points $(-1, 2)$ and $(3, 4)$.
    b. Find the slope of the line passing through the points $(-2, 3)$ and $(6, 8)$.
    c. Graph the two lines.
    d. Are the lines parallel?

20. a. Find the slope of the line passing through the points $(-2, -3)$ and $(3, 4)$.
    b. Find the slope of the line passing through the points $(-1, -2)$ and $(5, 5)$.
    c. Graph the two lines.
    d. Are the lines parallel?

21. a. Find the slope of the line passing through the points $(0, -7)$ and $(8, -5)$.
    b. Find the slope of the line passing through the points $(5, 7)$ and $(8, -5)$.
    c. Graph the two lines.
    d. Are the lines perpendicular (does the product of their slopes equal -1)?

22. a. Find the slope of the line passing through the points $(8, 0)$ and $(6, 6)$.
    b. Find the slope of the line passing through the points $(-3, 3)$ and $(6, 6)$.
    c. Graph the two lines.
    d. Are the lines perpendicular (does the product of their slopes equal -1)?

23. a. Find the slope of the line passing through the points $(-1, -2)$ and $(3, 3)$.
    b. Find the slope of the line passing through the points $(0, 4)$ and $(-6, -8)$.
    c. Graph the two lines.
    d. Are the lines perpendicular (does the product of their slopes equal -1?)

24. a. Find the slope of the line passing through the points $(-4, -6)$ and $(2, 1)$.
    b. Find the slope of the line passing through the points $(-3, 5)$ and $(1, -2)$.
    c. Graph the two lines.
    d. Are the lines perpendicular (does the product of their slopes equal -1)?

25. If one line has a slope of -2, what must the slope of a second line be if the two lines are perpendicular?

26. If one line has a slope of 3, what must the slope of a second line be if the two lines are perpendicular?

27. If one line has a slope of $2/3$, what must the slope of a second line be if the two lines are perpendicular?

28. If one line has a slope of $-3/5$, what must the slope of a second line be if the two lines are perpendicular?

29. The slope of a horizontal line is ? .

30. The slope of a vertical line is ? .

12.6 EQUATIONS OF STRAIGHT LINES

POINT-SLOPE FORM

In Section 12.5, we found the slope of a straight line by using the formula

$$m = \frac{y_2 - y_1}{x_2 - x_1} \quad (x_2 \neq x_1)$$

Let us say we know that a line goes through the point (2, 3) and has a slope of 2.

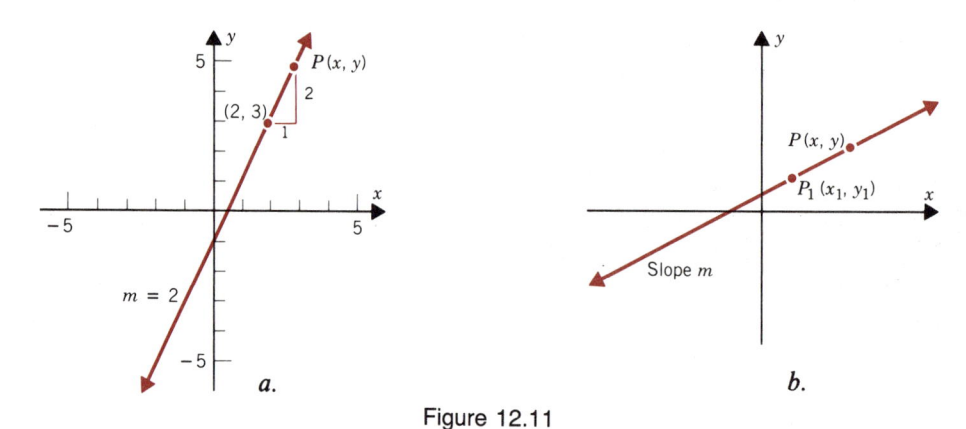

FIRST-DEGREE EQUATIONS AND INEQUALITIES IN TWO VARIABLES

Figure 12.11

If we denote *any* other point on the line as $P(x, y)$ (See Figure 12.11a), by the slope formula

$$m = 2 = \frac{y - 3}{x - 2}$$

from which

$$2(x - 2) = y - 3$$
$$2x - 4 = y - 3$$
$$2x - 1 = y$$

or

$$y = 2x - 1 \qquad (1)$$

Thus, Equation (1) is the equation of the line that goes through the point (2, 3) and has a slope of 2.

In general let us say we know a line passes through a point $P_1(x_1, y_1)$ and has slope m. If we denote *any* other point on the line as $P(x, y)$ (see Figure 12.11b), by the slope formula

$$m = \frac{y - y_1}{x - x_1}$$

from which

$$m(x - x_1) = y - y_1$$

or

$$y - y_1 = m(x - x_1) \qquad (2)$$

Equation (2) is called the **point-slope form** for a linear equation. In Equation (2), m, x_1 and y_1 are known and x and y are variables that represent the coordinates of *any* point on the line. Thus, whenever we know the slope of a line and a point on the line, we can find the equation of the line by using Equation (2).

Example 1 A line has slope -2 and passes through point (2, 4). Find the equation of the line.

Solution Substitute -2 for m and (2, 4) for (x_1, y_1) in Equation (2)

12.6 EQUATIONS OF STRAIGHT LINES

$$y - 4 = -2(x - 2)$$

where y_1 value, slope, and x_1 value are indicated.

$$y - 4 = -2x + 4$$
$$y = -2x + 8$$

Thus, a line with slope -2 that passes through the point $(2, 4)$ has the equation $y = -2x + 8$. We could also write the equation in equivalent forms $y + 2x = 8$, $2x + y = 8$, or $2x + y - 8 = 0$.

SLOPE-INTERCEPT FORM

Now consider the equation of a line with slope m and y intercept b as shown in Figure 12.12. Substituting 0 for x_1 and b for y_1 in the point-slope form of a linear equation, we have

$$y - b = m(x - 0)$$
$$y - b = mx$$

or

$$y = mx + b \qquad (3)$$

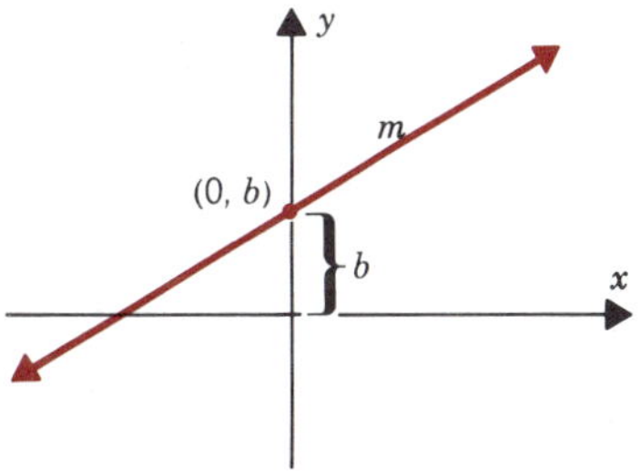

Figure 12.12

Equation (3) is called the **slope-intercept form** for a linear equation. The slope and y intercept can be obtained directly from an equation in this form.

Example 2 If a line has the equation

$$y = -2x + 8$$

where -2 is the slope and 8 is the y intercept, then the slope of the line must be -2 and the y interecept must be 8. Similarly, the graph

$$y = -3x + 4$$

has a slope -3 and a y intercept 4; and the graph of

$$y = \frac{1}{4}x - 2$$

has a slope $1/4$ and a y intercept -2.

If an equation is not written in $x = mx + b$ form and we want to know the slope and/or the y intercept, we rewrite the equation by solving for y in terms of x.

Example 3 Find the slope and y intercept of $2x - 3y = 6$.

Solution We first solve for y in terms of x by adding $-2x$ to each number.

$$2x - 3y - 2x = 6 - 2x$$
$$-3y = 6 - 2x$$

FIRST-DEGREE EQUATIONS AND INEQUALITIES IN TWO VARIABLES

Now dividing each member by -3, we have

$$\frac{-3y}{-3} = \frac{-2x}{-3} + \frac{6}{-3}$$

$$y = \frac{2}{3}x - 2$$

Comparing this equation with the form $y = mx + b$, we note that the slope m (the coefficient of x) equals $2/3$, and the y intercept equals -2.

EXERCISES 12.6

Find the equation of the line with the given slope and passing through the given point. Write the equation in the form $y = mx + b$. See Example 1.

Complete Example

$m = 3;\ (-2, 4)$

$y - \underline{\quad}_{(1)} = 3[x - (\underline{\quad}_{(2)})]$

Substitute -2 for x_1, 4 for y_1, and 3 for m in the point-slope form $y - y_1 = m(x - x_1)$

$y - 4 = 3(x + 2)$

Remove parentheses and simplify.

$y - 4 = 3x + \underline{\quad}_{(3)}$

$y = \underline{\quad}_{(4)}$

(1) 4   (2) -2 (3) 6   (4) $3x + 10$

1. $m = 3;\ (-1, 1)$
2. $m = -2;\ (1, -1)$
3. $m = -2;\ (-3, 4)$
4. $m = -3;\ (-2, 4)$
5. $m = 0;\ (-2, 1)$
6. $m = 0;\ (-3, 2)$
7. $m = \frac{1}{2};\ (4, -3)$
8. $m = \frac{-1}{2};\ (2, -6)$
9. $m = -4;\ (-2, 3)$
10. $m = 5;\ (2, -5)$
11. $m = \frac{2}{3};\ (-3, 1)$
12. $m = \frac{-2}{3};\ (-6, 2)$

Find the equation of the line passing through the two given points. Write the equation in the form $y = mx + b$.

Complete Example

$(-1, 3),\ (2, 6)$

Using the slope formula, first find the slope.

$$m = \frac{y_2 - y_1}{x_2 - x_1} = \frac{6 - 3}{2 - (\underline{\quad}_{(1)})}$$

408

12.6 EQUATIONS OF STRAIGHT LINES

$$m = \frac{3}{3} = 1$$

Use the point-slope form for a linear equation with $m = 1$ and (x_1, y_1) equal to $(-1, 3)$ or $(2, 6)$. We use $(-1, 3)$.

$$y - \underline{}_{(2)} = 1[x - (-1)]$$

Remove parentheses and simplify.

$$y - 3 = x + \underline{}_{(3)}$$

$$y = \underline{}_{(4)}$$

(1) -1 (2) 3 (3) 1 (4) $x + 4$

13. $(-2, 4), (1, 7)$ 14. $(2, 5), (4, 9)$ 15. $(-1, 3), (0, 7)$ 16. $(1, 2), (0, -6)$

17. $(0, 0), (3, 5)$ 18. $(0, 0), (-2, -7)$ 19. $(2, 4), (-3, 4)$ 20. $(1, 3), (-2, 3)$

21. $(6, 4), (-2, 5)$ 22. $(5, -2), (-1, 3)$

In Exercises 23–24, find the slope and y intercept of the graph of the equation. See Examples 2 and 3.

Complete Example

$$2x + 3y = 6$$

$$2x + 3y + (-2x) = 6 + (-2x)$$

Solve for y in terms fo x.

$$3y = \underline{}_{(1)} + 6$$

Divide each member by 3.

$$\frac{3y}{3} = \frac{-2x}{3} + \frac{6}{3}$$

$$y = \frac{-2}{3}x + 2$$

Compare with slope-intercept form $y = mx + b$.

Slope: $m = \underline{}_{(2)}$; y intercept: $\underline{}_{(3)}$

(1) $-2x$ (2) $\frac{-2}{3}$ (3) 2

23. $y = 5x - 2$ 24. $y = 3x + 4$ 25. $y = -4x + 3$ 26. $y = -6x - 3$

27. $3x + y = 4$ 28. $4x + y = -5$ 29. $6x + 3y = 5$ 30. $4x + 2y = 7$

31. $5x + 4y = -3$ 32. $2x - 3y = 6$ 33. $2x - 3y = 6$ 34. $x - 2y = -7$

FIRST-DEGREE EQUATIONS AND INEQUALITIES IN TWO VARIABLES

35.
   a. What is the slope of the line $y = -3x + 2$?
   b. What is the slope of a line parallel to $y = -3x + 2$?
   c. A line is parallel to the line $y = -3x + 2$ and passes through the point $(1, 3)$. What is its equation?

36.
   a. What is the slope of the line $y = 2x + 1$?
   b. What is the slope of a line parallel to $y = 2x + 1$?
   c. A line is parallel to the line $y = 2x + 1$ and passes through the point $(-2, 1)$. What is its equation?

37.
   a. What is the slope of the line $y = -2x + 4$?
   b. What is the slope of a line perpendicular to $y = -2x + 4$?
   c. A line is perpendicular to the line $y = -2x + 4$ and passes through the point $(4, 2)$. What is its equation?

38.
   a. What is the slope of the line $y = -3x - 2$?
   b. What is the slope of a line perpendicular to $y = -3x - 2$?
   c. A line is perpendicular to the line $y = -3x - 2$ and passes through the point $(6, -2)$. What is its equation?

12.7 DIRECT VARIATION

A special case of a first-degree equation in two variables is given by

$$y = kx \quad (k \text{ is a constant})$$

Such a relationship is called a **direct variation.** We say that the variable y *varies directly as* x.

Example 1 We know that the pressure P in a liquid *varies directly* as the depth d below the surface of the liquid. We can state this relationship in symbols as

$$P = kd$$

In a direct variation, if we kow a set of conditions on the two variables, and if we further know another value for one of the variables, we can find the value of the second variable for this new set of conditions.

In the above example, we can solve for the constant k to obtain

$$k = \frac{P}{d}$$

Since the ratio $\frac{P}{d}$ is constant for each set of conditions, we can use a proportion to solve problems involving direct variation.

Example 2 If pressure P varies directly as depth d, and $P = 40$ when $d = 10$, find P when $d = 15$.

Solution Since the ratio $\frac{P}{d}$ is constant, we can substitute values for P and d and obtain the proportion

$$k = \frac{40}{10} = \frac{P}{15}$$

from which

$$40 \cdot 15 = 10 \cdot P$$

$$\frac{40 \cdot 15}{10} = P$$

$$60 = P$$

Thus, $P = 60$ when $d = 15$.

12.7 DIRECT VARIATION

EXERCISES 12.7
Write an equation expressing the relationship between the variables using k as the constant of variation. See Example 1.

Complete Examples

a. The distance (d) that a car travels at a constant speed varies directly with the time (t).

Ans. $d = $ _____
(1)

b. The sales tax (T) on a certain item varies directly as the cost (C) of the item.

Ans. $T = $ _____
(2)

(1) kt (2) kC

1. y varies directly with x.
2. P varies directly with R.
3. R varies directly with L.
4. I varies directly with d.
5. The cost c in dollars of a certain item varies directly with the weight w in kilograms.
6. The current I in an electric circuit varies directly with the voltage E.
7. The tension T on a spring varies directly with the distance s it is stretched.
8. The pressure P exerted on a liquid at a given point varies directly with the depth d of the point beneath the surface of the liquid.

Solve. See Example 2.

Complete Example

If y varies directly as x, and $y = 10$ when $x = 15$, find y if $x = 24$.

$y = $ _____ (1)   Write a relationship between y and x.

$\dfrac{y}{x} = k$ Solve for k. Divide each member by x.

$\dfrac{10}{15} = \dfrac{y}{\underline{\quad(2)\quad}}$ Since the ratio y/x is a constant, ratios are equal for different sets of conditions. Substitute appropriate values in each of two ratios to obtain a proportion.

Solve for y.

$10(24) = \underline{\quad(3)\quad} y$

$\dfrac{10(24)}{15} = y$

$y = \underline{\quad(4)\quad}$

(1) kx (2) 24   (3) 15   (4) 16

FIRST-DEGREE EQUATIONS AND INEQUALITIES IN TWO VARIABLES

9. If y varies directly with x, and $y = 6$, when $x = 4$, find y when $x = 14$.

10. If z varies directly with x, and $z = 9$ when $x = 12$, find z when $x = 3$.

11. If R varies directly with L, and $R = 42$ when $L = 30$, find R when $= 45$.

12. If I varies directly with d, and $I = 640$ when $d = 120$, find I when $d = 600$.

13. The distance d in miles that a car travels varies directly with the time t in hours. If $d = 96$ when $t = 2$, find the distance traveled in $3\frac{1}{2}$ hours.

14. The cost C in dollars of a certain vitamin varies directly with the weight w in kilograms. If $C = 60$ when $w = 24$, find the cost of 54 kilograms.

15. The current I in an electric circuit varies directly with the voltage E. If $I = 4.2$ when $E = 60$, find E when $I = 6.6$.

16. The tension T on a spring varies directly as the distance s it is stretched. If $T = 54$ when $s = 12$, find s if $T = 20$.

17. The pressure P exerted by a liquid at a given point varies directly with the depth d of the point beneath the surface of the liquid. If a certain liquid exerts a pressure of 20 pounds per square foot at a depth of 6 feet, what would be the pressure at 9 feet?

18. What would be the pressure of the liquid in exercise 17 at a depth of 15 feet?

19. The circumference C of a circle varies directly with the diameter D. What is the effect on the circumference if the diameter is doubled in size?

20. What is the effect on the diameter of a circle if the circumference is reduced to one-fourth of its original length? (See Exercise 19.)

12.8 INEQUALITIES IN TWO VARIABLES

In Section 12.3 and 12.4, we graphed equations in two variables. In this section we will graph inequalities in two variables. For example, consider the inequality

$$y \leq -x + 6$$

The solutions are ordered pairs of numbers that "satisfy" the inequality. That is, (a, b) is a solution of the inequality if the inequality is a true statement after we substitute a for x and b for y.

Example 1 Determine if the given ordered pair is a solution of $y \leq -x + 6$.

    a. (1, 1)   b. (2, 5)

Solution The ordered pair (1, 1) is a solution because, when 1 is substituted for x and 1 is substituted for y, we get

$$(1) \leq (1) + 6, \quad \text{or} \quad 1 \leq 5$$

which is a true statement. On the other hand, (2, 5) is *not* a solution because when 2 is substituted for x and 5 is substituted for y, we obtain

$$(5) \leq -(2) + 6, \quad \text{or} \quad 5 \leq 4$$

which is a false statement.

12.8 INEQUALITIES IN TWO VARIABLES

To graph the inequality $y \leq -x + 6$, we first graph the equation $y = x + 6$ shown in Figure 12.13. Notice that (3, 3), (3, 2), (3, 1), (3, 0), and so on, associated with the points that are on or below the line, are all solutions of the inequality $y \leq -x + 6$, whereas (3, 4), (3, 5), and (3, 6), associated with points above the line are not solutions of the inequality. In fact, all ordered pairs associated with points on or below the line are solutions of $y \leq -x + 6$. Thus, *every* point on or below the line is in the graph. We represent this by shading the region below the line (see Figure 12.14).

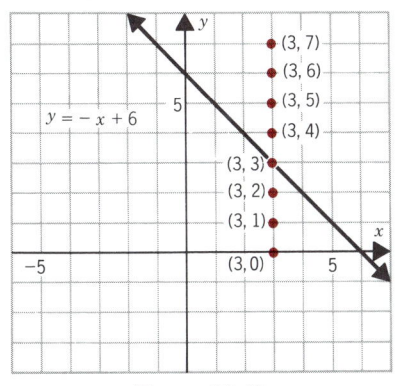

Figure 12.13

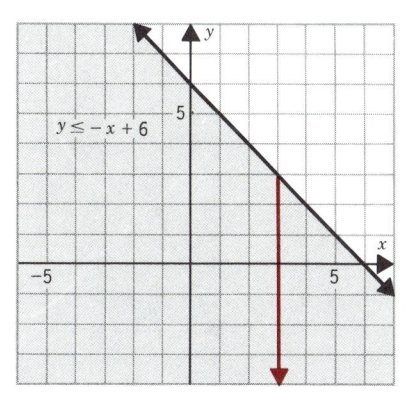

Figure 12.14

In general, to graph a first-degree inequality in two variables of the form $Ax + By \leq C$ or $Ax + By \geq C$, we first graph the equation $Ax + By = C$ and then determine which half-plane (a region above or below the line) contains the solutions. We then shade this half-plane. We can always determine which half-plane to shade by selecting a point (not on the line of the equation $Ax + By = C$) and testing to see if the ordered pair associated with the point is a solution of the given inequality. If so, we shade the half-plane containing the test point; otherwise, we shade the other half-plane. Often, (0, 0) is a convenient test point.

Example 2 Graph $2x + 3y \geq 6$.

Solution We first graph the line $2x + 3y = 6$ (see graph *a*). Using the origin as a test point, we determine whether (0, 0) is a solution of $2x + 3y \geq 6$. Since the statement

$$2(0) + 3(0) \geq 6$$

is false, (0, 0) is not a solution and we shade the half-plane that does not contain the origin (see graph *b*).

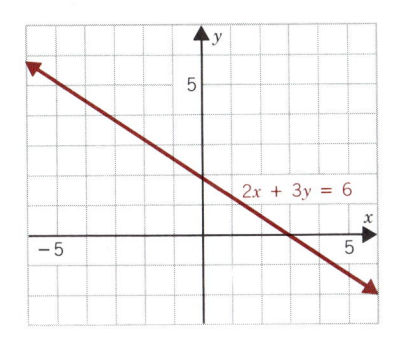

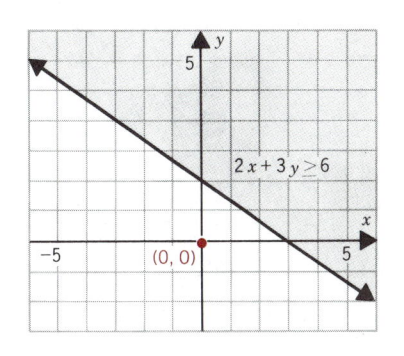

FIRST-DEGREE EQUATIONS AND INEQUALITIES IN TWO VARIABLES

When the line $Ax + By = C$ passes through the origin, $(0, 0)$ is not a valid test point since it is on the line.

Example 3 Graph $y \geq 2x$.

Solution We begin by graphing the line $y = 2x$ (see graph *a*). Since the line passes through the origin, we must choose another point not on the line as our test point. We will use $(0, 1)$. Since the statement

$$(1) \geq 2(0)$$

is true, $(0, 1)$ is a solution and we shade the half-plane that contains $(0, 1)$. (see graph *b*.)

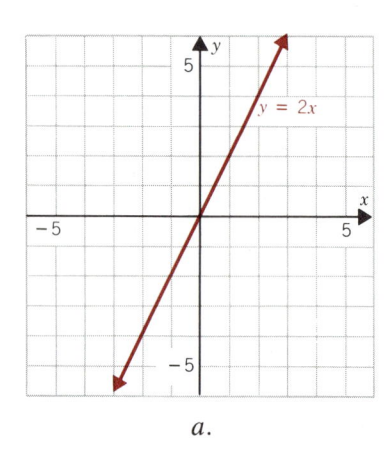

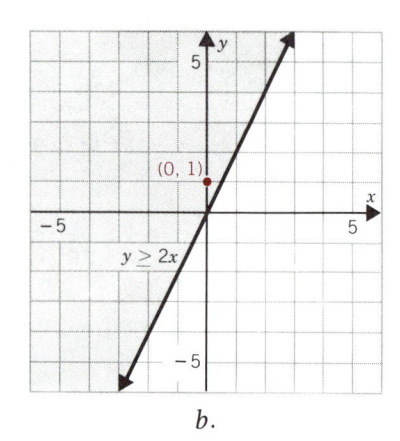

a.

b.

If the inequality symbol is $<$ or $>$, the points on the graph of $Ax + By = C$ are not solutions of the inequality. We then use a dashed line for the graph of $Ax + By = C$. An example is shown in the exercises.

EXERCISES 12.8

Determine if the given ordered pair is a solution of the inequality. See Example 1.

Complete Example

a. $y < 4x$; $(1, -3)$

b. $2x + y \geq 4$; $(2, -2)$

Substitute 1 for x and -3 for y.

Substitute 2 for x and -2 for y.

$-3 < 4(\underline{})$
$\qquad\qquad(1)$

$2(2) + (\underline{}) \geq 4$
$\qquad\qquad\quad(3)$

$-3 < 4$ is true.

$4 + (-2) \geq 4$

Ans. $(1, -3)$ $\underline{\text{is/is not}}$ a solution.
$\qquad\qquad\quad(2)$

$2 \geq 4$ is false.

Ans. $(2, -2)$ $\underline{\text{is/is not}}$ a solution.
$\qquad\qquad\qquad(4)$

(1) 1   (2) is   (3) -2 (4) is not

12.8 INEQUALITIES IN TWO VARIABLES

1. $y < x$; (0, 1)
2. $y > x$; (1, 3)
3. $y < x + 3$; (−1, 2)
4. $y \geq x - 5$; (3, 2)
5. $x + y < -2$; (−2, −1)
6. $x + y > -1$; (−1, 3)
7. $2x + y \leq 1$; (−1, −1)
8. $2x - y > 3$; (−2, −4)
9. $x < 4y + 1$; (0, −3)
10. $x < 3y + 5$; (2, 0)
11. $0 \leq x + 4y$; (−1, 1)
12. $2 > x + 2y$; (0, −2)

Graph each inequality. See Examples 2 and 3.

Complete Example

$-3x + 2y < 6$

Graph the equation $-3x + 2y = 6$. Since the inequality symbol is $<$ rather than $\leq$, use a dashed line.

Test (0, 0): $-3(\underline{}) + 2(\underline{}) < 6$
                      (1)   (2)

$0 + 0 < 6$ is true.

Shade the half-plane containing (0, 0).

(3)

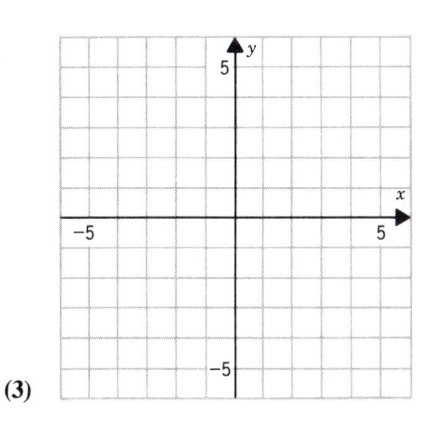

(1) 0   (2) 0   (3) [graph showing dashed line with shaded region, labeled $-3x + 2y < 6$]

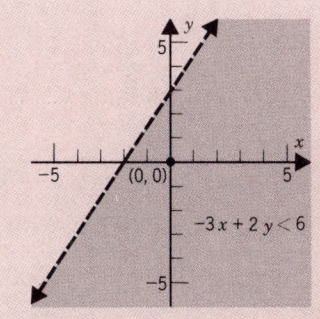

13. $y - 3x \leq 5$
14. $y + 3x \leq 6$
15. $3x - 4y \geq 12$
16. $2x + 5y \geq 10$
17. $y + 3x < 6$
18. $2x - y < 4$
19. $y > 3x + 2$
20. $y > -2x + 1$
21. $y < x - 3$
22. $y < -x - 3$
23. $6x + 4y \leq 12$
24. $3x + 2y \leq 12$
25. $y \leq 2$
26. $y \leq -1$
27. $x > -3$
28. $x > 2$
29. $y \leq 2x$
    (*Hint:* (0, 0) is not a good test point.)
30. $y \geq -2x$
    (*Hint:* (0, 0) is not a good test point.)
31. $y < 3x$
32. $y > 3x$
33. $y > 4x$
34. $y < -4x$

FIRST-DEGREE EQUATIONS AND INEQUALITIES IN TWO VARIABLES

CHAPTER TWELVE REVIEW EXERCISES

1. Given $I = 0.05P$, the yearly interest (I) on an investment equals the rate (0.05) times the principal (P).
   a. Which symbols are constants?
   b. Which symbols are variables?
   c. If the principal increases, what happens to the interest?

2. Solve $2x - y = 4$ for y in terms of x.

3. Solve $2y - 3x = 6$ for y in terms of x.

4. If $y = 2x + 1$, find the solutions with specified first components.
   a. (3, ?)   b. (−2, ?)   c. (0, ?)   d. $(-\frac{1}{2}, ?)$

5. If $f(x) = x - 5$, find
   a. $f(4)$ b. $f(-2)$ c. $f(0)$ d. $f(-6)$

6. Graph the following ordered pairs on a set of rectangular axes.
   a. (3, 4)   b. (−2, 3)   c. (3, −2)   d. (0, 4)

In Exercises 7–10, graph each equation.

7. $x + y = 3$
8. $2y - x = 4$
9. $x = 3$
10. $3x + 2y = 6$

11. Where does the graph of $x - y = 8$ cross the x-axis? The y-axis?

12. Find the slope of the line containing the points (2, 3) and (−4, 5).

13. Is the line passing through the points (2, 4) and (−1, 3) parallel to the line passing through (5, −1) and (−3, 2)?

14. What is the equation of a line with slope −2 and passing through the point (−2, −5)? Write your answer in $y = mx + b$ form.

15. What is the equation of a line passing through the points (−2, 3) and (1, 5)? Write your answer in $y = mx + b$ form.

16. What is the slope and y-intercept of the graph of the equation $2y - 5x = 0$?

17. If y varies directly with x, and $y = 20$ when $x = 6$, find x when $y = 44$.

In Exercises 18–20, graph each inequality.

18. $2x + 3y \geq 6$
19. $x \leq -2$
20. $y > -3x$

13 SYSTEMS OF LINEAR EQUATIONS

13.1 GRAPHICAL SOLUTIONS

Often, we want to find a single ordered pair that is a solution to two different linear equations. One way to obtain such an ordered pair is by graphing the two equations on the same set of axes and determining the coordinates of the point where they intersect.

Example 1 Graph the equations

$$x + y = 5$$
$$x - y = 1$$

on the same set of axes and determine the ordered pair that is a solution for each equation.

Solution Using the intercept method of graphing, we find that two ordered pairs that are solutions of $x + y = 5$ are

(0, 5)   and   (5, 0)

And two ordered pairs that are solutions of $x - y = 1$ are

(0, −1)   and   (1, 0)

The graphs of the equation are shown. The point of intersection is (3, 2). Thus, (3, 2) should satisfy each equation. In fact,

3 + 2 = 5   and   3 − 2 = 1

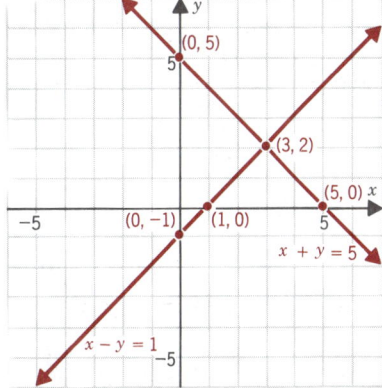

In general, graphical solutions are only approximate. We will develop methods for exact solutions in later sections.

417

SYSTEMS OF LINEAR EQUATIONS

Linear equations considered together in this fashion are said to form a **system of equations.** As in the above example, the solution of a system of linear equations can be a single ordered pair. The components of this ordered pair satisfy *each* of the two equations.

Some systems have no solutions, while others have an infinite number of solutions. If the graphs of the equations in a system do not intersect—that is, if the lines are parallel (see Figure 13.1a)—the equations are said to be **inconsistent**, and there is no ordered pair that will satisfy both equations. If the graphs of the equations are the same line (see Figure 13.1b), the equations are said to be **dependent**, and each ordered pair which satisfies one equation will satisfy both equations. Notice that when a system is inconsistent, the slopes of the lines are the same but the y intercepts are different. When a system is dependent, the slopes and y intercepts are the same.

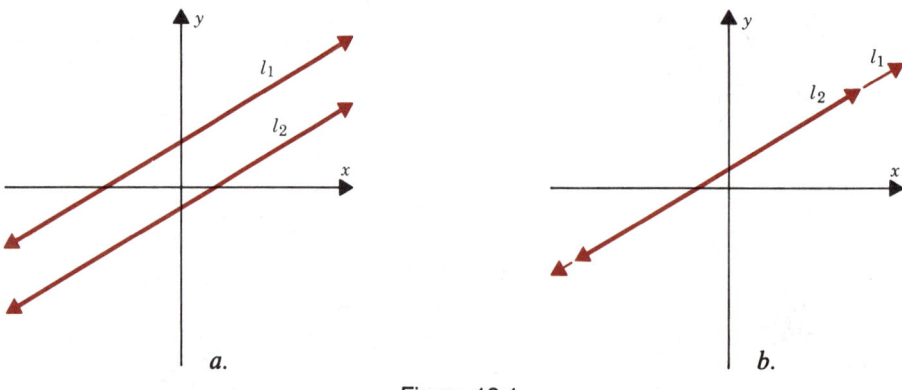

Figure 13.1

In our work we will be primarily interested in systems that have one and only one solution and that are said to be consistent and independent. The graph of such a system is shown in the solution of Example 1.

EXERCISES 13.1

Find the solution of each system by graphical methods. If the system is inconsistent or dependent, so state. See Example 1.

Complete Example

$4x - 2y = 6$

$x + y = 3$

Find the intercepts.

For $4x - 2y = 6$, if $x = 0$, $y = -3$; if $y = 0$, $x = \underline{}$.
(1)

For $x + y = 3$, if $x = 0$, $y = 3$; if $y = 0$, $x = \underline{}$.
(2)

Graph the equations.

13.1 GRAPHICAL SOLUTIONS

(3)

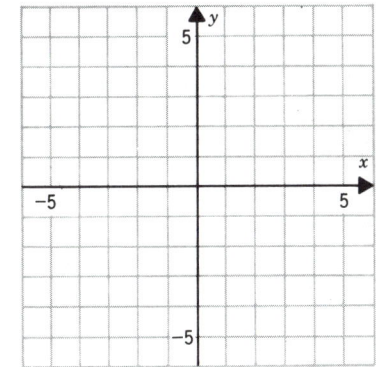

The graphs of the two equations intersect at the point corresponding to (2, 1).

(1) $\dfrac{3}{2}$ (2) 3   (3)

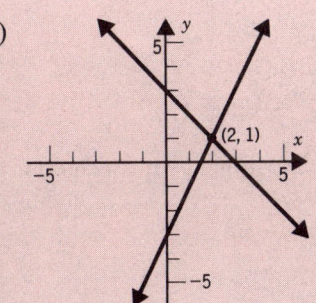

1. $x - y = 2$
   $x + y = 8$

2. $x - y = -2$
   $3x - y = 8$

3. $2x - y = 0$
   $2x - 3y = -4$

4. $y + 3x = 0$
   $4x - 3y = 13$

5. $x - 3y = 0$
   $2x - y = -5$

6. $2y + x = 0$
   $y - x = 6$

7. $x + y = 6$
   $x - y = 2$

8. $5y - x = 6$
   $-y + x = 4$

9. $y = x$
   $y = 4 - x$

10. $y = -x$
    $y = 3 - 2x$

11. $y - x = 1$
    $y + x = -5$

12. $4x - 4y = 0$
    $x + y = 2$

13. $2x - y + 5 = 0$
    $x - 3y = -20$

14. $2x + 7y - 12 = 0$
    $x + 5y = 0$

15. $5x - 2y + 8 = 0$
    $3x + y + 7 = 0$

16. $7y - 2x + 6 = 0$
    $8y - 5x - 4 = 0$

17. $x + y = 4$
    $2x + 2y = 8$

18. $x - 2y = 3$
    $2x - 4y = 6$

19. $x + 3y = 5$
    $2x + 6y = 5$

20. $x - 3y = -4$
    $-2 + 6y = 5$

21. $3x - y = 1$
    $6x = 2y + 2$

22. $-x + 2y = 3$
    $3x = 6y - 9$

23. $x - 2y = 4$
    $-4y = 5 - 2x$

24. $x + y = 7$
    $-2y = 5 + 2x$

419

SYSTEMS OF LINEAR EQUATIONS

13.2 SOLVING SYSTEMS BY ADDITION I

We can solve systems of equations algebraically. What is more, the solutions we obtain by algebraic methods are exact.

The system in the following example is the system we considered in Section 13.1 on page 417.

Example 1   Solve

$$x + y = 5 \tag{1}$$
$$x - y = 1 \tag{2}$$

Solution   We can obtain an equation in one variable by adding Equations (1) and (2)

$$x + y = 5 \tag{1}$$
$$\underline{x - y = 1} \tag{2}$$
$$2x \phantom{{}-y} = 6$$

Solving the resulting equation for x yields

$$2x = 6, \quad x = 3$$

We can now substitute 3 for x in either Equation (1) or Equation (2) to obtain the corresponding value of y. In this case, we have selected Equation (1) and obtain

$$(3) + y = 5$$
$$y = 2$$

Thus, the solution is $x = 3$, $y = 2$; or $(3, 2)$.

Notice that we are simply applying the addition property of equality so we can obtain an equation containing a single variable. The equation in one variable, together with either of the original equations, then forms an equivalent system whose solution is easily obtained.

In the above example, we were able to obtain an equation in one variable by adding Equations (1) and (2) because the terms $+y$ and $-y$ are the negatives of each other. Sometimes, it is necessary to multiply each member of one of the equations by -1 so that terms in the same variable will have opposite signs.

Example 2   Solve

$$2a + b = 4 \tag{3}$$
$$a + b = 3 \tag{4}$$

Solution   We begin by multiplying *each member* of Equation (4) by -1, to obtain

$$2a + b = 4 \tag{3}$$
$$-a - b = -3 \tag{4'}$$

where $+b$ and $-b$ are negatives of each other.

The symbol $'$, called "prime," indicates an equivalent equation; that is, an equation that has the same solutions as the original equation. Thus, Equation (4') is equivalent to Equation (4). Now adding Equations (3) and (4'), we get

13.2 SOLVING SYSTEMS BY ADDITION I

$$2a + b = 4 \quad (3)$$
$$\underline{-a - b = -3} \quad (4')$$
$$a\phantom{{}+b} = 1$$

Substituting 1 for a in Equation (3) or Equation (4) [say, Equation (4)], we obtain

$$1 + b = 3$$
$$b = 2$$

and our solution is $a = 1$, $b = 2$ or $(1, 2)$. When the variables are a and b, the ordered pair is given in the form (a, b).

EXERCISES 13.2
Solve each system.

Complete Example (see Example 1.)

$$3x + y = 8 \quad (1)$$
$$3x - y = 4 \quad (2)$$

Obtain an equation in one variable by adding Equations (1) and (2). Solve the resulting equation for x.

$$3x + y = 8$$
$$\underline{3x - y = 4}$$
$$6x\phantom{{}+y} = 12$$

$$x = \underline{}_{(1)}$$

Substitute 2 for x in Equation (1). (We could have used Equation (2) as well.) Solve for y.

$$3(\underline{}_{(2)}) + y = 8$$

$$y = \underline{}_{(3)}$$

Ans. $x = 2$, $y = 2$; or $(\underline{}_{(4)})$

---

(1) 2   (2) 2   (3) 2   (4) (2, 2)

---

1. $x + y = 5$
   $x - y = 1$

2. $x + y = 10$
   $x - y = 4$

3. $x + y = 3$
   $-x + y = 5$

4. $-x + y = 7$
   $x + y = 5$

5. $x + 2y = 10$
   $x - 2y = 2$

6. $2x + y = 3$
   $-2x + y = -1$

7. $3x + 2y = 5$
   $-2x - 2y = -4$

8. $5x - 3y = -1$
   $3x + 3y = 9$

9. $3x + 3y = 15$
   $3x - 3y = 27$

10. $6x - y = 4$
    $2x + y = 4$

11. $7x - 3y = -10$
    $x + 3y = 2$

12. $3x - 2y = 11$
    $3x + 2y = 19$

421

SYSTEMS OF LINEAR EQUATIONS

Complete Example

$4a + 2b = 3$ (1)
$2a + 2b = 1$ (2)

Multiply each member of Equation (2) by -1 to obtain (2′). Add Equations (1) and (2′). Solve for a.

$4a + 2b = 3$
$-2a - 2b = -1$ (2′)
$2a = 2$
$a = \underline{\qquad}$
 (1)

Substitute 1 for a in Equation (2). (We could have used Equation (1) as well.) Solve for b.

$2(\underline{\qquad}) + 2b = 1$.
  (2)
$2b = -1$
$b = \underline{\qquad}$
 (3)

Ans. $a = 1$, $b = \dfrac{-1}{2}$; or $\left(\underline{\qquad} \right)$
  (4)

---

(1) 1   (2) 1   (3) $\dfrac{-1}{2}$ (4) $\left(1, \dfrac{-1}{2}\right)$

---

13. $2a + b = 3$
    $a + b = 2$

14. $a + b = 6$
    $3a + b = 10$

15. $a + b = 7$
    $a + 3b = 11$

16. $a + 2b = -1$
    $3a + 2b = 1$

17. $3a + 2b = 1$
    $2a + 2b = 0$

18. $6a + 5b = 6$
    $-4a + 5b = -4$

19. $4a + 2b = 4$
    $3a + 2b = 8$

20. $4a + 2b = 4$
    $6a + 2b = 8$

21. $a + 3b = 2$
    $2a + 3b = 7$

22. $3a - 4b = 19$
    $-2a - 4b = -6$

23. $3a - 4b = 0$
    $3a + 2b = 18$

24. $3a - 2b = 5$
    $-a - 2b = -15$

25. $x + y = -1$
    $2x + y = -5$

26. $2x + 3y = -5$
    $2x - 4y = 16$

27. $3x + 4y = -10$
    $x + 4y = -6$

28. $4x - 4y = -4$
    $4x - 3y = -3$

29. $7x + 6y = 12$
    $9x + 6y = 12$

30. $3a + 2b = -12$
    $a + 2b = -4$

13.2 SOLVING SYSTEMS BY ADDITION I

Complete Example

$$\frac{x}{6} + \frac{y}{4} = \frac{3}{2} \quad (1)$$

$$\frac{2x}{3} - \frac{y}{2} = 0 \quad (2)$$

Remove fractions by multiplying each equation by LCD of fractions in that equation.

$$(\cancel{12})\overset{2}{\frac{x}{\cancel{6}}} + (\cancel{12})\overset{3}{\frac{y}{\cancel{4}}} = (\cancel{12})\overset{6}{\frac{3}{\cancel{2}}} \quad (1)$$

$$(\cancel{6})\overset{2}{\frac{2x}{\cancel{3}}} - (\cancel{6})\overset{3}{\frac{y}{\cancel{2}}} = (6)0 \quad (2)$$

$$2x + 3y = 18 \quad (1')$$
$$\underline{4x - 3y = 0} \quad (2')$$
$$6x = 18$$

Simplify Equations (1) and (2). Solve for x.

$$x = \underline{}_{(1)}$$

Substitute 3 for x in Equation (2'). Solve for y.

$$4(\underline{}_{(2)}) - 3y = 0$$

$$-3y = -12$$

$$y = \underline{}_{(3)}$$

Ans. $x = 3$, $y = 4$; or ($\underline{}_{(4)}$)

---

(1) 3   (2) 3   (3) 4   (4) (3, 4)

---

31. $y + \frac{1}{2}x = 0$
$\frac{1}{3}y - \frac{1}{3}x = 2$

32. $\frac{a}{5} + \frac{2b}{5} = 2$
$\frac{a}{2} - b = 1$

33. $\frac{2x}{3} + y = 3$
$\frac{x}{2} - \frac{y}{4} = \frac{5}{4}$

34. $x - \frac{2}{3}y = -4$
$\frac{x}{4} + \frac{y}{2} = -1$

35. $\frac{a}{4} - \frac{b}{3} = 0$
$\frac{a}{2} + \frac{b}{3} = 3$

36. $a + \frac{b+1}{3} = 0$
$\frac{5a}{2} - \frac{b+8}{4} = -5$

423

SYSTEMS OF LINEAR EQUATIONS

13.3 SOLVING SYSTEMS BY ADDITION II

As we saw in Section 13.2, solving a system of equations by addition depends on one of the variables in both equations having coefficients that are the negatives of each other. If this is not the case, we can find equivalent equations that do have variables with such coefficients.

Example 1 Solve the system

$$-5x + 3y = -11 \qquad (1)$$
$$-7x - 2y = -3 \qquad (2)$$

Solution If we multiply each member of Equation (1) by 2 and each member of Equation (2) by 3, we obtain the equivalent system

$$(2)(-5x) + (2)(3y) = (2)(-11)$$
$$(3)(-7x) - (3)(2y) = (3)(-3)$$

or

$$-10x + 6y = -22 \qquad (1')$$
$$-21x - 6y = -9 \qquad (2')$$

Now, adding Equations (1') and (2'), we get

$$-31x = -31$$
$$x = 1$$

Substituting 1 for x in Equation (1) yields

$$-5(1) + 3y = -11$$
$$3y = -6$$
$$y = -2$$

The solution is $x = 1$, $y = -2$ or $(1, -2)$.

Note that in Equations (1) and (2), the terms involving variables are in the left-hand member and the constant term is in the right-hand member. We will refer to such arrangements as the **standard form** for systems. It is convenient to arrange systems in standard form before proceeding with their solution. For example, if we want to solve the system

$$3y = 5x - 11 \qquad (3)$$
$$-7x = 2y - 3 \qquad (4)$$

we would first write the system in standard form by adding   to each member of Equation (3) and by adding   to each member of Equation (4). Thus, we get

$$-5x + 3y = -11$$
$$-7x - 2y = -3$$

and we can now proceed as shown above.

13.3 SOLVING SYSTEMS BY ADDITION II

EXERCISES 13.3
Solve. See Example 1.

Complete Example

$3x + 2y = 11$ (1)

$5x - 4y = 11$ (2)

Multiply each member of Equation (1) by 2. Add Equations (1′) and (2).

$6x + 4y = 22$ (1′)

$5x - 4y = 11$ (2)

$11x = 33$

$x = \underline{}$ **(1)**

Solve for x.

$3(\underline{}_{(2)}) + 2y = 11$

Substitute 3 for x in Equation (1) and solve for y.

$2y = 2$

$y = \underline{}_{(3)}$

Ans. $x = 3$, $y = 1$; or $\underline{}_{(4)}$

(1) 3   (2) 3   (3) 1   (4) (3, 1)

1. $3x + 2y = 7$
   $x + y = 3$

2. $2x - 3y = 8$
   $x + y = -1$

3. $2x - y = 2$
   $3x + 2y = 10$

4. $a - 4b = 9$
   $3a + 2b = 13$

5. $3a - b = -5$
   $2a + 3b = -7$

6. $-x + 3y = -1$
   $-6x + y = -6$

7. $3a - 3b = -3$
   $-6a + 2b = 14$

8. $3x - 6y = 6$
   $x - 2y = 3$

9. $5x + 3y = 19$
   $2x - y = 12$

10. $5x - 3y = 32$
    $2x + 6y = -16$

11. $3x - 5y = -1$
    $x + 2y = 18$

12. $3x + 3y = 0$
    $6x + 9y = -6$

Complete Example

$2y = 11 - 3x$ (1)

$5x = 11 + 4y$ (2)

Add $3x$ to each member of Equation (1) and add $-4y$ to each member of Equation (2) to arrange system in standard form.

$\underline{}_{(1)} = 11$

$\underline{}_{(2)} = 11$

Solve as described in the previous example.

(1) $3x + 2y$ (2) $5x - 4y$

425

SYSTEMS OF LINEAR EQUATIONS

13. $3x + 2y = 7$
    $x = y - 1$

14. $3x + 9 = -2y$
    $x = -3y + 25$

15. $a + 5b = 0$
    $-3a + 10 = 10b$

16. $x = 8 - 2y$
    $2x = 6 + y$

17. $8b - 3a = 5$
    $a - b = 0$

18. $6x = 22 + 2y$
    $8x = 33 - y$

19. $8a = 4b + 4$
    $3a = 2b + 3$

20. $x + 3y = 35$
    $0 = 2x - y$

21. $5b = 3a + 8$
    $2b = -a + 1$

22. $x - y = 15$
    $2x = -3y$

23. $3x - 2y = 3$
    $2x = y + 2$

24. $2x = 2y + 2$
    $4x = 5 + 4y$

25. $2x + 3y = -1$
    $3x + 5y = -2$

26. $3x - 2y = 13$
    $7x + 3y = 15$

27. $2x = 3y - 1$
    $3x + 4y = 24$

28. $5x - 2y = 0$
    $2x - 3y = -11$

29. $4b - a = 8$
    $2b + a = 10$

30. $3b - 4a = 3$
    $2b = 4a - 2$

31. $2a + 3b = 0$
    $5a - 2b = -19$

32. $8x - 7y = 0$
    $7x = 8y + 15$

Complete Example

$$\frac{x+1}{2} - \frac{2y-1}{6} = \frac{7}{6} \quad (1)$$

$$\frac{2x-1}{4} - \frac{y-3}{4} = 1 \quad (2)$$

Express each equation without fractions by multiplying by LCD of fractions in each equation.

$$(\cancel{6})^3\frac{x+1}{\cancel{2}} - (\cancel{6})\frac{2y-1}{\cancel{6}} = (\cancel{6})\frac{7}{\cancel{6}} \quad (1)$$

$$(\cancel{4})\frac{2x-1}{\cancel{4}} - (\cancel{4})\frac{y-3}{\cancel{4}} = (4)1 \quad (2)$$

$$3(x+1) - (2y-1) = \underline{}_{(1)}$$

$$(\underline{}_{(2)}) - (y-3) = 4$$

$$3x - 2y = 3 \quad (1')$$

$$2x - y = \underline{}_{(3)} \quad (2')$$

Write system in standard form.

$$3x - 2y = 3 \quad (1')$$
$$\underline{-4x + 2y = -4} \quad (2'')$$
$$-x = -1$$

$$x = \underline{}_{(4)}$$

Multiply Equation (2') by -2, and solve for x,

$$3(1) - 2y = 3$$
$$-2y = 0$$
$$y = \underline{}_{(5)}$$

Substitute 1 for x in Equation (1'). Solve for y.

Ans. $x = 1$, $y = 0$; or $(1, 0)$

---

(1) 7   (2) $2x - 1$ (3) 2   (4) 1   (5) 0

33. $\dfrac{5a}{4} + b = \dfrac{11}{2}$

$a + \dfrac{b}{3} = 3$

34. $2a - \dfrac{5b}{2} = 13$

$\dfrac{a}{3} + \dfrac{b}{5} = \dfrac{14}{5}$

35. $\dfrac{x}{4} + \dfrac{y}{5} = 1$

$\dfrac{2x}{9} - \dfrac{y}{9} = -2$

36. $\dfrac{5x}{8} + y = \dfrac{1}{4}$

$\dfrac{5x}{4} - \dfrac{3y}{2} = 4$

37. $\dfrac{x}{3} - \dfrac{y}{2} = 1$

$\dfrac{2x+3}{5} - \dfrac{y}{2} = 2$

38. $\dfrac{2x+1}{7} + \dfrac{3y+2}{5} = \dfrac{1}{5}$

$\dfrac{3x-2}{2} + \dfrac{y+4}{4} = 4$

13.4 SOLVING SYSTEMS BY SUBSTITUTION

In Sections 13.2 and 13.3, we solved systems of first-degree equations in two variables by the addition method. Another method, called the **substitution method**, can also be used to solve such systems.

Example 1 Solve the system

$$-2x + y = 1 \qquad (1)$$
$$x + 2y = 17 \qquad (2)$$

Solution Solving Equation (1) for y in terms of x, we obtain

$$y = 2x + 1 \qquad (1')$$

We can now substitute $2x + 1$ for y in Equation (2) to obtain

$$x + 2(2x + 1) = 17$$
$$x + 4x + 2 = 17$$
$$5x = 15$$
$$x = 3$$

Substituting 3 for x in Equation (1'), we have

$$y = 2(3) + 1 = 7$$

Thus, the solution of the system is $x = 3$, $y = 7$; or $(3, 7)$.

In the above example, it was easy to express y explicitly in terms of x using Equation (1). But we also could have used Equation (2) to write x explicitly in terms of y

$$x = -2y + 17 \qquad (2')$$

Now substituting $-2y + 17$ for x in Equation (1), we get

$$-2(-2y + 17) + y = 1$$
$$4y - 34 + y = 1$$
$$5y = 35$$
$$y = 7$$

SYSTEMS OF LINEAR EQUATIONS

Substituting 7 for y in Equation (2'), we have

$$x = -2(7) + 17 = 3$$

The solution of the system is again (3, 7).

Note that the substitution method is useful if we can easily express one variable in terms of the other variable.

EXERCISES 13.4
Solve each system using the substitution method. See Example 1.

Complete Example

$$2y + 3x = 7 \quad (1)$$
$$y = x + 1 \quad (2)$$

Substitute $x + 1$ for y in Equation (1).

$$2(\underline{\quad\quad}) + 3x = 7$$
$$\quad\quad (1)$$

Solve for x.

$$2x + 2 + 3x = 7$$
$$5x = 5$$
$$x = \underline{\quad\quad}$$
$$\quad\quad (2)$$

Substitute 1 for x in Equation (2).

$$y = (1) + 1 = \underline{\quad\quad}$$
$$\quad\quad\quad\quad (3)$$

Ans. $x = 1, \quad y = 2;$ or $\underline{\quad\quad}$
$$\quad\quad\quad\quad\quad\quad\quad\quad (4)$$

(1) $x + 1$ (2) 1   (3) 2   (4) (1, 2)

1. $y = 2x$
   $3x + y = 10$

2. $y = 3x$
   $x + 2y = 14$

3. $y = x - 2$
   $2x + y = 7$

4. $y = x - 4$
   $x + 4y = 9$

5. $2x + 3y = 4$
   $y = 2x + 4$

6. $x + 2y = -1$
   $y = x - 5$

7. $x - y = 6$
   $y = x - 6$

8. $x - 2y = 1$
   $y = 2x - 5$

9. $3x - 2y = 10$
   $y = -x$

10. $2x + y = -3$
    $y = -x$

11. $\dfrac{x}{2} - 2y = \dfrac{17}{2}$
    $y = x - 5$

12. $\dfrac{x}{3} + \dfrac{y}{2} = \dfrac{3}{2}$
    $y = 2x - 5$

13.5 APPLICATIONS USING TWO VARIABLES

Complete Example

$$x - 10 = 3y \quad (1)$$
$$2x + 3y = -7 \quad (2)$$
$$x = \underline{} \quad (1')$$
$$\quad\quad (1)$$

Since x has a coefficient 1 in Equation (1), solve for x in terms of y.

$$2(\underline{}) + 3y = -7$$
$$\quad (2)$$

Substitute $3y + 10$ for x in Equation (2).

$$6y + 20 + 3y = -7$$
$$9y = -27$$
$$y = \underline{}$$
$$\quad\quad (3)$$

Solve for y.

$$x = 3(-3) + 10 = \underline{}$$
$$\quad\quad\quad\quad\quad\quad (4)$$

Substitute -3 for y in Equation $(1')$.

Ans. $x = 1$, $y = -3$; or $(1, -3)$

---

(1) $3y + 10$ (2) $3y + 10$ (3) -3 (4) 1

---

13. $x = y + 2$
     $2x - y = 7$

14. $x = y - 4$
     $3x - 7y = -8$

15. $x + 1 = 3y$
     $2x - 3y = 4$

16. $x - 3 = 2y$
     $5x + 5y = 0$

17. $x + 5 = 2y$
     $2x = 1 - 7y$

18. $x - 3 = 5y$
     $3x = 2y - 4$

19. $\dfrac{x}{4} + y = \dfrac{5}{2}$
     $x = y$

20. $\dfrac{x}{2} + \dfrac{y}{3} = \dfrac{1}{2}$
     $x = -y$

13.5 APPLICATIONS USING TWO VARIABLES

If two variables are related by a single first-degree equation, there are infinitely many ordered pairs that are solutions of the equation. But if the two variables are related by two independent first-degree equations, there can be only one ordered pair that is a solution of both equations. Therefore, *to solve problems using two variables, we must represent two independent relationships using two equations.* We can often solve problems more easily by using a system of equations than by using a single equation involving one variable. We will follow the six steps outlined on page 264, with minor modifications as shown in the next example.

Example 1 The sum of two numbers is 26. The larger number is 2 more than three times the smaller number. Find the numbers.

Solution **Steps 1–2** We represent what we want to find as *two* word phrases. Then, we represent the word phrases in terms of *two* variables.

Smaller number: x
Larger number: y

SYSTEMS OF LINEAR EQUATIONS

Step 3 A sketch is not applicable.

Step 4 Now we must write *two* equations representing the conditions stated.

The sum of two numbers is 26.
$$x + y = 26$$

The larger number is two more than three times the smaller number.
$$y = 2 + 3 \cdot x$$

Step 5 To find the numbers, we solve the system

$$x + y = 26 \quad (1)$$
$$y = 2 + 3x \quad (2)$$

Since Equation (2) shows y explicitly in terms of x, we will solve the system by the substitution method. Substituting $2 + 3x$ for y in Equation (1), we get

$$x + (2 + 3x) = 26$$
$$4x = 24$$
$$x = 6$$

Substituting 6 for x in Equation (2), we get

$$y = 2 + 3(6) = 20$$

Step 6 The smaller number is 6 and the larger number is 20.

EXERCISES 13.5

In each of the following exercises, represent two independent conditions of the problem by a system of equations using two variables. Then, solve the system. Follow the six steps outlined on page 264.

Complete Example

A 12-foot board is cut into two parts so that one part is 2 feet longer than the other. How long is each part?

Step 1–2 Represent the two quantities you want to find as two word phrases. Represent these quantities in terms of *two* variables.

The longer part: x
The shorter part: y

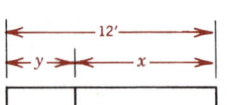

Step 3 A sketch may be helpful.

Step 4 Write two equations representing the conditions in the problem.

$$x + y = \underline{\qquad} \quad (1)$$
$$x = y + \underline{\qquad} \quad (2)$$

430

13.5 APPLICATIONS USING TWO VARIABLES

Step 5 Solve the system using the substitution method. Substitute $y + 2$ for x in Equation (1).

$$y + 2 + y = 12$$
$$2y = 10$$
$$y = \underline{}_{(3)}$$

Substitute 5 for y in Equation (2).

$$x = 5 + 2 = \underline{}_{(4)}$$

Step 6 The longer part is 7 feet; the shorter part is 5 feet.

---

(1) 12   (2) 2   (3) 5   (4) 7

---

1. The sum of two numbers is 25 and their difference is 9. What are the numbers?

2. The sum of two numbers is 21 and their difference is 13. What are the numbers?

3. A 20-meter board is cut into two pieces, one of which is 2 meters longer than the other. How long is each piece?

4. A 30-meter board is cut into two pieces, one of which is 6 meters shorter than the other. How long is each piece?

5. Two packages weighed together total 28 kilograms. One of the packages weighs 8 kilograms less than twice the other. How much does each weigh?

6. Two packages weighed together total 45 kilograms. One of the packages weighs 11 kilograms more than the other. How much does each weigh?

7. A car and trailer together sold for $12,000. The car was valued at $5000 more than the trailer. What was the value of each separately?

8. A guitar and amplifier together cost $356. The amplifier cost $20 more than two times the guitar. What was the price of each?

9. It took 24 working hours to paint the outside walls and trim of a house. If it took 6 more hours to paint the trim than to paint the walls, how long did it take to paint each?

10. An ice cream cone and a cookie together cost 72 cents. The ice cream cone costs a dime more than the cookie. What is the cost of each?

11. A certain fishing spot is located 45 kilometers from town. Part of the distance can be driven in a car, but part of it must be traveled on foot. If it is possible to drive 19 more kilometers than must be walked, how far must be walked?

12. Two trains left towns A and B, which are 240 kilometers apart, at the same time and proceeded toward each other on parallel tracks. At the time they met, the train from town A had traveled 10 kilometers farther than the train from town B. How many kilometers from town A were the trains when they met?

13. A freight train is made up of 92 cars, not counting the engine and its caboose. These cars are partly flat cars and partly box cars. There are 28 more flat cars than box cars. How many flat cars are there in the train?

14. A man bought a toy train and a doll as presents for his children. He paid a total of $55. If the train cost $1 less than three times the doll, what was the price of each?

15. The sum of two numbers is 24. One half of one number is 3 more than the other number. What are the numbers?

16. The difference of two numbers is 13. If the smaller number is 2 more than one fourth of the larger, what are the numbers?

SYSTEMS OF LINEAR EQUATIONS

17. There were 7672 votes cast in a recent election. The winning candidate received 12 votes more than her opponent. How many votes did each candidate receive?

18. At a recent election, the winning candidate received 122 votes more than his opponent. If there were a total of 10,764 votes cast, how many votes did each candidate receive?

19. A collection of 34 coins consists of dimes and quarters. How many coins of each kind are in the collection if the total value is $5.50?

20. A collection of 42 coins consists of nickels and dimes. How many coins of each kind are in the collection if the total value if $3.60?

21. A sum of $3600 is invested, part at 8% and the remainder at 12%. Find the amount of each investment if the interest on each investment is the same.

22. The total income from two investments is $750. One investment yields 9% and the second investment yields 10%. How much is invested at each rate if the total investment was $8000?

CHAPTER THIRTEEN REVIEW EXERCISES

Find the solution of each system by graphical methods. If the system is inconsistent or dependent, so state.

1. $x + y = 5$
   $2x - y = 4$

2. $x - y = 5$
   $2x + 3y = 5$

3. $3x + y = -4$
   $6x + 2y = -8$

4. $2x = 3y - 1$
   $4x = 24 + 6y$

Find the solution of each system by using the addition method.

5. $x - y = 3$
   $x + y = 5$

6. $2x + y = -6$
   $x - y = -3$

7. $x - 3y = 1$
   $x + 2y = 1$

8. $4 = 2x - y$
   $-y = x + 7$

9. $4y - 3x = -5$
   $x - 7 = -20y$

10. $8y = -3x - 1$
    $6 + 2x = -8y$

11. $\dfrac{y}{6} = \dfrac{-x}{2}$
    $7y + 2x = 19$

12. $\dfrac{2x}{3} + \dfrac{y}{2} = \dfrac{-19}{6}$
    $\dfrac{x+3}{4} = \dfrac{-y}{10}$

Find the solution of each system by using the addition method.

13. $y = 3x - 5$
    $y - x = -1$

14. $y = 2x + 1$
    $2x + 3y = -21$

15. $x - y = 6$
    $3x - 4y = 16$

16. $4x + y = 5$
    $8x - 2y = -2$

432

Solve.

17. Two packages together weigh 84 kilograms, and one of the packages weighs 20 kilograms more than the other. How much does each weigh?
18. A collection of 25 coins consists of dimes and quarters. How many coins of each kind are in the collection if the total value is $3.55?
19. The sum of two numbers is -40, and their difference is -8. What are the numbers?
20. One number is eight more than four times the other, and their sum is -2. What are the numbers?

REVIEW OF FACTORING

The following exercises review processes we studied earlier. The ability to factor will be helpful to you in Chapter 14. If you have difficulty in factoring these expressions, you should review the appropriate sections in Chapter 9.

Factor each polynomial completely.

1. $5x + 10y$
2. $3x^2 + 6x - 3$
3. $-2x^2 - 4$
4. $4x^2 - 8x$
5. $4y^2 - 8y - 8$
6. $3x^2 - 6x + x^3$
7. $x^2 + 5x + 6$
8. $y^2 - 6y + 9$
9. $y^2 - 7y - 8$
10. $y^2 + 2y - 35$
11. $2x^2 + 6x - 20$
12. $3y^2 + 9y - 12$
13. $2y^2 - y - 3$
14. $6y^2 + y - 1$
15. $6x^2 - 13x + 6$
16. $3x^2 - 8x - 35$
17. $6x^2 + 5x - 4$
18. $8y^2 - 2y - 1$
19. $4y^2 + 6y + 2$
20. $6x^2 + 21x + 9$
21. $18x^2 - 42x - 16$
22. $80x^2 - 10x - 25$
23. $4x^2 - 10x - 6$
24. $18y^2 - 3y - 6$
25. $x^2 + 2ax + a^2$
26. $x^2 - 2ax + a^2$
27. $y^2 + 6by + 9b^2$
28. $y^2 - 4by + 4b^2$
29. $x^2 + 8ax + 16a^2$
30. $y^2 - 10by + 25b^2$
31. $x^2 - 16$
32. $y^2 - 36$
33. $y^2 - 4$
34. $x^2 - 49$
35. $4x^2 - 25$
36. $9y^2 - 81$
37. $12y^2 - 48$
38. $50x^2 - 32$

14 QUADRATIC EQUATIONS

A **quadratic equation** is an equation of the form
$$ax^2 + bx + c = 0 \qquad (1)$$
where a, b, and c are constants and $a \neq 0$. To make quadratic equations easier to work with, we will regard Equation (1) as the **standard form** for such equations. Note that in standard form, the right-hand member is 0 and the terms in the left-hand member are in order of descending powers of the variable. If all three terms are present, as in
$$2x^2 + 3x - 1 = 0 \quad \text{and} \quad 3x^2 - 2x + 1 = 0$$
the equation is called a **complete quadratic equation.** If either $b = 0$ or $c = 0$, that is, if one of the last two terms is missing, as in
$$x^2 - 3 = 0, \quad x^2 + 4x = 0, \quad \text{and} \quad 4x^2 = 0$$
the equation is called an **incomplete quadratic equation.** In this chapter, we will study one method of solving quadratic equations.

14.1 SOLVING EQUATIONS IN FACTORED FORM

As we learned earlier, a solution of an equation is a number that, when substituted for the variable, results in a true statement. Now, suppose we have an equation such as
$$(x - 3)(x - 2) = 0$$
The equation says that the product of two numbers, $(x - 3)$ and $(x - 2)$, is 0. To find solutions of the equation, we use the following principle.

> **If the product of two factors is 0, at least one of the factors is 0.**

435

QUADRATIC EQUATIONS

In symbols,

If $ab = 0$, then $a = 0$ or $b = 0$ or both a and b equal 0

Example 1 Solve $(x - 3)(x - 2) = 0$.

Solution The equation
$$(x - 3)(x - 2) = 0.$$
will be true only if
$$x - 3 = 0 \quad \text{or} \quad x - 2 = 0$$
For what value of x will $x - 3 = 0$? For $x = 3$. For what value of x will $x - 2 = 0$? For $x = 2$. Thus, the solutions are $x = 2$ or $x = 3$.

To solve an equation in factored form where one member is 0:
1. Set each factor equal to 0.
2. Solve each of the resulting equations.

Example 2 Solve $(2x - 5)(3x + 1) = 0$.

Solution The equation
$$(2x - 5)(3x + 1) = 0.$$
will be true only if
$$2x - 5 = 0 \quad \text{or} \quad 3x + 1 = 0$$
$$2x = 5 \qquad\qquad 3x = -1$$
$$x = \frac{5}{2} \qquad\qquad x = \frac{-1}{3}$$

Thus, the solutions are $x = 5/2$ or $x = -1/3$.

Sometimes the values of the variable for which one or both factors equal zero can be determined by inspection. We should do this whenever possible.

EXERCISES 14.1
For what value of x will each of the following expressions equal 0?

Complete Example (see Example 1)

$x + 3$

$x + 3 = $ _____
        (1)

$x = $ _____
   (2)

Determine the value by inspection or set $x + 3$ equal to 0 and solve the resulting equation.

(1) 0   (2) -3

14.1 SOLVING EQUATIONS IN FACTORED FORM

1. $x - 7$
2. $x - 2$
3. $3x - 6$
4. $2x + 8$
5. $3x - 1$
6. $4x + 1$
7. $5x + 15$
8. $5x - 5$

For what values of x will each of the following products equal 0?

Complete Example (see Example 1)

$(x - 4)(x + 2)$

$x - 4 = 0$ or $x + 2 = \underline{\quad}$ (1)

$x = \underline{\quad}$ (2) $x = \underline{\quad}$ (3)

Determine values by inspection or set each factor equal to 0 and solve the resulting equations.

(1) 0   (2) 4   (3) −2

9. $(x - 3)(x - 5)$
10. $(x + 2)(x + 5)$
11. $x(x + 4)$
12. $x(x - 6)$
13. $(2x - 6)(3x + 9)$
14. $(5x + 5)(4x - 8)$
15. $(3x + 1)(2x - 3)$
16. $(4x + 3)(3x - 2)$
17. $2x(5x + 4)(2x - 7)$
18. $3x(x - 5)(6x + 9)$

Solve each of the following equations. See Example 2.

Complete Example

$(x - 3)(x + 4) = 0$

$x - 3 = 0$ or $x + 4 = \underline{\quad}$ (1)

$x = \underline{\quad}$ (2) $x = \underline{\quad}$ (3)

(1) 0   (2) 3   (3) −4

19. $(x - 2)(x - 3) = 0$
20. $(x + 2)(x - 4) = 0$
21. $y(y - 4) = 0$
22. $p(p - 7) = 0$
23. $(r + 3)(r) = 0$
24. $(x + 6)(2x) = 0$
25. $(x - 2)(x + 3) = 0$
26. $(x + 5)(x - 5) = 0$
27. $(x - 1)(x + 8) = 0$
28. $(x + 3)(x + 7) = 0$
29. $(t + 2)(t - 3) = 0$
30. $(x + 1)(x - 3) = 0$
31. $u(u - 4) = 0$
32. $x(x + 3) = 0$
33. $(b + 4)(b - 3) = 0$

QUADRATIC EQUATIONS

34. $(b - 7)(b - 1) = 0$ **35.** $(2x - 3)(4x + 3) = 0$ **36.** $(3u - 1)(u + 1) = 0$

37. $(3y - 2)(3y + 2) = 0$ **38.** $(x + 8)(3x - 8) = 0$ **39.** $2z(2z + 3) = 0$

40. $r(3r + 7) = 0$

Complete Example

$(x + 1)(x - 3)(2x - 1) = 0$

$x + 1 = 0$ or $x - 3 = 0$ or $2x - 1 = \underline{}$
                                                                                                       (1)

$x = \underline{}$; $x = \underline{}$; $x = \dfrac{1}{2}$
    (2)   **(3)**

---

(1) 0   **(2)** -1 **(2)** 3

---

41. $(x - 3)(x - 2)(x - 1) = 0$ **42.** $(x + 6)(x + 5)(x + 4) = 0$

43. $x(x + 2)(x - 1) = 0$ **44.** $x(x - 3)(x + 2) = 0$

45. $2x(x + 4)(x - 3) = 0$ **46.** $x(x + 1)(x - 1) = 0$

47. $x(2x + 1)(2x - 1) = 0$ **48.** $(2x - 3)(3x + 2)(2x + 2) = 0$

14.2 SOLVING QUADRATIC EQUATIONS BY FACTORING I

We saw in the preceding section that we can solve a quadratic equation if one member is in factored form and the other member is zero.

> **To solve a quadratic equation by factoring:**
> 1. Write the equation in standard form.
> 2. Factor the left-hand member.
> 3. Set each factor equal to zero.
> 4. Solve each of the resulting equations.

Example 1   Solve $x^2 - 2x = 24$.

Solution   We first write the equation in standard form as

$$x^2 - 2x - 24 = 0$$

Then, factoring the left-hand member, we obtain
$$(x - 6)(x + 4) = 0$$
Setting each factor equal to 0, we have
$$x - 6 = 0 \quad \text{or} \quad x + 4 = 0$$
from which
$$x = 6, \quad x = -4$$

If the left-hand member is not factorable, we must use other methods to solve the equation. These methods are discussed in Chapter 16.

We can check solutions for quadratic equations the same way we check solutions for linear equations—by direct substitution in the original equation. For the above example,
$$(6)^2 - 2(6) = 36 - 12 = 24$$
and
$$(-4)^2 - 2(-4) = 16 + 8 = 24$$

EXERCISES 14.2
Solve by factoring. See Example 1.

Complete Example

$$x^2 = 5x$$ Write in standard form.

$$x^2 - \underline{\qquad}_{(1)} = 0$$ Factor left-hand member.

$$x(\underline{\qquad}_{(2)}) = 0$$ Determine solutions by inspection on set each factor equal to 0 and solve the resulting equations.

$$x = 0 \quad \text{or} \quad x - 5 = 0$$

$$x = 0 \qquad x = \underline{\qquad}_{(3)}$$

(1) $5x$ (2) $x - 5$ (3) 5

1. $x^2 + 3x = 0$
2. $x^2 - 2x = 0$
3. $2y^2 - 5y = 0$
4. $3y^2 - 7y = 0$

5. $2y^2 = 9y$
6. $4y^2 = 3y$
7. $4x^2 = 16x$
8. $5x^2 = 10x$

QUADRATIC EQUATIONS

Complete Example

$$x^2 = 25$$
$$x^2 - 25 = \underline{}_{(1)}$$ Write in standard form.

$$(x - 5)(\underline{}_{(2)}) = 0$$ Factor left-hand member.

$$x - 5 = 0 \quad \text{or} \quad x + 5 = \underline{}_{(3)}$$ Determine solutions by inspection or set each factor equal to 0 and solve the resulting equations.

$$x = \underline{}_{(4)} \quad x = -5$$

(1) 0   (2) $x + 5$ (3) 0   (4) 5

9. $x^2 - 1 = 0$ 10. $x^2 - 36 = 0$ 11. $x^2 - 4 = 0$ 12. $x^2 - 9 = 0$

13. $x^2 - 16 = 0$ 14. $x^2 - 100 = 0$ 15. $x^2 = 64$ 16. $x^2 = 49$

17. $3x^2 = 27$ 18. $2x^2 = 32$ 19. $5x^2 = 45$ 20. $7x^2 = 63$

Complete Example

$$x^2 - 4x - 5 = 0$$
$$(x - 5)(\underline{}_{(1)}) = 0$$ Factor left-hand member.

$$x - 5 = 0 \quad \text{or} \quad x + 1 = \underline{}_{(2)}$$ Determine solutions by inspection or set each factor equal to 0 and solve the resulting equations.

$$x = \underline{}_{(3)} \quad x = -1$$

(1) $x + 1$ (2) 0   (3) 5

21. $x^2 - 3x + 2 = 0$ 22. $x^2 + 3x + 2 = 0$ 23. $y^2 + 4y + 4 = 0$ 24. $y^2 - 8y + 12 = 0$

25. $y^2 - 3y - 4 = 0$ 26. $y^2 - 3y - 10 = 0$ 27. $x^2 + 4x - 21 = 0$ 28. $x^2 + x - 42 = 0$

29. $x^2 + 5x - 14 = 0$ 30. $x^2 + 8x + 15 = 0$ 31. $x^2 + 12x + 36 = 0$ 32. $x^2 + 14x + 49 = 0$

14.2 SOLVING QUADRATIC EQUATIONS BY FACTORING I

Complete Example

$$2x^2 - 6x = 8$$

$$2x^2 - 6x - 8 = \underline{}_{(1)}$$

$$2(x^2 - 3x - \underline{}_{(2)}) = 0$$

$$2(x - 4)(\underline{}_{(3)}) = 0$$

$$x - 4 = 0 \quad \text{or} \quad x + 1 = \underline{}_{(4)}$$

$$x = \underline{}_{(5)} \quad x = -1$$

Write equation in standard form.

Completely factor left-hand member.

Determine solutions by inspection or set each factor containing a variable equal to 0 and solve the resulting equations. The constant 2 has no effect on the solution.

(1) 0   (2) 4   (3) $x + 1$ (4) 0   (5) 4

---

33. $2x^2 - 10x = 12$
34. $3x^2 - 6x = -3$
35. $4y^2 - 12y = 16$
36. $4y^2 - 24y = 28$

37. $2y^2 + 2y = 60$
38. $3y^2 + 6y = 45$
39. $3x^2 - x = 4$
40. $4x^2 + 4x = 3$

41. $6x^2 = 11x - 3$
42. $4x^2 = 4x + 3$
43. $4x^2 = 4x - 1$
44. $12x^2 = 8x + 15$

45. $12x - 9 = 4x^2$
46. $x + 15 = 2x^2$

Complete Example

$$x(x + 2) = 8$$

$$x^2 + \underline{}_{(1)} = 8$$

$$x^2 + 2x - 8 = \underline{}_{(2)}$$

$$(x + 4)(\underline{}_{(3)}) = 0$$

$$x + 4 = 0 \quad \text{or} \quad x - 2 = \underline{}_{(4)}$$

$$x = \underline{}_{(5)} \quad x = 2$$

Remove parentheses.

Write in standard form.

Factor left-hand member.

Determine solutions by inspection or set each factor equal to 0 and solve the resulting equations.

(1) $2x$ (2) 0   (3) $x - 2$ (4) 0   (5) -4

441

QUADRATIC EQUATIONS

47. $y(2y - 3) = -1$ **48.** $x(x + 2) = 3$ **49.** $2(x^2 - 1) = 3x$

50. $y(y - 2) = 6 - y$ **51.** $x(x + 2) - 3x - 2 = 0$ **52.** $2y(y - 2) = y + 3$

Complete Example

$(x - 4)(x + 3) = -10$ Multiply the factors in the left-hand member.

$x^2 - x - \underline{\quad(1)\quad} = -10$ Write in standard form.

$x^2 - x - 2 = 0$ Factor left-hand member.

$(x - 2)(\underline{\quad(2)\quad}) = 0$ Determine solutions by inspection or set each factor equal to 0 and solve the resulting equations.

$x - 2 = 0 \quad \text{or} \quad x + 1 = \underline{\quad(3)\quad}$

$x = \underline{\quad(4)\quad} \quad x = -1$

(1) 12   (2) $x + 1$ (3) 0   (4) 2

53. $(x - 2)(x + 1) = 4$ **54.** $(x - 5)(x + 1) = -8$

55. $(x - 2)(x - 1) = 1 - x$ **56.** $(x + 3)^2 = 2x + 14$

57. $(2x + 5)(x - 4) = -18$ **58.** $(2x - 1)(x - 2) = -1$

59. $(6x + 1)(x + 1) = 4$ **60.** $(x - 2)(x + 1) = x(2 - x)$

14.3 SOLVING QUADRATIC EQUATIONS BY FACTORING II

In Section 11.7, we "cleared" a first-degree equation of fractions by multiplying each member by the LCD of the fractions. We can use the same procedure for second-degree equations.

Example 1 Solve $\dfrac{x^2}{2} + \dfrac{5x}{4} = 3$.

14.3 SOLVING QUADRATIC EQUATIONS BY FACTORING II

Solution We first multiply each member by 4 and then rewrite the equation in standard form

$$(4)\frac{x^2}{2} + (4)\frac{5x}{4} = (4)3$$

$$2x^2 + 5x = 12$$

$$2x^2 + 5x - 12 = 0$$

Then, factoring the left-hand member, we have

$$(2x - 3)(x + 4) = 0$$

from which

$$2x - 3 = 0 \quad \text{or} \quad x + 4 = 0$$

$$2x = 3 \qquad\qquad x = -4$$

$$x = \frac{3}{2}$$

Thus, the solutions are $x = 3/2$ or $x = -4$.

Recall from Section 11.7 that if we multiply each member of an equation by an expression containing a variable, we must make sure that the expression does not equal 0.

Example 2 Solve $\dfrac{7}{x - 3} - \dfrac{3}{x - 4} = \dfrac{1}{2}$.

Solution We multiply each member by the LCD $2(x - 3)(x - 4)$. Note that x cannot be 3 or 4 since $(x - 3)(x - 4)$ equals 0 for these values of x. Thus,

$$2(x - 3)(x - 4)\frac{7}{(x - 3)} - 2(x - 3)(x - 4)\frac{3}{(x - 4)} = 2(x - 3)(x - 4)\frac{1}{2}$$

$$14(x - 4) - 6(x - 3) = (x - 3)(x - 4)$$

Now we remove parentheses to obtain

$$14x - 56 - 6x + 18 = x^2 - 7x + 12$$

$$8x - 38 = x^2 - 7x + 12$$

Writing the equation in standard form yields

$$x^2 - 15x + 50 = 0$$

and factoring the left-hand member, we get

$$(x - 10)(x - 5) = 0$$

Lastly, we can determine solutions by inspection or set each factor equal to 0 and solve

$$x - 10 = 0 \quad \text{or} \quad x - 5 = 0$$

$$x = 10 \qquad\qquad x = 5$$

Thus, the solutions are $x = 10$ or $x = 5$.

QUADRATIC EQUATIONS

EXERCISES 14.3
Solve

Complete Example (see Example 1)

$$\frac{15}{2}x^2 - \frac{10}{3} = 0$$

$$(\overset{3}{\cancel{6}})\frac{15}{\cancel{2}}x^2 - (\overset{2}{\cancel{6}})\frac{10}{\cancel{3}} = (6)0$$

Multiply each term by the LCD 6.

$$45x^2 - \underline{\qquad}_{(1)} = 0$$

Completely factor left-hand member.

$$5(9x^2 - \underline{\qquad}_{(2)}) = 0$$

$$5(3x - 2)(\underline{\qquad}_{(3)}) = 0$$

Determine solutions by inspection or set each factor containing a variable equal to 0 and solve the resulting equations. The constant 5 has no effect on the solution.

$$3x - 2 = 0 \quad \text{or} \quad 3x + 2 = 0$$

$$3x = 2 \qquad\qquad 3x = -2$$

$$x = \frac{2}{3} \qquad\qquad x = \underline{\qquad}_{(4)}$$

---

(1) 20 (2) 4 (3) $3x + 2$ (4) $\frac{-2}{3}$

---

1. $x^2 - \frac{1}{9} = 0$
2. $\frac{1}{3}x^2 - \frac{4}{3} = 0$
3. $\frac{2}{3}x^2 - \frac{3}{2} = 0$
4. $\frac{5}{2}y^2 - 10 = 0$

5. $\frac{x^2}{2} = 8$
6. $3x^2 = \frac{75}{4}$
7. $\frac{x^2}{2} + x = 0$
8. $\frac{x^2}{3} - 2x = 0$

9. $\frac{x^2}{4} + \frac{x}{2} = 0$
10. $\frac{x^2}{18} + \frac{x}{3} = 0$
11. $\frac{x^2}{5} = x$
12. $\frac{x^2}{6} = \frac{x}{2}$

Complete Example (see Example 1)

$$y^2 = \frac{13}{6}y - 1$$

$$(6)y^2 = (\cancel{6})\frac{13}{\cancel{6}}y - (6)1$$

Multiply each member by LCD 6.

$$6y^2 = \underline{\qquad}_{(1)} - 6$$

Write in standard form; factor left-hand member.

$$6y^2 - 13y + 6 = 0$$

$$(2y - 3)(\underline{\qquad}_{(2)}) = 0$$

444

14.3 SOLVING QUADRATIC EQUATIONS BY FACTORING II

$2y - 3 = 0$ or $3y - 2 = $ ___(3)___

$2y = 3 \qquad\qquad 3y = 2$

$y = $ ___ $\qquad\qquad y = \dfrac{2}{3}$

(4)

Determine solutions by inspection or set each factor equal to 0 and solve the resulting equations.

---

(1) $13y$ (2) $3y - 2$ (3) 0 (4) $\dfrac{3}{2}$

---

13. $\dfrac{2}{3}x^2 + \dfrac{1}{3}x - 2 = 0$

14. $\dfrac{3}{4}x^2 + \dfrac{5}{2}x - 2 = 0$

15. $x^2 + 3x + \dfrac{9}{4} = 0$

16. $\dfrac{3}{2}x^2 - \dfrac{1}{4}x - \dfrac{1}{2} = 0$

17. $4x^2 + 13x + \dfrac{15}{2} = 0$

18. $\dfrac{1}{3}x^2 - \dfrac{5}{2}x + 3 = 0$

19. $\dfrac{x^2}{2} + x = \dfrac{15}{2}$

20. $x - 1 = \dfrac{x^2}{4}$

21. $\dfrac{x^2}{3} + x = \dfrac{-2}{3}$

22. $\dfrac{21}{2} + 2y = \dfrac{y^2}{2}$

23. $\dfrac{x^2}{6} + \dfrac{x}{3} = \dfrac{1}{2}$

24. $\dfrac{x^2}{15} = \dfrac{x}{5} + \dfrac{2}{3}$

Complete Example (see Example 2)

$\dfrac{1}{8x^2} - \dfrac{13}{24x} = -\dfrac{1}{2}$

$(24x^2)^{\;3}\dfrac{1}{8x^2} - (24x^2)^{\;x}\dfrac{13}{24x} = -(24x^2)^{\;12}\dfrac{1}{2}$

$3 - $ ___(1)___ $= -12x^2$

___(2)___ $- 13x + 3 = 0$

$(4x - 3)($ ___(3)___ $) = 0$

$4x - 3 = 0$ or $3x - 1 = 0$

$4x = 3 \qquad\qquad 3x = 1$

$x = \dfrac{3}{4} \qquad\qquad x = $ ___(4)___

Multiply each term by LCD $24x^2$. Note, x cannot be 0.

Write in standard form.

Factor left-hand member.

Determine solutions by inspection or set each factor equal to 0 and solve the resulting equations.

---

(1) $13x$ (2) $12x^2$ (3) $3x - 1$ (4) $\dfrac{1}{3}$

445

QUADRATIC EQUATIONS

25. $x + \dfrac{1}{x} = 2$

26. $\dfrac{x}{4} - \dfrac{3}{4} = \dfrac{1}{x}$

27. $1 - \dfrac{2}{x} = \dfrac{15}{x^2}$

28. $\dfrac{1}{2} + \dfrac{1}{2x} = \dfrac{1}{x^2}$

29. $1 + \dfrac{1}{x(x-1)} = \dfrac{3}{x}$

30. $\dfrac{4}{x} - 3 = \dfrac{5}{2x+3}$

31. $\dfrac{14}{x-6} - \dfrac{6}{x-8} = \dfrac{1}{2}$

32. $\dfrac{12}{x-3} + \dfrac{12}{x+4} = 1$

33. $\dfrac{2}{x-3} - \dfrac{6}{x-8} = -1$

34. $\dfrac{4}{x-2} - \dfrac{7}{x-3} = \dfrac{2}{15}$

35. $\dfrac{4}{x-1} - \dfrac{4}{x+2} = \dfrac{3}{7}$

36. $\dfrac{3}{x+6} - \dfrac{2}{x-5} = \dfrac{5}{4}$

14.4 APPLICATIONS

A variety of word problems lead to quadratic equations. Again, we will follow the six steps outlined on page 264 when solving word problems. However, because most quadratic equations have two solutions, it is important to check both results in the original problem to make sure they fulfill the physical conditions stated in the problem.

Example 1 An object is thrown off a building 48 feet high. The object's height h above the ground, at a particular time t, is given by

$$h = 48 + 32t - 16t^2 \qquad (1)$$

Determine the number of seconds t it would take the object to strike the ground.

Solution **Steps 1–2** Time to strike ground: t

Step 3 A sketch is helpful to visualize the problem.

Step 4 When the object hits the ground, $h = 0$. Thus, allowing $h = 0$ in Equation (1) yields

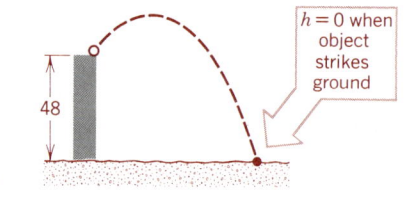

$$0 = 48 + 32t - 16t^2$$

Step 5 Solving for t, we get

$$16t^2 - 32t - 48 = 0$$

Dividing each member by 16 produces

$$t^2 - 2t - 3 = 0$$

Factoring, we have

$$(t - 3)(t + 1) = 0$$

Then

$$t - 3 = 0 \quad \text{or} \quad t + 1 = 0$$
$$t = 3 \qquad\qquad t = -1$$

Step 6 In this case, -1 does not meet the physical requirements of the problem, since time t must be positive. Thus, the object would strike the ground 3 seconds after it was thrown.

EXERCISES 14.4

Solve each word problem. See Example 1.

Complete Example

The square of an integer 7 less than eight times the integer. Find the integer.

Steps 1–2 The integer: x
Step 3 A sketch is not applicable.
Step 4 $x^2 = 8x - 7$
Step 5 $x^2 - 8x + \underline{\quad\quad}_{(1)} = 0$

$(x - 7)(\underline{\quad\quad}_{(2)}) = 0$

$x - 7 = 0 \quad \text{or} \quad x - 1 = 0$

$x = \underline{\quad\quad}_{(3)} \qquad x = 1$

Step 6 Since 1 and 7 *both* meet the conditions of the problem, *both* are valid solutions. The integer is either 1 or $\underline{\quad\quad}_{(4)}$.

(1) 7   (2) $x - 1$ (3) 7   (4) 7

---

1. The square of an integer is equal to five times the integer. Find the integer.

2. If three times the square of a certain integer is increased by the integer itself, the sum is 10. What is the integer?

3. Find two consecutive positive integers whose product is 72.

4. Find two consecutive positive integers whose product is 132.

5. The square of a positive integer increased by twice the square of the next consecutive integer gives 66. Find the integer.

6. The square of a positive integer is 79 less than twice the square of the next consecutive integer. Find the integers.

7. The distance h above the ground of a certain projectile launched upward from the top of a 160-foot building is given by the equation $h = 160 + 48t - 16t^2$, where t is in seconds. Find the time at which the projectile will strike the ground.

8. In Exercise 7, find the time at which the projectile is again at 160 feet above the ground.

9. The cost C of producing a certain radio set is related to the number of hours t it takes to manufacture the set by $C = 8t^2 - 32t - 16$. How many hours would be devoted to producing a set at a cost of $80?

10. In Exercise 9, how many hours would be devoted to producing a set at a cost of $24?

QUADRATIC EQUATIONS

Complete Example

The sum of a certain integer and twice its reciprocal is 19/3. Find the integer.

Steps 1–2 The integer: x $\left(\text{the reciprocal of } x \text{ is } \dfrac{1}{x}\right)$

Step 3 A sketch is not applicable.

Step 4 $$x + 2\left(\dfrac{1}{x}\right) = \dfrac{19}{3}$$

Step 5 $$(3x)x + (3\cancel{x})2\left(\dfrac{1}{\cancel{x}}\right) = (\cancel{3}x)\dfrac{19}{\cancel{3}}$$

$$3x^2 + \underline{\quad\quad}_{(1)} = 19x$$

$$3x^2 - 19x + 6 = 0$$

$$(3x - 1)(\underline{\quad\quad}_{(2)}) = 0$$

$3x - 1 = 0$ or $x - 6 = 0$

$3x = 1$ $\quad\quad\quad x = \underline{\quad\quad}_{(3)}$

$x = \dfrac{1}{3}$

Step 6 Since 1/3 is not an integer, it does not meet the conditions of the original problem. The integer is $\underline{\quad\quad}_{(4)}$.

---

(1) 6   (2) $x - 6$ (3) 6   (4) 6

---

11. The sum of a certain number and its reciprocal is $\tfrac{17}{4}$. What is the number?

12. The sum of a certain number and twice its reciprocal is $\tfrac{9}{2}$. What is the number?

13. The sum of the reciprocals of two consecutive odd integers is $\tfrac{8}{15}$. What are the integers?

14. The sum of the reciprocals of two consecutive even integers is $\tfrac{5}{12}$. Find the integers.

15. The sum of the reciprocal of a positive integer and twice the reciprocal of the next consecutive integer is $\tfrac{5}{6}$. What are the integers?

16. Twice the reciprocal of a positive integer is subtracted from three times the reciprocal of the next successive integer and the difference is $\tfrac{2}{21}$. What are the integers?

Complete Example

A boat travels 18 miles downstream and back in $4\tfrac{1}{2}$ hours. If the speed of the current is 3 miles per hours, what is the speed of the boat in still water?

Steps 1–2 Rate of boat in still water: x

14.4 APPLICATIONS

Step 3   Make a table showing distance, rate, and time.

	d	r	t = d/r
Downstream	18	x + 3	$\dfrac{18}{x+3}$
Upstream	18	x − 3	$\dfrac{18}{x-3}$

Step 4   Write an equation showing that the sum of the times is $4\frac{1}{2}$ or $\frac{9}{2}$ hours.

$$\frac{18}{x+3} + \frac{18}{x-3} = \underline{}_{(1)}$$

Step 5 $2(\cancel{x+3})(x-3)\dfrac{18}{\cancel{x+3}} + 2(x+3)(\cancel{x-3})\dfrac{18}{\cancel{x-3}} = \cancel{2}(x+3)(x-3)\dfrac{9}{\cancel{2}}$

$$36(x-3) + 36(x+3) = 9(x+3)(x-3)$$

$$36x - 108 + 36x + 108 = \underline{}_{(2)}$$

$$-9x^2 + 72x + \underline{}_{(3)} = 0$$

$$-9(x^2 - 8x - 9) = 0$$

$$9(x-9)(\underline{}_{(4)}) = 0$$

$$x - 9 = 0 \quad \text{or} \quad x + 1 = 0$$

$$x = 9 \qquad\qquad x = \underline{}_{(5)}$$

Step 6   Since −1 does not meet the conditions of the problem, only the positive solution is used. The boat travels 9 miles per hour in still water.

---

(1) $\dfrac{9}{2}$ (2) $9x^2 - 81$ (3) 81   (4) $x + 1$ (5) −1

---

17. A motor boat travels 24 miles downstream on a river and returns to its starting place. If the speed of the current is 2 miles per hour, and the round trip takes 5 hours, what is the speed of the boat in still water?

18. A crew rows a boat 6 miles downstream and then rows back to its starting place. If the speed of the current is 2 miles per hour, and the total trip takes 4 hours, how fast would the crew row in still water?

19. A man drove 180 miles from town A to town B and returned to town A. If he drove 15 miles per hour faster on the return trip than he did on the initial trip, and the initial trip took one hour longer, what was the man's speed on each trip?

20. A plane flew 480 miles at a certain speed, then increased its speed by 20 miles per hour and continued on the same course. After having flown a distance of 840 miles in a total of 5 hours, the plane landed. What was its original speed?

QUADRATIC EQUATIONS

CHAPTER FOURTEEN REVIEW EXERCISES

In Exercises 1–12, solve for x or y.

1. a. $(x - 2)(x + 5) = 0$ b. $y(y + 3) = 0$
2. a. $x^2 - 2x = 0$ b. $3x^2 = 6x$
3. a. $x^2 - 49 = 0$ b. $3y^2 = 27$
4. a. $y^2 - 4y - 5 = 0$ b. $y^2 - 7y - 18 = 0$
5. a. $x^2 - 2x = 3$ b. $x^2 = 4x - 4$
6. a. $2x^2 + 5x = 12$ b. $2x^2 = 3 - 5x$
7. a. $3x^2 + 5x = 2$ b. $3x^2 = 14x - 8$
8. a. $y(y - 6) = 16$ b. $x(x + 2) = 8$
9. a. $(x + 5)(x - 8) = -36$ b. $(x + 1)(x - 2) = 4$
10. a. $\dfrac{x^2}{4} - 9 = 0$ b. $3x^2 = \dfrac{27}{25}$
11. a. $\dfrac{y^2}{6} + \dfrac{y}{6} = 2$ b. $\dfrac{y}{2} - 1 = \dfrac{y^2}{16}$
12. a. $\dfrac{15}{y^2} + \dfrac{2}{y} = 1$ b. $\dfrac{2}{x} - \dfrac{1}{6} = \dfrac{2}{x + 2}$

13. The difference of two positive numbers is 7 and their product is 60. Find the numbers.
14. The sum of two positive numbers is 11 and their product is 30. Find the numbers.
15. The sum of the reciprocals of two consecutive integers is $\frac{11}{30}$. Find the integers.

15 RADICAL EXPRESSIONS

15.1 RADICALS

In this chapter, we will study a new kind of number—numbers that will enable us to solve additional equations.

SQUARE ROOTS

A **square root** of a positive number a is a number whose square is a. For example,

a square root of 9 is 3 because $3^2 = 9$

a square root of 4 is 2 because $2^2 = 4$

and

a square root of 16 is -4 because $(-4)^2 = 16$

Notice that

$$(-2)^2 = 4 \quad \text{and} \quad (2)^2 = 4$$

Thus, -2 and 2 are both square roots of 4. In fact, since $(a)(a) = a^2$ and $(-a)(-a) = a^2$, every positive number has two square roots, one positive and one negative.

Example 1 Find the two square roots of

a. 16
b. $\dfrac{1}{4}$

Solutions a. 4 and -4 since

$(4)(4) = 16$ and

$(-4)(-4) = 16.$

b. $\dfrac{1}{2}$ and $\dfrac{-1}{2}$ since

$\left(\dfrac{1}{2}\right)\left(\dfrac{1}{2}\right) = \dfrac{1}{4}$ and

$\left(\dfrac{-1}{2}\right)\left(\dfrac{-1}{2}\right) = \dfrac{1}{4}.$

451

RADICAL EXPRESSIONS

RADICAL SIGN

We use a special symbol, $\sqrt{}$, called a **radical sign**, to denote the positive or **principal square root** of a positive number. That is, for all positive numbers a, $\sqrt{a}$ (read "the square root of a") is the positive number whose square is a. The number a is called the **radicand**. In symbols,

$$\sqrt{a} \cdot \sqrt{a} = (\sqrt{a})^2 = a$$

where $\sqrt{a}$ and a are positive.

Example 2

$\sqrt{4} = 2$ because $2^2 = 4$
$\sqrt{9} = 3$ because $3^2 = 9$

and

$\sqrt{16} = 4$ because $4^2 = 16$

In the special case of zero,

$\sqrt{0} = 0$ because $0^2 = 0$

NEGATIVE SQUARE ROOTS

To represent the negative square root of a, we use the symbol $-\sqrt{a}$.

Example 3 a. $-\sqrt{4} = -2$ b. $-\sqrt{9} = -3$ c. $-\sqrt{16} = -4$

Using radical notation, we represent the two square roots of a positive number a by $\sqrt{a}$ and $-\sqrt{a}$. The symbol $\pm\sqrt{a}$ is sometimes used to denote both the positive and negative square roots of a.

Example 4 a. $\pm\sqrt{16} = \pm 4$ b. $\pm\sqrt{81} = \pm 9$ c. $\pm\sqrt{36} = \pm 6$

THE RADICAND

Since the square of any number is positive or zero, a symbol such as $\sqrt{-4}$ is not meaningful, because there is no number among the numbers we are studying whose square is -4. In general, $\sqrt{a}$ represents one of the numbers we are studying only if $a \geq 0$. In this book, *we will assume that all variables and expressions in radicands that involve the square root symbol represent nonnegative numbers.*

Example 5 For $\sqrt{x}$, we assume that $x \geq 0$;

for $\sqrt{x-2}$, we assume that $x - 2 \geq 0$, or $x \geq 2$;

for $\sqrt{x+2}$, we assume that $x + 2 \geq 0$, or $x \geq -2$.

With the above agreement that radicands are positive or zero, we can rewrite some radical expressions that contain variables as equivalent expressions without radical notation.

Example 6

$\sqrt{4x^2} = 2x$ because $(2x)^2 = 4x^2$

and

$\sqrt{16x^6} = 4x^3$ because $(4x^3)^2 = 16x^6$

EXERCISES 15.1

For Exercises 1–6, find the two square roots of the given number. See Example 1.

Complete Examples

a. 25

b. $\dfrac{4}{9}$

Ans. 5, ____(1)____

Ans. $\dfrac{2}{3}$, ____(2)____

(1) -5 (2) $\dfrac{-2}{3}$

1. 81
2. 49
3. 121
4. 100
5. $\dfrac{16}{25}$
6. $\dfrac{4}{25}$

Find each square root. See Examples 2–4.

Complete Examples

a. $\sqrt{49} = $ ____(1)____

b. $-\sqrt{\dfrac{4}{81}} = $ ____(2)____

c. $\pm\sqrt{\dfrac{4}{25}} = $ ____(3)____

d. $\sqrt{0} = $ ____(4)____

(1) 7   (2) $\dfrac{-2}{9}$ (3) $\pm\dfrac{2}{5}$ (4) 0

7. $\sqrt{16}$
8. $\sqrt{36}$
9. $-\sqrt{81}$
10. $-\sqrt{121}$
11. $\pm\sqrt{144}$
12. $\pm\sqrt{225}$
13. $\sqrt{9}$
14. $\sqrt{1}$
15. $\sqrt{\dfrac{1}{36}}$
16. $-\sqrt{\dfrac{1}{4}}$
17. $\pm\sqrt{\dfrac{4}{9}}$
18. $\pm\sqrt{\dfrac{9}{25}}$

RADICAL EXPRESSIONS

State the restriction on the variable in each expression. See Example 5.

Complete Example

$\sqrt{x+1}$

$x + 1 \geq$ ___(1)___ or $x \geq$ ___(2)___ $x + 1$ must be nonnegative.

Ans. $x \geq -1$.

---

(1) 0   (2) −1

---

19. $\sqrt{x-3}$ 20. $\sqrt{x-5}$ 21. $\sqrt{x+7}$

22. $\sqrt{x+4}$ 23. $\dfrac{1}{\sqrt{x-6}}$ 24. $\dfrac{1}{\sqrt{x-4}}$

Find each square root. See Example 6.

Complete Examples

a. $\sqrt{y^6} = $ ___(1)___   b. $-\sqrt{49x^2y^6} = $ ___(2)___   c. $\pm\sqrt{(c+d)^2} = $ ___(3)___

---

(1) y^3 (2) $-7xy^3$ (3) $\pm(c+d)$

---

25. $\sqrt{x^2}$ 26. $\sqrt{y^4}$ 27. $\sqrt{4x^2}$ 28. $\sqrt{a^2b^2}$

29. $-\sqrt{a^2c^4}$ 30. $-\sqrt{9x^6y^6}$ 31. $\pm\sqrt{36a^6}$ 32. $\pm\sqrt{100x^{10}}$

33. $\sqrt{121a^2b^2}$ 34. $\sqrt{(x+y)^2}$ 35. $-\sqrt{(a+b)^2}$ 36. $-\sqrt{4(x+y)^2}$

37. $\sqrt{\dfrac{a^2}{b^2}}$ 38. $-\sqrt{\dfrac{b^4}{100}}$ 39. $\pm\sqrt{\dfrac{9}{x^2y^2}}$ 40. $\pm\sqrt{\dfrac{4x^2}{y^2}}$

15.2 IRRATIONAL NUMBERS

If the numerator of an arithmetic fraction is an integer and the denominator is a nonzero integer, the fraction represents a **rational number**. For example,

$$\frac{2}{3}, \quad \frac{-4}{7}, \quad \text{and} \quad \frac{15}{8}$$

are rational numbers. Note that all integers are rational numbers, since any integer can be expressed as the quotient of itself and 1. That is,

$$-2 = \frac{-2}{1}, \quad 6 = \frac{6}{1}, \quad \text{and} \quad 0 = \frac{0}{1}$$

IRRATIONAL NUMBERS

As we saw in Section 10.1, we can associate rational numbers with points on the number line. But there are other numbers that can be associated with points on a number line that cannot be expressed as the quotient of two integers. Such numbers are called **irrational numbers**. A detailed study of irrational numbers is beyond the scope of this book. We will simply observe that any radical whose radicand is not the square of a rational number represents an irrational number.

Example 1
$$\sqrt{2}, \quad \sqrt{3}, \quad \sqrt{5}, \quad \text{and} \quad \sqrt{\frac{5}{7}}$$

are irrational numbers, but

$$\sqrt{4}, \quad \sqrt{\frac{9}{25}}, \quad \text{and} \quad \sqrt{16}$$

are rational numbers, since they can be written as 2, 3/5, and 4, respectively—that is, the quotient of two integers

As noted in Section 15.1, every positive number has two square roots. These square roots may be irrational.

Example 2 The two square roots of 7 are $\sqrt{7}$ and $-\sqrt{7}$;

the two square roots of 11 are $\sqrt{11}$ and $-\sqrt{11}$;

the two square roots of 15 are $\sqrt{15}$ and $-\sqrt{15}$.

SQUARE ROOT TABLE

We cannot represent irrational numbers exactly by common fractions or decimal fractions. However, we can approximate irrational numbers to any desired degree of accuracy. To do this, we can use either a table of square roots (see inside back cover) or a hand-held calculator that has square root capability.

Note in the table of square roots or by using a calculator, the square roots of 1, 4, 9, 16, 25, 36, 49, 64, 81, and 100 are the rational numbers are 1, 2, 3, 4, 5, 6, 7, 8, 9, and 10, respectively. The square roots of all other integers between 1 and 100 are irrational numbers, and the entries shown for these numbers in the table are only approximations to their true value. For example, $\sqrt{2}$ is approximately equal to 1.414, and $\sqrt{3}$ is approximately equal to 1.732. The symbol $\approx$ is often used for the phrase "is approximately equal to." Thus,

$$\sqrt{2} \approx 1.414 \quad \text{and} \quad \sqrt{3} \approx 1.732$$

In studying operations with radicals in this and the following sections, we assume that all laws valid for operations with rational numbers also hold for irrational numbers and that the symbols for the fundamental operations are unchanged. For example, the expression $2\sqrt{3}$ represents the product of 2 and $\sqrt{3}$. The expression $4 + \sqrt{7}$ represents the sum of 4 and $\sqrt{7}$. Both $2\sqrt{3}$ and $4 + \sqrt{7}$ are irrational.

Example 3 Find decimal approximations for

a. $2\sqrt{3}$ b. $4 + \sqrt{7}$

Round off answers to two decimal places.

RADICAL EXPRESSIONS

Solution Using the table of square roots or a calculator, we find that $\sqrt{3} \approx 1.732$ and $\sqrt{7} \approx 2.646$.

a. $2\sqrt{3} \approx 2(1.732)$
    ≈ 3.464
    ≈ 3.46

b. $4 + \sqrt{7} \approx 4 + (2.646)$
    ≈ 6.646
    ≈ 6.65

REAL NUMBERS

Both rational and irrational numbers are called **real numbers**. The real numbers completely fill the number line. In Section 8.5, we used a number line to graph inequalities in which the variable represented an integer. Now we can graph inequalities in which the variable represents a real number.

Example 4 Graph $x > 3$ and $x \geq 3$ where x represents a real number.

Solution If x represents a real number, the graph of $x > 3$ is shown in figure *a* and the graph of $x \geq 3$ is shown in figure *b*. The colored line on the number line represents *an infinite set of points*.

Notice that we use an open dot when we do not want to include the number in the graph (figure *a*) and a closed dot when we do (figure *b*).

EXERCISES 15.2

For Exercises 1–6, write the two square roots of the given number. Use radical notation for exact values. See Example 1.

Complete Examples

a. 14

Ans. $\sqrt{14}$, _____(1)

b. 18

Ans. $\sqrt{18}$, _____(2)

(1) $-\sqrt{14}$ (2) $-\sqrt{18}$

1. 10
2. 8
3. 17
4. 19
5. 22
6. 27

15.2 IRRATIONAL NUMBERS

Which of the following numbers are rational and which are irrational? If the number is rational, express it as an integer or as a quotient of two integers. See Example 2.

Complete Examples

a. $\sqrt{\frac{4}{9}}$ is _____(1)_____

since $\sqrt{\frac{4}{9}} = $ _____(2)_____

b. $\sqrt{3}$ is _____(3)_____

---

(1) rational (2) $\frac{2}{3}$ (3) irrational

---

7. 6
8. 8
9. $\sqrt{2}$
10. $\sqrt{4}$
11. $\sqrt{6}$
12. $\sqrt{9}$
13. $\sqrt{25}$
14. $\sqrt{100}$
15. $3\sqrt{16}$
16. $\sqrt{7}$
17. $-\sqrt{13}$
18. $-2\sqrt{100}$
19. $\sqrt{\frac{4}{9}}$
20. $-\sqrt{\frac{16}{25}}$
21. $-\sqrt{\frac{2}{3}}$
22. $\sqrt{\frac{4}{5}}$
23. $1 + \sqrt{4}$
24. $3 + \sqrt{4}$
25. $2 + \sqrt{5}$
26. $1 + \sqrt{3}$

Using the table of square roots or a calculator, find a decimal approximation for each of the following. Round off answers to two decimal places. If the digit in the third decimal place is 5, round to the next higher digit in the second decimal place. See Example 3.

Complete Examples

a. $\sqrt{23} \approx 4.796$

$\approx$ _____(1)_____

b. $-2\sqrt{46} \approx -2 ($ _____(2)_____ $)$

$= -13.564$

$\approx$ _____(3)_____

---

(1) 4.80 (2) 6.782 (3) −13.56

---

27. $\sqrt{57}$
28. $\sqrt{83}$
29. $\sqrt{3}$
30. $\sqrt{17}$
31. $\sqrt{5}$
32. $\sqrt{92}$
33. $-\sqrt{26}$
34. $-\sqrt{54}$
35. $2\sqrt{3}$
36. $3\sqrt{2}$
37. $-5\sqrt{3}$
38. $-6\sqrt{5}$
39. $\frac{1}{3}\sqrt{18}$
40. $\frac{1}{4}\sqrt{48}$
41. $-\frac{1}{5}\sqrt{75}$
42. $-\frac{2}{3}\sqrt{21}$

457

RADICAL EXPRESSIONS

Complete Examples

a. $3 + 2\sqrt{2} \approx 3 + 2(\underline{})$
   (1)
   $\approx 3 + 2.828$
   $= 5.828$
   $\approx \underline{}$
   (2)

b. $\dfrac{3 + \sqrt{3}}{2} \approx \dfrac{3 + 1.732}{\underline{}}$
   (3)
   $= \dfrac{4.732}{2}$
   $= 2.366$
   $\approx \underline{}$
   (4)

(1) 1.414   (2) 5.83   (3) 2   (4) 2.37

43. $1 + \sqrt{3}$
44. $2 - \sqrt{5}$
45. $3 - \sqrt{2}$
46. $5 + \sqrt{5}$
47. $5 + 3\sqrt{7}$
48. $-3 + 2\sqrt{6}$
49. $-7 - 3\sqrt{28}$
50. $-6 + 2\sqrt{35}$
51. $\dfrac{3 + 2\sqrt{2}}{2}$
52. $\dfrac{7 - 5\sqrt{5}}{3}$
53. $\dfrac{6 - 2\sqrt{3}}{5}$
54. $\dfrac{7 + 3\sqrt{3}}{2}$
55. $\sqrt{3} - \sqrt{2}$
56. $\sqrt{5} - \sqrt{7}$
57. $3\sqrt{3} - 2\sqrt{5}$
58. $2\sqrt{2} - 5\sqrt{5}$

Graph each set of numbers on a separate number line. Estimate the location of graphs of numbers between integers.

Complete Example

$\sqrt{13}, 4, \sqrt{7}$ From the table of square roots or by using a calculator, $\sqrt{13} \approx 3.606$ and $\sqrt{7} \approx 2.646$.

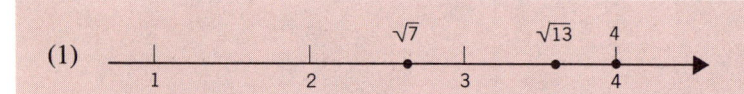

(1) [number line from 1 to 4]

(1) [number line showing √7 between 2 and 3, √13 and 4 between 3 and 4]

59. $7, 8, \sqrt{55}$
60. $\sqrt{21}, -\sqrt{25}, 6$
61. $\sqrt{3}, \sqrt{5}, -\sqrt{7}$
62. $\sqrt{9}, -\sqrt{16}, \sqrt{25}$
63. $\sqrt{1}, -\sqrt{2}, \sqrt{3}$
64. $-\sqrt{6}, \sqrt{8}, \sqrt{10}$
65. $-\sqrt{4}, \sqrt{3}, -\sqrt{2}$
66. $\sqrt{2}, 0, -\sqrt{2}$
67. $-\sqrt{40}, \sqrt{30}, -\sqrt{20}$
68. $-\sqrt{1}, 0, \sqrt{1}$
69. $\sqrt{35}, -\sqrt{16}, -\sqrt{37}$
70. $\sqrt{21}, \sqrt{27}, \sqrt{30}$

Graph the inequalities. All variables represent real numbers. See Example 4.

Complete Examples

a. $x > -3$

b. $x \leq 2$

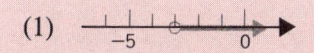

(1) [number line showing x > -3, open circle at -3, arrow right]

(2) [number line showing x ≤ 2, closed circle at 2, arrow left]

(1) [number line showing x > -3, open circle at -3, arrow right]

(2) [number line showing x ≤ 2, closed circle at 2, arrow left]

| 71. $x > 5$ | 72. $x > -2$ | 73. $x \leq 4$ | 74. $x \leq -2$ |
| 75. $x > -2$ | 76. $x > -4$ | 77. $x \leq 1$ | 78. $x \leq 3$ |

15.3 SIMPLIFYING RADICAL EXPRESSIONS I

We consider a radical expression to be in *simplest form* if no prime factor of the radicand occurs more than once. We can determine whether a radical is in simplest form by examining the prime factors of the radicand. For example, $\sqrt{78}$ is in simplest form because none of the factors is repeated when we completely factor the radicand, $\sqrt{13 \cdot 3 \cdot 2}$. On the other hand, $\sqrt{20}$ is not in simplest form because the factor 2 occurs more than once when we completely factor the radicand, $\sqrt{2 \cdot 2 \cdot 5}$. We shall now consider a way to simplify a radical that has a repeated factor in the radicand.

First note that

$$\sqrt{4 \cdot 9} = \sqrt{36} = 6$$

and

$$\sqrt{4}\,\sqrt{9} = 2 \cdot 3 = 6$$

Thus,

$$\sqrt{4 \cdot 9} = \sqrt{4} \cdot \sqrt{9}$$

In general,

The square root of a product is equal to the product of the square roots of its factors.

In symbols,

$$\sqrt{ab} = \sqrt{a}\,\sqrt{b} \quad (a, b \geq 0)$$

Simplifying radicals is easiest if we first express the radicand in completely factored form.

459

RADICAL EXPRESSIONS

Example 1 Simplify $\sqrt{216}$.

Solution We first express 216 in completely factored form
$$\sqrt{216} = \sqrt{2 \cdot 2 \cdot 2 \cdot 3 \cdot 3 \cdot 3}$$
Now, using the fact that $\sqrt{ab} = \sqrt{a}\sqrt{b}$ and $\sqrt{a^2} = a$, we can write
$$\sqrt{216} = \sqrt{2 \cdot 2 \cdot 2 \cdot 3 \cdot 3 \cdot 3}$$
$$= \sqrt{2^2}\sqrt{3^2}\sqrt{2 \cdot 3}$$
$$= 2 \cdot 3\sqrt{6}$$
$$= 6\sqrt{6}$$

In practice, it is convenient to group the repeated factors by two's and then directly simplify the square roots.

Example 2

$$\sqrt{216} = \sqrt{\underbrace{2 \cdot 2} \cdot \underbrace{3 \cdot 3} \cdot 2 \cdot 3} \quad \text{— Group by two's.}$$
$$= \sqrt{\underbrace{2 \cdot 2}}\sqrt{\underbrace{3 \cdot 3}}\sqrt{2 \cdot 3}$$
$$= \quad 2 \quad \cdot \quad 3 \quad \sqrt{2 \cdot 3}$$
$$= 6\sqrt{6}$$

We can apply the same procedure to radical expressions when the radicands contain variables.

Example 3 a. $\sqrt{x^6} = \sqrt{x \cdot x \cdot x \cdot x \cdot x \cdot x}$ Group by two's.
$$= \sqrt{x \cdot x}\sqrt{x \cdot x}\sqrt{x \cdot x}$$
$$= \quad x \quad\quad x \quad\quad x \quad = x^3$$

b. $\sqrt{x^5} = \sqrt{x \cdot x \cdot x \cdot x \cdot x}$ Group by two's.
$$= \sqrt{x \cdot x}\sqrt{x \cdot x}\sqrt{x}$$
$$= \quad x \quad\quad x \quad \sqrt{x} \quad = x^2\sqrt{x}$$

You may notice that when the radicand contains a variable with an even exponent, we obtain the square root by dividing the exponent by 2. Similarly, when the radicand has a variable with an odd exponent, we can simplify by factoring the variable factor into two factors, one having an even exponent and the other having an exponent of 1.

Example 4 a. $\sqrt{x^6} = x^3$ (6 ÷ 2)   b. $\sqrt{x^{10}} = x^5$ (10 ÷ 2)

c. $\sqrt{x^7} = \sqrt{x^6}\sqrt{x} = x^3\sqrt{x}$ d. $\sqrt{x^{11}} = \sqrt{x^{10}}\sqrt{x} = x^5\sqrt{x}$

By simplifying radical expressions, we can extend the scope of the table of square roots.

15.3 SIMPLIFYING RADICAL EXPRESSIONS I

Example 5 Approximate $\sqrt{216}$.

Solution $\sqrt{216}$ is not available in the table. But we can approximate $\sqrt{216}$ by observing that

$$\sqrt{216} = \sqrt{2 \cdot 2 \cdot 2 \cdot 3 \cdot 3 \cdot 3}$$
$$= 2 \cdot 3\sqrt{6}$$
$$\approx 6(2.449) = 14.694$$

Common Error   Note that

$$\sqrt{16 + 9} \neq \sqrt{16} + \sqrt{9}$$

since

$$\sqrt{16 + 9} = \sqrt{25} = 5$$

and

$$\sqrt{16} + \sqrt{9} = 4 + 3 = 7$$

In general,

$$\sqrt{a + b} \neq \sqrt{a} + \sqrt{b}$$

EXERCISES 15.3
Simplify.

Complete Examples (see Examples 1 and 2)

a. $\sqrt{24}$

$= \sqrt{2 \cdot 2 \cdot \underline{}}$ (1)

$= \;\; 2 \;\sqrt{2 \cdot 3}$

$= \underline{}$ (2)

b. $\sqrt{1575}$

$= \sqrt{5 \cdot 5 \cdot 3 \cdot 3 \cdot \underline{}}$ (3)

$= \;\; 5 \;\cdot\; 3 \;\sqrt{7}$

$= \underline{}$ (4)

Factor radicand completely and group factors by two's.

(1) 3   (2) $2\sqrt{6}$ (3) 7   (4) $15\sqrt{7}$

1. $\sqrt{8}$
2. $\sqrt{12}$
3. $\sqrt{18}$
4. $\sqrt{49}$
5. $-\sqrt{20}$
6. $-\sqrt{27}$
7. $-\sqrt{72}$
8. $-\sqrt{24}$
9. $\sqrt{64}$
10. $\sqrt{162}$
11. $\sqrt{288}$
12. $\sqrt{84}$
13. $\sqrt{125}$
14. $\sqrt{450}$
15. $-\sqrt{1080}$
16. $-\sqrt{882}$
17. $\pm\sqrt{720}$
18. $\pm\sqrt{588}$
19. $\pm\sqrt{1944}$
20. $\pm\sqrt{1125}$

RADICAL EXPRESSIONS

Complete Examples (see Examples 3 and 4)

a. $\sqrt{x^4}$

Method I
$$\sqrt{x^4} = \sqrt{\underbrace{x \cdot x} \cdot \underbrace{x \cdot x}}$$
$$= x \cdot x$$
$$= \underline{}$$
$$(1)$$

Method II
$$\sqrt{x^4} = x^{4 \div 2}$$
$$= \underline{}$$
$$(2)$$

b. $\sqrt{y^9}$

Method I
$$\sqrt{y^9} = \sqrt{\underbrace{y \cdot y} \cdot \underbrace{y \cdot y} \cdot \underbrace{y \cdot y} \cdot \underbrace{y \cdot y} \cdot y}$$
$$= y \cdot y \cdot y \cdot y \sqrt{y}$$
$$= \underline{}$$
$$(3)$$

Method II
$$\sqrt{y^9} = \sqrt{y^8}\sqrt{y}$$
$$= y^{8 \div 2}\sqrt{y}$$
$$= \underline{}$$
$$(4)$$

---

(1) x^2 (2) x^2 (3) $y^4\sqrt{y}$ (4) $y^4\sqrt{y}$

---

21. $\sqrt{x^3}$ 22. $\sqrt{y^5}$ 23. $\sqrt{y^7}$ 24. $\sqrt{x^{10}}$

25. $-\sqrt{x^{11}}$ 26. $-\sqrt{x^{13}}$ 27. $\sqrt{x^6}$ 28. $-\sqrt{x^4}$

29. $\pm\sqrt{x^8}$ 30. $\pm\sqrt{x^{15}}$ 31. $\sqrt{x^{12}}$ 32. $\sqrt{x^{14}}$

Complete Examples (see Examples 1–4)

a. $\sqrt{12y^3}$
$$= \sqrt{\underbrace{2 \cdot 2} \cdot \underbrace{y \cdot y} \cdot 3 \cdot y}$$
$$= 2 \cdot y \sqrt{3 \cdot y}$$
$$= \underline{}$$
$$(1)$$

b. $-\sqrt{20x^2y^3}$
$$= -\sqrt{\underbrace{2 \cdot 2} \cdot \underbrace{x \cdot x} \cdot \underbrace{y \cdot y} \cdot 5 \cdot y}$$
$$= -\ 2 \cdot x \cdot y \sqrt{5 \cdot y}$$
$$= \underline{}$$
$$(2)$$

---

(1) $2y\sqrt{3y}$ (2) $-2xy\sqrt{5y}$

---

33. $\sqrt{4x^2}$ 34. $\sqrt{8x^2}$ 35. $\sqrt{9x^3}$ 36. $\sqrt{12x^3}$

37. $-\sqrt{24y^5}$ 38. $-\sqrt{121x^5}$ 39. $\sqrt{64y^4}$ 40. $\sqrt{36x^5}$

41. $\sqrt{49x^7}$ 42. $\sqrt{16x^2}$ 43. $\pm\sqrt{32x^3}$ 44. $\pm\sqrt{72y^3}$

45. $\sqrt{80x^2}$ 46. $\sqrt{98y^3}$ 47. $-\sqrt{64x}$ 48. $-\sqrt{3x^2}$

49. $\pm\sqrt{5y^3}$ 50. $\pm\sqrt{7x^2}$ 51. $\sqrt{48x^2y}$ 52. $\sqrt{20x^2y^2}$

15.3 SIMPLIFYING RADICAL EXPRESSIONS I

53. $\sqrt{25x^3y^2}$ **54.** $\sqrt{50xy^2}$ **55.** $-\sqrt{45a^4b}$ **56.** $-\sqrt{40x^5y^2}$

57. $\sqrt{\dfrac{9}{16}x^2y^2}$ **58.** $\sqrt{\dfrac{4}{9}x^3y}$ **59.** $\pm\sqrt{\dfrac{25}{36}y^2}$ **60.** $\pm\sqrt{\dfrac{1}{4}ab^2c^3}$

Complete Examples

a. $3\sqrt{4x^3}$

$= 3\sqrt{2 \cdot 2 \cdot x \cdot x \cdot x}$

$= 3 \cdot 2 \cdot x \sqrt{x}$

$= \underline{}$
       (1)

b. $\dfrac{3x}{y}\sqrt{18x^3y^3}$

$= \dfrac{3x}{y}\sqrt{3 \cdot 3 \cdot x \cdot x \cdot y \cdot y \cdot 2 \cdot x \cdot y}$

$= \dfrac{3x}{y} \cdot 3 \cdot x \cdot y \sqrt{2 \cdot x \cdot y}$

$= \underline{}$
       (2)

---

(1) $6x\sqrt{x}$ (2) $9x^2\sqrt{2xy}$

---

61. $2\sqrt{x^2}$ **62.** $3\sqrt{x^2y}$ **63.** $3\sqrt{4x}$ **64.** $4\sqrt{5x^2}$

65. $7\sqrt{49y^3}$ **66.** $-4\sqrt{16x^2}$ **67.** $2x\sqrt{x^2y}$ **68.** $3x\sqrt{9x^3}$

69. $-\dfrac{1}{3}\sqrt{9a^3}$ **70.** $\dfrac{1}{5}\sqrt{25y^3}$ **71.** $\pm\dfrac{1}{2}x\sqrt{16x^4}$ **72.** $\pm\dfrac{2a}{3}\sqrt{36a^3b^3}$

Use the table of square roots to approximate each expression. Round off answers to two decimal places. See Example 5.

Complete Examples

a. $\sqrt{243}$

$= \sqrt{3 \cdot 3 \cdot 3 \cdot 3 \cdot 3}$

$= 3 \cdot 3 \sqrt{3}$

$\approx 9(1.732)$

$= 15.588$

$\approx \underline{}$
   (1)

b. $3 + 2\sqrt{200}$

$= 3 + 2\sqrt{2 \cdot 2 \cdot 5 \cdot 5 \cdot 2}$

$= 3 + 2 \cdot 2 \cdot 5 \sqrt{2}$

$= 3 + 20\sqrt{2}$

$\approx 3 + 20(\underline{})$
              (2)

$= 3 + 28.28 = \underline{}$
                    (3)

---

(1) 15.59   (2) 1.414   (3) 31.28

463

RADICAL EXPRESSIONS

73. $\sqrt{108}$
74. $\sqrt{162}$
75. $\sqrt{275}$
76. $\sqrt{207}$
77. $3 + \sqrt{125}$
78. $24 - \sqrt{176}$
79. $5 - \sqrt{300}$
80. $11 - \sqrt{242}$
81. $-2\sqrt{243}$
82. $-5\sqrt{120}$
83. $6 + 3\sqrt{104}$
84. $1 + 2\sqrt{128}$

85. Use a numerical example to show that $\sqrt{x} + \sqrt{9}$ is not equivalent to $\sqrt{x + 9}$.

86. Use a numerical example to show that $\sqrt{y} + \sqrt{16}$ is not equivalent to $\sqrt{y + 16}$.

15.4 SIMPLIFYING RADICAL EXPRESSIONS II

SUMS AND DIFFERENCES

Recall from Section 6.2 that we add like terms by adding their numerical coefficients; that is,

$$2r + 3r = 5r$$

where r represents any number. In particular, if r represents an irrational number, say $\sqrt{2}$, we have

$$2\sqrt{2} + 3\sqrt{2} = 5\sqrt{2}$$

Thus, we may add radical expressions by adding their numerical coefficients, provided the radicands involved are identical. If the radicands differ, we can only indicate addition, for example, $3\sqrt{2} + 4\sqrt{3}$ cannot be written as a single term. As before, if there is no numerical coefficient before a radical, it is understood that the coefficient is 1. These same ideas apply to differences.

Example 1

a. $4\sqrt{3} + \sqrt{3} = 4\sqrt{3} + 1\sqrt{3}$
$= 5\sqrt{3}$

b. $5\sqrt{x} + 3\sqrt{x} = 8\sqrt{x}$

c. $5\sqrt{2} - 2\sqrt{2} = 3\sqrt{2}$

d. $\sqrt{7} - 4\sqrt{7} = 1\sqrt{7} - 4\sqrt{7}$
$= -3\sqrt{7}$

It is a good idea to write radicals in simplest form before attempting to combine like terms.

Example 2

a. $\sqrt{20} + 2\sqrt{45} = \sqrt{2 \cdot 2 \cdot 5} + 2\sqrt{3 \cdot 3 \cdot 5}$
$= 2\sqrt{5} + 2 \cdot 3\sqrt{5}$
$= 2\sqrt{5} + 6\sqrt{5} = 8\sqrt{5}$

b. $3\sqrt{4x} - 2\sqrt{x} = 3\sqrt{2 \cdot 2x} - 2\sqrt{x}$
$= 3 \cdot 2\sqrt{x} - 2\sqrt{x}$
$= 6\sqrt{x} - 2\sqrt{x} = 4\sqrt{x}$

15.4 SIMPLIFYING RADICAL EXPRESSIONS II

PRODUCTS AND FACTORS

Products involving radicals can also be written in equivalent forms. From the distributive property,

$$a(\sqrt{3} + \sqrt{2}) = a\sqrt{3} + a\sqrt{2}$$

where a may be either rational or irrational. And by the symmetric property of equality, we can reverse the multiplication to obtain

$$a\sqrt{3} + a\sqrt{2} = a(\sqrt{3} + \sqrt{2})$$

where the right-hand member is in factored form. For example,

Example 3 a. $3(\sqrt{2} + 5) = 3\sqrt{2} + 3 \cdot 5$ b. $3\sqrt{2} + 15 = 3\sqrt{2} + 3 \cdot 5$
$\phantom{3(\sqrt{2} + 5)} = 3\sqrt{2} + 15$ $\phantom{3\sqrt{2} + 15} = 3(\sqrt{2} + 5)$

EXERCISES 15.4
Simplify.

Complete Example (see Example 1)
a. $\sqrt{5} + 3\sqrt{5}$
$= \underline{}$
$ (1)$

b. $\sqrt{5} - 3\sqrt{5}$
$= \underline{}$
$ (2)$

(1) $4\sqrt{5}$ (2) $-2\sqrt{5}$

1. $\sqrt{3} + 2\sqrt{3}$
2. $\sqrt{7} - 3\sqrt{7}$
3. $3\sqrt{5} - 2\sqrt{5}$
4. $8\sqrt{5} - 2\sqrt{5} + 3\sqrt{5}$
5. $2\sqrt{3} - 4\sqrt{3} + 2\sqrt{3}$
6. $\sqrt{5} - 3\sqrt{5} + 7\sqrt{5}$

Complete Examples (see Example 2)
a. $5\sqrt{2} - \sqrt{8} + \sqrt{12}$
$= 5\sqrt{2} - \underline{}_{(1)} + 2\sqrt{3}$
$= \underline{}_{(2)} + 2\sqrt{3}$

b. $2\sqrt{3a} + \sqrt{27a} - 2\sqrt{12a}$
$= 2\sqrt{3a} + \underline{}_{(3)} - 4\sqrt{3a}$
$= \underline{}_{(4)}$

Simplify radicals.
Combine like terms.

(1) $2\sqrt{2}$ (2) $3\sqrt{2}$ (3) $3\sqrt{3a}$ (4) $\sqrt{3a}$

465

RADICAL EXPRESSIONS

7. $2\sqrt{3} + \sqrt{27}$
8. $\sqrt{8} + \sqrt{18}$
9. $\sqrt{50} - 2\sqrt{32}$
10. $2\sqrt{6} - 2\sqrt{24} + \sqrt{54}$
11. $\sqrt{12} + 2\sqrt{27} - 3\sqrt{48}$
12. $\sqrt{20} + \sqrt{45} - 2\sqrt{80}$
13. $3\sqrt{2} - 4\sqrt{3} + \sqrt{2}$
14. $2\sqrt{3} - \sqrt{4} + 3\sqrt{3}$
15. $\sqrt{3} + 2\sqrt{12} + \sqrt{18}$
16. $\sqrt{36} - 2\sqrt{32} + \sqrt{49}$
17. $3\sqrt{144} - 4\sqrt{49} + 3\sqrt{24}$
18. $\sqrt{12} - \sqrt{27} + 2\sqrt{8}$
19. $\sqrt{4a} + \sqrt{9a}$
20. $\sqrt{12a} - \sqrt{3a}$
21. $2\sqrt{x} + 2\sqrt{25x}$
22. $3\sqrt{2x} - \sqrt{8x}$
23. $\sqrt{16b^3} - b\sqrt{25b} + 3b\sqrt{b}$
24. $\sqrt{xy^2} + 2\sqrt{xy^2} - \sqrt{4xy^2}$

Express without parentheses. See Example 3.

Complete Examples

a. $2(3 + 4\sqrt{2})$
   $= 6 + \underline{}_{(1)}$

b. $a(4\sqrt{3} - 6\sqrt{a})$
   $= 4a\sqrt{3} - \underline{}_{(2)}$

Apply distributive property.

(1) $8\sqrt{2}$ (2) $6a\sqrt{a}$

25. $4(\sqrt{3} + 1)$
26. $2(3 - \sqrt{2})$
27. $-5(6 + \sqrt{7})$
28. $-2(\sqrt{6} - 3)$
29. $4(\sqrt{2} - \sqrt{3})$
30. $3(\sqrt{3} + \sqrt{7})$
31. $3(1 + 3\sqrt{a})$
32. $2(4\sqrt{a} - 5)$
33. $x(\sqrt{x} + 2\sqrt{y})$
34. $y(\sqrt{xy} - \sqrt{y})$
35. $xy(y\sqrt{x} + 2)$
36. $xy(4 - x\sqrt{x})$
37. $2(\sqrt{2} - 3\sqrt{3} - 5)$
38. $-3(6 + \sqrt{5} - 2\sqrt{3})$
39. $-(\sqrt{a} + \sqrt{b} - \sqrt{c})$
40. $-(2\sqrt{a} - \sqrt{b} + 2\sqrt{c})$

Simplify radical expressions where possible and factor. See Example 3.

Complete Examples

a. $8\sqrt{3} - 10$
   $= 2(4\sqrt{3} - \underline{}_{(1)})$

b. $\sqrt{12} + 4$
   $= \underline{}_{(2)} + 4$
   $= 2(\underline{}_{(3)} + 2)$

c. $y\sqrt{x} - y^2\sqrt{y}$
   $= y(\sqrt{x} - \underline{}_{(4)})$

(1) 5   (2) $2\sqrt{3}$ (3) $\sqrt{3}$ (4) $y\sqrt{y}$

466

41. $2 + 2\sqrt{3}$
42. $6 - 3\sqrt{2}$
43. $4\sqrt{2} - 12$
44. $3\sqrt{7} - 3$
45. $4\sqrt{5} + 8$
46. $8 + 32\sqrt{5}$
47. $8 - 32\sqrt{5}$
48. $4 + 2\sqrt{3}$
49. $6 + 24\sqrt{2}$
50. $5\sqrt{5} - 10$
51. $3 + \sqrt{18}$
52. $2 - \sqrt{32}$
53. $4 - 2\sqrt{8}$
54. $6 + 2\sqrt{27}$
55. $21 + \sqrt{18}$
56. $6 + \sqrt{72}$
57. $4 + \sqrt{16y}$
58. $3 - 2\sqrt{9y}$
59. $3x\sqrt{x} - 6x\sqrt{y}$
60. $2y^2\sqrt{x} - 8y\sqrt{y}$
61. $3\sqrt{x^2y} - 9x\sqrt{x}$
62. $4\sqrt{x^3y} + 6x\sqrt{x}$
63. $2\sqrt{y^3} - 6\sqrt{y^4}$
64. $3\sqrt{x^4} + 6\sqrt{x^5}$

15.5 FRACTIONS INVOLVING RADICAL EXPRESSIONS

The properties of fractions we considered in Chapters 10 and 11 also apply to fractions that contain radical expressions.

Example 1 Reduce $\dfrac{8 - \sqrt{80}}{4}$.

Solution We first simplify the radical expression to get

$$\frac{8 - 4\sqrt{5}}{4}$$

We then factor the numerator and use the fundamental principle of fractions to obtain

$$\frac{\cancel{4}(2 - \sqrt{5})}{\cancel{4}} = 2 - \sqrt{5}$$

As in Chapter 11, fractions must have common denominators before we can write the sum or difference as a single fraction.

Example 2 Write as a single fraction.

a. $\dfrac{\sqrt{2}}{3} + \dfrac{x}{2}$

b. $4 - \dfrac{2\sqrt{3}}{5}$

Solutions a. We find the LCD 6 and build each fraction to obtain

$$\frac{(2)\sqrt{2}}{(2)3} + \frac{(3)x}{(3)2} = \frac{2\sqrt{2} + 3x}{6}$$

b. We find the LCD 5 and build each fraction to obtain

$$\frac{(5)4}{(5)1} - \frac{2\sqrt{3}}{5} = \frac{20 - 2\sqrt{3}}{5}$$

RADICAL EXPRESSIONS

EXERCISES 15.5
Reduce the fractions. See Example 1.

Complete Example

$$\frac{-2 - \sqrt{72}}{4}$$

$$= \frac{-2 - 6\sqrt{2}}{4} \qquad \text{Simplify radicals.}$$

$$= \frac{2(-1 - 3\sqrt{2})}{\underline{\qquad(1)\qquad}} \qquad \text{Factor numerator and simplify.}$$

$$= \underline{\qquad(2)\qquad}$$

(1) 4 (2) $\dfrac{-1 - 3\sqrt{2}}{2}$

1. $\dfrac{4 + 6\sqrt{3}}{2}$ 2. $\dfrac{3 - 3\sqrt{2}}{3}$ 3. $\dfrac{6 - 2\sqrt{5}}{2}$ 4. $\dfrac{9 - 3\sqrt{5}}{3}$

5. $\dfrac{-2 + \sqrt{8}}{2}$ 6. $\dfrac{-6 + \sqrt{54}}{3}$ 7. $\dfrac{4 + 3\sqrt{12}}{2}$ 8. $\dfrac{2 - \sqrt{8}}{4}$

9. $\dfrac{3 + \sqrt{18}}{6}$ 10. $\dfrac{8 + \sqrt{32}}{16}$ 11. $\dfrac{5 - \sqrt{75}}{10}$ 12. $\dfrac{16 - 2\sqrt{48}}{16}$

13. $\dfrac{2x - \sqrt{8}}{2x}$ 14. $\dfrac{6x - \sqrt{18}}{6x}$ 15. $\dfrac{x - \sqrt{x^3}}{x}$ 16. $\dfrac{2x + \sqrt{3x^3}}{x}$

17. $\dfrac{4y - 2\sqrt{y^3}}{2y}$ 18. $\dfrac{6y + 3\sqrt{2y^3}}{6y}$

Write each sum or difference as a single fraction. See Example 2.

Complete Examples

a. $\dfrac{5}{7} + \dfrac{\sqrt{3}}{7}$ b. $\dfrac{2}{3} - \dfrac{\sqrt{7}}{3}$

$= \underline{\qquad(1)\qquad}$ $= \underline{\qquad(2)\qquad}$ Add or subtract numerators.

(1) $\dfrac{5 + \sqrt{3}}{7}$ (2) $\dfrac{2 - \sqrt{7}}{3}$

468

15.5 FRACTIONS INVOLVING RADICAL EXPRESSIONS

19. $\dfrac{2}{3} + \dfrac{\sqrt{2}}{3}$ 20. $\dfrac{5}{2} - \dfrac{\sqrt{3}}{2}$ 21. $\dfrac{\sqrt{3}}{5} - \dfrac{1}{5}$ 22. $\dfrac{\sqrt{2}}{7} + \dfrac{1}{7}$

23. $\dfrac{2\sqrt{10}}{3} - \dfrac{\sqrt{3}}{3}$ 24. $\dfrac{\sqrt{17}}{5} - \dfrac{3\sqrt{7}}{5}$ 25. $\dfrac{\sqrt{11}}{a} + \dfrac{1}{a}$ 26. $\dfrac{\sqrt{5}}{b} - \dfrac{3}{b}$

Complete Examples

a. $\dfrac{2}{3} + \dfrac{5\sqrt{7}}{6}$

$= \dfrac{(2)2}{(\underline{})3} + \dfrac{5\sqrt{7}}{6}$

$= \dfrac{}{(2)}$

b. $\dfrac{1}{2} - \dfrac{\sqrt{3}}{3}$

$= \dfrac{(3)1}{(3)2} - \dfrac{\sqrt{3}(2)}{3(\underline{})}$

$= \dfrac{}{(4)}$

Build fractions with LCD 6.

Add or subtract numerators.

(1) 2 (2) $\dfrac{4 + 5\sqrt{7}}{6}$ (3) 2 (4) $\dfrac{3 - 2\sqrt{3}}{6}$

27. $\dfrac{5}{4} + \dfrac{3\sqrt{2}}{2}$ 28. $\dfrac{1}{10} - \dfrac{2\sqrt{3}}{5}$ 29. $\dfrac{1}{2a} + \dfrac{\sqrt{3}}{6a}$ 30. $\dfrac{\sqrt{5}}{3b} - \dfrac{5}{6b}$

31. $\dfrac{2}{5} + \dfrac{\sqrt{3}}{3}$ 32. $\dfrac{3}{7} - \dfrac{\sqrt{2}}{2}$ 33. $\dfrac{\sqrt{3}}{4} - \dfrac{1}{3}$ 34. $\dfrac{\sqrt{5}}{2} - \dfrac{2}{3}$

35. $\dfrac{2\sqrt{3}}{3} - \dfrac{\sqrt{2}}{2}$ 36. $\dfrac{3\sqrt{5}}{4} + \dfrac{\sqrt{3}}{5}$ 37. $4 + \dfrac{3\sqrt{2}}{2}$ 38. $2 - \dfrac{\sqrt{2}}{3}$

39. $\dfrac{\sqrt{2}}{5} + 1$ 40. $\dfrac{2\sqrt{2}}{3} - 1$ 41. $\dfrac{3\sqrt{3}}{2} + 3$ 42. $\dfrac{3\sqrt{7}}{2} - 3$

43. $\dfrac{\sqrt{2}}{x} + \dfrac{1}{2x}$ 44. $\dfrac{\sqrt{3}}{x} - \dfrac{1}{4x}$ 45. $\dfrac{3}{4} - 2\sqrt{y}$ 46. $\dfrac{4}{5} + 3\sqrt{y}$

47. $\dfrac{\sqrt{x}}{2} + \dfrac{\sqrt{y}}{3}$ 48. $\dfrac{2\sqrt{y}}{3} - \dfrac{\sqrt{x}}{4}$

RADICAL EXPRESSIONS

15.6 PRODUCTS OF RADICAL EXPRESSIONS

In Section 15.3, we saw that

$$\sqrt{ab} = \sqrt{a}\sqrt{b} \quad (a, b \geq 0)$$

By the symmetric property of equality,

$$\sqrt{a}\sqrt{b} = \sqrt{ab} \quad (a, b \geq 0)$$

Stated in words:

> **The product of two square roots is equal to the square root of the product of the radicands.**

We can use this property to simplify products involving radicals.

Example 1 Simplify $\sqrt{6x}\sqrt{2xy}$.

Solution We first write the product as

$$\sqrt{6x \cdot 2xy}$$

Then, we simplify as in Section 15.3 to obtain

$$\sqrt{6x \cdot 2xy} = \sqrt{2 \cdot 2 \cdot x \cdot x \cdot 3 \cdot y}$$
$$= 2 \cdot x \cdot \sqrt{3 \cdot y}$$
$$= 2x\sqrt{3y}$$

We can use the above property together with the distributive property to simplify radical expressions containing parentheses. If the product contains two binomials, we can use the FOIL method.

Example 2 a. $\sqrt{3}(\sqrt{3} - 2) = \sqrt{3} \cdot \sqrt{3} - \sqrt{3} \cdot 2$
$$= 3 - 2\sqrt{3}$$

b. $(\sqrt{3} - 1)(\sqrt{3} + 2) = \sqrt{3} \cdot \sqrt{3} + \sqrt{3} \cdot 2 - 1 \cdot \sqrt{3} - 1 \cdot 2$
$$= 3 + 2\sqrt{3} - \sqrt{3} - 2$$
$$= 1 + \sqrt{3}$$

15.6 PRODUCTS OF RADICAL EXPRESSIONS

EXERCISES 15.6
Simplify.

Complete Examples (see Example 1)

a. $\sqrt{2}\sqrt{3}$
$= \sqrt{2 \cdot 3}$
$= \sqrt{\underline{}}$ (1)

b. $\sqrt{2x}\sqrt{10xy}$
$= \sqrt{2x \cdot 10xy}$
$= \sqrt{2 \cdot 2 \cdot x \cdot x \cdot 5 \cdot y}$
$= 2 \cdot \underline{}_{(2)} \sqrt{5 \cdot y}$
$= \underline{}_{(3)}$

(1) 6   (2) x (3) $2x\sqrt{5y}$

1. $\sqrt{3}\sqrt{5}$
2. $\sqrt{2}\sqrt{7}$
3. $\sqrt{3}\sqrt{10}$
4. $\sqrt{5}\sqrt{13}$
5. $\sqrt{3}\sqrt{6}$
6. $\sqrt{2}\sqrt{10}$
7. $\sqrt{8}\sqrt{2}$
8. $\sqrt{27}\sqrt{3}$
9. $\sqrt{2x}\sqrt{3x}$
10. $\sqrt{5a}\sqrt{3a}$
11. $\sqrt{2xy}\sqrt{6xy^2}$
12. $\sqrt{6x^2}\sqrt{3x^2y}$
13. $\sqrt{8a}\sqrt{18a}$
14. $\sqrt{12b}\sqrt{32b}$
15. $\sqrt{10x^2}\sqrt{15y}$
16. $\sqrt{18a^2}\sqrt{6b}$

Complete Examples (see Example 1)

a. $2\sqrt{6}\sqrt{8}$
$= 2\sqrt{48}$
$= 2\sqrt{2 \cdot 2 \cdot 2 \cdot 2 \cdot \underline{}_{(1)}}$
$= 2 \cdot 2 \cdot \underline{}_{(2)} \sqrt{3}$
$= \underline{}_{(3)}$

b. $(\sqrt{x})(2\sqrt{xy})$
$= 2\sqrt{x \cdot x \cdot \underline{}_{(4)}}$
$= 2 \cdot x \sqrt{y}$
$= \underline{}_{(5)}$

(1) 3   (2) 2   (3) $8\sqrt{3}$ (4) y (5) $2x\sqrt{y}$

17. $\sqrt{2}\sqrt{5}\sqrt{3}$
18. $\sqrt{5}\sqrt{3}\sqrt{7}$
19. $\sqrt{5}\sqrt{10}\sqrt{2}$
20. $\sqrt{6}\sqrt{3}\sqrt{2}$
21. $(2\sqrt{3})(\sqrt{2})(\sqrt{9})$
22. $(5\sqrt{5})(3\sqrt{10})(\sqrt{4})$
23. $(2\sqrt{x})(3\sqrt{x})(\sqrt{x})$
24. $(x\sqrt{2})(x\sqrt{3})(\sqrt{6})$
25. $(a\sqrt{b})(b\sqrt{c})(c\sqrt{a})$
26. $(b\sqrt{a})(a\sqrt{b})(a\sqrt{ab})$
27. $(x\sqrt{x})(\sqrt{x^2})(\sqrt{x^3})$
28. $(a^2\sqrt{a})(2a\sqrt{a})(a\sqrt{a^2})$

471

RADICAL EXPRESSIONS

Complete Examples (see Example 2)

a. $\sqrt{3}(2 + \sqrt{2})$
$= (\sqrt{3})(2) + (\sqrt{3})(\sqrt{2})$
$= 2\sqrt{3} + \underline{}_{(1)}$

b. $\sqrt{3}(\sqrt{6} - \sqrt{15})$
$= (\sqrt{3})(\sqrt{6}) - (\sqrt{3})(\sqrt{15})$
$= \sqrt{18} - \sqrt{\underline{}_{(2)}}$
$= 3\sqrt{2} - \underline{}_{(3)}$

---

(1) $\sqrt{6}$ (2) 45 or $3 \cdot 3 \cdot 5$ (3) $3\sqrt{5}$

---

29. $\sqrt{2}(3 + \sqrt{3})$
30. $\sqrt{3}(5 + \sqrt{5})$
31. $\sqrt{3}(\sqrt{6} + 2)$
32. $\sqrt{2}(\sqrt{6} + 3)$

33. $\sqrt{5}(4 + \sqrt{10})$
34. $\sqrt{3}(2 - \sqrt{15})$
35. $\sqrt{3}(\sqrt{2} + \sqrt{6})$
36. $\sqrt{5}(\sqrt{3} - \sqrt{10})$

37. $\sqrt{3}(\sqrt{3} + \sqrt{2})$
38. $\sqrt{5}(\sqrt{5} + \sqrt{3})$
39. $\sqrt{2}(\sqrt{10} - \sqrt{2})$
40. $\sqrt{3}(\sqrt{3} + \sqrt{15})$

Complete Example (see Example 2)

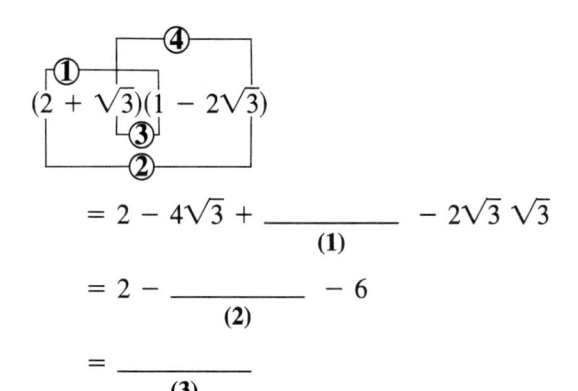

$(2 + \sqrt{3})(1 - 2\sqrt{3})$

$= 2 - 4\sqrt{3} + \underline{}_{(1)} - 2\sqrt{3}\sqrt{3}$ Apply the FOIL method.

$= 2 - \underline{}_{(2)} - 6$ Simplify.

$= \underline{}_{(3)}$

---

(1) $\sqrt{3}$ (2) $3\sqrt{3}$ (3) $-4 - 3\sqrt{3}$

---

41. $(3 + \sqrt{2})(1 - \sqrt{2})$
42. $(2 - \sqrt{2})(3 + \sqrt{2})$
43. $(\sqrt{5} - 1)(\sqrt{5} + 3)$

44. $(\sqrt{7} + 3)(\sqrt{7} - 5)$
45. $(2 + \sqrt{3})(2 - \sqrt{3})$
46. $(3 + \sqrt{2})(3 - \sqrt{2})$

47. $(3 - 2\sqrt{5})(3 + 2\sqrt{5})$
48. $(4 - 3\sqrt{6})(4 + 3\sqrt{6})$
49. $(2\sqrt{3} + \sqrt{5})(\sqrt{3} - 2\sqrt{5})$

50. $(2\sqrt{5} - 3\sqrt{2})(\sqrt{5} + \sqrt{2})$
51. $(3\sqrt{7} - 2\sqrt{5})(2\sqrt{7} + 3\sqrt{5})$
52. $(5\sqrt{6} - 2\sqrt{3})(\sqrt{6} - \sqrt{3})$

15.7 QUOTIENTS OF RADICAL EXPRESSIONS

Observe that

$$\sqrt{\frac{36}{9}} = \sqrt{4} = 2$$

and

$$\frac{\sqrt{36}}{\sqrt{9}} = \frac{6}{3} = 2$$

Thus,

$$\frac{\sqrt{36}}{\sqrt{9}} = \sqrt{\frac{36}{9}}$$

Stated in words,

> **The quotient of two square roots is equal to the square root of the quotient of the radicands.**

In general,

$$\frac{\sqrt{a}}{\sqrt{b}} = \sqrt{\frac{a}{b}}, \quad (a \geq 0, b > 0) \tag{1}$$

For example, $\dfrac{\sqrt{7}}{\sqrt{2}} = \sqrt{\dfrac{7}{2}}$ and $\sqrt{\dfrac{3}{x}} = \dfrac{\sqrt{3}}{\sqrt{x}}$. Furthermore, the expressions in (1) above can be written in another form which we call the simplest form of a quotient.

> A quotient involving radical expressions is in simplest form if:
> 1. No radical expressions are contained in the denominators of fractions.
> 2. The radicand does not contain a fraction.

Example 1   Which quotients are in simplest form?

   a. $\sqrt{\dfrac{3}{2}}$ b. $\dfrac{\sqrt{6}}{2}$ c. $\dfrac{5x}{\sqrt{3}}$ d. $\dfrac{\sqrt{15x}}{3}$

Solution   Expressions b and d are in simplest form. Expression a contains a fraction in the radicand and expression c contains a radical in the denominator; thus, a and c are not in simplest form.

When quotients involving radical expressions are not in simplest form, we can sometimes use Equation (1) to rewrite the expression in simplest form.

Example 2   Write in simplest form.

   a. $\dfrac{\sqrt{6}}{\sqrt{3}}$ b. $\dfrac{\sqrt{4x^2}}{\sqrt{2x}}$

RADICAL EXPRESSIONS

Solutions Using Equation (1), we have

a. $\dfrac{\sqrt{6}}{\sqrt{3}} = \sqrt{\dfrac{6}{3}}$
   $= \sqrt{2}$

b. $\dfrac{\sqrt{4x^2}}{\sqrt{2x}} = \sqrt{\dfrac{4x^2}{2x}}$
   $= \sqrt{2x}$

To simplify quotients involving radical expressions, we may need to use the fundamental principle of fractions as well as Equation (1).

Example 3 Write in simplest form.

a. $\dfrac{\sqrt{2}}{\sqrt{3}}$

b. $\sqrt{\dfrac{2}{3a}}$

Solutions a. We multiply the numerator and denominator by $\sqrt{3}$ to obtain

$$\dfrac{\sqrt{2}\,\sqrt{3}}{\sqrt{3}\,\sqrt{3}} = \dfrac{\sqrt{6}}{3}$$

b. We first use Equation (1) to obtain

$$\sqrt{\dfrac{2}{3a}} = \dfrac{\sqrt{2}}{\sqrt{3a}}$$

Now multiplying the numerator and denominator by $\sqrt{3a}$, we obtain

$$\dfrac{\sqrt{2}\,\sqrt{3a}}{\sqrt{3a}\,\sqrt{3a}} = \dfrac{\sqrt{6a}}{3a}$$

In Example 3, we rewrote each quotient so that the denominator was free of radicals. This process is sometimes called rationalizing the denominator. We rationalize the denominator of a fraction by building to a fraction in which the denominator is a perfect square, and then we simplify.

Example 4 Rationalize the denominator and simplify.

a. $\dfrac{\sqrt{12}}{\sqrt{3x}}$

b. $\dfrac{2\sqrt{5}}{\sqrt{8}}$

Solutions a. We multiply the numerator and denominator by $\sqrt{3x}$ to obtain

$$\dfrac{\sqrt{12}\,\sqrt{3x}}{\sqrt{3x}\,\sqrt{3x}} = \dfrac{\sqrt{36x}}{3x}$$

Now, we simplify $\sqrt{36x}$ to get

$$\dfrac{\sqrt{36x}}{3x} = \dfrac{\overset{2}{\cancel{6}}\sqrt{x}}{\cancel{3}x} = \dfrac{2\sqrt{x}}{x}$$

b. Since $8 \cdot 2 = 16$ and 16 is a perfect square, we multiply numerator and denominator by $\sqrt{2}$ to obtain

$$\dfrac{2\sqrt{5}\,\sqrt{2}}{\sqrt{8}\,\sqrt{2}} = \dfrac{2\sqrt{10}}{\sqrt{16}}$$

15.7 QUOTIENTS OF RADICAL EXPRESSIONS

Now, simplifying yields

$$\frac{2\sqrt{10}}{\sqrt{16}} = \frac{\cancel{2}\sqrt{10}}{\cancel{4}_2} = \frac{\sqrt{10}}{2}$$

To see one advantage of the rationalized form, we will compute a decimal approximation for $\frac{\sqrt{2}}{\sqrt{3}}$. If we approach this problem directly, we obtain

$$\frac{\sqrt{2}}{\sqrt{3}} \approx \frac{1.414}{1.732}$$

and arrive at a problem in long division. But if we first rationalize the denominator, we can arrive at a simpler division process as shown in Example 5.

Example 5 Find a decimal approximation to two decimal places, for $\frac{\sqrt{2}}{\sqrt{3}}$.

Solution First, rationalize the denominator to get

$$\frac{\sqrt{2}\,\sqrt{3}}{\sqrt{3}\,\sqrt{3}} = \frac{\sqrt{6}}{3}$$

Using the table, we have

$$\frac{\sqrt{6}}{3} \approx \frac{2.449}{3} = 0.816$$

$$= 0.82$$

EXERCISES 15.7

Rewrite each fraction so that no radical appears in the denominator, and no fraction appears in the radicand. That is, write in simplest form.

Complete Examples (see Examples 1 and 2)

a. $\dfrac{\sqrt{12}}{\sqrt{3}} = \sqrt{\dfrac{12}{3}}$

$= \sqrt{\underline{}}_{(1)}$

$= \underline{}_{(2)}$

b. $\dfrac{3\sqrt{6a}}{\sqrt{2a}} = 3\sqrt{\dfrac{6a}{2a}}$

$= 3\sqrt{\underline{}}_{(3)}$

(1) 4   (2) 2   (3) 3

475

RADICAL EXPRESSIONS

1. $\dfrac{\sqrt{18}}{\sqrt{2}}$
2. $\dfrac{\sqrt{8}}{\sqrt{2}}$
3. $\dfrac{\sqrt{75}}{\sqrt{3}}$
4. $\dfrac{\sqrt{80}}{\sqrt{5}}$
5. $\dfrac{\sqrt{27a}}{\sqrt{3a}}$

6. $\dfrac{\sqrt{28a}}{\sqrt{7a}}$
7. $\dfrac{\sqrt{8a}}{\sqrt{2}}$
8. $\dfrac{\sqrt{12a}}{\sqrt{3}}$
9. $\dfrac{\sqrt{15b}}{\sqrt{5}}$
10. $\dfrac{\sqrt{21b}}{\sqrt{7b}}$

11. $\dfrac{\sqrt{ab}}{\sqrt{a}}$
12. $\dfrac{\sqrt{7abc}}{\sqrt{bc}}$
13. $\dfrac{2\sqrt{14bc}}{\sqrt{2c}}$
14. $\dfrac{3\sqrt{8a}}{\sqrt{2}}$
15. $\dfrac{\sqrt{2}\,\sqrt{3}}{\sqrt{6}}$

16. $\dfrac{\sqrt{6}\,\sqrt{8}}{\sqrt{3}}$
17. $\dfrac{\sqrt{a}\,\sqrt{ab}}{\sqrt{b}}$
18. $\dfrac{\sqrt{3a}\,\sqrt{6a}}{\sqrt{2}}$
19. $\dfrac{\sqrt{ab}\,\sqrt{3b}}{\sqrt{a}}$
20. $\dfrac{\sqrt{3}\,\sqrt{10}}{\sqrt{6}}$

Complete Examples (see Example 3)

a. $\dfrac{\sqrt{3}}{\sqrt{a}} = \dfrac{\sqrt{3}\,\sqrt{a}}{\sqrt{a}\,\underline{\quad\quad}}$
 $\phantom{\dfrac{\sqrt{3}}{\sqrt{a}}}\hspace{2.2em}$ (1)

$\phantom{\dfrac{\sqrt{3}}{\sqrt{a}}}= \dfrac{\sqrt{3a}}{\underline{\quad\quad}}$
$\phantom{\dfrac{\sqrt{3}}{\sqrt{a}}}\hspace{1.6em}$ (2)

b. $\sqrt{\dfrac{1}{2}} = \dfrac{\sqrt{1}}{\underline{\quad\quad}}$
$\phantom{\sqrt{\dfrac{1}{2}}}\hspace{1.2em}$ (3)

$\phantom{\sqrt{\dfrac{1}{2}}} = \dfrac{1\,\sqrt{2}}{\sqrt{2}\,\sqrt{2}}$

$\phantom{\sqrt{\dfrac{1}{2}}} = \underline{\quad\quad}$
$\phantom{\sqrt{\dfrac{1}{2}}}\hspace{1.2em}$ (4)

---

(1) $\sqrt{a}$ (2) a (3) $\sqrt{2}$ (4) $\dfrac{\sqrt{2}}{2}$

---

21. $\dfrac{5}{\sqrt{2}}$
22. $\dfrac{5}{\sqrt{3}}$
23. $\dfrac{2}{\sqrt{x}}$
24. $\dfrac{5}{\sqrt{x}}$

25. $\dfrac{a}{\sqrt{b}}$
26. $\dfrac{x}{\sqrt{y}}$
27. $\sqrt{\dfrac{1}{3}}$
28. $\sqrt{\dfrac{1}{5}}$

29. $\sqrt{\dfrac{2}{a}}$
30. $\sqrt{\dfrac{2a}{b}}$
31. $\sqrt{\dfrac{3a}{b}}$
32. $\sqrt{\dfrac{5b}{a}}$

15.7 QUOTIENTS OF RADICAL EXPRESSIONS

Complete Examples (see Example 4)

a. $\dfrac{\sqrt{18}}{\sqrt{2x}} = \dfrac{\sqrt{18}\,\sqrt{2x}}{\sqrt{2x}\,\sqrt{\underline{}}}$
(1)

$= \dfrac{\sqrt{36x}}{\underline{}}$
(2)

$= \dfrac{6\sqrt{x}}{2x}$

$= \underline{}$
(3)

b. $\dfrac{6\sqrt{y}}{\sqrt{27}} = \dfrac{6\sqrt{y}\,\sqrt{3}}{\sqrt{27}\,\sqrt{3}}$

$= \dfrac{6\sqrt{3y}}{\sqrt{\underline{}}}$
(4)

$= \dfrac{6\sqrt{3y}}{9}$

$= \underline{}$
(5)

---

(1) $2x$ (2) $2x$ (3) $\dfrac{3\sqrt{x}}{x}$ (4) 81 (5) $\dfrac{2\sqrt{3y}}{3}$

---

33. $\dfrac{\sqrt{18}}{\sqrt{2x}}$ 34. $\dfrac{\sqrt{8}}{\sqrt{2y}}$ 35. $\dfrac{\sqrt{75}}{\sqrt{3y}}$ 36. $\dfrac{\sqrt{80}}{\sqrt{5x}}$

37. $\dfrac{a\sqrt{2}}{\sqrt{a}}$ 38. $\dfrac{b\sqrt{3}}{\sqrt{b}}$ 39. $\dfrac{4\sqrt{3x}}{\sqrt{8}}$ 40. $\dfrac{9\sqrt{5x}}{\sqrt{27}}$

41. $\dfrac{a\sqrt{32}}{\sqrt{2a}}$ 42. $\dfrac{b\sqrt{21}}{\sqrt{3b}}$ 43. $\dfrac{4y\sqrt{3x}}{\sqrt{4y}}$ 44. $\dfrac{9x\sqrt{2y}}{\sqrt{27x}}$

Write in simplest form. See Examples 3–4.

Complete Examples

a. $\sqrt{\dfrac{20}{3}} = \dfrac{\sqrt{20}}{\sqrt{\underline{}}}$
(1)

$= \dfrac{2\sqrt{5}\,\sqrt{3}}{\sqrt{3}\,\underline{}}$
(2)

$= \dfrac{2\sqrt{15}}{\underline{}}$
(3)

b. $\sqrt{\dfrac{4}{3x}} = \dfrac{\sqrt{4}}{\sqrt{3x}}$

$= \dfrac{2\sqrt{3x}}{\sqrt{3x}\,\underline{}}$
(4)

$= \dfrac{2\sqrt{3x}}{\underline{}}$
(5)

---

(1) 3 (2) $\sqrt{3}$ (3) 3 (4) $\sqrt{3x}$ (5) $3x$

477

RADICAL EXPRESSIONS

45. $\sqrt{\dfrac{8}{3}}$ 46. $\sqrt{\dfrac{18}{5}}$ 47. $\sqrt{\dfrac{9}{2}}$ 48. $\sqrt{\dfrac{12}{7}}$

49. $\sqrt{\dfrac{72}{5}}$ 50. $\sqrt{\dfrac{98}{3}}$ 51. $\sqrt{\dfrac{50}{2x}}$ 52. $\sqrt{\dfrac{75}{3y}}$

53. $\sqrt{\dfrac{24}{3x}}$ 54. $\sqrt{\dfrac{32}{4y}}$ 55. $\sqrt{\dfrac{x^3}{xy}}$ 56. $\sqrt{\dfrac{y^5}{xy^2}}$

Rationalize each denominator and find a decimal approximation (round off answers to two decimal places). See Example 5.

Complete Examples

a. $\sqrt{\dfrac{5}{3}} = \dfrac{\sqrt{5}\sqrt{3}}{\sqrt{3}\sqrt{3}}$

$\phantom{\sqrt{\dfrac{5}{3}}} = \dfrac{\sqrt{15}}{\underline{}}$
(1)

$\phantom{\sqrt{\dfrac{5}{3}}} \approx \dfrac{3.873}{3}$

$\phantom{\sqrt{\dfrac{5}{3}}} = \underline{} \approx 1.29$
(2)

b. $4\sqrt{\dfrac{1}{3}} = \dfrac{4\sqrt{1}\sqrt{3}}{\sqrt{3}\,\underline{}}$
(3)

$\phantom{4\sqrt{\dfrac{1}{3}}} = \dfrac{4\sqrt{3}}{3}$

$\phantom{4\sqrt{\dfrac{1}{3}}} \approx \dfrac{4(1.732)}{3}$

$\phantom{4\sqrt{\dfrac{1}{3}}} = 2.309 \approx \underline{}$
(4)

(1) 3   (2) 1.291   (3) $\sqrt{3}$ (4) 2.31

57. $\sqrt{\dfrac{1}{7}}$ 58. $\sqrt{\dfrac{3}{5}}$ 59. $\dfrac{3}{\sqrt{2}}$ 60. $\dfrac{2}{\sqrt{3}}$

61. $\dfrac{3}{\sqrt{5}}$ 62. $\dfrac{2}{\sqrt{6}}$ 63. $3\sqrt{\dfrac{1}{3}}$ 64. $5\sqrt{\dfrac{1}{5}}$

CHAPTER FIFTEEN REVIEW EXERCISES

1. Which of the following are irrational numbers?

$\sqrt{6},\ -\sqrt{9},\ \sqrt{4},\ \sqrt{\dfrac{9}{25}},\ -\sqrt{\dfrac{2}{3}}$

2. Using the table of square roots, find a decimal approximation for
   a. $\sqrt{93}$ b. $2\sqrt{47}$ c. $\dfrac{1}{4}\sqrt{32}$

CHAPTER FIFTEEN REVIEW EXERCISES

3. a. Locate the numbers 8, $\sqrt{80}$, and $\frac{19}{2}$ on a number line.
   b. Graph $x \leq 3$ where x represents a real number.

Simplify each expression in Exercises 4–8.

4. a. $\sqrt{72}$ b. $-\sqrt{90}$ c. $\sqrt{175}$
5. a. $\sqrt{y^4}$ b. $\sqrt{x^{15}}$ c. $\sqrt{x^3 y^7}$
6. a. $x\sqrt{27}$ b. $3\sqrt{3x^2}$ c. $\frac{1}{3}\sqrt{27x^2}$
7. a. $\sqrt{7} + 3\sqrt{7}$ b. $4\sqrt{6} - 3\sqrt{6} + \sqrt{6}$ c. $7\sqrt{2} - 3\sqrt{2} + \sqrt{2}$
8. a. $2\sqrt{12} - 4\sqrt{27}$ b. $\sqrt{9x} - \sqrt{4x}$ c. $\sqrt{x^2 y} - 3\sqrt{4x^2 y} + 7x\sqrt{y}$

9. Express without parentheses.
   a. $3(\sqrt{y} - 6)$ b. $y(\sqrt{xy} - 2y)$ c. $\sqrt{3}(\sqrt{2} - \sqrt{6})$

10. Simplify radicals where possible and factor.
    a. $2 - \sqrt{8}$ b. $\sqrt{27} - 3\sqrt{5}$ c. $3x^2 - \sqrt{5x^4}$

Simplify each expression.

11. a. $\dfrac{3 - 2\sqrt{27}}{3}$ b. $\dfrac{6 - 3\sqrt{12}}{3}$ c. $\dfrac{4 - \sqrt{32}}{4}$
12. a. $\dfrac{3}{2} - \dfrac{\sqrt{3}}{2}$ b. $\dfrac{\sqrt{15}}{5} + \dfrac{2\sqrt{15}}{5}$ c. $\dfrac{2}{3} - \dfrac{\sqrt{3}}{6}$
13. a. $\dfrac{\sqrt{5}}{3} - \dfrac{3}{2}$ b. $\dfrac{2\sqrt{3}}{5} - 1$ c. $\dfrac{3\sqrt{6}}{4} + 2$
14. a. $\sqrt{5}\sqrt{3}$ b. $\sqrt{16}\sqrt{32}$ c. $\sqrt{3x^2}\sqrt{6x}$
15. a. $\sqrt{5}\sqrt{2}\sqrt{15}$ b. $(2\sqrt{6})(\sqrt{3})(\sqrt{2})$ c. $(x\sqrt{3})(\sqrt{4x})$
16. a. $\sqrt{3}(2 - \sqrt{2})$ b. $\sqrt{7}(\sqrt{2} - \sqrt{14})$ c. $\sqrt{8}(\sqrt{2} - \sqrt{3})$
17. a. $(2 - \sqrt{3})(2 + \sqrt{3})$ b. $(\sqrt{5} - \sqrt{7})(\sqrt{5} + \sqrt{7})$ c. $(2 - \sqrt{3})(3 - 2\sqrt{3})$

Rationalize the denominator and simplify.

18. a. $\dfrac{\sqrt{27}}{\sqrt{3}}$ b. $\dfrac{\sqrt{12a}}{\sqrt{2}}$ c. $\dfrac{\sqrt{3a}\sqrt{ab}}{\sqrt{b}}$
19. a. $\sqrt{\dfrac{3}{5}}$ b. $\sqrt{\dfrac{12}{5}}$ c. $\sqrt{\dfrac{3}{2x}}$
20. a. $\dfrac{\sqrt{85}}{\sqrt{5x}}$ b. $\dfrac{\sqrt{3y}}{\sqrt{2x^2}}$ c. $\dfrac{4\sqrt{7x}}{\sqrt{2x}}$

479

16 SOLVING QUADRATIC EQUATIONS BY OTHER METHODS

In Chapter 14, we solved quadratic equations whose solutions are rational numbers by using factoring methods. Now that we have studied irrational numbers, we are ready to examine additional methods of solving quadratic equations.

16.1 EXTRACTION OF ROOTS

We can solve an equation of the form $x^2 - a = 0$ by first writing it in the form

$$x^2 = a \qquad (1)$$

We see that for a greater than or equal to zero, x is a number that, when multiplied by itself, yields a. Therefore, by the definition of a square root, x, must be a square root of a. Since a has two square roots, we have $\sqrt{a}$ and $-\sqrt{a}$ as solutions for Equation (1). That is,

$$(\sqrt{a})^2 = \sqrt{a}\,\sqrt{a} = a$$

and

$$(-\sqrt{a})^2 = (-\sqrt{a})(-\sqrt{a}) = a$$

We can use a special notation to represent both roots; that is, $x = \pm\sqrt{a}$.

Example 1 Solve $x^2 - 5 = 0$.

Solution By adding 5 to each member, we have

$$x^2 = 5$$

Therefore, $x = \pm\sqrt{5}$, and the solutions are $\sqrt{5}$ or $-\sqrt{5}$.

Sometimes, we must apply the multiplication and division properties of equality to obtain an equation in the form $x^2 = a$.

SOLVING QUADRATIC EQUATIONS BY OTHER METHODS

Example 2 Solve $\frac{2}{3}x^2 - 4 = 0$.

Solution Adding 4 to each member yields
$$\frac{2}{3}x^2 = 4$$
which is equivalent to
$$3\left(\frac{2}{3}\right)x^2 = 4(3)$$
$$2x^2 = 12$$
$$x^2 = 6$$
Therefore, $x = \pm\sqrt{6}$, and the solutions are $\sqrt{6}$ or $-\sqrt{6}$.

The method that we used in the above examples is sometimes referred to as **extraction of roots.** Note that to apply this method, we must get the squared term on one side of the equation and the constant term on the other.

We can also use extraction of roots to solve quadratic equations of the form
$$(x + k)^2 = d$$

Example 3 Solve $(x - 2)^2 = 9$.

Solution Extracting the roots, we obtain
$$x - 2 = \pm\sqrt{9} = \pm 3$$
Thus, $x - 2 = 3$ or $x - 2 = -3$. Solving for x yields the equations

$x - 2 = 3$ or $x - 2 = -3$
$x = 3 + 2$ $x = -3 + 2$
$x = 5$ $x = -1$

The solutions are $x = 5$, $x = -1$.

EXERCISES 16.1
Solve

Complete Example (see Examples 1 and 2).

$2y^2 = 72$
$y^2 = \underline{}$ (1) Write with y^2 as left-hand member.

$y = \pm\sqrt{36} = \pm\underline{}$ (2) Extract square root of each member. Simplify.

$y = 6$ or $y = \underline{}$ (3)

---

(1) 36   (2) 6   (3) −6

16.1 EXTRACTION OF ROOTS

1. $x^2 = 4$
2. $y^2 = 9$
3. $x^2 - 16 = 0$
4. $x^2 - 25 = 0$

5. $98 = 2x^2$
6. $12 = 3x^2$
7. $x^2 - 3 = 0$
8. $7 - z^2 = 0$

9. $y^2 - 10 = 0$
10. $3x^2 - 15 = 0$
11. $4x^2 - 24 = 0$
12. $7x^2 = 42$

13. $12 = z^2$
14. $24 = b^2$
15. $x^2 - 18 = 0$
16. $3x^2 - 24 = 0$

17. $3x^2 - 54 = 0$
18. $5x^2 - 100 = 0$

Complete Example (see Examples 1 and 2.)

$5x^2 - 3 = 2x^2 + 33$ Write with x^2 as left-hand member.

$3x^2 = \underline{}$
 (1)

$x^2 = 12$ Extract square root of each member.

$x = \pm\sqrt{12} = \underline{\phantom{\pm 2\sqrt{3}}}$
$\phantom{x = \pm\sqrt{12} = }$ (2)   Simplify radical expression.

$x = 2\sqrt{3}$ or $x = \underline{\phantom{-2\sqrt{3}}}$
$\phantom{x = 2\sqrt{3} \text{ or } x = }$ (3)

(1) 36   (2) $\pm 2\sqrt{3}$ (3) $-2\sqrt{3}$

19. $2x^2 - 4 = x^2$
20. $6x^2 + 3 = 4x^2 + 11$
21. $3 = x^2 - 2$

22. $4t^2 - 16 = 16$
23. $4y^2 - 10 = y^2 - 10$
24. $s^2 - 2 = 4 - s^2$

SOLVING QUADRATIC EQUATIONS BY OTHER METHODS

Complete Example (see Example 2)

$$\frac{1}{3}x^2 - \frac{3}{4} = 0$$

$$(\cancel{12}^4)\frac{1}{\cancel{3}}x^2 - (\cancel{12}^3)\frac{3}{\cancel{4}} = (12)0$$ 　　Multiply each member by LCD 12.

$$4x^2 - \underline{} = 0$$ 　　Write with x^2 as left-hand member.
$$(1)$$

$$4x^2 = \underline{}$$
$$(2)$$

$$x^2 = \frac{9}{4}$$ 　　Extract square root of each member.

$$x = \pm\sqrt{\frac{9}{4}} = \pm\frac{3}{2}$$

$$x = \frac{3}{2} \quad \text{or} \quad x = \underline{}$$
$$\phantom{x = \frac{3}{2} \quad \text{or} \quad x =}(3)$$

---

(1) 9　　　(2) 9　　　(3) $\frac{-3}{2}$

---

25. $\frac{1}{4}x^2 = 5$ 　　　　**26.** $\frac{2}{3}y^2 - 4 = 0$ 　　　　**27.** $\frac{2}{3}x^2 = 6$

28. $\frac{2x^2}{3} - 4 = \frac{x^2}{3}$ 　　**29.** $\frac{5x^2}{2} - 4 = 2x^2$ 　　**30.** $\frac{1}{2}x^2 - 4 = \frac{3}{2}$

Complete Example (see Example 3)

$$(x + 3)^2 = 25$$

$$x + 3 = \pm\sqrt{25} = \underline{}$$ 　　Extract square root of each member.
$$\phantom{x + 3 = \pm\sqrt{25} = }(1)$$ 　　Solve resulting first-degree equations.

$$x + 3 = 5 \quad \text{or} \quad x + 3 = \underline{}$$
$$\phantom{x + 3 = 5 \quad \text{or} \quad x + 3 = }(2)$$

$$x = \underline{} \quad \text{or} \quad x = -8$$
$$(3)$$

---

(1) ±5　　　(2) −5　　　(3) 2

16.1 EXTRACTION OF ROOTS

31. $(x-1)^2 = 4$ **32.** $(x+3)^2 = 9$ **33.** $(x-2)^2 = 25$ **34.** $(x+1)^2 = 36$

35. $(x-5)^2 = 1$ **36.** $(x+7)^2 = 1$ **37.** $(x-a)^2 = 25$ **38.** $(x+b)^2 = 4$

39. $(x-3)^2 = a^2$ **40.** $(x+5)^2 = b^2$ **41.** $(x-a)^2 = b^2$ **42.** $(x+a)^2 = b^2$

Complete Example (see Example 3)

$(x-2)^2 = 20$

$x - 2 = \pm\sqrt{20} = \underline{\qquad}$ (1)

$x - 2 = 2\sqrt{5}$ or $x - 2 = -2\sqrt{5}$

$x = 2 + 2\sqrt{5}$ or $x = \underline{\qquad}$ (2)

(1) $\pm 2\sqrt{5}$ (2) $2 - 2\sqrt{5}$

43. $(x+3)^2 = 2$ **44.** $(x-2)^2 = 3$ **45.** $(x+5)^2 = 5$ **46.** $(x-6)^2 = 7$

47. $(x+10)^2 = 8$ **48.** $(x-1)^2 = 12$ **49.** $(x-5)^2 = a$ **50.** $(x+2)^2 = a$

51. $(x+1)^2 = b$ **52.** $(x-7)^2 = b$ **53.** $(x-b)^2 = a$ **54.** $(x+b)^2 = a$

Solve each of the following equations for the indicated variable.

Complete Example

$K = \dfrac{v^2}{64},$ for v

$(64)K = (64)\dfrac{v^2}{64}$ Multiply each member by 64.

$64K = \underline{\qquad}$ (1)

$v^2 = 64K$ Extract the square root of each member and simplify.

$v = \pm\sqrt{64K} = \pm \underline{\qquad} \sqrt{K}$ (2)

$v = 8\sqrt{K}$ or $v = \underline{\qquad}$ (3)

(1) v^2 (2) 8 (3) $-8\sqrt{K}$

485

SOLVING QUADRATIC EQUATIONS BY OTHER METHODS

55. $x^2 - a = 0$, for x **56.** $b = x^2$, for x **57.** $\dfrac{x^2}{3} - b^2 a^3 = 0$, for x

58. $\dfrac{ax^2}{2} - b = 0$, for x **59.** $\dfrac{2y^2}{5} = \dfrac{b}{3}$, for y **60.** $\dfrac{y^2}{2} + a = \dfrac{2a}{3} + 2y^2$, for y

61. $s = \dfrac{1}{2} gt^2$, for t **62.** $V = \dfrac{1}{3} \pi r^2 h$, for r **63.** $A = 4\pi r^2$, for r

64. $C = bh^2 r$, for h **65.** $I = \dfrac{3k}{d^2}$, for d **66.** $F = \dfrac{k}{d^2}$, for d

16.2 COMPLETING THE SQUARE

So far, the methods we have used to solve quadratic equations apply to special cases only. Let us now develop a method that is applicable to any quadratic equation.

First, we will examine the squares of several binomials expressed as trinomials.

$$(x + 1)^2 = x^2 + x + x + 1 = x^2 + 2x + 1$$

$$(x - 2) = x^2 - 2x - 2x + 4 = x^2 - 4x + 4$$

$$\left(x + \dfrac{1}{3}\right)^2 = x^2 + \dfrac{1}{3}x + \dfrac{1}{3}x + \dfrac{1}{9} = x^2 + \dfrac{2}{3}x + \dfrac{1}{9}$$

$$\left(x - \dfrac{2}{5}\right)^2 = x^2 - \dfrac{2}{5}x - \dfrac{2}{5}x + \dfrac{4}{25} = x^2 - \dfrac{4}{5}x + \dfrac{4}{25}$$

In general, we have

$$(x + k)^2 = x^2 + kx + kx + k^2 = x^2 + 2kx + k^2$$

The coefficient of x is twice the second term of the binomial.

The constant term is the square of the second term of the binomial.

We want to use these observations to perform a process called **completing the square.** For example, if we have the expression $x^2 + 6x$, we want to find a value k^2, such that when we add k^2 to $x^2 + 6x$, we get the square of a binomial. That is, we want to find a value of k^2 such that

$$x^2 + 6x + k^2 = (x + k)^2$$

From the examples above, we know that 6 (the coefficient of x) must be twice k. Therefore, k must be $\dfrac{1}{2} \cdot 6$ or 3. Now, since k is 3, k^2 must be 3^2 or 9, and we can write

$$x^2 + 6x + 9 = (x + 3)^2$$

16.2 COMPLETING THE SQUARE

In general, if we are given an expression of the form $x^2 + bx$, we can *complete the square* by adding $\left[\frac{1}{2}b\right]^2$.

$$x^2 + bx + \left[\frac{1}{2}b\right]^2$$
Add to complete the square

We can now write this trinomial as the square of a binomial

$$\left(x + \frac{1}{2}b\right)^2$$

Using this procedure, we can write any quadratic equation in the form $(x + k)^2 = d$, which we learned to solve in the preceding section.

Example 1 Solve $x^2 - 6x - 7 = 0$.

Solution We first write the equation in the form

$$x^2 - 6x = 7$$

Now, by adding 9 [the *square* of $\frac{1}{2}(-6)$] to each member, we have

$$x^2 - 6x + 9 = 7 + 9$$

or

$$(x - 3)^2 = 16$$

We can now solve by extraction of roots to obtain

$$(x - 3)^2 = \pm\sqrt{16} = \pm 4$$
$$x - 3 = 4 \quad \text{or} \quad x - 3 = -4$$
$$x = 7 \quad \text{or} \quad x = -1$$

Thus, the solutions are 7 or -1.

In the event the coefficient on the second-degree term is not 1, we must divide each term in the equation by that coefficient before proceeding.

Example 2 Solve $2x^2 - 3x - 9 = 0$.

Solution We begin by dividing each term by 2

$$\frac{2x^2}{2} - \frac{3x}{2} - \frac{9}{2} = \frac{0}{2}$$

$$x^2 - \frac{3}{2}x - \frac{9}{2} = 0$$

SOLVING QUADRATIC EQUATIONS BY OTHER METHODS

Now, we will add $\frac{9}{2}$ to each member to obtain

$$x^2 - \frac{3}{2}x = \frac{9}{2}$$

We can complete the square by adding $\frac{9}{16}$ [the *square* of $\frac{1}{2}\left(\frac{-3}{2}\right)$] to each member to get

$$x^2 - \frac{3}{2}x + \frac{9}{16} = \frac{9}{2} + \frac{9}{16}$$

We can now express the left-hand member as a perfect square and simplify the right-hand member. We obtain

$$\left(x - \frac{3}{4}\right)^2 = \frac{(8)9}{(8)2} + \frac{9}{16}$$

$$\left(x - \frac{3}{4}\right)^2 = \frac{81}{16}$$

Extracting the roots, we obtain

$$x - \frac{3}{4} = \pm\sqrt{\frac{81}{16}} = \pm\frac{9}{4}$$

$$x - \frac{3}{4} = \frac{9}{4} \quad \text{or} \quad x - \frac{3}{4} = \frac{-9}{4}$$

$$x = \frac{9}{4} + \frac{3}{4} \qquad\qquad x = \frac{-9}{4} + \frac{3}{4}$$

$$= \frac{12}{4} = 3 \qquad\qquad = \frac{-6}{4} = \frac{-3}{2}$$

Thus, the solutions are 3 or $\frac{-3}{2}$.

The method of completing the square is summarized below.

To solve an equation by completing the square:

1. Write the equation in standard form.
2. If the coefficient of the second-degree term is different from 1, divide each term in the equation by this coefficient.
3. Write the equation with the constant term in the right-hand member.
4. Add to each member the square of one half the coefficient of the first-degree term.
5. Rewrite the equation with the left-hand member expressed as a perfect square; simplify the right-hand member.
6. Solve by extraction of roots.

To become familiar with the method of completing the square, solve all equations in the following exercises by this method even though many can be solved by factoring, which is usually easier.

16.2 COMPLETING THE SQUARE

EXERCISES 16.2
Solve by completing the square.

Complete Example (see Example 1)

$$x^2 - x - 2 = 0$$

Rewrite equation with constant term in the right-hand member.

$$x^2 - x = 2$$

Square one-half the coefficient of x and add to each member; $\left[\frac{1}{2}(-1)\right]^2 = \frac{1}{4}$.

$$x^2 - x + \underline{}_{(1)} = 2 + \frac{1}{4}$$

Rewrite left-hand member as a perfect square; simplify right-hand member.

$$\left(x - \underline{}_{(2)}\right)^2 = \frac{(4)2}{(4)1} + \frac{1}{4} = \underline{}_{(3)}$$

Extract square root of each member.

$$x - \frac{1}{2} = \pm\sqrt{\frac{9}{4}} = \pm\frac{3}{2}$$

Solve resulting first-degree equations.

$$x - \frac{1}{2} = \frac{3}{2} \quad \text{or} \quad x - \frac{1}{2} = \frac{-3}{2}$$

$$x = \frac{1}{2} + \frac{3}{2} \qquad\qquad x = \frac{1}{2} + \frac{-3}{2}$$

$$x = \underline{}_{(4)} \quad \text{or} \quad x = \underline{}_{(5)}$$

(1) $\frac{1}{4}$ (2) $\frac{1}{2}$ (3) $\frac{9}{4}$ (4) 2 (5) -1

1. $x^2 + 4x - 12 = 0$
2. $x^2 - 2x - 15 = 0$
3. $z^2 - 2z + 1 = 0$
4. $y^2 - y - 6 = 0$

5. $y^2 + y - 20 = 0$
6. $x^2 - x - 20 = 0$
7. $x^2 + 3x + 2 = 0$
8. $u^2 + 5u + 6 = 0$

9. $z^2 - 3z - 4 = 0$
10. $y^2 + 9y + 20 = 0$
11. $r^2 - 3r - 10 = 0$
12. $p^2 - 5p + 4 = 0$

13. $x^2 - 2x - 1 = 0$
14. $y^2 + 4y = 4$
15. $z^2 = 3z + 3$
16. $s^2 + 1 = -3s$

17. $t^2 = 3 - t$
18. $-x^2 - 6x = 1$

489

SOLVING QUADRATIC EQUATIONS BY OTHER METHODS

Complete Example (See Example 2.)

$$3x^2 - 2x - 5 = 0$$

Divide each term by 3, the coefficient of x^2, and rewrite with constant in the right-hand member.

$$x^2 - \frac{2}{3}x = \underline{}_{(1)}$$

Square one-half the coefficient of x and add to each member; $\left[\frac{1}{2}\left(\frac{-2}{3}\right)\right]^2 = \frac{1}{9}$.

$$x^2 - \frac{2}{3}x + \underline{}_{(2)} = \frac{5}{3} + \frac{1}{9}$$

Rewrite the left-hand member as a perfect square; simplify the right-hand member.

$$\left(x - \frac{1}{3}\right)^2 = \frac{(3)5}{(3)3} + \frac{1}{9} = \underline{}_{(3)}$$

Extract square root of each member.

$$x - \frac{1}{3} = \pm\sqrt{\frac{16}{9}} = \pm\frac{4}{3}$$

Solve resulting equations.

$$x - \frac{1}{3} = \frac{4}{3} \quad \text{or} \quad x - \frac{1}{3} = \frac{-4}{3}$$

$$x = \frac{1}{3} + \frac{4}{3} \qquad x = \frac{1}{3} + \frac{-4}{3}$$

$$= \underline{}_{(4)} \qquad = \underline{}_{(5)}$$

(1) $\frac{5}{3}$ (2) $\frac{1}{9}$ (3) $\frac{16}{9}$ (4) $\frac{5}{3}$ (5) -1

19. $4x^2 + 4x - 3 = 0$ 20. $4y^2 - 4y = 3$ 21. $2x^2 = 2 - 3x$

22. $6z^2 + 6 = 13z$ 23. $2t^2 - t - 15 = 0$ 24. $1 - r = 6r^2$

16.3 QUADRATIC FORMULA

Solving the general quadratic equation

$$ax^2 + bx + c = 0, \quad a \neq 0$$

by completing the square, we can obtain a formula expressing the solutions of the equation in terms of the coefficients a, b, and c. We can then solve any quadratic equation by simply substituting the numerical coefficients of the terms in the formula and evaluating the result.

16.3 QUADRATIC FORMULA

We complete the square in the general quadratic equation as follows:

$$ax^2 + bx + c = 0$$

$$ax^2 + bx = -c$$

$$x^2 + \frac{b}{a}x = \frac{-c}{a}$$

$$x^2 + \frac{b}{a}x + \frac{b^2}{4a^2} = \frac{-c}{a} + \frac{b^2}{4a^2}$$

$$\left(x + \frac{b}{2a}\right)^2 = \frac{-c(4a)}{a(4a)} + \frac{b^2}{4a^2}$$

$$\left(x + \frac{b}{2a}\right)^2 = \frac{b^2 - 4ac}{4a^2}$$

$$x + \frac{b}{2a} = \pm\sqrt{\frac{b^2 - 4ac}{4a^2}}$$

$$x = \frac{-b}{2a} \pm \frac{\sqrt{b^2 - 4ac}}{2a}$$

$$x = \frac{-b \pm \sqrt{b^2 - 4ac}}{2a}$$

The last equation is called the **quadratic formula.** Since this formula was developed from the quadratic equation in standard form, we should write any quadratic equation in standard form before attempting to determine values for a, b, and c to substitute in the formula. Furthermore, the sign on the coefficient must be substituted with the coefficient.

Example 1 Indicate the values of a, b, and c for the equation

$$-x + 3x^2 = 4$$

Solution We write the equation in the standard form

$$3x^2 - x - 4 = 0$$

from which

$$a = 3, \quad b = -1, \quad \text{and} \quad c = -4$$

If a quadratic equation has fractional coefficients, it is generally helpful to clear the equation of fractions before proceeding.

Example 2 Indicate the values of a, b, and c for the equation

$$\frac{2}{3} - \frac{1}{2}x = -2x^2$$

Solution We multiply each term by LCD 6 to get

$$(\overset{2}{\cancel{6}})\frac{2}{3} - (\overset{3}{\cancel{6}})\frac{1}{2}x = (6)(-2x^2)$$

$$4 - 3x = -12x^2$$

SOLVING QUADRATIC EQUATIONS BY OTHER METHODS

from which
$$12x^2 - 3x + 4 = 0$$
and
$$a = 12, \quad b = -3, \quad \text{and} \quad c = 4$$

We can use the quadratic formula to solve quadratic equations now that we know how to identify a, b, and c.

Example 3 Solve $3x^2 - x - 4 = 0$.

Solution From Example 1, $a = 3$, $b = -1$, and $c = -4$. We can now solve for x by substituting these values into the quadratic formula.

$$x = \frac{-b \pm \sqrt{b^2 - 4ac}}{2a}$$

$$x = \frac{-(-1) \pm \sqrt{(-1)^2 - 4(3)(-4)}}{2(3)}$$

$$= \frac{1 \pm \sqrt{1 - (-48)}}{6} = \frac{1 \pm \sqrt{49}}{6} = \frac{1 \pm 7}{6}$$

Simplifying, we obtain

$$x = \frac{1+7}{6} \quad \text{or} \quad x = \frac{1-7}{6}$$

$$x = \frac{8}{6} = \frac{4}{3} \qquad x = \frac{-6}{6} = -1$$

In actual practice, we should use the quadratic formula only when easier methods (factoring or extraction of roots) fail. Many of the following exercises are easier to solve by other methods, but they are included to indicate the complete generality of the quadratic formula. These exercises should be solved by use of the formula.

EXERCISES 16.3

In Exercises 1–20, indicate the values for a, b, and c to be substituted in the quadratic formula.

Complete Examples (see Example 1)

a. $x^2 = x + 2$
$x^2 - x - 2 = 0$
$a = 1, \quad b = \underline{\quad}_{(1)}, \quad c = \underline{\quad}_{(2)}$

b. $2x^2 = x$
$2x^2 - x = 0$ Write in standard form.
$a = 2, \quad b = \underline{\quad}_{(3)}, \quad c = \underline{\quad}_{(4)}$

---

(1) -1 (2) -2 (3) -1 (4) 0

---

1. $x^2 - 3x + 2 = 0$
2. $y^2 + 5y + 4 = 0$
3. $x^2 - x - 30 = 0$
4. $y^2 + 3y - 4 = 0$
5. $x^2 - 2x = 0$
6. $y^2 = 5y$
7. $4y^2 - 3 = 0$
8. $2y^2 - 1 = 0$

16.3 QUADRATIC FORMULA

9. $2x^2 = 7x - 6$
10. $6x^2 + x = 1$
11. $6x^2 = 5x - 1$
12. $3x^2 - 5 = 0$
13. $y^2 + 4 = 8y$
14. $x^2 = 7x$

Complete Example (see Example 2)

$$\frac{x}{3} = 4 - \frac{x^2}{2}$$

Multiply each term by LCD 6.

$$(\overset{2}{\cancel{6}})\frac{x}{\cancel{3}} = (6)4 - (\overset{3}{\cancel{6}})\frac{x^2}{\cancel{2}}$$

$$\underline{} = 24 - \underline{}$$
$\quad\ \ (1) \qquad\qquad\qquad (2)$

Write in standard form.

$3x^2 + 2x - 24 = 0$

$a = 3, \quad b = \underline{}, \quad c = \underline{}$
$\qquad\qquad\quad\ \ (3) \qquad\qquad\ (4)$

---

(1) $2x$ (2) $3x^2$ (3) 2 (4) -24

---

15. $x^2 = x + \frac{1}{2}$
16. $x^2 = \frac{15}{4} - x$
17. $2x^2 - 1 + \frac{7}{3}x = 0$
18. $y^2 + 1 = \frac{13}{6}y$
19. $\frac{9}{4}y^2 + \frac{3}{2}y - 2 = 0$
20. $\frac{x^2}{3} = \frac{x}{2} + \frac{3}{2}$

Solve by use of the quadratic formula. See Example 3.

Complete Example

$x^2 - x - 6 = 0$

$x = \dfrac{-b \pm \sqrt{b^2 - 4ac}}{2a}$

Substitute 1 for a, -1 for b, and -6 for c.

$x = \dfrac{-(-1) \pm \sqrt{(-1)^2 - 4(1)(-6)}}{2(1)}$

Perform indicated operations.

$x = \dfrac{1 \pm \sqrt{1 + 24}}{\underline{}}$
$\qquad\qquad\ (1)$

Simplify.

$x = \dfrac{1 \pm \sqrt{25}}{2} = \dfrac{1 \pm 5}{2}$

$x = \dfrac{1 + 5}{2} = \underline{}$ or $x = \dfrac{1 - 5}{2} = \underline{}$
$\qquad\qquad\quad\ \ (2) \qquad\qquad\qquad\qquad\qquad (3)$

---

(1) 2 (2) 3 (3) -2

SOLVING QUADRATIC EQUATIONS BY OTHER METHODS

21. $x^2 - 3x + 2 = 0$
22. $y^2 + 5y + 4 = 0$
23. $z^2 - 4z - 12 = 0$
24. $x^2 - x - 30 = 0$
25. $x^2 + 2x - 15 = 0$
26. $y^2 + 3y - 4 = 0$
27. $x^2 + 3x - 1 = 0$
28. $y^2 + 5y + 5 = 0$
29. $y^2 - 3y - 2 = 0$
30. $x^2 + x - 1 = 0$

Complete Example

$x^2 - 9 = 0$

$x = \dfrac{-b \pm \sqrt{b^2 - 4ac}}{2a}$ — Substitute 1 for a, 0 for b, and -9 for c.

$x = \dfrac{-(0) \pm \sqrt{(0)^2 - 4(1)(-9)}}{2(1)}$ — Perform indicated operations.

$x = \dfrac{\pm\sqrt{36}}{\underset{(1)}{}} = \dfrac{\pm 6}{2}$ — Simplify.

$x = \dfrac{+6}{2} = \underline{}_{(2)}$ or $x = \dfrac{-6}{2} = \underline{}_{(3)}$

(1) 2 (2) 3 (3) −3

31. $x^2 - 2x = 0$ (Hint: $c = 0$.)
32. $x^2 - 4 = 0$ (Hint: $b = 0$.)
33. $y^2 = 5y$
34. $z^2 = 9$
35. $7x = x^2$
36. $16 = y^2$
37. $z^2 - 3z = 0$
38. $x^2 = 1$
39. $4y^2 - 3 = 0$
40. $2y^2 - 1 = 0$

Complete Example

$2x^2 = 2 - 3x$

$2x^2 \times 3x - 2 = 0$ — Write in standard form.

$x = \dfrac{-b \pm \sqrt{b^2 - 4ac}}{2a}$ — Substitute 2 for a, 3 for b, and -2 for c.

$x = \dfrac{-(3) \pm \sqrt{(3)^2 - 4(2)(-2)}}{2(2)}$ — Perform indicated operations.

$x = \dfrac{-3 \pm \sqrt{9 + 16}}{\underset{(1)}{}}$ — Simplify.

$x = \dfrac{-3 \pm \sqrt{25}}{4} = \dfrac{3 \pm 5}{4}$

$x = \dfrac{-3 + 5}{4} = \underline{}_{(2)}$ or $x = \dfrac{-3 - 5}{4} = \underline{}_{(3)}$

(1) 4 (2) $\dfrac{1}{2}$ (3) −2

16.4 GRAPHING QUADRATIC EQUATIONS IN TWO VARIABLES

41. $2x^2 = 7x - 6$
42. $5 = 6y - y^2$
43. $6x^2 + x = 1$
44. $-z = 3 - 2z^2$

45. $6x^2 - 13x - 5 = 0$
46. $6x^2 = 5x - 1$
47. $x^2 = 2x + 1$
48. $x^2 = 2x + 4$

49. $y^2 - 4y - 2 = 0$
50. $z^2 + 4 = 8z$
51. $2x^2 - 3x - 1 = 0$
52. $3x^2 - x - 1 = 0$

Complete Example

$$\frac{x^2}{3} = \frac{1}{3} - \frac{x}{2}$$

Multiply by LCD 6.

$$(\overset{2}{\cancel{6}})\frac{x^2}{\cancel{3}} = (\overset{2}{\cancel{6}})\frac{1}{\cancel{3}} - (\overset{3}{\cancel{6}})\frac{x}{\cancel{2}}$$

$$\underline{} = \underline{} - \underline{}$$
$\quad$ (1) $\qquad\qquad$ (2) $\qquad\qquad$ (3)

Write in standard form. Solve as shown in preceding Complete Example.

(1) $2x^2$ $\qquad$ (2) 2 $\qquad$ (3) $3x$

53. $x^2 = \frac{15}{4} - x$
54. $2x^2 - 1 + \frac{7}{3}x = 0$
55. $y^2 + 1 = \frac{13}{6}y$

56. $\frac{9}{4}y^2 + \frac{3}{2}y - 2 = 0$
57. $\frac{1}{3}x^2 = \frac{1}{2}x + \frac{3}{2}$
58. $\frac{3}{5}x^2 - x - \frac{2}{5} = 0$

16.4 GRAPHING QUADRATIC EQUATIONS IN TWO VARIABLES

In Chapter 12, we learned how to graph first-degree equations in two variables. We can now use the same procedure to graph second-degree equations of the form

$$y = ax^2 + bx + c$$

SOLUTIONS OF EQUATIONS IN TWO VARIABLES

As before with first-degree equations in two variables, solutions of Equation (1) are ordered pairs.

Example 1 Find a second component such that each ordered pair satisfies the equation $y = x^2 - 3x + 1$.

a. $(-1, ?)$
b. $(0, ?)$
c. $(1, ?)$
d. $(2, ?)$
e. $(3, ?)$
f. $(4, ?)$

495

SOLVING QUADRATIC EQUATIONS BY OTHER METHODS

Solutions To find the second component *y*, we substitute the given value of *x* into the expression $x^2 - 3x + 1$.

a. For $x = -1$, $y = (-1)^2 - 3(-1) + 1 = 5$; $(-1, 5)$
b. For $x = 0$, $y = (0)^2 - 3(0) + 1 = 1$; $(0, 1)$
c. For $x = 1$, $y = (1)^2 - 3(1) + 1 = -1$; $(1, -1)$
d. For $x = 2$, $y = (2)^2 - 3(2) + 1 = -1$; $(2, -1)$
e. For $x = 3$, $y = (3)^2 - 3(3) + 1 = 1$; $(3, 1)$
f. For $x = 4$, $y = (4)^2 - 3(4) + 1 = 5$; $(4, 5)$

GRAPHS OF SECOND-DEGREE EQUATIONS IN TWO VARIABLES

The graph of a second-degree equation in two variables is not a straight line. Hence, we must plot *more than two points* to determine the graph. The graph as shown in Figure 16.1 is called a **parabola**. The significant features of a parabola include the maximum (high) or minimum (low) point of the curve and the points where the curve intersects the axes. We should therefore choose ordered pairs for graphing that will display these features. The ability to determine the necessary number of ordered pairs and to choose the proper first components comes with experience.

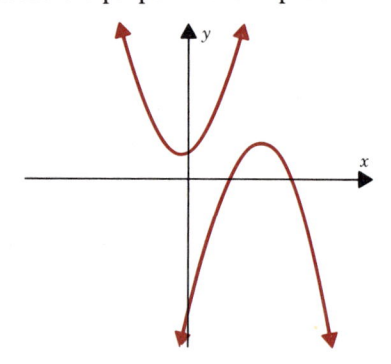

Figure 16.1

Example 2 Graph $y = x^2 - 3x + 1$. For *x* components, use all integers between -2 and 5.

Solution In Example 1, we determined solutions for this equation for *x* values between -2 and 5. We can list these solutions in a table. We then graph these ordered pairs and connect them with a smooth curve.

x	y
-1	5
0	1
1	-1
2	-1
3	1
4	5

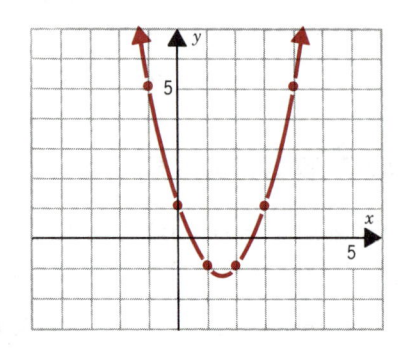

16.4 GRAPHING QUADRATIC EQUATIONS IN TWO VARIABLES

EXERCISES 16.4

In Exercises 1–6, find a second component such that each of the ordered pairs satisfies the equation $y = x^2 - 2x - 3$.

Complete Example (see Example 1.)

$(-3, ?)$

$y = (\underline{})^2 - 2(\underline{}) - 3$
   (1)   (2)

Substitute -3 for x in the equation $y = x^2 - 2x - 3$.

Simplify.

$= 9 + \underline{} - 3$
          (3)

$= \underline{}$
    (4)

Ans. $(-3, 12)$

(1) -3 (2) -3 (3) 6 (4) 12

1. $(0, ?)$
2. $(1, ?)$
3. $(-1, ?)$
4. $(2, ?)$
5. $(-2, ?)$
6. $(3, ?)$

In Exercises 7–12, find a second component such that each of the ordered pairs satisfies the equation $y = x^2 + x - 2$.

7. $(0, ?)$
8. $(-1, ?)$
9. $(1, ?)$
10. $(2, ?)$
11. $(-2, ?)$
12. $(-3, ?)$

In Exercises 13–18, find a second component such that each of the ordered pairs satisfies the equation $y = x^2 - 7x + 12$.

13. $(0, ?)$
14. $(1, ?)$
15. $(2, ?)$
16. $(3, ?)$
17. $(4, ?)$
18. $(5, ?)$

19. Graph the ordered pairs obtained in Exercises 1–6 and connect the points with a smooth curve.

20. Graph the ordered pairs obtained in Exercises 7–12 and connect the points with a smooth curve.

SOLVING QUADRATIC EQUATIONS BY OTHER METHODS

In Exercises 21–28, graph each equation. For x components, use all integers between the given numbers. See Example 2.

Complete Example

$$y = x^2 - 2x + 1 \quad (-2 \text{ and } 4)$$

Using a table, determine ordered pairs that are solutions to the equation. Graph these ordered pairs and connect them with a smooth curve.

x	y
−1	4
0	__(1)__
1	0
2	1
3	__(2)__

(3)

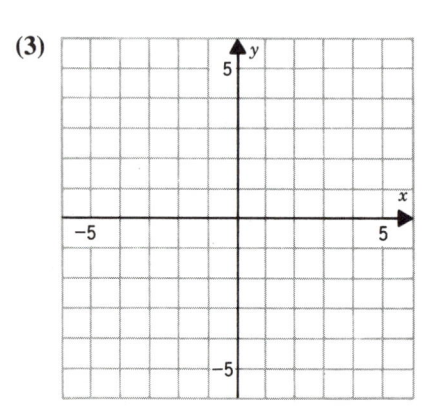

(1) 1   (2) 4   (3)

21. $y = x^2 - 2x$, $(-3 \text{ and } 5)$
22. $y = x^2 - 4$, $(-3 \text{ and } 3)$
23. $y = x^2 + 2x$, $(-5 \text{ and } 3)$
24. $y = x^2 + 1$, $(-3 \text{ and } 3)$
25. $9 - x^2 = y$, $(-4 \text{ and } 4)$
26. $3x - x^2 = y$, $(-2 \text{ and } 5)$
27. $y = x^2 - 5x + 4$, $(-2 \text{ and } 6)$
28. $y = x^2 + x - 6$, $(-4 \text{ and } 3)$

29. In Exercises 21–24, estimate from the graph the values of x for which y is 0.

30. In Exercises 25–28, estimate from the graph the values of x for which y is 0.

Graph each equation.

31. $y = x^2$
32. $y = x^2 + 3$
33. $y = 9 - x^2$
34. $y = 4 - x^2$
35. $x^2 - 2x - 3 = y$
36. $x^2 + 2x - 3 = y$

CHAPTER SIXTEEN REVIEW EXERCISES

Solve for x, y, or z by any method.

1. a. $x^2 - 25 = 0$ b. $3x^2 - 27 = 0$
2. a. $6x^2 - 42 = 0$ b. $\dfrac{2y^2}{3} = 4$
3. a. $(z - 2)^2 = 9$ b. $(z + 3)^2 = 1$
4. a. $(x - 7)^2 = 16$ b. $(x - a)^2 = c^2$
5. a. $(x + 3)^2 = a$ b. $(y + a)^2 = 4$
6. a. $x^2 + 3x - 4 = 0$ b. $y^2 = 3y + 3$
7. a. $2x^2 + 4x + 2 = 0$ b. $y^2 - y = 2$

8. Solve by completing the square: $y^2 - 4y = 5$.

In Exercises 9–11, solve each equation using the quadratic formula.

9. $\dfrac{x^2}{4} = \dfrac{15}{4} - \dfrac{x}{2}$ 10. $x^2 + 3x + 1 = 0$ 11. $\dfrac{x^2}{4} + 1 = \dfrac{13}{12}x$

12. Graph: $y = x^2 + 3x$.
13. Graph: $y = x^2 + 3x - 4$.
14. Graph: $y = x^2 - x - 6$.
15. In Exercise 14, estimate from the graph the values of x for which y is 0.

SOLVING QUADRATIC EQUATIONS BY OTHER METHODS

SUMMARY: PART II

PROPERTIES OF ADDITION AND MULTIPLICATION

$a + b = b + a$
$a \cdot b = b \cdot a$ *commutative properties*

$(a + b) + c = a + (b + c)$
$(a \cdot b) \cdot c = a \cdot (b \cdot c)$ *associative properties*

$a \cdot (b + c) = a \cdot b + a \cdot c$ *distributive property*

PROPERTIES OF FRACTIONS

$$\frac{a}{b} = a \cdot \frac{1}{b}$$

$$\frac{a}{b} = \frac{-a}{-b} = -\frac{-a}{b} = -\frac{a}{-b}$$

$$\frac{-a}{b} = \frac{a}{-b} = -\frac{a}{b} = -\frac{-a}{-b}$$

$$\frac{a}{b} = \frac{a \cdot c}{b \cdot c}$$

$$\frac{a}{b} \cdot \frac{c}{d} = \frac{ac}{bd} \quad \text{and} \quad a\left(\frac{b}{c}\right) = \frac{ab}{c}$$

$$\frac{a}{b} \div \frac{c}{d} = \frac{a}{b} \cdot \frac{d}{c}$$

$$\frac{a}{c} + \frac{b}{c} = \frac{a + b}{c} \quad \text{and} \quad \frac{a}{c} - \frac{b}{c} = \frac{a}{c} + \frac{-b}{c} = \frac{a - b}{c}$$

$$\frac{a}{b} + \frac{d}{c} = \frac{(c)a}{(c)b} + \frac{d(b)}{c(b)} = \frac{ca + db}{bc}$$

$$\text{If} \quad \frac{a}{b} = \frac{c}{d}, \quad \text{then} \quad ad = bc$$

PROPERTIES OF EXPONENTS

$a^n = a \cdot a \cdot a \cdot \ldots \cdot a$ (n factors, n a natural number)

$\dfrac{a^m}{a^n} = a^{m-n}$, if m is greater than n; $\dfrac{a^m}{a^n} = \dfrac{1}{a^{n-m}}$, if n is greater than m

$a^0 = a;$ $a^{-n} = \dfrac{1}{a^n}$

PROPERTIES OF EQUALITY

If $a = b$, then $b = a$ *symmetric property*

SUMMARY: PART II

The following equations are equivalent:

$$a = b, \quad a + c = b + c, \quad a - c = b - c$$
$$\frac{a}{c} = \frac{b}{c} \ (c \neq 0), \quad \text{and} \quad a \cdot c = b \cdot c \ (c \neq 0)$$

PROPERTIES OF INEQUALITY

$a < b$, $a + c < b + c$, and $a - c < b - c$ are equivalent inequalities.

If $c < 0$, then $a < b$ and $\frac{a}{c} < \frac{b}{c}$ are equivalent inequalities.

If $c < 0$, then $a < b$, $ac > bc$, and $\frac{a}{c} > \frac{b}{c}$ are equivalent inequalities.

QUADRATIC FORMULA

$$x = \frac{-b \pm \sqrt{b^2 - 4ac}}{2a}$$

can be used to solve any quadratic equation, where a, b, and c are the coefficients in the quadratic equation $ax^2 + bx + c = 0$.

PROPERTIES OF GRAPHS

The graph of a first-degree equation in two variables is a line in the plane.

The *slope* of a line containing the points $P_1(x_1, y_1)$ and $P_2(x_2, y_2)$ is given by

$$m = \frac{y_2 - y_1}{x_2 - x_1} \ (x_2 \neq x_1)$$

Two lines are parallel if they have the same slope ($m_1 = m_2$).
Two lines are perpendicular if the product of their slopes is -1 ($m_1 \cdot m_2 = -1$).
The **point-slope form** of a line with slope m and passing through the point (x_1, y_1) is

$$y - y_1 = m(x - x_1)$$

The *slope-intercept form* of a line with slope m and y intercept b is

$$y = mx + b$$

The graph of a first-degree inequality in two variables is a half-plane.

The graph of

$$y = ax^2 + bx + c$$

is a **parabola**.

PROPERTIES OF RADICALS

$$\sqrt{a}\sqrt{a} = (\sqrt{a})^2 = a$$
$$\sqrt{ab} = \sqrt{a}\sqrt{b} \quad \text{or} \quad \sqrt{a}\sqrt{b} = \sqrt{ab}$$
$$\sqrt{\frac{a}{b}} = \frac{\sqrt{a}}{\sqrt{b}} \quad \text{or} \quad \frac{\sqrt{a}}{\sqrt{b}} = \sqrt{\frac{a}{b}}$$

SOLVING QUADRATIC EQUATIONS BY OTHER METHODS

CUMULATIVE REVIEW FOR PART II

1. Write $24x^3y^2$ in completely factored form.
2. If $x = -2$ and $y = -4$, find the value of $\dfrac{x^2 + 2x - 4}{y}$.
3. For what value of x is $\dfrac{3x - 2}{x - 1}$ meaningless?
4. Simplify: $\dfrac{6a^3b^2}{3ab^2}$.
5. Solve for x: $3x - a = \dfrac{2ax - a^2}{a}$.
6. Solve the system: $3x + 4y = 11$
   $2x - y = 0$.
7. Represent $\dfrac{3}{x + y} + \dfrac{4}{x^2 + xy}$ as a single fraction.
8. Find three ordered pairs that are solutions of $2x - y = 4$.
9. Simplify: $3x\sqrt{x^2y} + 2\sqrt{x^4y}$.
10. The sum of two numbers is a. If one number is x, the other is __?__.
11. How many solutions has the equation $2x^2 - 8 = 0$?
12. What conclusions can be drawn concerning a and b if $ab = 0$?
13. The sum of two numbers is 18. If one number is -9, the other is __?__.
14. One number is four times another. If their sum is 40, find the numbers.
15. The sum of two numbers is 16, and their product is 63. Find the numbers.
16. Simplify: $\dfrac{4^2 - 2^2}{3} - \dfrac{4^2 + 2^2}{4}$.
17. If $a = 0$, $b = -1$, and $c = 2$, find the value of $a(b^2 + c^2)$.
18. Write $-3, \dfrac{3}{8}, \dfrac{3}{7}, 3, 0, \dfrac{-5}{2}$, and $\dfrac{-5}{3}$ in order from smallest to largest.
19. Factor: $x^3 - 3x^2 + 2x$.
20. Multiply: $(2x - 3)(x + 4)$.
21. Write $-\dfrac{x + 2}{-3}$ in standard form.
22. Simplify: $(x - 3)(x + 2) - (x^2 - 3x)$.
23. Represent $\dfrac{3}{x^2 + x} - \dfrac{2}{x + 1}$ as a single fraction.
24. Express $\dfrac{3}{x - 1}$ as a fraction with a denominator of $x^2 + 2x - 3$.
25. Simplify: $\dfrac{3x^2 - x - 2}{2x^2 + x - 3}$.
26. A positive number has __?__ square roots.
27. If $a = -4$, $b = 2$, and $c = 0$, find the value of $\dfrac{2a - 4}{3 - bc}$.
28. Multiply: $(-3ab)(2c)(b^2)$.
29. Represent $\dfrac{a - b}{2} + \dfrac{a - 3b}{3}$ as a single fraction.

CUMULATIVE REVIEW FOR PART II

30. Reduce to lowest terms: $\dfrac{3a^2x - 3ax}{24a^2x^2 + 60ax^2}$.
31. Divide: $(x^4 + 3x^3 - x^2)$ by (x^2).
32. Factor: $4x^2 - 12x + 9$.
33. Solve: $3(x - 2) - 7 = 5(2x + 5) - 3$.
34. Solve: $\dfrac{y - 6}{5} = \dfrac{12y + 3}{5} - 3y + 3$.
35. The sum of two numbers is 146. If n represents the smaller of the two numbers, represent the larger number in terms of n.
36. Write in terms of x, the number of cents in x quarters.
37. Solve the system: $\dfrac{x}{2} - \dfrac{y}{3} = 1$
    $x - \dfrac{2y}{3} = 2$.
38. Write in scientific notation.
    a. 3,270,000   b. 0.0000649
39. Solve: $2x^2 = 5x + 3$.
40. The sum of two numbers is 154. If the larger number is eight less than twice the smaller, what are the numbers?
41. What is the slope of the line with the equation $3x + 2y = 5$?
42. Simplify: $\sqrt{18} - 2\sqrt{5} - \sqrt{20} + \sqrt{50}$.
43. Simplify: $(2 - \sqrt{3})(2 + \sqrt{3})$.
44. Rationalize the denominator and simplify: $\dfrac{\sqrt{4x}\sqrt{9x}}{\sqrt{6x}}$.
45. Rationalize the denominator and simplify: $\dfrac{3\sqrt{2y}}{\sqrt{3y}}$.
46. Solve the system $2x - 3 = y$ by algebraic methods.
    $3x + 2y = 15$
47. Solve: $\dfrac{7}{8} = \dfrac{42}{y + 4}$.
48. Find the equation of a line passing through the points (1, 1) and (3, 5).
49. Solve $3y^2 - 2y = y$.
50. If S varies directly with T, and $S = 4.2$ when $T = 12.3$, find T when $S = 12.6$.

State if the two expressions are equivalent. If they are not equivalent, use a numerical example to show that they are not equivalent.

51. $-\dfrac{2x - 1}{3}$; $\dfrac{-2x - 1}{3}$ 52. $-a^2$; $(-a)^2$
53. $\dfrac{3 + 2x}{2x}$; 3   54. $\dfrac{2x + 4}{2}$; $x + 4$
55. $3 + 4x$; $7x$ 56. $3 - (x - 2)$; $3 - x + 2$
57. $\dfrac{2x + 5}{2}$; $x + 5$ 58. $\dfrac{x}{2} + \dfrac{x}{2}$; x
59. $-(5 - y)$; $-5 - y$ 60. $x^4 \cdot x^3$; x^{12}
61. $-\dfrac{3 - x}{2}$; $\dfrac{-3 + x}{2}$ 62. $3x^{-1}$; $\dfrac{1}{3x}$

SOLVING QUADRATIC EQUATIONS BY OTHER METHODS

63. $(x + 3)^2$; $x^2 + 9$
64. $\dfrac{4 + y}{8 + y}$; $\dfrac{1}{2}$
65. $\dfrac{4 + x}{4}$; x
66. $x^2 \cdot x^3$; x^5
67. $(xy)^0$; 0
68. $(\sqrt{x} + \sqrt{y})^2$; $x + y$
69. $\dfrac{2x + 6}{2}$; $x + 3$
70. $-\dfrac{x + 2}{3}$; $\dfrac{-x + 2}{3}$
71. $\dfrac{3 + 2x}{3}$; $2x$
72. $\dfrac{x}{2} + \dfrac{y}{2}$; $\dfrac{x + y}{4}$
73. $-2x^2$; $4x^2$
74. $5 - 2\sqrt{x}$; $3\sqrt{x}$
75. $(4x)^{-1}$; $\dfrac{1}{4x}$
76. $\sqrt{x + y}$; $\sqrt{x} + \sqrt{y}$
77. $x - y$; $y - x$
78. $(2x + 3) \cdot \dfrac{1}{2x}$; 3
79. $\dfrac{x^2 - 1}{x - 1}$; x
80. $(4 + x)^0$; 1

PART III
GEOMETRY

17 ELEMENTS OF GEOMETRY

17.1 INTRODUCTORY CONCEPTS

Just as numbers are basic elements of arithmetic and algebra, *points, lines,* and *planes* are basic elements of geometry. These objects and the symbols used to identify them are listed in Table 17.1.

TABLE 17.1

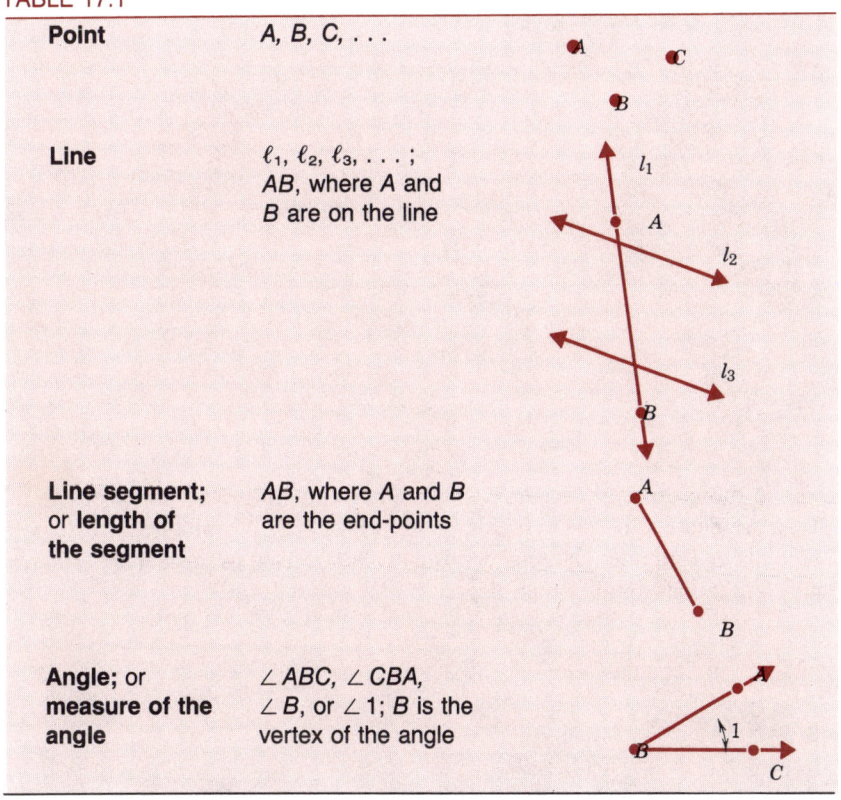

Point	A, B, C, ...	
Line	$\ell_1, \ell_2, \ell_3, \ldots$; AB, where A and B are on the line	
Line segment; or **length of the segment**	AB, where A and B are the end-points	
Angle; or **measure of the angle**	$\angle ABC$, $\angle CBA$, $\angle B$, or $\angle 1$; B is the vertex of the angle	

ELEMENTS OF GEOMETRY

In Figure 17.1, angles 1 and 2 share a common vertex and a common side but no interior points between them. Such angles are called **adjacent** angles. In this case, $\angle ABD = \angle 1 + \angle 2$. If $\angle 1 = \angle 2$, then BC is called the **bisector** of $\angle ABD$.

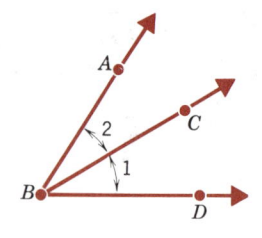

Figure 17.1

MEASURES OF ANGLES

A protractor such as that shown in Figure 17.2 is used to measure angles. The measures of several angles are shown in the figure. You should visualize the semicircle from A to C as being divided into 180 equal parts, each of which is called a **degree** and denoted by the symbol °.

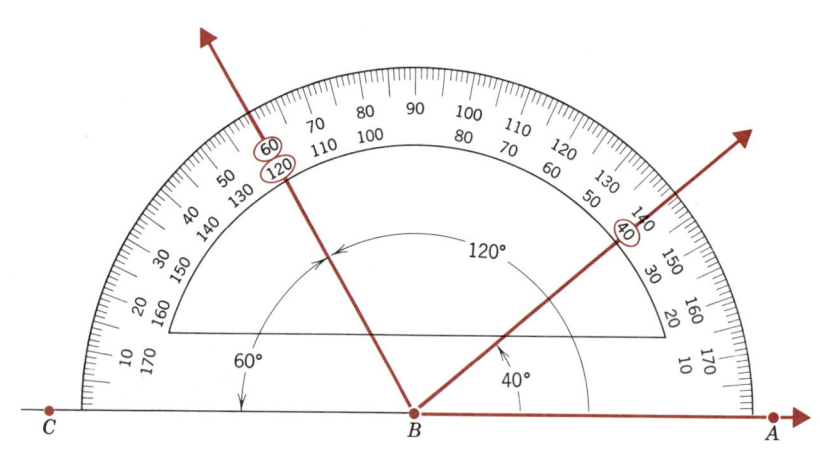

Figure 17.2

To determine more precise measures, smaller units of **minutes** and **seconds** are used. These are denoted by the symbols ′ and ″. There are 60 minutes in 1 degree and 60 seconds in 1 minute. Decimal forms can also be used as measures of angles that require gradations smaller than degree units.

Example 1 Express 28°24′ in decimal form

Solution Since 60′ = 1°,

17.1 INTRODUCTORY CONCEPTS

$$24' = \frac{24}{60}(1°) = \frac{2}{5}(1°) = 0.4°$$

Thus $28°24' = 28.4°$.

Example 2 Express $30.4°$ in degrees and minutes.

Solution Since $1° = 60'$,

$$0.4° = 0.4(60') = 24'$$

Thus $30.4° = 30°24'$.

Two angles are called **complementary** if their sum is $90°$. Two angles are called **supplementary** if their sum is $180°$.

Example 3 In Figure a, $\angle BAC$ and $\angle CAD$ are complementary since their sum is $20° + 70° = 90°$. In Figure b, $\angle A$ and $\angle B$ are complementary since their sum is $30° + 60° = 90°$, and $\angle B$ and $\angle C$ are supplementary since their sum is $60° + 120° = 180°$.

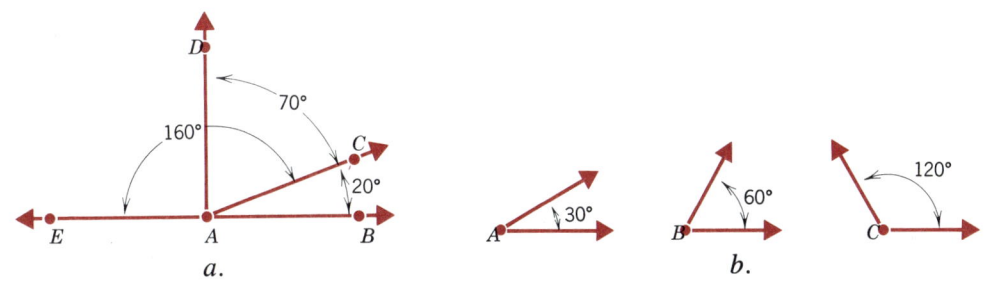

NAMES OF ANGLES

Angles are named according to their measures, as shown in Table 17.2.

TABLE 17.2

Name of Angle	Measure of Angle
Acute angle:	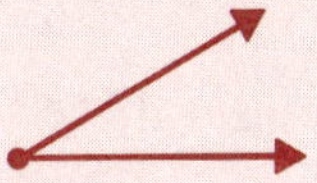 Less than $90°$
Right angle:	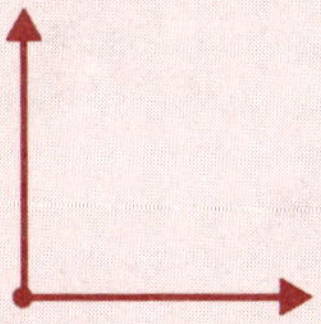 Equals $90°$
Obtuse angle:	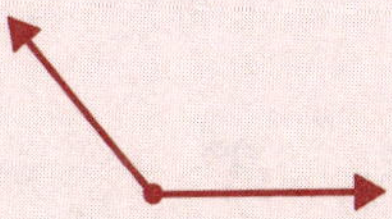 Between $90°$ and $180°$
Straight angle:	 Equals $180°$

ELEMENTS OF GEOMETRY

EXERCISES 17.1
Express each measure in decimal form. See Example 1.

Complete Examples

a. 24°12′

$12' = \frac{12}{60}(1°) = \frac{1}{5}(1°)$

= _____ (1)

Ans. 24°12′ = _____ (2)

b. 16°45′

$45' = \frac{45}{60}(1°) = \frac{3}{4}(1°)$

= _____ (3)

Ans. 16°45′ = _____ (4)

---

(1) 0.2°   (2) 24.2°   (3) 0.75°   (4) 16.75°

---

1. 24′
2. 15′
3. 6°36′
4. 28°48′

5. 15°6′
6. 48°3′
7. 107°9′
8. 221°18′

Express each measure in degrees and minutes. See Example 2.

Complete Examples

a. 12.7°

0.7° = 0.7(60′)

= _____ (1)

Ans. 12.7° = _____ (2)

b. 14.55°

0.55° = 0.55(60′)

= _____ (3)

Ans. 14.55° = _____ (4)

---

(1) 42′   (2) 12°42′   (3) 33′   (4) 14°33′

---

9. 0.4°
10. 0.6°
11. 5.9°
12. 21.3°

13. 15.35°
14. 29.45°
15. 154.65°
16. 240.85°

510

17.1 INTRODUCTORY CONCEPTS

State the measure of a. the complement and b. the supplement of an angle with the given measure. See Example 3.

Complete Examples

a. 46°12′

The complement of 46°12′ is
90° − 46°12′ = 89°60′ − 46°12′ = _____(1)_____ .

The supplement of 46°12′ is
180° − 46°12′ = _____(2)_____ .

b. 57.3°

The complement of 57.3° is
90° − 57.3° = _____(3)_____ .

The supplement of 57.3° is
180° − 57.3° = _____(4)_____ .

---

(1) 43°48′   (2) 133°48′   (3) 32.7°   (4) 122.7°

---

17. 28°
18. 54°
19. 86°10′
20. 44°54′

21. 27.4°
22. 88.6°
23. 10.42°
24. 22.88°

In the figure AD and CE are straight lines, OB bisects ∠AOC, ∠3 is a right angle, and the measure of ∠6 is 30°.

a. Find the measure of each angle.
b. Specify whether the angle is an acute, right, obtuse, or straight angle.

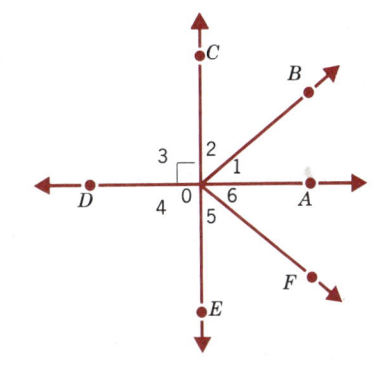

25. ∠AOC
26. ∠BOC
27. ∠BOF
28. ∠DOE

29. ∠BOE
30. ∠BOD
31. ∠DOF
32. ∠COE

33. Name the angle complementary to ∠1.
34. Name the angle complementary to ∠6.

35. Name the angle supplementary to ∠1.
36. Name the angle supplementary to ∠5.

ELEMENTS OF GEOMETRY

Solve. Refer to page 264 for the six steps used to solve word problems.

Complete Example

The measure of the angle is equal to four-fifths that of its complement. Find the measure of each angle.

Steps 1–2 Measure of the larger angle: x

Measure of the smaller angle: _____
(1)

Step 3 Not applicable.

Step 4 The sum of the measures of complementary angles equals 90°.

$$x + \frac{4}{5}x = \underline{\qquad}$$
(2)

Step 5 Solve the equation.

$$x + \frac{4}{5}x = 90.$$

$$5\left(x + \frac{4}{5}x\right) = (90)5$$

$$5x + \underline{\qquad} = 450$$
(3)

$$9x = 450$$

$$x = \underline{\qquad}$$
(4)

Step 6 The measures of the angles are 50° and $\frac{4}{5}(50)$ or _____ .
(5)

---

(1) $\frac{4}{5}x$ (2) 90   (3) $4x$ (4) 50   (5) 40

---

37. The measure of an angle is equal to one-fifth that of its complement. Find the measure of each angle.

38. The measure of an angle is equal to four-fifths that of its supplement. Find the measure of each angle.

39. The measure of an angle is equal to 30° less than twice its supplement. Find the measure of each angle.

40. The measure of an angle is equal to 30° more than three times its complement. Find the measure of each angle.

17.2 INTERSECTING LINES AND PARALLEL LINES

Nonadjacent angles formed by two intersecting lines are called **vertical angles** and have the same measure.

Example 1 In the figure, $\angle 1$ and $\angle 3$ are vertical angles, and so are $\angle 2$ and $\angle 4$.

17.2 INTERSECTING LINES AND PARALLEL LINES

PERPENDICULAR LINES

If two intersecting lines form angles of 90°, the lines are said to be **perpendicular.** The symbol ⊥ means "is perpendicular to."

Example 2 In the figure, $\ell_1 \perp \ell_2$.

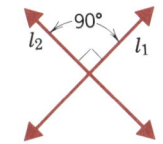

PARALLEL LINES

If two lines in a plane do not intersect, the lines are said to be **parallel.** This relationship is expressed in symbols as $\ell_1 \parallel \ell_2$.

Example 3 In the figure, $\ell_1 \parallel \ell_2$.

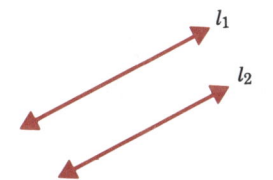

A line that intersects two or more other lines is called a **transversal** of those lines. When two lines are intersected by a transversal, certain pairs of the angles formed are given special names. Those angles occupying the same position relative to the transversal are called **corresponding angles.** Angles on opposite sides of the transversal have names involving the word **alternate** and are referred to as **alternate exterior** or **alternate interior** angles, depending on the position.

Example 4 In the figure, ℓ_1 is a *transversal* of ℓ_2 and ℓ_3. Also,

$\angle 1$ and $\angle 5$, $\angle 2$ and $\angle 6$,
$\angle 3$ and $\angle 7$, $\angle 4$ and $\angle 8$ are *corresponding angles*.

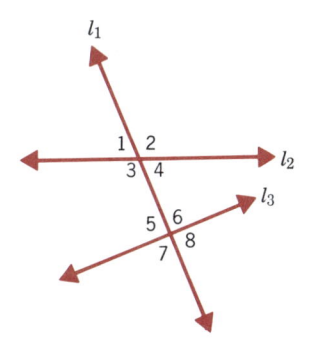

In addition, $\angle 1$ and $\angle 8$, and $\angle 2$ and $\angle 7$ are called **alternate exterior angles,** while $\angle 4$ and $\angle 5$, and $\angle 3$ and $\angle 6$ are called **alternate interior angles.**

513

ELEMENTS OF GEOMETRY

If two lines cut by a transversal are parallel, corresponding angles are equal, alternate interior angles are equal, and alternate exterior angles are equal.

Example 5 In the figure, $l_1 \parallel l_2$ because the angles are corresponding angles,

$$\angle 1 = \angle 5, \quad \angle 3 = \angle 7, \quad \angle 2 = \angle 6, \quad \text{and} \quad \angle 4 = \angle 8;$$

because the angles are alternate interior angles,

$$\angle 3 = \angle 6 \quad \text{and} \quad \angle 4 = \angle 5;$$

and because the angles are alternate exterior angles,

$$\angle 1 = \angle 8 \quad \text{and} \quad \angle 2 = \angle 7.$$

Furthermore, because they are vertical angles,

$$\angle 1 = \angle 4, \quad \angle 2 = \angle 3, \quad \angle 5 = \angle 8, \quad \text{and} \quad \angle 6 = \angle 7.$$

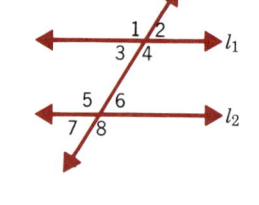

BISECTORS

A point or line (or line segment) that separates a given line segment into two equal parts is called a **bisector** of the line segment. A line (or line segment) that separates a given angle into two parts is called an angle bisector.

Example 6
a. In Figure a, if $AB = BC$, the point B, line ℓ_1, and line segment DE are bisectors of line segment AC.

b. In Figure b, if $\angle 1 = \angle 2$, then ℓ_1 is the bisector of $\angle ABC$.

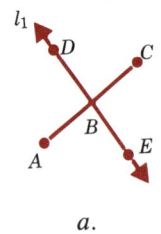

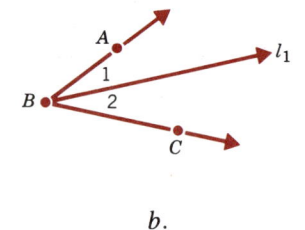

a.

b.

EXERCISES 17.2
Name pairs of angles that meet each of the following conditions. See Examples 1–4.

1. Vertical angles.
2. Alternate exterior angles.
3. Corresponding angles.
4. Alternate interior angles.

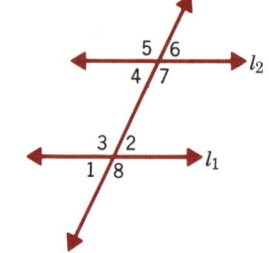

514

17.2 INTERSECTING LINES AND PARALLEL LINES

In the figure above, if $\ell_1 \parallel \ell_2$ and $\angle 1 = 40°$, find the measure of each angle. See Example 5.

Complete Example

$\angle 8$ is complementary/supplementary to $\angle 1$.

$\qquad\qquad$ (1)

Therefore, the measure of $\angle 8$ is $180° - 40° = $ _____ .
$\qquad\qquad\qquad\qquad\qquad\qquad$ (2)

---

(1) supplementary (2) 140°

---

5. $\angle 2$ 6. $\angle 3$ 7. $\angle 4$

8. $\angle 5$ 9. $\angle 6$ 10. $\angle 7$

Given straight lines AD and BE, and that CO $\perp$ DA and $\angle AOB = 35°$. Find the measure of each angle. See Example 5.

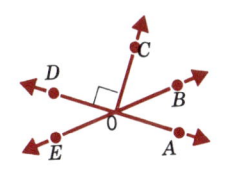

Complete Example

Since $\angle EOD$ and $\angle AOB$ are vertical angles,

$\angle EOD = $ _____ .
$\qquad\qquad$ (1)

Since $\angle EOB$ is a straight angle,

$\angle EOB = $ _____ .
$\qquad\qquad$ (2)

---

(1) 35° (2) 180°

---

11. $\angle COD$ 12. $\angle EOA$ 13. $\angle COA$

14. $\angle COB$ 15. $\angle BOD$ 16. $\angle COE$

ELEMENTS OF GEOMETRY

Given that $\ell_1 \parallel \ell_2$ and $\ell_4 \perp \ell_1$, and that the measure of $\angle 1$ is $x°$. Find the measures of each angle in terms of x. See Examples 1–5.

17. $\angle 2$
18. $\angle 4$
19. $\angle 5$
20. $\angle 6$

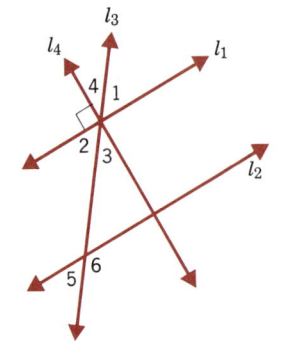

Given that $\ell_1 \parallel \ell_2$, $\ell_3 \parallel \ell_4$, and $\angle 6 = x°$. Find the measure of each angle. See Examples 1–5.

21. $\angle 11$
22. $\angle 5$
23. $\angle 10$
24. $\angle 2$
25. $\angle 14$
26. $\angle 1$

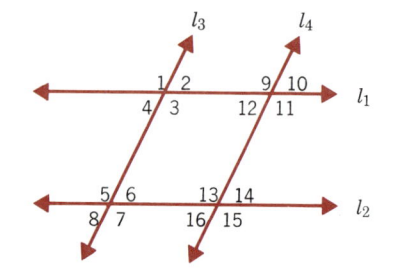

Given that $\ell_1 \parallel \ell_2$, ℓ_4 bisects the angle formed by ℓ_2 and ℓ_3, and that $\angle 2 = 48°$. Find the measure of each angle. See Examples 4–6.

27. $\angle 5$
28. $\angle 6$
29. $\angle 11$
30. $\angle 14$

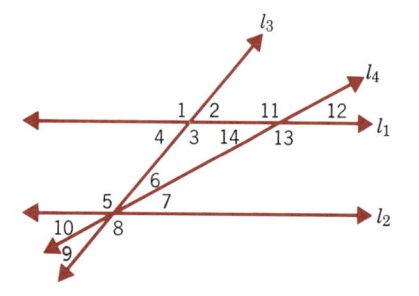

17.3 TRIANGLES

When three points that do not lie on the same line are joined by three line segments, the resulting figure is called a **triangle**. Triangles are sometimes named in terms of the lengths of their sides, as shown in Table 17.3.

17.3 TRIANGLES

TABLE 17.3

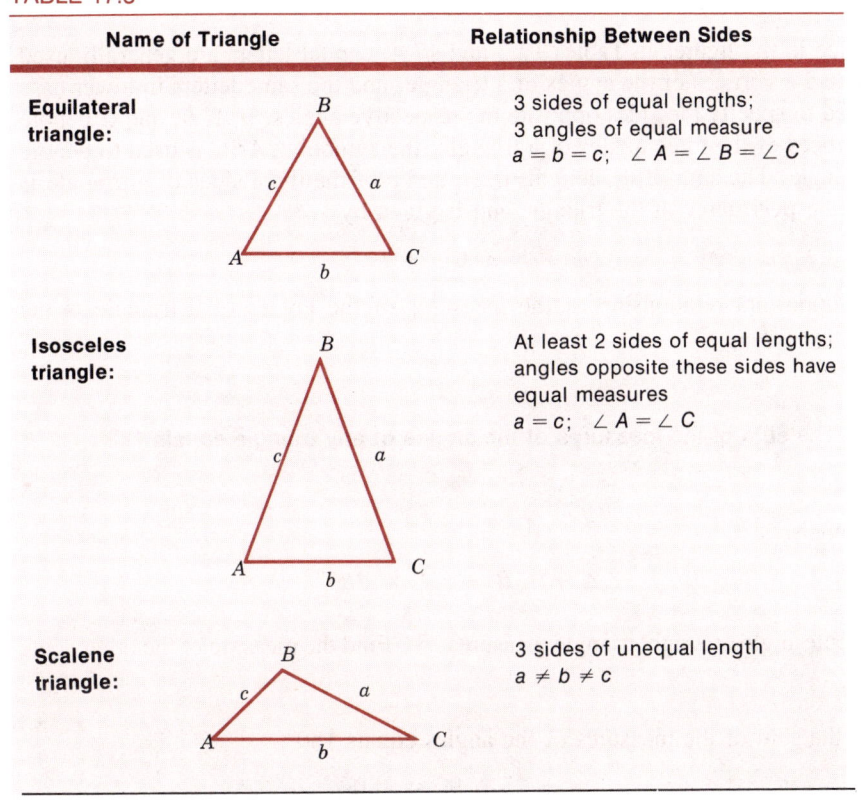

Name of Triangle		Relationship Between Sides
Equilateral triangle:		3 sides of equal lengths; 3 angles of equal measure $a = b = c$; $\angle A = \angle B = \angle C$
Isosceles triangle:		At least 2 sides of equal lengths; angles opposite these sides have equal measures $a = c$; $\angle A = \angle C$
Scalene triangle:		3 sides of unequal length $a \neq b \neq c$

Triangles are also named in terms of the measures of their angles, as shown in Table 17.4.

TABLE 17.4

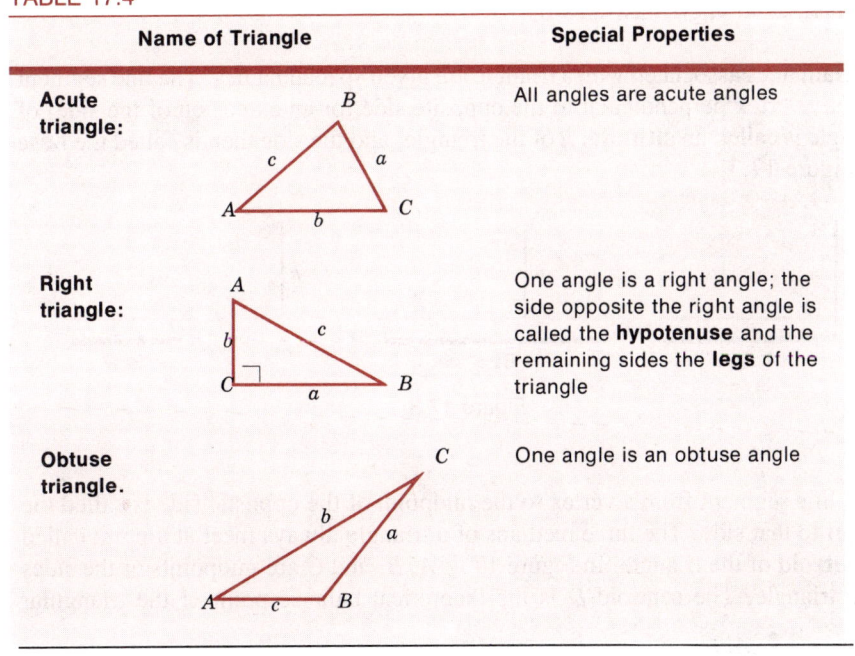

Name of Triangle		Special Properties
Acute triangle:		All angles are acute angles
Right triangle:		One angle is a right angle; the side opposite the right angle is called the **hypotenuse** and the remaining sides the **legs** of the triangle
Obtuse triangle.		One angle is an obtuse angle

517

ELEMENTS OF GEOMETRY

PROPERTIES OF TRIANGLES

As noted in the figures in Tables 17.3 and 17.4, capital letters are generally used to name the vertices of the angles of a triangle, and the same letters in lower case are used to specify the sides opposite the respective angles. If *A*, *B*, and *C* denote the vertices of the three angles of a triangle, the symbol $\triangle ABC$ is used to denote the triangle. The sum of the lengths *a*, *b*, and *c* of the three sides of a triangle is called the **perimeter** of the triangle and is given by

$$P = a + b + c.$$

An important relationship of the measures of the angles of a triangle is the following:

> **The sum of the measures of the angles of any triangle equals 180°.**

In symbols,

$$\angle A + \angle B + \angle C = 180°.$$

Example 1 One acute angle of a right triangle measures 23°. Find the measure of the other acute angle, *x*.

Solution Since the sum of the measures of the angles equals 180°,

$$x + 25 + 90 = 180$$

Solving for *x* yields

$$x + 115 = 180$$
$$x = 65$$

The other acute angle measures 65°.

Certain lines associated with a triangle are given special names. The line segment from any vertex perpendicular to the opposite side (or an extension of the side) of a triangle is called an **altitude**, *h* of the triangle, and the side then is called the **base** (see Figure 17.3).

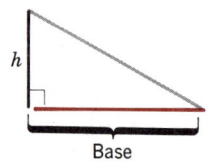

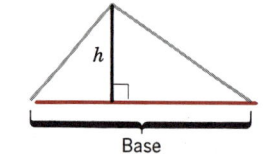

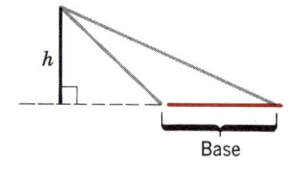

Figure 17.3

The line segment from a vertex to the midpoint of the opposite side is called the **median** to that side. The three medians of a triangle always meet at a point called the **centroid** of the triangle. In Figure 17.4, *A*, *B*, and *C* are midpoints of the sides of the triangle. The centroid *D* is the theoretical balance point of the triangular region.

518

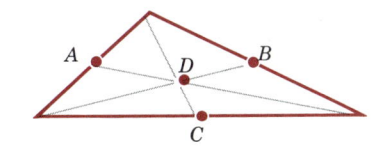

Figure 17.4

The *area* of a triangle is given by the following:

The *area* of a triangle is equal to one-half the product of a base and the corresponding altitude.

In symbols,

$$A = \frac{1}{2}bh$$

where b is the length of any base and h is the length of the corresponding altitude. The area is in square units. Thus, if b and h are in centimeters, the area is in square centimeters.

Example 2 Find the perimeter and area of triangle ABC.

Solution The perimeter is

$$P = 10 + 5 + 5\sqrt{3}$$
$$= 15 + 5\sqrt{3}$$

The area is $A = \frac{1}{2}bh$ where $b = 10$ and $h = \frac{5\sqrt{3}}{2}$. Thus,

$$A = \frac{1}{2}(10)\left(\frac{5\sqrt{3}}{2}\right) = \frac{25\sqrt{3}}{2}$$

PYTHAGOREAN THEOREM

The following relationship concerning the sides of a right triangle is particularly important and has many applications.

In any right triangle, the square of the length of the hypotenuse is equal to the sum of the squares of the lengths of the legs.

In symbols,

$$c^2 = a^2 + b^2$$

where c is the length of the hypotenuse, and a and b are the lengths of the legs.

ELEMENTS OF GEOMETRY

Example 3 In right triangle ABC,
$$c^2 = 5^2 + 2^2$$
$$= 25 + 4 = 29$$ 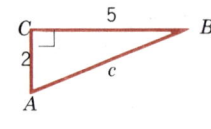
from which
$$c = \sqrt{29}$$

Because of the importance of the foregoing relationship, it is usually named and known as the **Pythagorean theorem** in honor of the Greek mathematician, Pythagoras.

EXERCISES 17.3

Classify each triangle in reference to sides and reference to angles. See Tables 17.3 and 17.4.

1. (triangle with sides 3, 3, 3)
2. (right triangle with legs 4, 4 and hypotenuse $4\sqrt{2}$) 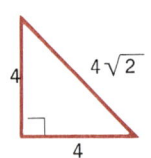
3. (triangle with sides 3, 4, 5)
4. (triangle with sides 8, 6, 5)
5. (triangle with sides 4, 5, 5)
6. (triangle with sides 4, 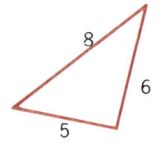 5, 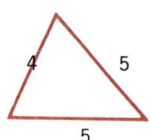 6)

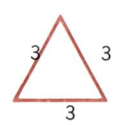

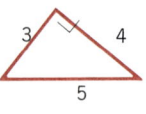

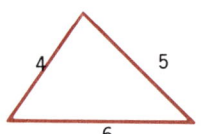

Given that $\ell_1 \parallel \ell_2$. Use the information given in the figure and the fact that the sum of the angles of a triangle equals 180° to find the measure of each angle. See Example 1.

7. $\angle 3$
8. $\angle 4$
9. $\angle 1$
10. $\angle 2$
11. $\angle 5$
12. $\angle 6$

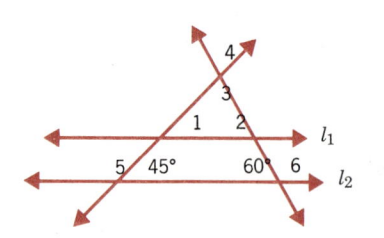

(figure shows two parallel lines ℓ_1 and ℓ_2 with a triangle; angles 45° and 60° marked)

520

17.3 TRIANGLES

Given that BF ⊥ ED, EB ⊥ AD, ∠1 = 30°, ∠2 = 45°, and BC = CD. See Example 1.

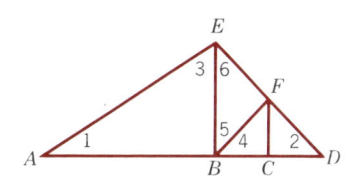

13. Name the altitude to base ED of △EBD.
14. Name the altitude to base AD of △AED.
15. Name a median of △BDF.
16. Name the hypotenuse of △BFE.
17. Name the hypotenuse of △ABE.
18. Name the hypotenuse of △BED.
19. Find the measure of ∠3.
20. Find the measure of ∠6.
21. Find the measure of ∠5.
22. Find the measure of ∠4.

Find the area and perimeter of each triangle. See Example 2.

23. △ABD
24. △BDE
25. △BEC
26. △BDC

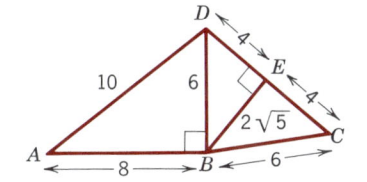

Use the Pythagorean theorem to find the lengths of each of the following line segments designated by x. See Example 3.

27.

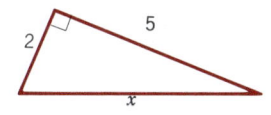

28.

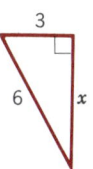

29.

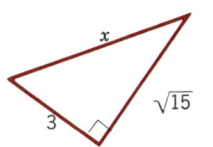

30.

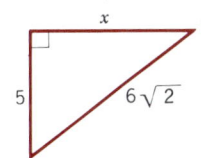

521

ELEMENTS OF GEOMETRY

Solve.

Complete Example

The perimeter of a triangle is 30 centimeters. If the second side is 2 centimeters longer than the first, and the first is two thirds of the third, find the length of each side.

Steps 1–2   Length of the third side: x

Length of the first side: _____
                                 (1)

Length of the second side: $\frac{2}{3}x + 2$

Step 3   A figure is helpful.

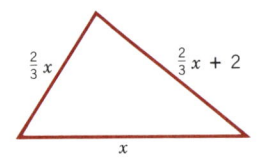

Step 4   Write an equation expressing the fact that the sum of the lengths of the sides of the triangle equals 30.

$$(x) + \left(\frac{2}{3}x\right) + \left(\frac{2}{3}x + 2\right) = 30$$

Step 5
$$3\left(x + \frac{2}{3}x + \frac{2}{3}x + 2\right) = (30)3$$

$$3x + 2x + 2x + 6 = 90$$

$$7x = \underline{}$$
$$(2)$$

$$x = \underline{}$$
$$(3)$$

Step 6   The lengths of the sides are 12, $\frac{2}{3}(12)$ or 8, and $\frac{2}{3}(12) + 2$ or _____ centimeters.
(4)

---

(1) $\frac{2}{3}x$ (2) 84   (3) 12   (4) 10

---

31. The length of one side of a triangle is two-thirds the length of each of the other sides, which are equal. How long is each of the sides if the perimeter is 16 inches?

32. One angle of a triangle is 90°. If two-thirds of one acute angle is added to one-half of the second acute angle, the result is 50°. Find the number of degrees in each acute angle.

33. If one-half of one of the acute angles of a right triangle is equal to seven-fourths of the other, how large is each acute angle?

34. If one of the equal angles in an isosceles triangle is two-fifths of the vertex angle, how large is each angle in the triangle?

522

17.4 CONGRUENT TRIANGLES; SIMILAR TRIANGLES

Complete Example

A 25-foot ladder is placed against a wall so that its foot is 7 feet from the foot of the wall. How far up the wall does the ladder extend?

Steps 1–2 Height up the wall: a

Step 3 A figure is helpful.

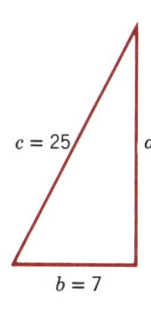

Step 4 Substitute 7 for b and 25 for c in $a^2 + b^2 = c^2$.

$$a^2 + 7^2 = (\underline{}_{(1)})^2$$

Step 5 $a^2 + \underline{}_{(2)} = 625$

$$a^2 = \underline{}_{(3)}$$

$$a = \pm\sqrt{576}$$

$$a = \pm\underline{}_{(4)}$$

Step 6 For the length of a ladder, 24 is the only meaningful solution.

---

(1) 25 (2) 49 (3) 576 (4) 24

---

35. How long must a wire be to stretch from the top of a 40-meter telephone pole to a point on the ground 30 meters from the foot of the pole?

36. How high on a building will a 25-foot ladder reach if its foot is 15 feet from the wall against which the ladder is to be placed?

37. If a 30-meter pine tree casts a shadow of 30 meters, how far is it from the tip of the shadow to the top of the tree?

38. One leg of a right triangle is 6 centimeters longer than the other, and the hypotenuse has a length of 6 centimeters less than twice that of the shorter leg. Find the lengths of the sides of the right triangle.

17.4 CONGRUENT TRIANGLES; SIMILAR TRIANGLES

CONGRUENT TRIANGLES

If two triangles have the same size and shape so that one triangle can in theory be superimposed exactly on the other, the triangles are called **congruent triangles.** The parts that coincide are called **corresponding parts** of the triangles. The symbol ≅ means "is congruent to."

ELEMENTS OF GEOMETRY

Example 1 In the figure below, triangles *ABC* and *DEF* are congruent. We write $\triangle ABC \cong \triangle DEF$. Since corresponding parts have equal measure.

$$\angle A = \angle D, \quad \angle B = \angle E, \quad \angle C = \angle F$$

and

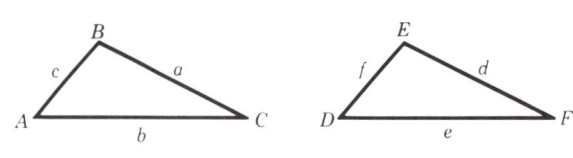

$$a = d, \quad b = e, \quad c = f$$

SIMILAR TRIANGLES

If two triangles have the same shape (corresponding angles are equal, but the relative sides do not necessarily have the same length), the triangles are called **similar triangles.**

The symbol ~ means "is similar to." Note that if two triangles are congruent, then they are also similar.

Example 2 In the figure below, if $\angle A = \angle D$, $\angle B = \angle E$, and $\angle C = \angle F$, the triangles are similar. We write $\triangle ABC \sim \triangle DEF$.

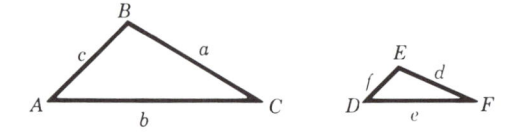

Since, for any triangle the sum of the angles is 180°, if two angles of two triangles have equal measures, then the third angles also have equal measures. Thus:

> **If two angles of one triangle have the same measures as two angles of another, the triangles are similar.**

The following relationship is a property of similar triangles.

> **The lengths of corresponding sides of similar triangles are proportional.**

Example 3 The two triangles in Example 2 are similar. Hence,

$$\triangle ABC \sim \triangle DEF$$

and

$$\frac{a}{d} = \frac{b}{e} = \frac{c}{f}$$

17.4 CONGRUENT TRIANGLES; SIMILAR TRIANGLES

EXERCISES 17.4
List the corresponding parts of each pair of congruent triangles. See Example 1.

Complete Example

△ADC ≅ △BEC

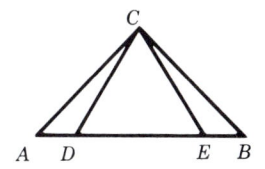

∠A and ∠B,   ∠ADC and ──(1)──,   ∠ACD and ──(2)──

AD and EB,   AC and ──(3)──,   CD and ──(4)──

(1) ∠BEC   (2) ∠BEC   (3) BC   (4) CE

1. △ACD ≅ △BCD

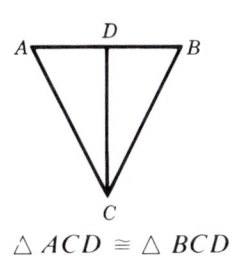

2. △ADF ≅ △BEF

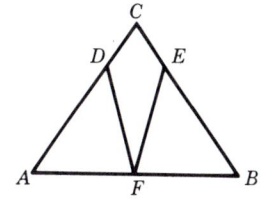

3. △AEC ≅ △BDC

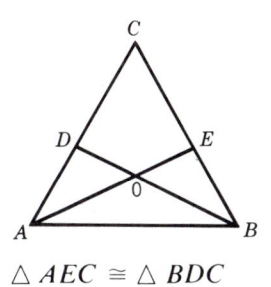

4. △ADO ≅ △CEO

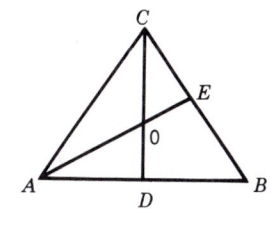

5. △ACF ≅ △BCD

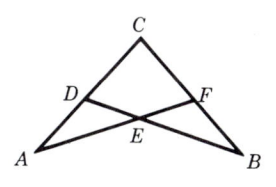

6. △ABC ≅ △EAD

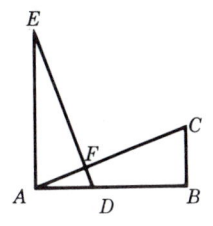

525

ELEMENTS OF GEOMETRY

In Exercises 7–16 the lengths of certain sides of two similar triangles are given. In each case, find the missing lengths. See Examples 2 and 3.

Complete Example
$\triangle ABC \sim \triangle DEF$

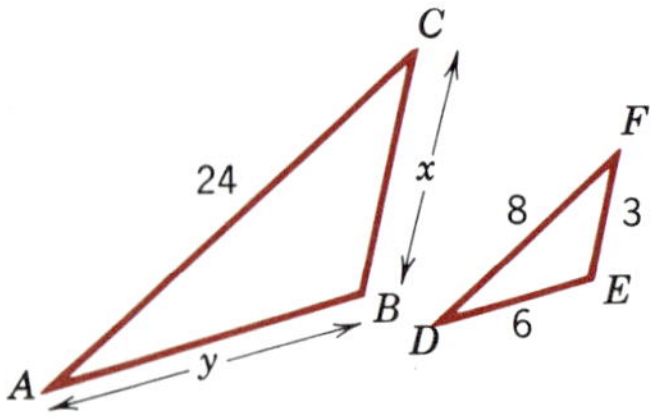

To find x, we set up the proportion

$$\frac{x}{3} = \frac{8}{\underline{\qquad(1)\qquad}}$$

Solving for x yields

$8x = \underline{\qquad(2)\qquad}$; $\quad x = 9$

To find y, we set up the proportion

$$\frac{y}{6} = \frac{24}{\underline{\qquad(3)\qquad}}$$

Solving for y yields

$8y = \underline{\qquad(4)\qquad}$; $\quad y = 18$

---

(1) 24   (2) 72   (3) 8   (4) 144

7.

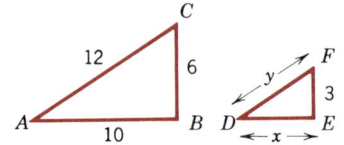

$\triangle ABC \sim \triangle DEF$

8.

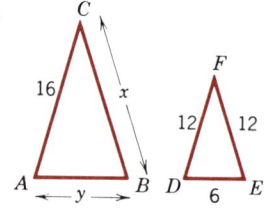

$\triangle ABC \sim \triangle DEF$

9.

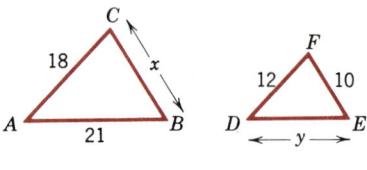

$\triangle ABC \sim \triangle DEF$

10.

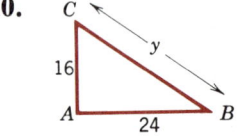

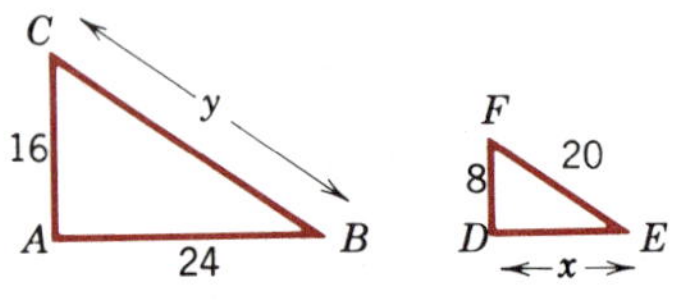

$\triangle ABC \sim \triangle DEF$

17.4 CONGRUENT TRIANGLES; SIMILAR TRIANGLES

Complete Example

$\triangle ABC \sim \triangle ADE$

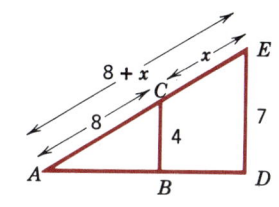

Since $\triangle ABC \sim \triangle ADE$, the ratios of the lengths of corresponding sides are equal. Thus, $\dfrac{ED}{CB} = \dfrac{AE}{AC}$.
To find x, set up the proportion

$$\dfrac{7}{4} = \dfrac{8+x}{\underline{\quad(1)\quad}}$$

Solving for x yields

$$56 = 4(8 + x)$$
$$56 = 32 + \underline{\quad(2)\quad}$$
$$24 = 4x$$
$$\underline{\quad(3)\quad} = x$$

(1) 8 (2) $4x$ (3) 6

---

11. $\triangle ABC \sim \triangle ADE$

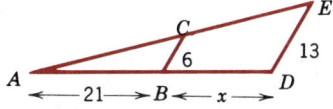

12. $\triangle ABC \sim \triangle ADE$

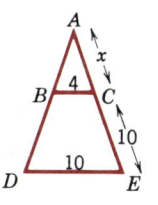

13. $\triangle ABC \sim \triangle EDC$

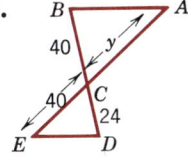

14. $\triangle ACD \sim \triangle CBD$

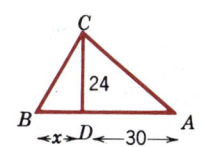

15. $\triangle ABC \sim \triangle CBD$

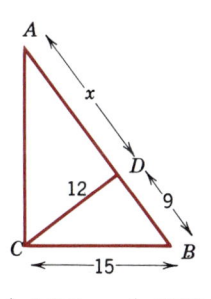

16. $\triangle ABC \sim \triangle EDC$

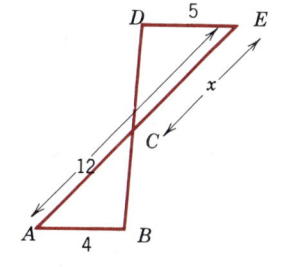

ELEMENTS OF GEOMETRY

17.5 QUADRILATERALS

The word **quadrilateral** is used to refer to any geometric figure having four sides. Table 17.5 shows some quadrilaterals having special names, and lists some properties of each.

TABLE 17.5

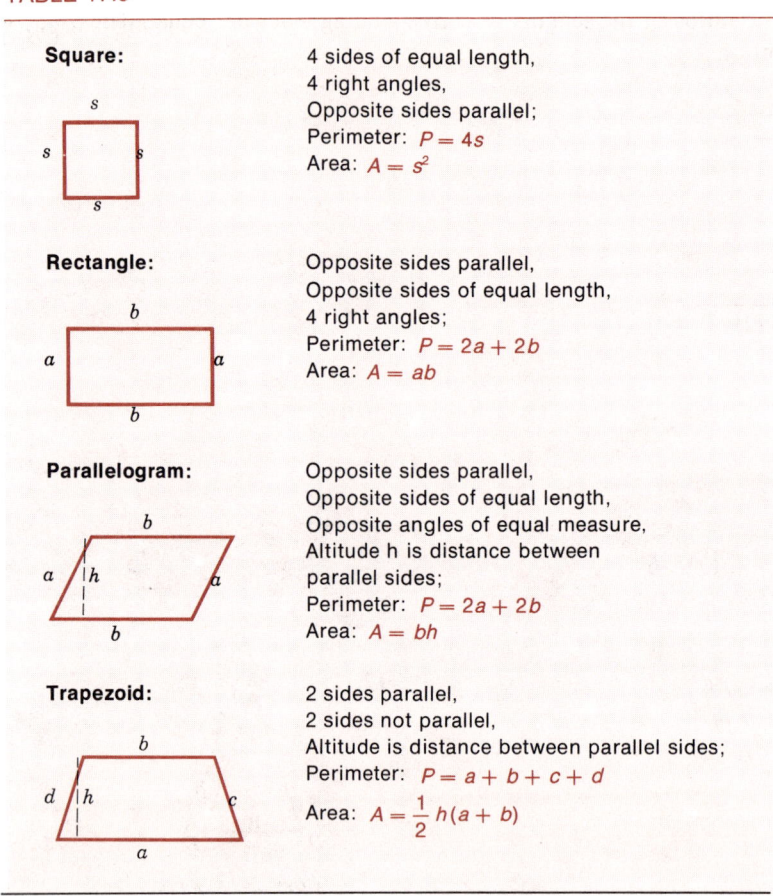

Square: 4 sides of equal length,
4 right angles,
Opposite sides parallel;
Perimeter: $P = 4s$
Area: $A = s^2$

Rectangle: Opposite sides parallel,
Opposite sides of equal length,
4 right angles;
Perimeter: $P = 2a + 2b$
Area: $A = ab$

Parallelogram: Opposite sides parallel,
Opposite sides of equal length,
Opposite angles of equal measure,
Altitude h is distance between parallel sides;
Perimeter: $P = 2a + 2b$
Area: $A = bh$

Trapezoid: 2 sides parallel,
2 sides not parallel,
Altitude is distance between parallel sides;
Perimeter: $P = a + b + c + d$
Area: $A = \frac{1}{2} h(a + b)$

Note from Table 17.5 that some quadrilaterals can be identified by more than one name. For example, a square is a rectangle and also a parallelogram. However, the word "square" would generally be used to describe this quadrilateral because squares have some properties that not all rectangles and parallelograms have.

In a quadrilateral, a line segment, other than a side, whose end points are vertices, is a **diagonal** of the quadrilateral. In Figure 17.5, the diagonal from A to C is shown by a dashed line.

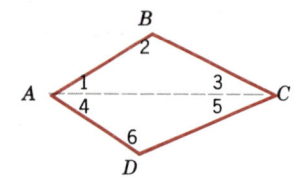

Figure 17.5

17.5 QUADRILATERALS

Since the sum of the measures of a triangle is 180°, we note from Figure 17.5 that

$$\angle 1 + \angle 2 + \angle 3 = 180° \quad \text{and} \quad \angle 4 + \angle 5 + \angle 6 = 180°$$

Therefore it follows that the sum of the measures of the angles of a quadrilateral equals 360°. In Figure 17.5,

$$\angle A + \angle B + \angle C + \angle D = 360°$$

where $\angle A = \angle 1 + \angle 4$ and $\angle C = \angle 3 + \angle 5$.

Example 1 Referring to Figure 17.5 above, if $\angle A = 50°$, $\angle B = 120°$, and $\angle C = 40°$, then

$$50 + 120 + 40 + \angle D = 360$$
$$210 + \angle D = 360$$
$$\angle D = 150°$$

EXERCISES 17.5
For each figure a. give the most appropriate name, b. give all other possible names, and c. give the measure of each angle whose measure is not given. See Table 17.5 and Example 1.

Complete Example

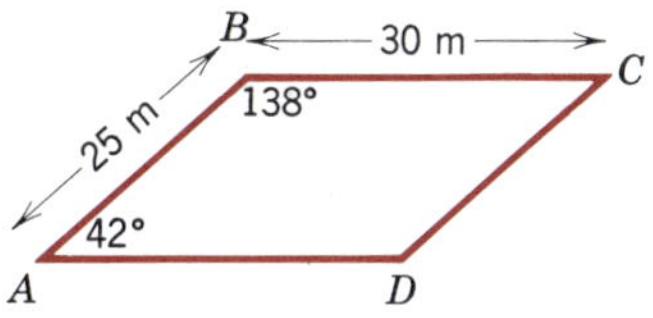

a. $ABCD$ is a _____ .
    (1)
b. It can also be called a quadrilateral.
c. $\angle A = \angle C =$ _____ ; $\angle B = \angle D =$ _____ .
                    (2)   (3)

(1) parallelogram   (2) 42°   (3) 138°

1.

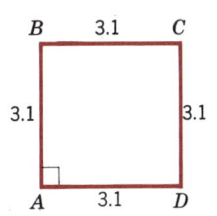

2.

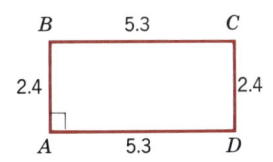

3. 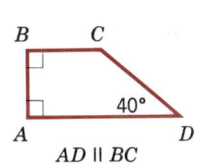 $AD \parallel BC$

4.

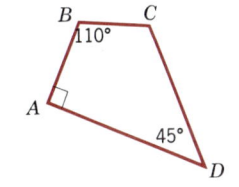

529

ELEMENTS OF GEOMETRY

Use the measures shown in the figure to find each of the following

5. Perimeter of EFGH.
6. Perimeter of ABHG.
7. Area of EFGH.
8. Area of ABHG.
9. Measure of ∠GFH.
10. Measure of ∠HDE.
11. Measure of ∠FHD.
12. Measure of ∠FED.
13. Area of BHDC.
14. Area of AGDC.

FH ∥ ED

Solve.

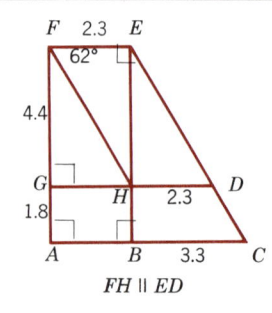

Complete Example

Find the length of the diagonal of a rectangle whose length is 6 meters and whose width is 4 meters.

Steps 1–2   length of diagonal: d.

Step 3

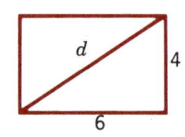

Step 4   Use the Pythagorean theorem $c^2 = a^2 + b^2$.

$d^2 = (4)^2 + (6)^2$

Step 5 $d^2 = 16 +$ _____ (1) $=$ _____ (2)

$d = \pm$ _____ (3) $= \pm 2\sqrt{13}$

Step 6   The only meaningful answer for the length of the diagonal is _____ (4) .

(1) 36   (2) 52   (3) $\pm\sqrt{52}$ (4) $2\sqrt{13}$

15. Find the length of the diagonal of a rectangle whose length is 4 centimeters and whose width is 3 centimeters.

16. Find the length of the diagonal of a rectangle whose length is 12 meters and whose width is 5 meters.

17. A baseball diamond is a square whose sides are 90 feet in length. Find the straight line distance from home plate to second base. (Use the table of square roots and find the length to the nearest foot.)

18. Find the length of the diagonal of a square whose side is a millimeters in length.

19. Find the length of a rectangle whose width is 5 inches and whose diagonal is 13 inches long.

20. Find the length of a rectangle whose diagonal is 20 meters long and whose width is 12 meters.

17.5 QUADRILATERALS

Complete Example

The length of a rectangle is 2 meters greater than the width. The diagonal is 10 meters long. Find dimensions of the rectangle.

Steps 1–2   Width: x
Length: _____
       (1)

Step 3

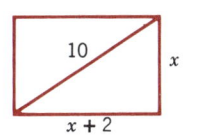

$x + 2$, x, 10 (diagonal)

Step 4   Substitute x for a, $x + 2$ for b, and 10 for c in $a^2 + b^2 = c^2$

$$(x)^2 + (x + 2)^2 = (\underline{})^2$$
$$(2)$$

Step 5
$$x^2 + x^2 + 4x + 4 = 100$$
$$2x^2 + 4x - 96 = 0$$
$$2(x^2 + 2x - 48) = 0$$
$$2(x - 6)(\underline{}) = 0$$
$$(3)$$

$x - 6 = 0 \qquad x + 8 = 0$
$x = 6 \qquad\quad x = \underline{}$
(4)

Step 6   6 is the only meaningful solution for the width; the length is $6 + 2$ or 8 meters.

---

(1) $x + 2$ (2) 10   (3) $x + 8$ (4) -8

---

21. The length of a rectangle is 3 meters greater than the width, and the diagonal is 15 meters in length. Find the dimensions of the rectangle.

22. The width of a rectangle is 7 centimeters less than the length, and the diagonal is 13 centimeters in length. Find the dimensions of the rectangle.

23. The width of a rectangle is 3 millimeters less than the length, and the square of the length of the diagonal is 29 millimeters. Find the dimensions of the rectangle.

24. The length of a rectangle is 5 inches greater than the width, and the square of the length of the diagonal is 73 inches. Find the dimensions of the rectangle.

25. Find the dimensions of a rectangle whose area is 12 square meters if the length is 4 meters greater than the width.

26. Find the dimensions of a rectangle whose area is 28 square feet if the length is 3 feet greater than the width.

27. If the sum of the length and the width of a rectangle is 12 centimeters, and its area is 42 square centimeters, what are the dimensions of the rectangle?

28. If the sum of the length and the width of a rectangle is 21 inches, and its area is 108 square inches, what are the dimensions of the rectangle?

29. If the perimeter of a rectangle is 40 meters, and its area is 96 square meters, what are the dimensions of the rectangle?

30. If the perimeter of a rectangle is 52 inches, and its area is 168 square inches, what are the dimensions of the rectangle?

ELEMENTS OF GEOMETRY

17.6 CIRCLES

A **circle** is a figure in a plane each of whose points is located the same (constant) distance from a point within called its **center.**

Table 17.6 shows some geometric facts about circles where π is an irrational number approximately equal to 3.14.

TABLE 17.6

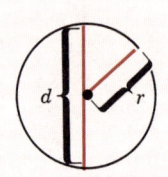

Radius, r: the distance between the center and any point on the circle (or, sometimes, the segment joining these points)

Diameter, d: the length of a segment containing the center and with endpoints on the circle (or, sometimes, the segment itself)

Circumference: the distance around a circle (its perimeter)

$$C = 2\pi r, \qquad C = \pi d$$

Area: $A = \pi r^2, \qquad A = \frac{1}{4}\pi d^2$

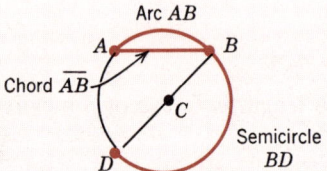

Arc: any part of a circle

Chord: any segment with endpoints on a circle

Semicircle: each of the arcs determined by a diameter

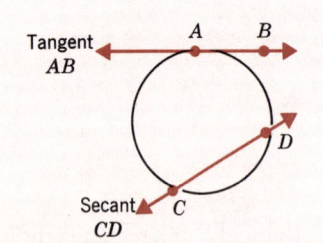

Tangent: a line touching a circle at any one point

Secant: a line containing a chord of a circle

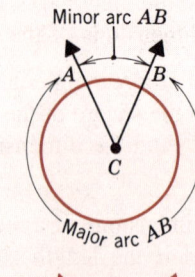

Central angle ACB: an angle with vertex at the center of a circle
Intercepted arc AB: arc determined by an angle
Minor arc AB: arc with length less than a semicircle
Major arc AB: arc with length greater than a semicircle

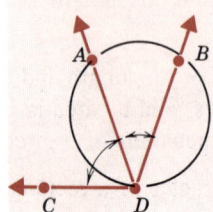

Inscribed angles $\angle ADB$ and $\angle ADC$: an angle with vertex on a circle and sides contained in secants (or tangents) of the circle

17.6 CIRCLES

Arcs of circles are sometimes measured in terms of the *central angle* intercepting them. Thus a central angle measuring 46° intercepts an arc measuring 46°. It is also true that the measure of the arc intercepted by an *inscribed angle* equals twice the measure of the angle.

Example 1 In the figure, $\angle ACB$ is a central angle measuring 46°. Therefore arc $AB = 46°$ and measures twice $\angle ADB$. Therefore $\angle ADB = 23°$.

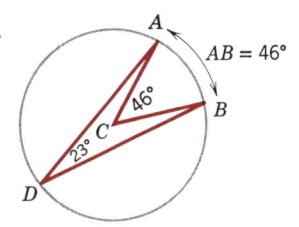

EXERCISES 17.6

Exercises 1–14 refer to the figure, where the center of the circle shown is O. Name each line, line segment, or arc. See Table 17.6.

1. Line EF.
2. Segment OB.
3. Segment AC.
4. Segment EC.
5. Arc EBC.
6. Arc BC.

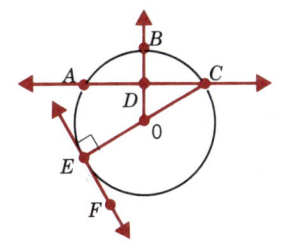

Assume that $\angle OCD$ measures 30°, and that radius OC is 2 centimeters long. Find each of the following. See Example 1.

7. The measure of $\angle COD$.

8. The measure of arc AE.

9. The measure of arc BC.

10. The length of segment OD.

11. The length of segment AC.

12. The length of segment BD.

533

ELEMENTS OF GEOMETRY

Find the area (to the nearest tenth of a unit) for each shaded region. All arcs pictured are circular arcs. (Use 3.14 for π.)

Complete Example

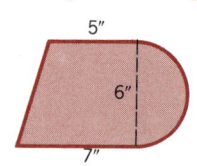

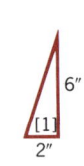

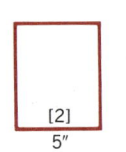

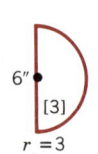

Divide figure into 3 parts.

Find area of each.

$A_{[1]} = \dfrac{(6)(2)}{2} = $ _____ ; $A_{[2]} = (5)(6) = $ _____ ; $A_{[3]} = \dfrac{3.14}{2}($ _____ $)^2 = 14.13$
    (1)   (2)   (3)

$A = 6 + 30 + 14.13 = $ _____
                        (4)

The area is 50.13 square centimeters.

(1) 6   (2) 30   (3) 3   (4) 50.13

13.

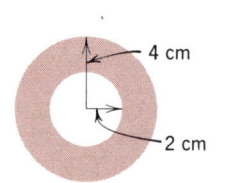

14.

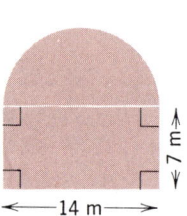

15.

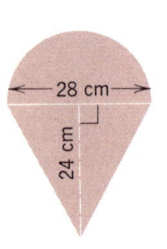

16.

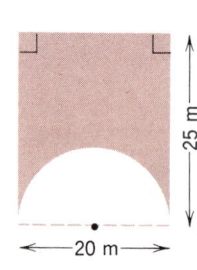

17.

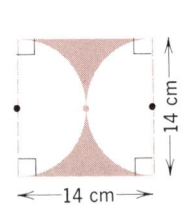

18.

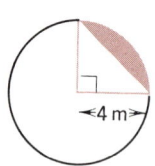

19–22. Find the perimeter to the nearest tenth of a unit of each shaded figure in Exercises 17–20. (Use 3.14 for π.)

Solve

23. A circular manhole cover has a radius of 28 centimeters. The cover contains 20 holes, each with a diameter of 2 centimeters. To the nearest tenth of a square centimeter, what is the net surface area of one side of the cover?

24. Find to the nearest tenth of a square millimeter the interior cross-sectional area of a circular pipe with outside diameter of 35 millimeters and wall thickness of 3.5 millimeters.

534

25. How long must a piece of steel bar be if it is to mold the rim of a wheel 140 centimeters in diameter?

26. A point on the outer surface of a tire is located 35 centimeters from the center of the wheel of a car. To the nearest revolution, how many revolutions will the wheel make in traveling 1 kilometer?

27. Water in a main pipe 12 inches in diameter flows into other pipes having 3 inch diameters. How many such pipes can be supplied by the main if, in general, the total cross-sectional area of the small pipes must equal that of the main?

28. What to the nearest tenth of an inch, is the radius of a circle if the circumference exceeds the radius by 37 inches?

17.7 SOLIDS

Many common geometric figures determine solids. Table 17.7 below shows some of the more important geometric solids together with certain of their properties.

Because solids occupy three-dimensional space, it is necessary to consider cubic units of measurement as well as linear and square units. Thus, lengths, widths, radii, etc., are specified in linear units, surface area in square units, and volume in cubic units.

TABLE 17.7

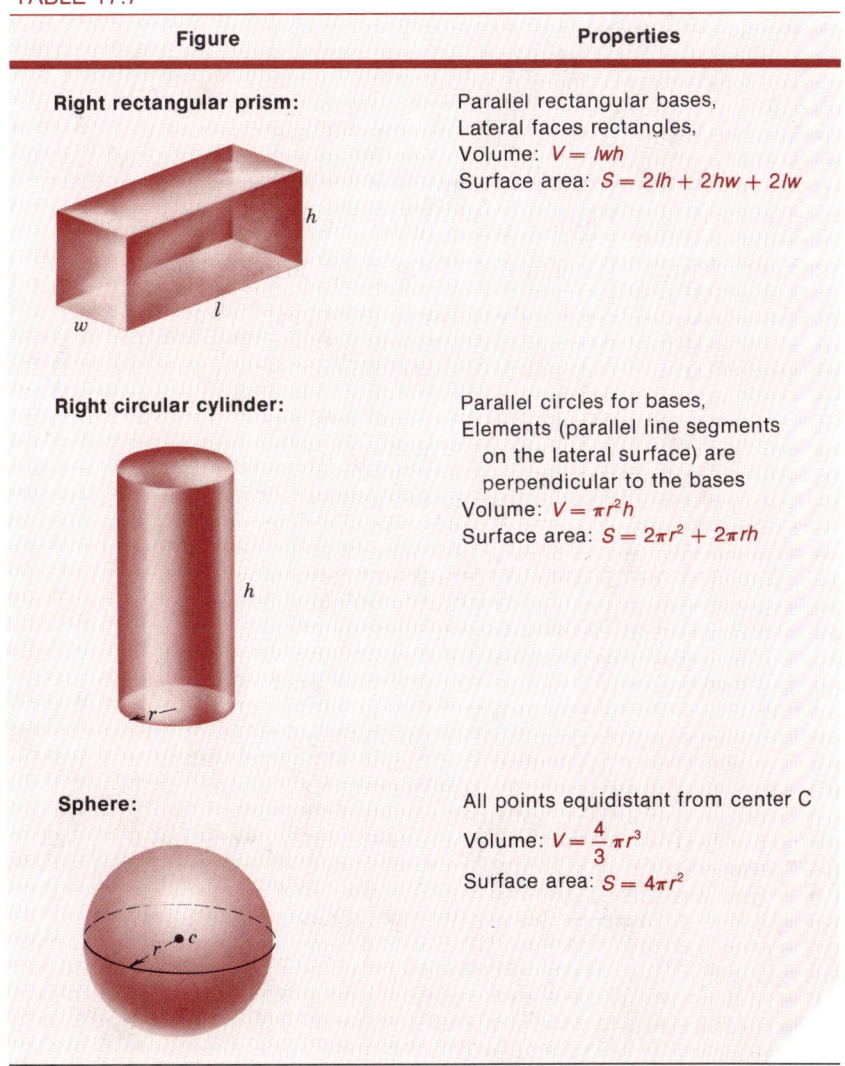

Figure	Properties
Right rectangular prism:	Parallel rectangular bases, Lateral faces rectangles, Volume: $V = lwh$ Surface area: $S = 2lh + 2hw + 2lw$
Right circular cylinder:	Parallel circles for bases, Elements (parallel line segments on the lateral surface) are perpendicular to the bases Volume: $V = \pi r^2 h$ Surface area: $S = 2\pi r^2 + 2\pi rh$
Sphere:	All points equidistant from center C Volume: $V = \frac{4}{3}\pi r^3$ Surface area: $S = 4\pi r^2$

ELEMENTS OF GEOMETRY

EXERCISES 17.7

In each exercise, express your answers to the nearest tenth of a unit.
Find the volume and total surface area of each solid. See Table 17.7.

Complete Example

Rectangular prism: dimensions of base, 5 centimeters by 10 centimeters; the height is 22 centimeters.

A sketch is usually helpful.

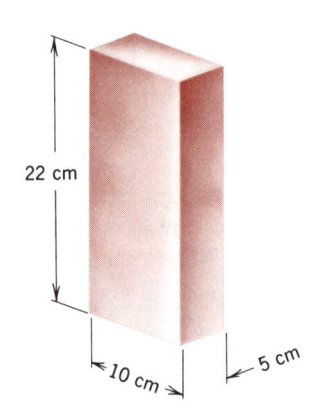

$V = lwh$

$= 10(5)(\underline{\quad\quad}) = \underline{\quad\quad}$
$\qquad\qquad$ (1) $\qquad\qquad$ (2)

$S = 2lh + 2hw + 2lw$

$= 2(10)(22) + 2(22)(\underline{\quad\quad}) + 2(5)(\underline{\quad\quad})$
$\qquad\qquad\qquad\qquad$ (3) $\qquad\qquad\qquad$ (4)

$= 440 + 220 + 100 = \underline{\quad\quad}$
$\qquad\qquad\qquad\qquad\qquad$ (5)

Ans. $V = 1100$ cubic centimeters; $S = 760$ square centimeters

(1) 22   (2) 1100   (3) 5   (4) 10   (5) 760

1. Rectangular prism; dimensions of base, 6 feet by 15 feet; the height is 18 feet.
2. Rectangular prism; dimensions of base, 4.2 inches by 7.3 inches; the height is 14.5 inches.
3. Right circular cylinder; radius of base is 24 centimeters; the height is 16 centimeters.
4. Right circular cylinder; radius of base is 7 inches; the height is 18 inches.
5. Sphere; the radius is 6 centimeters.
6. Sphere; the radius is 3 feet.

Find the volume of the pictured solid (Use 3.14 for π.)

7.

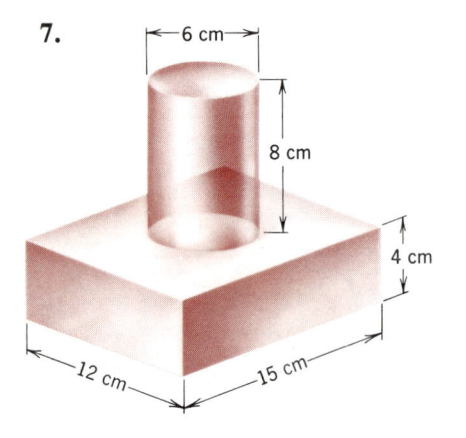

8.

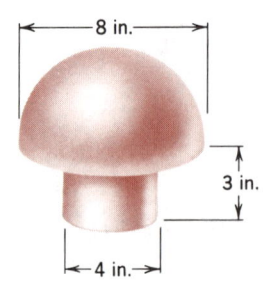

536

17.7 SOLIDS

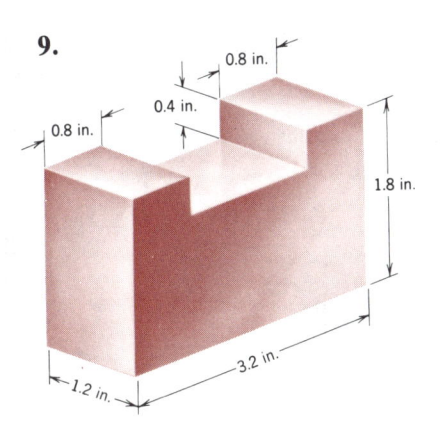

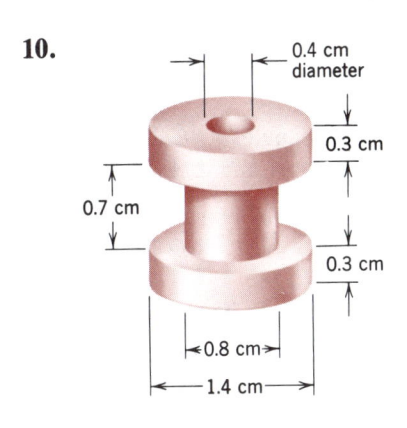

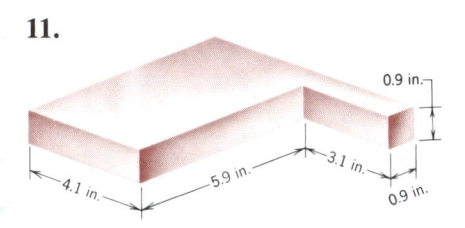

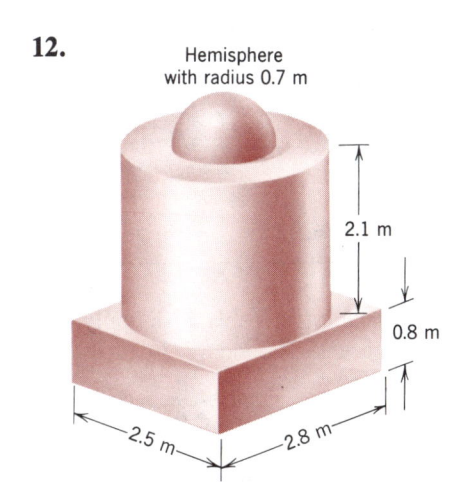

9. [figure: stepped rectangular block with dimensions 0.8 in., 0.4 in., 0.8 in., 1.8 in., 3.2 in., 1.2 in.]

10. [figure: spool-shaped solid with 0.4 cm diameter hole, 0.3 cm, 0.3 cm, 0.7 cm, 0.8 cm, 1.4 cm]

11. [figure: L-shaped slab with dimensions 4.1 in., 5.9 in., 3.1 in., 0.9 in., 0.9 in.]

12. [figure: cylinder on rectangular base topped with hemisphere of radius 0.7 m; cylinder 2.1 m, base 0.8 m, 2.5 m, 2.8 m]

Solve

13. A gallon of gasoline occupies 231 cubic inches. How many gallons of gasoline can be stored in a cylindrical tank 12 feet long and $3\frac{1}{2}$ feet in diameter?

14. What is the total surface area of the tank described in Problem 13?

15. Butane gas is stored in a spherical tank with diameter 7 meters. How many cubic meters will the tank hold?

16. If it costs $0.08 per square meter to paint the tank described in Problem 15, how much will it cost to paint the entire tank?

17. A cylindrical axle is made from steel that weighs 0.8 pound per cubic inch. If the cross-sectional diameter of the axle is 1.74 inches and the axle is 5.9 feet long, how much does the axle weigh?

18. If it costs $0.08 per square inch to plate the axle described in Problem 17, how much will it cost to plate the entire axle?

ELEMENTS OF GEOMETRY

17.8 TRIGONOMETRIC RATIOS

Consider *right* triangles ABC and ADE in Figure 17.6. These triangles are similar because $\angle A = \angle A$ and $\angle ACB = \angle AED$. Hence, their sides are proportional:

$$\frac{BC}{DE} = \frac{AB}{AD}, \quad \frac{AC}{AE} = \frac{AB}{AD}, \quad \text{and} \quad \frac{BC}{DE} = \frac{AC}{AE}.$$

These proportions can be written as

$$\frac{BC}{AB} = \frac{DE}{AD}, \quad \frac{AC}{AB} = \frac{AE}{AD}, \quad \text{and} \quad \frac{BC}{AC} = \frac{DE}{AE}.$$

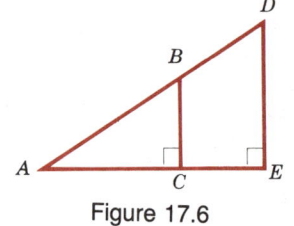

Figure 17.6

and suggest that the ratios of the sides of a triangle depend entirely on the angle A. These ratios, which involve the length of the hypotenuse and the lengths of the sides, are given the special names listed in Table 17.8.

TABLE 17.8
RATIOS IN RIGHT TRIANGLES

Sine of angle A	$\sin A = \dfrac{\text{length of side opposite } \angle A}{\text{length of hypotenuse}}$
Cosine of angle A	$\cos A = \dfrac{\text{length of side adjacent } \angle A}{\text{length of hypotenuse}}$
Tangent of angle A	$\tan A = \dfrac{\text{length of side opposite } \angle A}{\text{length of side adjacent } \angle A}$

The relationships in Table 17.8 apply to right triangles in any position. For example, in Figure 17.7a and 17.7b,

$$\sin A = \frac{a}{c}, \quad \cos A = \frac{b}{c}, \quad \text{and} \quad \tan A = \frac{a}{b}$$

and in Figure 17.7c,

$$\sin B = \frac{b}{c}, \quad \cos B = \frac{a}{c}, \quad \text{and} \quad \tan B = \frac{b}{a}$$

(a)   (b)   (c)

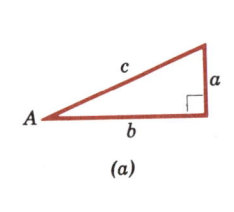

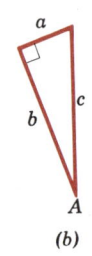

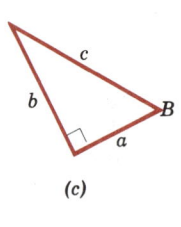

Figure 17.7

Example 1   For the right triangle shown, find the $\sin A$, $\cos A$, and $\tan A$.

Solution

$$\sin A = \frac{\text{length of side opposite}}{\text{length of hypotenuse}} = \frac{x}{7}$$

$$\cos A = \frac{\text{length of side adjacent}}{\text{length of hypotenuse}} = \frac{3}{7}$$

$$\tan A = \frac{\text{length of side opposite}}{\text{length of side adjacent}} = \frac{x}{3}$$

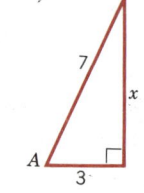

17.8 TRIGONOMETRIC RATIOS

To find values for trigonometric ratios for an angle A, we can use Table 17.9 below or a scientific calculator. For our purpose in this section, we will use the table.

Example 2   Find the values for the trigonometric ratios for an angle measuring 54°.

Solution   We simply locate 54° in one of the columns labeled "Angle," and read in the columns directly to the right,

$$\sin 54° = 0.8090, \quad \cos 54° = 0.5878, \quad \text{and} \quad \tan 54° = 1.3764$$

If we know the sine, cosine, or tangent of an angle A, we can use Table 17.9 to find the angle A.

TABLE 17.9
VALUES OF THE TRIGONOMETRIC RATIOS

Angle	Sine	Cosine	Tangent	Angle	Sine	Cosine	Tangent
1°	.0175	.9998	.0175	46°	.7193	.6947	1.0355
2°	.0349	.9994	.0349	47°	.7314	.6820	1.0724
3°	.0523	.9986	.0524	48°	.7431	.6691	1.1106
4°	.0698	.9976	.0699	49°	.7547	.6561	1.1504
5°	.0872	.9962	.0875	50°	.7660	.6428	1.1918
6°	.1045	.9945	.1051	51°	.7771	.6293	1.2349
7°	.1219	.9925	.1228	52°	.7880	.6157	1.2799
8°	.1392	.9903	.1405	53°	.7986	.6018	1.3270
9°	.1564	.9877	.1584	54°	.8090	.5878	1.3764
10°	.1736	.9848	.1763	55°	.8192	.5736	1.4281
11°	.1908	.9816	.1944	56°	.8290	.5592	1.4826
12°	.2079	.9781	.2126	57°	.8387	.5446	1.5399
13°	.2250	.9744	.2309	58°	.8480	.5299	1.6003
14°	.2419	.9703	.2493	59°	.8572	.5150	1.6643
15°	.2588	.9659	.2679	60°	.8660	.5000	1.7321
16°	.2756	.9613	.2867	61°	.8746	.4848	1.8040
17°	.2924	.9563	.3057	62°	.8829	.4695	1.8807
18°	.3090	.9511	.3249	63°	.8910	.4540	1.9626
19°	.3256	.9455	.3443	64°	.8988	.4384	2.0503
20°	.3420	.9397	.3640	65°	.9063	.4226	2.1445
21°	.3584	.9336	.3839	66°	.9135	.4067	2.2460
22°	.3746	.9272	.4040	67°	.9205	.3907	2.3559
23°	.3907	.9205	.4245	68°	.9272	.3746	2.4751
24°	.4067	.9135	.4452	69°	.9336	.3584	2.6051
25°	.4226	.9063	.4663	70°	.9397	.3420	2.7475
26°	.4384	.8988	.4877	71°	.9455	.3256	2.9042
27°	.4540	.8910	.5095	72°	.9511	.3090	3.0777
28°	.4695	.8829	.5317	73°	.9563	.2924	3.2709
29°	.4848	.8746	.5543	74°	.9613	.2756	3.4874
30°	.5000	.8660	.5774	75°	.9659	.2588	3.7321
31°	.5150	.8572	.6009	76°	.9703	.2419	4.0108
32°	.5299	.8480	.6249	77°	.9744	.2250	4.3315
33°	.5446	.8387	.6494	78°	.9781	.2079	4.7046
34°	.5592	.8290	.6745	79°	.9816	.1908	5.1446
35°	.5736	.8192	.7002	80°	.9848	.1736	5.6713
36°	.5878	.8090	.7265	81°	.9877	.1564	6.3138
37°	.6018	.7986	.7536	82°	.9903	.1392	7.1154
38°	.6157	.7880	.7813	83°	.9925	.1219	8.1443
39°	.6293	.7771	.8098	84°	.9945	.1045	9.5144
40°	.6428	.7660	.8391	85°	.9962	.0872	11.4301
41°	.6561	.7547	.8693	86°	.9976	.0698	14.3007
42°	.6691	.7431	.9004	87°	.9986	.0523	19.0811
43°	.6820	.7314	.9325	88°	.9994	.0349	28.6363
44°	.6947	.7193	.9657	89°	.9998	.0175	57.2900
45°	.7071	.7071	1.0000	90°	1.0000	.0000	

ELEMENTS OF GEOMETRY

Example 3   If $\sin A = 0.5446$, find A.

Solution   We are given $\sin A = 0.5446$ and wish to find the measure of angle A, we locate 0.5446 in one of the columns labeled "Sine" and read 33° in the column labeled "Angle" to the left. Thus,

$$\angle A = 33°.$$

Example 4   If $\cos A = 0.8900$, find A.

Solution   If we wish to find the measure of an angle A for which $\cos A = 0.8900$, a number not listed in the cosine column of the table, we simply find in the cosine column the entries between which 0.8900 lies. Thus 0.8900 lies between the table entries 0.8910 and 0.8829, so that A has a measure between 27° and 28°. Since 0.8900 is closer to 0.8910 than to 0.8829, we would use the approximation 27° for the measure of angle A.

EXERCISES 17.8
For each triangle, list sin A, cos A, and tan A. See Table 17.9 and Example 1.

Complete Example

$\sin A = \dfrac{\text{length of side opposite } \angle A}{\text{length of hypotenuse}} = \underline{\hspace{2cm}}$ (1)

$\cos A = \dfrac{\text{length of side adjacent } \angle A}{\text{length of hypotenuse}} = \underline{\hspace{2cm}}$ (2)

$\tan A = \dfrac{\text{length of side opposite } \angle A}{\text{length of side adjacent } \angle A} = \underline{\hspace{2cm}}$ (3)

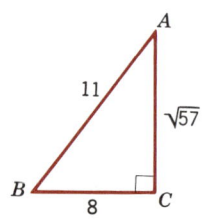

(1) $\dfrac{8}{11}$ (2) $\dfrac{\sqrt{57}}{11}$ (3) $\dfrac{8}{\sqrt{57}}$

1. Triangle with A, B, C; $AB = 10$, $BC = 6$, $AC = 8$, right angle at C.

2. Triangle with A, B, C; $CA = 5$, $BC = 12$, $AB = 13$, right angle at C.

3. Triangle with A, B, C; $AB = \sqrt{130}$, $AC = 7$, $BC = 9$, right angle at C.

4. Triangle with B, C, A; $BC = 3$, $AB = \sqrt{73}$, $AC = 8$, right angle at C.

5. Triangle with C, B, A; $CB = 14$, $AC = \sqrt{29}$, $AB = 15$, right angle at C.

6. Triangle with C, A, B; $CB = 7$, $CA = \sqrt{51}$, $AB = 10$, right angle at C.

17.8 TRIGONOMETRIC RATIOS

Determine the indicated trigonometric ratios. The listed sides are shown in the figure.

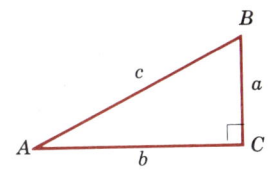

Complete Example

$a = 7$ and $c = 10$. Find $\sin \angle B$ and $\tan \angle B$

First use the Pythagorean theorem, $c^2 = a^2 + b^2$, to find b.
Substituting 7 for a and ___(1)___ for c

$$10^2 = 7^2 + b^2$$
$$b^2 = 100 - \underset{(2)}{\underline{\qquad}} = 51$$
$$b = \pm \underset{(3)}{\underline{\qquad}} ; \text{ the only meaningful answer is } \sqrt{51}.$$

$\sin \angle B = \dfrac{b}{c} = \dfrac{\sqrt{51}}{\underset{(4)}{\underline{\qquad}}}$ and $\tan \angle B = \dfrac{b}{a} = \dfrac{\sqrt{51}}{\underset{(5)}{\underline{\qquad}}}$

---

(1) 10   (2) 49   (3) $\sqrt{51}$ (4) 10   (5) 7

---

7. $a = 5$ and $b = 7$. Find $\sin A$ and $\sin B$.
8. $a = 4$ and $b = 9$. Find $\tan A$ and $\cos B$.
9. $a = 2$ and $c = 7$. Find $\sin B$ and $\tan B$.
10. $b = 3$ and $c = 5$. Find $\cos A$ and $\tan A$.
11. $b = 4$ and $c = 7$. Find $\cos B$ and $\tan B$.
12. $a = 2$ and $c = 6$. Find $\sin A$ and $\cos B$.

Use Table 17.9, page 539, to find the specified value. See Example 2.

13. $\sin 38°$
14. $\tan 8°$
15. $\cos 85°$
16. $\tan 2°$
17. $\cos 47°$
18. $\tan 71°$
19. $\sin 44°$
20. $\cos 89°$

Use Table 17.9 to find a value for A to the nearest degree. See Examples 3 and 4.

21. $\cos A = 0.9135$
22. $\tan A = 2.0000$
23. $\tan A = 2.7475$
24. $\sin A = 0.2600$
25. $\sin A = 0.5592$
26. $\cos A = 0.0700$
27. $\cos A = 0.1390$
28. $\tan A = 0.1900$

ELEMENTS OF GEOMETRY

17.9 SOLVING TRIANGLES

If, for some given measures of angles and lengths of sides of a triangle we find the measures of the remaining angles and lengths of the remaining sides, we say we have **solved** the triangle. The trigonometric ratios in conjunction with the fact that the sum of the measures of a triangle equals 180° (see page 518) can be used to solve triangles and thus enable us to solve many practical problems. In this text we only consider the solution of right triangles.

Example 1 Solve the triangle.

Solution To find BC, we set up an equation that involves the sine ratio for the 24° angle because this equation includes BC and two known measures.

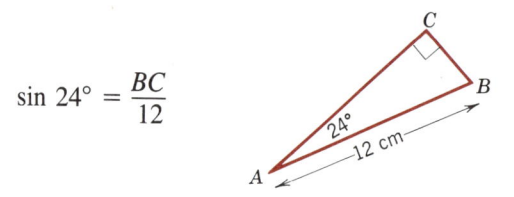

$$\sin 24° = \frac{BC}{12}$$

From Table 17.9, $\sin 24° = 0.4067$. We substitute 0.4067 for $\sin 24°$, and solve for BC.

$$0.4067 = \frac{BC}{12}$$

$$12(0.4067) = BC$$

$$4.9 = BC$$

To find AC, we set up an equation that involves the cosine ratio for the 24° angle because this equation includes AC and two known measures.

$$\cos 24° = \frac{AC}{12}$$

From Table 17.9, $\cos 24° = 0.9135$. We substitute 0.9135 for $\cos 24°$, and solve for AC.

$$0.9135 = \frac{AC}{12}$$

$$12(0.9135) = AC$$

$$11.0 = AC$$

To find $\angle B$, we use the fact that $\angle A + \angle B + \angle C = 180°$. Thus,

$$24° + \angle B + 90° = 180°,$$

and solving for $\angle B$, we have

$$B = 180° = (90° + 24°) = 180° - 114°$$

$$= 66°$$

Thus, to the nearest tenth, $BC = 4.9$ centimeters and $AC = 11.0$ centimeters; $\angle B = 66°$.

542

17.9 SOLVING TRIANGLES

EXERCISES 17.9
Find the value of x in each triangle to the nearest tenth. See Example 1.

Complete Example

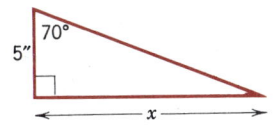

To find x, set up an equation that involves the tangent ratio for the 70° angle because this equation includes x and two known measures.

$$\tan 70° = \frac{x}{\underline{\quad(1)\quad}}$$

From Table 17.9, $\tan 70° = 2.7475$. Substitute 2.7475 for tan 70°, and solve for x.

$$2.7475 = \frac{x}{5}$$

$$(\underline{\quad(2)\quad})(2.7475) = x$$

$$\underline{\quad(3)\quad} = x$$

To the nearest tenth, $x = \underline{\quad(4)\quad}$ in.

(1) 5   (2) 5   (3) 13.7375   (4) 13.7

1. 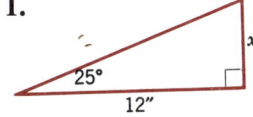
2. 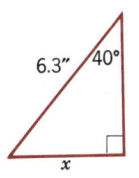
3. 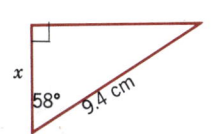
4. 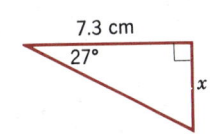
5. 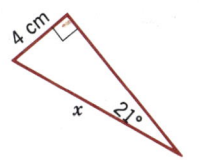
6.

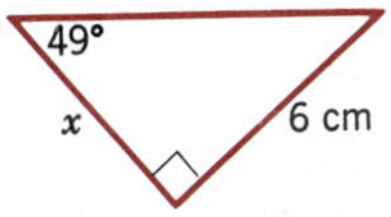

543

ELEMENTS OF GEOMETRY

Solve each right triangle. Express lengths of sides to the nearest tenth of a unit and measures of angles to the nearest degree. See Example 1.

7. [Triangle with A, B, C; right angle at C; AB-side labeled 6 cm between A and B at top; angle at A = 22°]

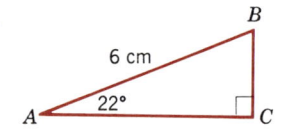

8. [Triangle with A, B, C; right angle at C; CB = 8 cm; angle at B = 25°]

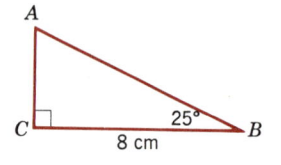

9. [Triangle with A, B, C; right angle at C; AB = 9.4 cm; angle at A = 44°]

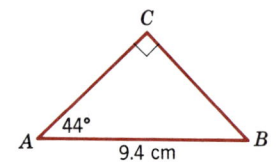

10. [Triangle with A, B, C; right angle at C; CB = 7.3 cm; angle at A = 56°]

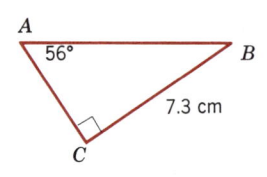

11. [Triangle with A, B, C; right angle at C; CA = 4.8"; angle at B = 32°]

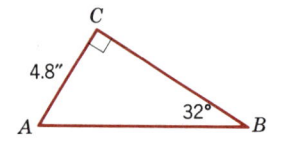

12. [Triangle with A, B, C; right angle at C; AB = 12.1"; angle at B = 48°]

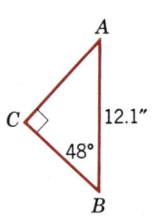

Solve each problem. Give measures of angles to the nearest degree and distances to the nearest tenth of a unit.

13. An airplane takes off at an angle of 12° to the ground. What is its height above the ground when it is above a point on the ground 5280 feet from the point of take off?

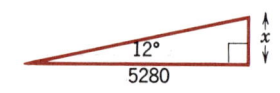

14. A ladder leaning against a wall is 6 meters long and makes an angle of 58° with the level ground. How far up the wall does the ladder reach?

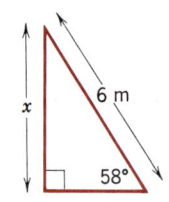

15. A guy wire supporting a pole makes an angle of 62° with the ground and is grounded 5.4 meters from the pole. How long is the wire?

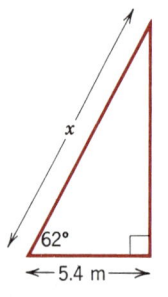

16. A road rises 80 meters for every 1000 meters along the road. What is the angle of inclination of the road?

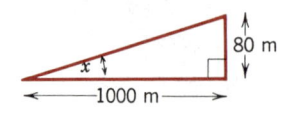

17. A mine shaft is inclined at an angle of 6° with the horizontal ground. How far directly below the ground is a miner after walking 300 feet into the tunnel?

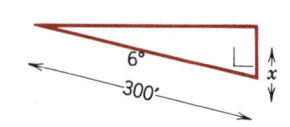

544

18. A boy "lets out" 900 feet of string attached to a kite. How high is the kite if the string is inclined at a 34° angle to the ground and we assume that the string is straight?

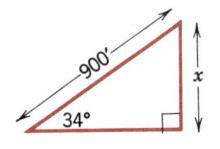

19. Each of the two equal angles of an isosceles triangle are 32° and each of the equal sides is 14.2 centimeters. What is the length of the altitude drawn to the side that is a common side to the equal angles?

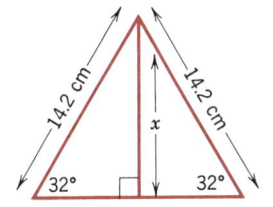

20. The altitude of an isosceles triangle drawn to the side common to each of two 64° angles is 12.8 centimeters. What is the length of one of the equal sides?

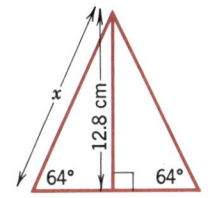

21. A searchlight is 180 meters from a building. What angle does the beam make with the ground if it lights up the building at a point 82 meters above the ground?

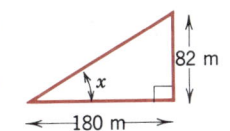

22. The span of a roof is 42 feet and its rise is 12 feet at the center of the span. What angle does the roof make with the horizontal?

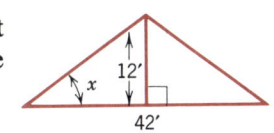

CHAPTER SEVENTEEN REVIEW EXERCISES

1. Express each measure in decimal form.
   a. 32°30′
   b. 42′

2. Express each measure in degrees and minutes.
   a. 4.75°
   b. 83.2°

3. Name (a) the complement and (b) the supplement of the angle whose measure is 47°21′.

In the figure for Exercises 4–6, $\ell_1 \parallel \ell_2$, $\ell_1 \perp \ell_4$, and $\angle 1 = 35°$.

4. Find the measure of $\angle 2$.
5. Find the measure of $\angle 3$.
6. Find the measure of $\angle 4$.

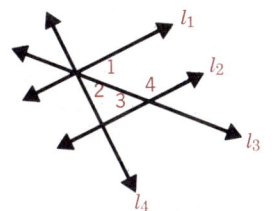

ELEMENTS OF GEOMETRY

In the figure for Exercises 7–11, AB = BC, BD ⊥ AC, and BD ⊥ EO.

7. Find ∠ABC if ∠A = 50°.
8. Find DC if BC = 6 and BD = 5.

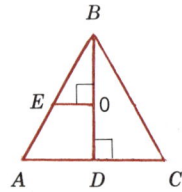

9. Find (a) the area and (b) the perimeter of △ABC using information in Exercise 8.
10. In the figure, △BOE ~ △BDA. If EB = 2, BD = 3, and AD = 2, find (a) AB, (b) EO, and (c) BO.
11. If DC = 4 and BC = 6, find (a) sin ∠DBC, (b) cos ∠C, and (c) tan ∠DBC.

In Exercises 12 and 13, use Table 17.9 to find the specified values.

12. a. cos 81° b. tan 29°
13. a. ∠A if sin A = 0.8900 b. ∠B if cos B = 0.9800

Solve.

15. A tree is 12 feet high. The shadow from the tree reaches a point on the horizontal ground 27 feet away from the tree. What is the angle of inclination of the sun?

16. Find the length of a diagonal of a rectangle whose length is 8 meters and whose width is 5 meters.

17. Find the length of the side of a rectangle whose width is 9 inches and whose diagonal is 13 inches.

18. What is the diameter of a circular mold that produces pipes with an outer circumference of 37.68 centimeters? (Use 3.14 for π.)

19. A cylindrical tank has a radius of 4 meters. If it can hold 400 cubic meters of liquid, how tall is the tank? (Uses 3.14 for π.)

20. A decorative spherical glass ball filled with colored water has a diameter of 5 centimeters. How many cubic centimeters of water will the ball hold? (Use 3.14 for π.)

ANSWERS AND SOLUTIONS

PART A: ANSWERS TO ODD-NUMBERED EXERCISES FOR EACH SECTION AND ALL ANSWERS FOR THE CUMULATIVE REVIEWS

EXERCISES 1.1

19.	7,707	21.	25,032	23.	100,670	25.	612,120
27.	98,765,432	29–35	omitted	37.	0.033	39.	9.0002
41.	1.0203	43.	0.27	45.	0.5499	47.	0.047
49.	0.0012354	51.	0.123408				

53. 0.55528, 0.55536, 0.55541

55. 0.802778, 0.802876, 0.802878

EXERCISES 1.2

1.a.	10	b.	14.8	c.	14.77	d.	14.774
3.a.	80	b.	76.3	c.	76.28	d.	76.283
5.a.	70	b.	69.9	c.	69.90	d.	69.899
7.a.	50	b.	45.9	c.	45.91	d.	45.910
9.a.	70	b.	71.0	c.	70.96	d.	70.960
11.a.	20	b.	20.0	c.	19.95	d.	19.951
13.a.	1.9	b.	1.91	c.	1.907		
15.a.	0.9	b.	0.92	c.	0.920		
17.a.	0.1	b.	0.10	c.	0.099		
19.a.	6.2	b.	6.17	c.	6.170		
21.	$1.29	23.	$8.40	25.	$47.29	27.	$58.00
29.	$38	31.	$510	33.	$601	35.	$1234
37.a.	2,100,000	b.	2,150,000	39.a.	2,000,000	b.	1,990,000

ANSWERS AND SOLUTIONS

EXERCISES 1.3

1. a. 15   b. 11   c. 10   d. 16   e. 13   f. 7   g. 16
3. a. 12   b. 16   c. 13   5. a. 13   b. 13   c. 14
7. 69   9. 569   11. 7.8   13. 79.5   15. 14.788
17. 74   19. 426   21. 14,791   23. 14,420   25. 18.46
27. 15,824   29. 165.41   31. 279,793   33. 600
35. 441.96   37. 18,770   39. 273.19   41. 1697
43. a. 19,645; 11,630; 6782; 18,376
    b. 8251; 9069; 11,173; 12,407; 15,533
    c. 56,433
45. 1.64 in.   47. 110   49. 6   51. 139   53. 2500   55. 6000

EXERCISES 1.4

1. a. 8   b. 5   c. 0   d. 8   e. 3   f. 0   g. 9
3. 17   5. 261   7. 5.2   9. 15.42   11. 47
13. 219   15. 6.17   17. 8.92   19. 159
21. 3.48   23. 43.789   25. 33,757   27. 2.568
29. $6.53   31. $13.92   33. 838   35. 58.33
37. 18.26   39. 64,832
41. a. 33,628 kwh   b. 52,448 kwh
    c. 38,998 kwh   d. 16,456 kwh
43. $609.05   45. Overdrawn $12.15   47. 1.392 in.
49. 0.719 in.   51. 63,856   53. 2375   55. 57.239

EXERCISES 1.5

1. 8   3. 18   5. 32   7. 68
9. 266   11. 39   13. 914   15. 158
17. 58   19. 46   21. 85   23. 160
25. 1298   27. 12   29. 763   31. 42
33. 1123   35. 38   37. 276   39. 3.772
41. 1.537   43. 8.134   45. 1.0576   47. 20.15
49. 5   51. 19   53. 9   55. 17
57. 6   59. 6

EXERCISES 1.6

1. $N + 107 = 118$ 3. $3.2 + N = 6.1$
5. $N = 248 + 76$ 7. $N - 82 = 49$
9. $9.87 - N = 7.42$ 11. $12{,}900 - 489 = N$
13. a. $43 + N = 126$ b. 83
15. a. $N - 20.3 = 37.6$ b. 57.9
17. a. $30.84 = N + 15.76$ b. 15.08
19. a. $1712 = N - 468$ b. 2180
21. a. $11{,}834 = 9275 + N$ b. 2559 yd
23. a. $49{,}320 = 26{,}942 + N$ b. 22,378 gal
25. a. $49{,}307 + N = 62{,}802$ b. 13,495 mi
27. a. $6382 + N = 6966$ b. $584
29. a. $N - 65 - 120 - 250 = 560$ b. $995
31. a. $812 - 75 - 236 - 49 - N = 417$ b. $35

PART A: ANSWERS

33.a. $1983.42 + 2604.67 + 3258.95 + N = 10{,}000$ b. $2152.96
35.a. $N - 627.55 - 849.42 - 763.31 - 935.77 = 3000$ b. $6176.05
37.a. $10 - 1.06 - 0.06 - 2.12 - 0.06 - 1.88 - 0.06 = N$ b. 4.76 in.
39.a. $250 + 115 + 87 + 106 - 95 - 180 = N$ b. 283 tons

EXERCISES 2.1

1.a. 10   b. 18   c. 28   d. 40   e. 54   f. 32
3.a. 35   b. 42   c. 56   d. 72   e. 0   f. 5
5. 84   7. 826   9. 111   11. 675   13. 168
15. 4774   17. 5544 19. 69,120 21. 25,296
23. 11,016 25. 4.08 27. 0.128 29. 0.0408
31. 0.12036 33. 3.25 35. 10.368 37. 1.7304
39. 0.00384 41.a. 0.6 b. 0.63 c. 0.626
43.a. 0.3 b. 0.31 c. 0.309
45.a. 0.4 b. 0.35 c. 0.353

EXERCISES 2.2

1. 0.895   3. 0.00895   5. 0.000895   7. 0.00895
9. 70,400   11. 70.4   13. 704   15. 4500   17. 3,500,000
19. 30,000   No answers 21–31   33. 8   35. 27   37. 16
39. 100   41. 243   43. 343   45. 1.728   47. 151.29

EXERCISES 2.3

1.a. 4   b. 3   c. 2   d. 1   e. 2   f. Undefined   g. 0
3.a. 4   b. 6   c. 8   d. 4   e. 0   f. Undefined   g. 1
5. 12   7. 341   9. 156   11. 115   13. 13
15. 168   17. 121 19. 45   21. 2.3   23. 4.51
25. 4.5   27. 12.34 29. 13.7 31. 15.3 33. 605
35. 230.4 37.a. 3.8 b. 3.79 39.a. 0.3 b. 0.33
41.a. 23.4 b. 23.36

EXERCISES 2.4

1. 40.7   3. 0.407   5. 0.386   7. 0.053   9. 0.00053
11. 70.99 13. 18   15. 143   17. 57   19. 409
21. 20   23. 100   25. 200

EXERCISES 2.5

1.a. $N = 19 \times 23$ b. 437 mi   3.a. $N = 58 \times 45$ b. 2610 words
5.a. $N = 125 \times 35$ b. 4375 lines   7.a. $N = 175 \times 12$ b. $2100
9. $478.80; $636.00; $3915.60; $12,573.72   11. $12490.38; $15591.96; $9849.96; $8109.99
13. $0.04   15. $0.19   17. $0.02
19. $0.75; $0.55; $0.50; $0.53   21.a. $5.18   b. $26.56
23.a. 671.4 mi   b. 2594.7 mi   25.a. $1.67   b. $10.33
27.a. $l = 35 \times 125$ b. 4375 lines   29.a. $l = \dfrac{3300}{25}$ b. 132 lines per minute

549

ANSWERS AND SOLUTIONS

31.a. $d = 12 \times 175$ b. $2100   33.a. $y = \frac{704}{16}$ b. 44 yards
35. $23.62   37. $32.51   39. $38.39

EXERCISES 2.6

1. 104
3. 6.4
5. 63.6
7. 3.12
9. 0.448
11. 46.25
13. 35
15. 51
17. 2.8
19. 4.7
21. 45.7
23. 36
25. 6.3
27. 9.8
29. 138
31. 1134
33. 20,160
35. 1,016,056
37. 338
39. 1234
41. 4781
43. 16,197,408
45. 12
47. 4
49. 5
51. 32
53. 5
55. 3
57. 10
59.a. $104 \times N = 66{,}454$ b. 639 items
61.a. $37{,}570 \times N = 30{,}769{,}830$ b. 819 items
63.a. $\frac{N}{276} = 4017$ b. 1,108,692
65.a. $N \times 9 = 9036$ b. 1004

EXERCISES 2.7

1. $\frac{5}{18}$
3. $\frac{17}{8}$
5. $\frac{8.87}{4.31}$
7. $7 \times 16 = 8 \times 14 = 112$
9. $3.5 \times 19 = 9.5 \times 7 = 66.5$
11. $12 \times 22.1 = 17 \times 15.6 = 265.2$
13. 3
15. 33
17. 99
19. 19.5
21. 13
23. 11
25. 8.4
27.a. $\frac{C}{42} = \frac{4}{7}$ b. 24¢   29.a. $\frac{C}{96} = \frac{20}{8}$ b. $2.40
31.a. $\frac{C}{14.97} = \frac{5}{3}$ b. $24.95   33.a. $\frac{C}{76.90} = \frac{8}{3}$ b. $205.07
35.a. $\frac{C}{6.24} = \frac{10}{3}$ b. $20.80   37.a. $\frac{N}{12} = \frac{54}{36}$ b. 18 in.
39.a. 11   b. 17   c. 420   d. 240
41. 165   43. 32 in.
45.a. 128   b. 160   c. 256   d. 7   e. 11

EXERCISES 3.1

1. Is a factor
3. Is not a factor
5. Is not a factor
7. Is a factor
9. Divisible by 2, 3, and 5
11. Divisible by 2 and 3
13. Divisible by 3 and 5
15. Divisible by 2 and 3
17. 2^2
19. 3^3
21. 2×3^3
23. 3^4
25. $2^4 \times 17$
27. $2^4 \times 11$
29. $2^4 \times 3 \times 11$
31. $2 \times 3 \times 5^3 \times 11$
33. 1409

EXERCISES 3.2

1. $\frac{4}{7}$
3. $\frac{15}{17}$
5. $\frac{5}{7}$
7. $\frac{7}{17}$
9. $\frac{5}{16}$
11. $\frac{2}{7}$
13. $\frac{3}{4}$
15. $\frac{1}{3}$
17. $\frac{4}{11}$
19. $\frac{3}{5}$
21. $\frac{9}{16}$
23. $\frac{19}{26}$

PART A: ANSWERS

25. $\dfrac{12}{24}$ 27. $\dfrac{21}{35}$ 29. $\dfrac{25}{40}$ 31. $\dfrac{21}{36}$

33. $\dfrac{55}{90}$ 35. $\dfrac{81}{144}$ 37. $\dfrac{48}{128}$ 39. $\dfrac{30}{105}$

41. $\dfrac{30}{108}$ 43. $\dfrac{15}{135}$ 45. $\dfrac{12}{12}$ 47. $\dfrac{80}{16}$

49. $\dfrac{4}{18} = \dfrac{2}{9}$ 51.a. $\dfrac{14}{36} = \dfrac{7}{18}$ b. $\dfrac{22}{36} = \dfrac{11}{18}$

EXERCISES 3.3

1. $\dfrac{8}{12}, \dfrac{9}{12}$ 3. $\dfrac{7}{28}, \dfrac{8}{28}$ 5. $\dfrac{4}{6}, \dfrac{1}{6}$ 7. $\dfrac{42}{6}, \dfrac{1}{6}$

9. $\dfrac{25}{30}, \dfrac{9}{30}$ 11. $\dfrac{25}{60}, \dfrac{6}{60}$ 13. $\dfrac{3}{12}, \dfrac{8}{12}, \dfrac{2}{12}$ 15. $\dfrac{14}{30}, \dfrac{1}{30}, \dfrac{12}{30}$

17. $\dfrac{10}{36}, \dfrac{4}{36}, \dfrac{3}{36}$ 19. $\dfrac{300}{60}, \dfrac{8}{60}, \dfrac{9}{60}$ 21. $\dfrac{144}{24}, \dfrac{4}{24}, \dfrac{15}{24}$ 23. $\dfrac{15}{72}, \dfrac{216}{72}, \dfrac{4}{72}$

25. $\dfrac{2}{11}, \dfrac{3}{11}, \dfrac{5}{11}$ 27. $\dfrac{1}{12}, \dfrac{5}{12}, \dfrac{7}{12}$ 29. $\dfrac{1}{12}, \dfrac{1}{6}, \dfrac{5}{24}$ 31. $\dfrac{5}{14}, \dfrac{3}{7}, \dfrac{10}{21}$

33. $\dfrac{7}{12}, \dfrac{2}{3}, \dfrac{3}{4}$ 35. $\dfrac{15}{32}, \dfrac{9}{16}, \dfrac{7}{8}$ 37. $\dfrac{7}{8}$ lb 39. $\dfrac{15}{32}$ in., $\dfrac{31}{64}$ in.

EXERCISES 3.4

1. $\dfrac{5}{7}$ 3. $\dfrac{1}{2}$ 5. $1\dfrac{3}{16}$ 7. $\dfrac{4}{5}$

9. $1\dfrac{5}{12}$ 11. $\dfrac{37}{40}$ 13. $\dfrac{7}{10}$ 15. $4\dfrac{11}{48}$

17. $\dfrac{17}{3}$ 19. $\dfrac{64}{5}$ 21. $\dfrac{341}{16}$ 23. $\dfrac{939}{32}$

25. $3\dfrac{2}{3}$ 27. $5\dfrac{3}{8}$ 29. $6\dfrac{11}{16}$ 31. $9\dfrac{29}{32}$

33. $7\dfrac{3}{5}$ 35. $11\dfrac{1}{2}$ 37. $18\dfrac{1}{4}$ 39. $4\dfrac{7}{8}$

41. $4\dfrac{49}{60}$ 43. $15\dfrac{5}{12}$ 45. $11\dfrac{5}{9}$ 47. $7\dfrac{7}{16}$

49. $39\dfrac{11}{16}$ 51. $34\dfrac{15}{16}$ 53. $1\dfrac{7}{8}$ in. 55. $46\dfrac{7}{8}$ tons

57. $73\dfrac{3}{8}$ 59. $49\dfrac{5}{8}$ 61. $104\dfrac{1}{4}$ tons 63. $50\dfrac{1}{4}$ tons

EXERCISES 3.5

1. $\dfrac{4}{7}$ 3. $\dfrac{1}{4}$ 5. $\dfrac{3}{32}$ 7. $\dfrac{1}{4}$

9. $\dfrac{1}{12}$ 11. $\dfrac{13}{40}$ 13. $\dfrac{5}{16}$ 15. $\dfrac{7}{8}$

17. $3\dfrac{3}{8}$ 19. $4\dfrac{7}{16}$ 21. $2\dfrac{1}{7}$ 23. $5\dfrac{5}{32}$

25. $9\dfrac{7}{64}$ 27. $18\dfrac{1}{6}$ 29. $6\dfrac{7}{8}$ 31. $3\dfrac{29}{32}$

551

ANSWERS AND SOLUTIONS

33. $3\frac{7}{12}$ 35. $5\frac{19}{24}$ 37. $6\frac{13}{20}$ 39. $8\frac{31}{36}$
41. $4\frac{3}{4}$ 43. $3\frac{11}{16}$ 45. $\frac{2}{5}$ 47. $15\frac{7}{16}$
49. $14\frac{13}{20}$ 51. $3\frac{23}{30}$ 53. $9\frac{3}{16}$ 55. $8\frac{1}{6}$ 57. $\frac{7}{16}$ in.
59. $749\frac{1}{2}$ gal; $688\frac{3}{4}$ gal; $633\frac{3}{8}$ gal; $561\frac{3}{8}$ gal; $477\frac{7}{8}$ gal; $367\frac{5}{8}$ gal

EXERCISES 3.6

1. $\frac{1}{12}$ 3. $\frac{1}{8}$ 5. $\frac{2}{15}$ 7. $\frac{5}{14}$
9. $\frac{1}{48}$ 11. $\frac{11}{96}$ 13. $\frac{1}{3}$ 15. $\frac{5}{144}$
17. $\frac{3}{25}$ 19. $16\frac{1}{2}$ 21. $7\frac{1}{7}$ 23. 100
25. $103\frac{1}{2}$ 27. 9   29. 6   31. 171
33. $9\frac{1}{4}$ 35. $22\frac{5}{6}$ 37. $\frac{1}{5}$ 39. $\frac{5}{3}$
41. $\frac{2}{7}$ 43. $\frac{9}{101}$ 45. $1\frac{1}{5}$ 47. $\frac{2}{15}$
49. $\frac{1}{18}$ 51. $19\frac{1}{2}$ 53. $\frac{5}{12}$ 55. 2
57. 2   59. $2\frac{1}{2}$ 61. $7\frac{1}{3}$ 63. $\frac{4}{15}$
65. a. $8\frac{5}{8}$ ft   b. $12\frac{15}{16}$ ft   c. $17\frac{31}{32}$ ft   d. $24\frac{11}{12}$ ft   e. $32\frac{79}{96}$ ft
67. a. $600   b. $186   c. $690   d. $1264

EXERCISES 3.7

1. a. 0.1   b. 0.13   c. 0.125   3. a. 0.2 b. 0.17 c. 0.167
5. a. 0.5   b. 0.47   c. 0.469   7. a. 0.8 b. 0.78 c. 0.778
9. a. 1.1   b. 1.06   c. 1.063   11. a. 1.2 b. 1.15 c. 1.154
13. 5.875   15. 3.75   17. 12.6   19. 15.375
21. 0.1   23. 0.001   25. 0.09   27. 0.013
29. 0.5   31. 0.75   33. 3.25   35. $5.33\frac{1}{3}$
37. $\frac{9}{10}$ 39. $\frac{21}{100}$ 41. $\frac{127}{1000}$ 43. $3\frac{9}{100}$
45. $4\frac{1}{2}$ 47. $2\frac{3}{4}$ 49. $\frac{1}{3}$ 51. $1\frac{2}{3}$

EXERCISES 4.1

1. 33%   3. 50.4%   5. 78.7%   7. 2.1%
9. 0.8%   11. 550%   13. 0.15   15. 0.004
17. 0.068   19. 0.027   21. 1.19   23. 2.59
25. 0.0325   27. 0.02375   29. 0.084   31. 0.005
33. 75%   35. 37.5%   37. 60%   39. 225%

41. 240%   43. 0.4%   45. $\frac{1}{4}$ 47. $\frac{5}{4}$
49. $\frac{9}{1000}$ 51. $\frac{3}{8}$

EXERCISES 4.2

1. 16	3. 24.15	5. 46.4	7. 64.8
9. 2.6	11. 1.92	13. 20	15. 57
17. 80%	19. 25%	21. 250%	23. 12%
25. 97.5%	27. 2.5%	29. 3.3%	31. 340
33. 133.3	35. 2666.7	37. 646.2	39. 727.3
41. $17.10	43. 12.5%	45. 14.8	47. 400
49. 35.7%	51. $17.85		

EXERCISES 4.3

1. $175   3. $365   5. $1920   7. $5040
9. $75.26   11. $84.40   13. $67.16   15. $119
17. $120   19. $257.60   21. $504   23. $312
25. 62.9%   27.a. 4.2%   b. 95.8%   29. 6.5%
31. A: 16%   B: 21.6%   C: 52.5%   D: 8%   F: 1.9%
33. 25% increase   35. 75% increase   37. 25% decrease   39. Less (salary is $351.90)
41. 99.99%   43.a. $846   b. $1034
45.a. $11.38   b. $13.63   47. 80   49. 525   51. 179
53. $2071.43   55. 70%   57. $858   59. 4366
61.a. 3649   b. 45,001   63. 30%

EXERCISES 4.4

1. 2.5   3. 0.25   5. 1.5   7.a. $112   b. $512
9.a. $378   b. $658   11.a. $1980   b. $9980
13.a. $1275   b. $3775   15. 10%; 6.7%
17. $24,216.22; $1860.47   19. $435.60
21. $5616   23. 17.8%   25. 11.2%
27.a. $141,052.63   b. 3 months

EXERCISES 5.1

1. 12 in.   3. 3 ft   5. 2 yd   7. 16 oz
9. 4 qt   11. 6 qt   13. 60 sec   15. 52 wk
17. 14 ft   19. 9 yd   21. 40,000 lb   23. 30 pt
25. 900 min   27. 126,720 ft   29. 126,720 in.   31. 80 gal
33. 4 hr   35. 272 pt   37. 768 qt   39. 21,600 min
41. 158 oz   43. 53 ft   45. 532 min   47. 23 qt
49. 4.19 pt   51. 7.94 lb   53. 1.79 mi   55. 1.88 da
57. 4.58 ft   59. 5.13 mi   61. 5.75 gal   63.a. 25,200 ft;   b. 4.8 mi
65.a. 1331 yd   b. 3993 ft   67. 1.2 mi   69. 2.66 in.

ANSWERS AND SOLUTIONS

EXERCISES 5.2

1.a. 22 ft 5 in.   b. 22.42 ft   3.a. 36 lb 7 oz   b. 36.44 lb
5.a. 49 ft 2 in.   b. 49.17 ft   7.a. 19 hr 20 min   b. 19.33 hr
9.a. 50 yd 1 ft   b. 50.33 yd   11. 557 mi
13.a. 8 da 3 hr   b. 8.13 da   15.a. 54 ft 5 in.   b. 54.42 ft
17. 1147 ft   19. 785 ft 4 in   21. 116 gal
23. 4 da   25. 19 hr 4 min or 1144 min   27. 11 lb 2 oz or 178 oz
29. 7 in   31. 60   33.a. 1024 fl oz   b. 8 gal
35. 4 ft 11 in   37. 3 ft 3 in or 39 in.   39. 40

EXERCISES 5.3

1. One-thousandth of a meter   3. One-hundredth of a gram
5. One-tenth of a liter   7. 10 grams
9. 100 liters   11. 1000 meters
13. 750 mm   15. 170 hm   17. 80.6 m   19. 9.37 cm   21. 37 dam
23. 270 km   25. 6.7 dg   27. 6.08 dag   29. 20,170 mg   31. 1.67 cl
33. 100.9 kl   35. 81,920 hl   37. 39,900 m   39. 8.88 g   41. 5630 mm
43. 0.06039 kl   45. 0.0001526 km   47. 68,780,000 mg
49. *a*   51. *a*   53. *c*   55. *c*

EXERCISES 5.4

1. 3.9 kg   3. 2449.4 g   5. 258.0 g   7. 11.9 cm
9. 4.4 m   11. 43.5 m   13. 25.4 km   15. 12.8 ℓ
17. 63.7 ℓ   19. 30.7 ℓ   21. 3.4 in   23. 30.5 ft
25. 108.2 yd   27. 347.4 m   29. 0.5 oz   31. 14.3 lb
33. 12.3 qt   35. 9.0 qt   37. 10.8 gal   39. 157.5 gal
41. 17,480.2 km   43. 1.4 in.   45. 7.8 km per liter   47. 6.2 km per liter
49. 20.5 mpg   51. 18.6 mpg

EXERCISES 5.5

1. $0.06 per oz   3. $4.07 per qt   5. $0.03 per oz
7. 28 fl oz   9. 1 gal   11. B
13.a. $1507   b. $107.64
15. A: $1373, $98.07;   B: $1478, $105.57;   A is the better buy
17.a. 270 C   b. 2.3 lb   19.a. 375 C   b. 4.8 lb.
21.a. 280 C   b. 9.6 lb   23. jogging   25. skiing
27. bicycling   29. 18,000 fewer C
31. Pat, 20; Fran, 24; Chris, 25
33. E. Adams, 77.7; M. Lopez, 78; L. Frye, 65
35. 94   37. 92   39. 100   41. 2.9
43. $160.46   45. $253.17
47.a. $3.12   b. $5.20   c. $6.19   d. $10.38
49.a. $119.29   b. $1.79

PART A: ANSWERS

CUMULATIVE REVIEW FOR PART I

1. $14.40
2. 73.53
3. 17 ft
4. $3\frac{5}{8}$
5. $7300
6. 6.2
7. 4 gal 2 qt
8. 3.501
9. $\frac{3}{10}$
10. 30
11. Two hundred thirty-four and seven hundredths
12. 9.1 kg
13. 37.5
14. $3.08
15. $\frac{7}{2}$
16. 179.7
17. $\frac{30}{24}$
18. 17.4 mi
19. 0.4%
20. 76.8
21. 50%
22. $33\frac{1}{3}\%$
23. 187.5 mi
24. 6.25%
25. larger box
26. 23,045
27. $\frac{19}{15} = 1\frac{4}{15}$
28. 1.7
29. 25.2
30. 25%
31. $2^3 \times 3 \times 5$
32. 17.22
33. 52.62
34. 3.2
35. 80%
36. $21.60
37. 60
38. 7.77
39. $5.94
40. $2\frac{1}{2}$
41. 204 words
42. 16.275 mi
43. Two hundred fifteen and thirty-five hundredths
44. $\frac{3}{5}$
45. 66.81
46. 20%
47. $233.20
48. 40.95
49. 40%
50. 24
51. 132
52. 2.4
53. 23.4
54. 3.2
55. 75%
56. $\frac{1}{4}$
57. $9\frac{3}{5}$
58. 45%
59. 2,500,000
60. 160 qt
61. $33\frac{1}{3}\%$
62. 27 ft
63. 27.0
64. 0.683
65. 16
66. 3.42
67. $\frac{7}{10}$
68. 0
69. 1
70. 100
71. 1
72. 24
73. 12
74. 7
75. 5

EXERCISES 6.1

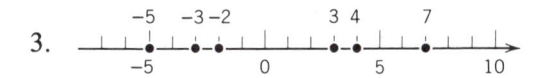

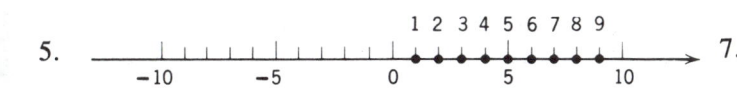

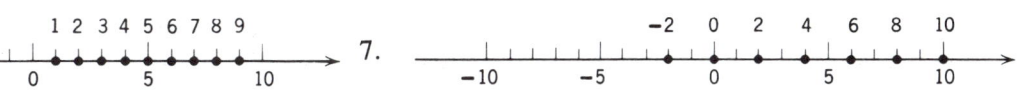

1. number line with points at −5, −3, −1, 0, 2, 5
3. number line with points at −5, −3, −2, 3, 4, 7
5. number line with points at 1, 2, 3, 4, 5, 6, 7, 8, 9
7. number line with points at −2, 0, 2, 4, 6, 8, 10
9. number line with point at −3

11. $b > c$
13. $a < d$
15. $b = b$
17. $0 < b$
19. $a < b$
21. $d > a$
23. <
25. <
27. >
29. >
31. <
33. <
35. >
37. 6
39. 10
41. −9
43. −6
45. <
47. =
49. >
51. >
53. <
55. >
57. Positive
59. Negative

EXERCISES 6.2

1. 7
3. −9
5. 4
7. −6
9. 8
11. 0
13. 10
15. −10
17. 5
19. −4
21. 5
23. 0
25. 8
27. −8
29. 5
31. 6
33. −3
35. −3
37. 3
39. 3

555

ANSWERS AND SOLUTIONS

EXERCISES 6.3

1.	6	3.	8	5.	3	7.	−3	9.	5	11. −5
13.	3	15.	−1	17.	−11	19.	−12	21.	2	23. 8
25.	0	27.	0	29.	4	31.	7	33.	−10	35. 3
37.	0	39.	5	41.	−3	43.	−6	45.	−6	47. 4

EXERCISES 6.4

1.	−15	3.	30	5.	0	7.	−8	9.	−15	11. −14
13.	0	15.	−24	17.	18	19.	0	21.	120	23. 8
25.	3^3	27.	$(-4)^2$	29.	$(-2)^3$	31.	2^5	33.	$5 \cdot 3^2$	35. $-1 \cdot 2^2 \cdot 7$
37.	3^3	39.	$-1 \cdot 2^6$	41.	−16	43.	16	45.	−12	47. 36
49.	2	51.	−16	53.	5	55.	−3	57.	0	59. Undefined
61.	−6	63.	−5	65.	−2	67.	0	69.	Undefined	

EXERCISES 6.5

1.	5	3.	14	5.	5	7.	2	9.	5
11.	32	13.	27	15.	20	17.	12	19.	22
21.	40	23.	49	25.	4	27.	2	29.	0
31.	14	33.	7	35.	5	37.	0	39.	1
41.	90	43.	−16	45.	0	47.	−16	49.	7

EXERCISES 7.1

1.	6^2	3.	x^3	5.	$3^2 y^2$	7.	$2x^2 y^3$
9.	$2a^2 bc^3$	11.	$(x - 3)^2$	13.	$3^2 + 2^3$	15.	$3x^2 + 5y^3$
17.	$(3a)^2 - b^2$	19.	$x^3 + x^2 y^2$	21.	$3a^2 - (3a)^2$	23.	$(x - y)^2 + y^2$
25.	$3 \cdot 3 \cdot xxyyy$	27.	$2 \cdot 2 \cdot 5 \cdot aabbc$	29.	$yy(3x)(3x)$	31.	$(5a)(5a)(2b)(2b)$
33.	$3 \cdot xx$	35.	$(3x)(3x)$	37.	$(a - 4)(a - 4)(a - 4)$		
39.	$yyy(3y + 4)(3y + 4)$						

41.a. Monomial   b. Coefficient of y^2 is 2
43.a. Binomial   b. Coefficient of x^2 is 5; coefficient of x is 3
45.a. Binomial   b. Coefficient of x^2 is 1; coefficient of x is 5
47.a. Monomial   b. Coefficient of y^5 is −1
49.a. Binomial   b. Coefficient of y^4 is 3; coefficient of x^2 is −2
51.a. Binomial   b. Coefficient of x is 1; coefficient of y is 1
53.a. Binomial   b. Coefficient of x^4 is 3; coefficient of y is 3
55.a. Trinomial   b. Coefficient of x^2 is 3;
                      Coefficient of y is 3;
                      Coefficient of z is −4
57. $5y^3$ has degree 3; y^2 has degree 2
59. y^4 has degree 4; $-3y$ has degree 1; 4 has degree 0
61. $4y^2$ has degree 2; $-2y$ has degree 1; 1 has degree 0
63. $4x^3$ has degree 3; $2x^2$ has degree 2; $2x$ has degree 1
65. 4   67. 5   69. 4

PART A: ANSWERS

EXERCISES 7.2

1. -8
3. 12
5. -4
7. 9
9. -48
11. 16
13. 10
15. -14
17. 0
19. 3
21. -2
23. -1
25. 1
27. -4
29. -12
31. 1
33. 3
35. -3
37. 5
39. 7
41. 1
43. -6
45. 0
47. -24
49. 5
51. 4
53. -9
55. 1
57. -18
59. $P(-1) = 4$; $P(2) = 7$
61. $Q(-2) = 9$; $Q(0) = 1$
63. $D(-3) = -6$; $D(3) = 12$

EXERCISES 7.3

1. $6y$
3. $9y^2$
5. $7x^2$
7. $6b^3$
9. $8xy$
11. $10xy^3$
13. $5x^2y + 2xy^2$
15. $3x + 6xy + 4y$
17. $2x$
19. $-3y$
21. 0
23. $14hk$
25. $7x^2 + x$
27. $3x^2 + 8c$
29. $5y + 4y^2$
31. $5x + 3y$
33. $3y + 3x$
35. $7xy + 3x$
37. x^2y
39. $5x + (-4x^2) + x^3$
41. $x^2yz + (-2xy^2z) + xyz^2$
43. $13x^3yz + (-2xyz) + (-10x^2yz)$
45. $9x + 2$
47. x
49. $4x^2 + 6x$
51. $5ab^2 + 6b$
53. $-6a^2b^2 + a^2 + b^2$

EXERCISES 7.4

1. $-3x$
3. $-2x$
5. $6y$
7. $4xy$
9. $-7x^3y$
11. $6y^2$
13. $6x$
15. $-4g$
17. $3ab^2$
19. $-3x$
21. $-11y^2$
23. $3a^2b$
25. $ab^2 + a^2b$
27. $3x^2y - 7xy^2$
29. $4x - 4$
31. $-4y^2 - 3y$
33. $-z^2 - 5z$
35. 0
37. $-x^2y - 4xy + xy^2$
39. $x^2y^2 - 3xy + 3$
41. $2xy^2 + 4xy - 2x$
43. $3x$
45. $x + 7$
47. $2a + 10b$
49. $4x + 2z$
51. $2a$
53. $-x + 2y + 6z$
55. $x - 11$
57. $-y^2 + 3$
59. $5y^2 + 4y$
61. $14y^2 + 6y - 4$

EXERCISES 7.5

1. x^5
3. y^{10}
5. $12y^5$
7. $12a^6$
9. $12a^4b^3$
11. $-10x^6y^4$
13. $6x^3$
15. y^7
17. $16a^3b^4$
19. a^3b^3
21. $a^5b^3c^3$
23. $-10a^2b^5c^4$
25. x^3
27. $3x^5$
29. $2y^3$
31. $6a^2b^2 - b^2$
33. $2y^4 - 6y^5$
35. $a^3 - a^2$
37. $4ab + 3b$
39. $8a^3b$

EXERCISES 7.6

1. $3x$
3. -4
5. Undefined
7. 0
9. $-x^3$
11. y^3
13. xy^2
15. x^3y
17. $-2xy^4$
19. $3xy^3$
21. -3
23. xy^2
25. -1
27. $-5x^2$
29. $6xy^2$
31. $-x$
33. x
35. $-3y$
37. $-4x^2$
39. 1

ANSWERS AND SOLUTIONS

41. $4xy$
43. $2hy^2$
45. -3
47. $-18y^2$
49. $3x - 3$
51. $2x$
53. $5y$
55. 0
57. $-5x$
59. 0
61. 0
63. $5y$
65. $4ab$

EXERCISES 8.1

1. No
3. Yes
5. Yes
7. -5
9. 1
11. 13
13. 5
15. -175
17. 0
19. $x = 3$
21. $z = 9$
23. 5
25. 7
27. -5
29. 1
31. -2
33. 3
35. 2
37. 0
39. 4
41. 3
43. 4
45. -4
47. -6
49. 2
51. -1
53. 3
55. 1
57. 0
59. 7
61. -4
63. 2
65. 20
67. -12
69. 20
71. 12
73. -20
75. -10
77. -36
79. 15
81. 10
83. 4
85. 6
87. 15

EXERCISES 8.2

1. 2
3. 0
5. 4
7. 8
9. 5
11. 1
13. 3
15. 2
17. -2
19. -7
21. -4
23. 18
25. 3
27. 2
29. 6
31. -2
33. -27
35. 10
37. -3
39. -2

EXERCISES 8.3

1. -9
3. 5
5. 2
7. 2
9. -32
11. 3
13. 2
15. $t = \dfrac{d}{r}$
17. $l = \dfrac{v}{wh}$
19. $d = \dfrac{c}{\pi}$
21. $r = \dfrac{d}{t}$
23. $h = \dfrac{v}{lw}$
25. $r = \dfrac{I}{pt}$
27. $g = \dfrac{v - k}{t}$
29. $m = \dfrac{Fd^2}{kM}$
31. $x = a$
33. $x = a$
35. $x = 3a$
37. $y = \dfrac{b}{a}$
39. $y = \dfrac{3a}{b}$
41. $y = \dfrac{5}{3a}$
43. $x = \dfrac{a^2}{c}$
45. $x = 2$
47. $x = \dfrac{bc}{a}$

EXERCISES 8.4

1. 7
3. -5
5. 8
7. 8
9. 4
11. 12, 14
13. 15, 17
15. $-12, -11, -10$
17. $-9, -7, -5$
19. 6, 8, 10
21. 60 centimeters, 84 centimeters
23. 32 feet, 44 feet
25. 4 feet, 8 feet, 12 feet

558

27. 12 feet, 15 feet, 24 feet
29. 2163, 2213
31. 94, 118
33. 10 kilograms
35. 21 minutes
37. 79 centimeters
39. 7 kilometers
41. 18 miles

EXERCISES 8.5

1. $<$
3. $>$
5. $<$
7. $x > 4$
9. $y < -15$
11. $x \geq 3$
13. $x > 25$
15. $y > -5$
17. $x \leq 12$
19. $y > 3$
21. $x < -21$

23. $x > 2$; 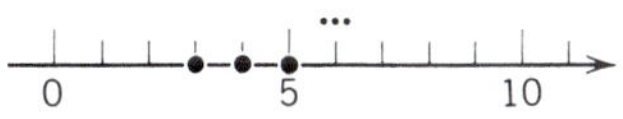
25. $x > -3$; 
27. $x \leq -6$; 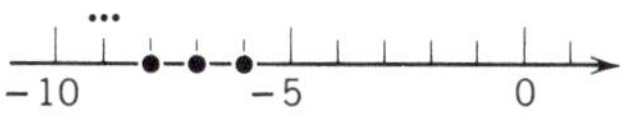
29. $x > 3$; 
31. $y > -5$; 
33. $y < -10$;
35. $x < 3$;
37. $x < 3$;
39. $x \geq 6$;

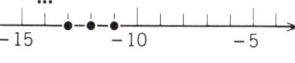

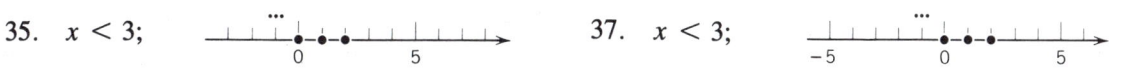

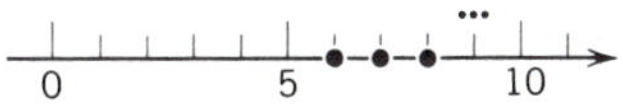

41. The smallest value could be 31.
43. The integer could be 10, 11, 12, and so on.
45. The shortest piece can be 9 feet or less.

EXERCISES 9.1

1. $3x - 12$
3. $10y - 10$
5. $-2x - 8$
7. $10a^2 + 6a$
9. $-b^2 + 2b$
11. $x^2y + xy^2$
13. $-2x^3 - 3x^2y$
15. $x^3 - 2x^2 + x$
17. $-y^3 + y^2 - 2y$
19. $4x^5 - 12x^4 + 16x^3$
21. $-y^6 + y^4 - y^3$
23. $-x^3y - x^2y^2 - xy^3$
25. $-a$
27. $2ax + x + a$
29. $-ax + ay - 2y$
31. $x^2 + 4x + 1$
33. $3x - 12y$
35. $3y^2 - 25y$
37. $ax^2 + 3ax - a$
39. $-4abx - aby + 2ab + b$
41. $-a - c$
43. $a - 2b + c$
45. $-3x - 2y + z$
47. $-1 + 3x - x^2$
49. 0
51. $-2x$

EXERCISES 9.2

1. $3(x + 2)$
3. $2(x - 3y)$
5. $2y(y - 1)$
7. $y(ay + 1)$
9. $3y(3ay + 2)$
11. $3(y^2 - y + 1)$
13. $a(x + y - z)$
15. $x(x - 3 + y)$
17. $2y(2y^2 - y + 1)$
19. $6axy(x - 3y + 4)$
21. $-a(a + b)$
23. $-x(1 + x)$
25. $-b(ac + a + c)$
27. $-3y(2y^2 + y + 1)$
29. $-x(1 - x + x^2)$
31. $-xy^2(y^3 + y^2 - 1)$
33. $d = k(1 + at)$
35. $S = kr^2(h + 1)$
37. $V = 2ga^2(D - d)$
39. $A = r^2(a + b + c)$

559

ANSWERS AND SOLUTIONS

EXERCISES 9.3

1. $x^2 + 7x + 12$
3. $y^2 - 2y - 3$
5. $a^2 + 7a + 10$
7. $b^2 - 2b - 8$
9. $x^2 + 9x + 8$
11. $y^2 - 8y + 7$
13. $a^2 + 8a + 16$
15. $b^2 - 25$
17. $x^2 + 2x + 1$
19. $y^2 - 1$
21. $4 - x^2$
23. $36 - y^2$
25. $x^2 - 4bx + 3b^2$
27. $x^2 + xy - 2y^2$
29. $x^2 + 4ax + 4a^2$
31. $y^2 - 36a^2$
33. $x^2 - t^2$
35. $x^2 + 8x + 16$
37. $x^2 - 14x + 49$
39. $x^2 - 2x + 1$
41. $x^2 + 4x + 4$
43. $a^2 - 2ab + b^2$
45. $2x^2 + 6x + 4$
47. $6y^2 + 60y + 150$
49. $6x^2 - 12x + 6$
51. $a^3 + 4a^2 - 5a$
53. $a^3 - 4a$
55. $xy^2 - 6xy + 9x$

EXERCISES 9.4

1. $(x + 2)(x + 3)$
3. $(x + 6)(x + 5)$
5. $(x + 9)(x + 5)$
7. $(y - 1)(y - 2)$
9. $(y - 7)(y - 9)$
11. $(x - 4)(x + 3)$
13. $(y + 5)(y - 4)$
15. $(a + 7)(a - 5)$
17. $(b - 20)(b + 1)$
19. $(a - 10)(a + 5)$
21. $(b - 9)(b + 5)$
23. $(y - 45)(y + 1)$
25. $2(x + 3)(x + 2)$
27. $y(y - 3)(y + 1)$
29. $5(c - 2)(c - 3)$
31. $4b(a + 6)(a - 3)$
33. $(x + 2a)(x + 2a)$
35. $(a - b)(a - 2b)$
37. $(s + 3a)(s + 2a)$
39. $(y + 5)(y + 2)$
41. $(x - 8)(x - 1)$
43. $(z - 8)(z - 4)$
45. $-(x - 3)(x + 7)$
47. $-(z - 12)(z + 2)$
49. $-(y - 9)(y + 2)$
51. Not factorable
53. Not factorable
55. Not factorable
57. $(x + 4y)(x + y)$
59. Not factorable

EXERCISES 9.5

1. $2x^2 + 7x + 3$
3. $3y^2 + y - 2$
5. $6y^2 + 11y + 3$
7. $20x^2 + 7x - 6$
9. $4x^2 - 4xy - 15y^2$
11. $8y^2 + 6xy - 9x^2$
13. $4x^2 + 4x + 1$
15. $25x^2 + 20x + 4$
17. $16y^2 + 40y + 25$
19. $x^2 - 4xy + 4y^2$
21. $9x^2 - 6xy + y^2$
23. $64x^2 + 48xy + 9y^2$
25. $4x^2 + 12xy + 9y^2$
27. $4x^2 - 9$
29. $36y^2 - 25$
31. $4x^2 - a^2$
33. $9x^2 - 4y^2$
35. $16x^2 - 49y^2$
37. $6x^2 - 16x - 6$
39. $12y^2 - 3$
41. $12x^2 - 60x + 75$
43. $2x^3 + x^2 - 10x$
45. $4x^3 - 4x^2 + x$
47. $9r^3 - r$

EXERCISES 9.6

1. Not factorable
3. Factorable
5. Not factorable
7. $(3a + 1)(a + 1)$
9. $(2x - 1)(x - 1)$
11. $(3b - 1)(3b - 1)$
13. $(2x - 1)(x - 3)$
15. $(4y - 1)(y - 1)$
17. $(8x + 5)(8x + 3)$
19. $(4y + 1)(y - 1)$
21. $(4a + 5)(a - 1)$
23. $(8x + 1)(2x - 5)$
25. $(16x - 1)(x + 5)$
27. $(2t + s)(t - 3s)$
29. $(3x - a)(x - 2a)$
31. $(2x + 3)(x - 1)$
33. $(2x - 3)(x + 1)$
35. $(3a + 1)(2a + 1)$
37. $(4x - 5)(4x + 1)$
39. $(4y + b)(y + b)$
41. $(2a + 5b)(2a + 3b)$
43. $2(3x + 1)(x + 1)$
45. $2(4y + 1)(y - 1)$
47. $9(2x - 3)(x + 1)$
49. $3y(3y + 1)(3y - 2)$
51. $3a(4b + a)(b + a)$
53. $2xy(5y + 4x)(5y - 2x)$

PART A: ANSWERS

EXERCISES 9.7

1. $(x + 3)(x - 3)$
3. $(x + 1)(x - 1)$
5. $(x + z)(x - z)$
7. $(xy + 4)(xy - 4)$
9. $(ax + 7b)(ax - 7b)$
11. $(6 + x)(6 - x)$
13. $(2b + 3)(2b - 3)$
15. $(5x + 4)(5x - 4)$
17. $(3 + 2x)(3 - 2x)$
19. $(9 + 2x)(9 - 2x)$
21. $(2a + 11b)(2a - 11b)$
23. $(5y + 7x)(5y - 7x)$
25. $(7ax + 12by)(7ax - 12by)$
27. $(2xy + 9)(2xy - 9)$
29. $(6ab + 1)(6ab - 1)$
31. $5(x + 1)(x - 1)$
33. $3x(x + 1)(x - 1)$
35. $2(x + 2y)(x - 2y)$
37. $3(ab + 2cd)(ab - 2cd)$
39. $4x^2(y + 2)(y - 2)$

EXERCISES 9.8

1. 7
3. -1
5. 3
7. -1
9. -2
11. 3
13. -5
15. 4
17. -5
19. 0
21. 2
23. -4
25. 3
27. 1
29. 4
31. 5
33. 1
35. 4
37. -3
39. -5

EXERCISES 9.9

1. a. $x + 4$ b. $5x$ c. $5(x + 4)$
3. $3(n + 6)$
5. $5(n - 8)$
7. a. $27 - n$ b. $3n$ c. $3(27 - n)$
9. a. $n + 16$ b. $5n$ c. $2(n + 16)$
11. a. $42 - n$ b. $4(42 - n)$ c. $5n$
13. a. $x + 2$ b. x^2 c. $(x + 2)^2$
15. a. $x + 1$ b. $3(x + 1)$ c. $x + 2$
17. a. $x + 1$ b. $(x + 1)^2$ c. $x(x + 1)$
19. 5, 3
21. 27, 24
23. 18, 7
25. 12, 14
27. 7, 21
29. 6, 8, 10
31. 9, 11
33. 9, 10, 11

EXERCISES 9.10

1. 10 nickels, 13 dimes
3. 9 pennies, 15 nickels, 3 dimes
5. 600 adults, 400 children
7. $1800 at 10%, $8000 at 11%
9. $18,000 at 8%, $16,000 at 9%
11. $1500 at 9%, $4500 at 12%
13. 16 grams
15. 40 kg
17. 15 quarts
19. 30 ounces

EXERCISES 10.1

1. $\dfrac{4}{7}$
3. $\dfrac{3x}{y}$
5. $\dfrac{7}{x - y}$
7. $\dfrac{x - 3}{4x + 1}$
9. $4 \cdot \dfrac{1}{7}$
11. $9 \cdot \dfrac{1}{5}$
13. $(x - 3) \cdot \dfrac{1}{4}$
15. $2 \cdot \dfrac{1}{x + 3}$

561

ANSWERS AND SOLUTIONS

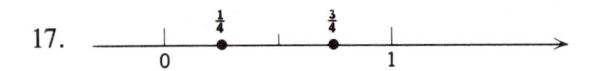

17. [number line showing 1/4 and 3/4 between 0 and 1]

19. [number line showing 1/2 and 5/2 between 0 and 3]

21. [number line showing −5/6 and 1/6 between −1 and 1]

23. [number line showing −5/2 and 5/4 between −3 and 3]

25. [number line showing −3, −3/4, 3/2 between −3 and 3]

27. [number line showing 2/5, 3/5, 4/5 between 0 and 1]

29. $\frac{3}{5}$
31. $\frac{2}{7}$
33. $\frac{-2}{5}$
35. $\frac{-a}{b}$
37. $\frac{-a}{b}$
39. $\frac{-x}{y}$
41. $\frac{7x}{8y}$
43. $-c$
45. $\frac{-(x+2)}{4}$
47. $\frac{x+5}{4}$
49. $\frac{-(2x-1)}{x+2}$
51. $\frac{x-3}{2}$
53. 0
55. 3

EXERCISES 10.2

1. $\frac{1}{4x}$
3. $\frac{-2}{3y^4}$
5. $\frac{-1}{2x}$
7. x^2y
9. $\frac{-1}{xy^2}$
11. $2x^2$
13. $\frac{3x^2}{2}$
15. $\frac{5bc^2}{4}$
17. $\frac{13a^2}{3c^2}$
19. $\frac{3}{4}$
21. $-4(x-y)$
23. 1
25. -2
27. $\frac{2}{x-a}$
29. $\frac{-1}{x+4}$
31. $\frac{1}{x+1}$
33. $\frac{1}{a-b}$
35. $\frac{a-b}{a+b}$
37. $\frac{a}{a+1}$
39. $\frac{x-2}{x-3}$
41. $\frac{a+3}{a-1}$
43. $4x^2y$
45. $\frac{-5y^2}{x}$
47. $\frac{x}{15y^2}$
49. b
51. a
53. b
55. a

EXERCISES 10.3

1. $2x - 1$
3. $y + 2$
5. $x + 3$
7. $3y^2 - 2y + 1$
9. $2y^2 - y + 3$
11. $4y - x + 1$
13. $3x^2 + 2x + \frac{-1}{3}$
15. $y + 2 + \frac{-1}{y}$
17. $3x^2 - 2 + \frac{-2}{3x^2}$
19. $y^2 - 3y + 2 + \frac{-1}{y}$
21. $y + 1 + \frac{1}{y}$
23. $x - 2 + \frac{3}{y}$
25. $x + 6$
27. $x + 1$
29. $2x + 1$
31. $x + 3$
33. $2x + 3$
35. $2x^2 - x + 1$
37. $x + 1 + \frac{-1}{x+2}$
39. $x - 2 + \frac{1}{x+5}$
41. $2x - 1 + \frac{-1}{x+1}$
43. $2x - 3 + \frac{-2}{2x+1}$
45. $x + 7$
47. $x - 6 + \frac{29}{x+6}$
49. $2x^2 + x + 1 + \frac{2}{x-1}$

PART A: ANSWERS

EXERCISES 10.4

1. $\dfrac{10}{6x}$ 3. $\dfrac{-12ab^2}{12b^3}$ 5. $\dfrac{-3x^2y}{3y^3}$ 7. $\dfrac{72}{36}$

9. $\dfrac{xy^2}{xy}$ 11. $\dfrac{3x^4y}{3x^2y}$ 13. $\dfrac{x+y}{2(x+y)}$ 15. $\dfrac{-2a(a+4)}{5(a+4)}$

17. $\dfrac{2a(a+3)}{a+3}$ 19. $\dfrac{3(x+y)}{(x-y)(x+y)}$ 21. $\dfrac{-3(x+1)}{(2x+1)(x+1)}$

23. $\dfrac{7a(b-3)}{(b+2)(b-3)}$ 25. $\dfrac{a^2}{a(a-3)}$ 27. $\dfrac{-3(x-y)}{(x+y)(x-y)}$

29. $\dfrac{y(y+2)}{(y-1)(y+2)}$ 31. $\dfrac{(x-1)(x+1)}{x^3-2x^2+x}$

EXERCISES 10.5

1. $\dfrac{1}{25}$ 3. $\dfrac{1}{x^6}$ 5. 1 7. $\dfrac{3}{64}$

9. $\dfrac{3}{1000}$ 11. $\dfrac{4}{x^2}$ 13. 2^{-3} 15. 5^{-2}

17. 2^{-2} or 4^{-1} 19. $x \cdot 5^{-2}$ or $x \cdot 25^{-1}$

21. $2 \cdot 10^{-2}$ 23. $x \cdot 10^{-3}$ 25. 10^2 27. 10^{-2}

29. 4.83×10^2 31. 7.2×10^{-2} 33. 4×10^3 35. 6.3×10^{-4}

37. 43,000 39. 0.00057 41. 82,340,000 43. 0.000008

45. 5.98×10^{27} 47. 3.0×10^8 49. 1.5×10^{-5}

EXERCISES 11.1

1. y 3. $\dfrac{4x^2}{5}$ 5. $4y$ 7. $\dfrac{-x}{y}$

9. $\dfrac{49rt}{4}$ 11. $\dfrac{-b}{az}$ 13. $\dfrac{3}{4}$ 15. 5

17. 1 19. $\dfrac{x-3}{x+7}$ 21. $\dfrac{3x-2}{3x+2}$ 23. $\dfrac{x+1}{x+3}$

25. $\dfrac{x-2}{2x(x+1)}$ 27. 1 29. $\dfrac{y-3}{y-6}$ 31. $\dfrac{2x}{3}$

33. $\dfrac{-2a}{5}$ 35. $\dfrac{3(a-b)}{4}$ 37. $\dfrac{-3(2x-y)}{5}$ 39. $\dfrac{3}{7}x$

41. $\dfrac{-5}{7}a$ 43. $\dfrac{5}{2}(a-b)$ 45. $\dfrac{1}{7}(x+y)$

EXERCISES 11.2

1. 1 3. $\dfrac{5}{16b^2}$ 5. $\dfrac{-x}{v}$ 7. $\dfrac{-x}{16y}$ 9. $\dfrac{b}{6a}$

11. $\dfrac{y^4}{3x}$ 13. $12y$ 15. $\dfrac{2y}{x}$ 17. $\dfrac{a+b}{2a}$ 19. $\dfrac{a}{2}$

563

ANSWERS AND SOLUTIONS

21. $\dfrac{5}{2}$
23. $\dfrac{5}{6x}$
25. $\dfrac{2x+y}{x+2y}$
27. $\dfrac{y-1}{y+7}$
29. $\dfrac{x-5}{x-2}$
31. $\dfrac{y^2+6y+8}{(y-5)(y-2)}$
33. $\dfrac{x^2-x-12}{(x-1)(2x+1)}$
35. $\dfrac{1}{y-1}$

EXERCISES 11.3

1. $\dfrac{2x+5y}{9}$
3. $\dfrac{5x-3}{11}$
5. $\dfrac{3-x}{5}$
7. $\dfrac{4}{a}$
9. $\dfrac{1}{b}$
11. $\dfrac{4}{x}$
13. $\dfrac{x}{y}$
15. $\dfrac{x-2}{2}$
17. $\dfrac{2x+y}{3x}$
19. $\dfrac{x}{a}$
21. x
23. $\dfrac{4x}{y}$
25. $\dfrac{x+6}{2}$
27. $\dfrac{a+3b}{a-b}$
29. $\dfrac{-a}{a+b}$
31. $\dfrac{3x+2y}{x+y}$
33. $\dfrac{1}{x+2y}$
35. $\dfrac{6a-b}{3}$
37. $3x+y$
39. $\dfrac{u}{2u-v}$
41. $\dfrac{x-1}{x+2}$
43. $\dfrac{4}{x-1}$
45. $\dfrac{1}{x-1}$
47. $\dfrac{x+5}{x}$

EXERCISES 11.4

1. x^2y
3. xyz
5. $24x^2y$
7. $(x+y)(x-y)$
9. $x^2(x+2)$
11. $(x+4)(x-1)^2$
13. $\dfrac{21x}{8}$
15. $\dfrac{5+3a}{ax}$
17. $\dfrac{6y+4x}{3xy}$
19. $\dfrac{yz+xz+xy}{xyz}$
21. $\dfrac{8y-5}{6}$
23. $\dfrac{4x+7}{6}$
25. $\dfrac{5x-y}{6x}$
27. $\dfrac{7x+3}{(x+3)(x-3)}$
29. $\dfrac{x^2+3xy-2y^2}{(x-y)(x+y)}$
31. $\dfrac{2x^2+x-2}{(x-2)(x+2)}$
33. $\dfrac{2y^2-4y+14}{(y-3)(y+2)}$
35. $\dfrac{y^2+4y-3}{(y+1)(y+1)(y-1)}$
37. $\dfrac{3x^2-3x}{(x+1)(x+2)(x-2)}$
39. $\dfrac{2x^2+2x-3}{x(x+3)(x+3)}$

EXERCISES 11.5

1. $\dfrac{2x-5y}{10}$
3. $\dfrac{3y-2x}{12}$
5. $\dfrac{5}{2x}$
7. $\dfrac{1}{3x}$
9. $\dfrac{2y-3x}{xy}$
11. $\dfrac{9y-5x}{6xy}$
13. $\dfrac{-x-4}{6}$
15. $\dfrac{4y-3}{6}$
17. $\dfrac{-4x-7}{6}$
19. $\dfrac{-4x+7}{6x}$
21. $\dfrac{x-5y}{6x}$
23. $\dfrac{6a^2-5ab+6b^2}{12ab}$
25. $\dfrac{3}{2(x+y)}$
27. $\dfrac{-2}{3(x+1)}$
29. $\dfrac{9}{4(2a+b)}$
31. $\dfrac{-6x}{(x+3)(x-3)}$
33. $\dfrac{38}{(3x-4)(5x+6)}$
35. $\dfrac{-3a-9}{(2a+1)(a-2)}$
37. $\dfrac{-8x}{(x+2)(x-2)}$
39. $\dfrac{-1}{(x+2)(x+3)}$
41. $\dfrac{a^2-7ab+2b^2}{(a+b)(a-b)}$
43. $\dfrac{3}{(x+1)(x+1)(x-2)}$
45. $\dfrac{x^2-x}{(x-2)(x+3)(x+5)}$
47. $\dfrac{2x^2+13x+6}{(x+1)(x+2)(x+2)}$

PART A: ANSWERS

EXERCISES 11.6

1. $\dfrac{3}{2}$
3. $\dfrac{5}{4}$
5. $\dfrac{8}{7}$
7. 6
9. $\dfrac{b}{3a}$
11. $\dfrac{5}{3xy}$
13. $\dfrac{5}{3}$
15. $\dfrac{5}{2x}$
17. $\dfrac{3}{2}$
19. 6
21. $\dfrac{1}{4}$
23. $\dfrac{1}{4}$
25. $\dfrac{1}{14}$
27. 5
29. $\dfrac{2ab - a^2}{2ab - b^2}$
31. $\dfrac{y^2 - 1}{y^2 + 1}$
33. $\dfrac{x}{4}$
35. $\dfrac{a}{11b}$
37. $\dfrac{ab - a}{2b + 3}$
39. $\dfrac{2x - 1}{x}$

EXERCISES 11.7

1. 1
3. 6
5. 12
7. 2
9. 3
11. $\dfrac{1}{2}$
13. 15
15. 5
17. 6
19. 5
21. 4
23. 3
25. No solution
27. $\dfrac{a + b}{2}$
29. -4
31. $\dfrac{4a - 3b}{a}$
33. $-2a - 6$
35. No solution
37. 3
39. 2
41. 15

EXERCISES 11.8

1. 5
3. 14, 15
5. 18
7. $240
9. 20°
11. 150 kg
13. 72
15. Slower driver: 40 mph, faster driver: 60 mph
17. $1\tfrac{3}{4}$ miles per hour
19. Slower man: 30 mph, faster man: 60 mph
21. Plane: 1080 mph, rocket: 3780 mph
23. Slower car: 25 mph, faster car: 75 mph

EXERCISES 12.1

1. a. d and t b. d increases   c. t d. d e. 12
3. a. (2, 1)   b. (−1, 4)
5. a. (2, −5)   b. (−2, 3)
7. a. (−3, 8)   b. (3, −1)
9. a. $\left(-4, \dfrac{-3}{2}\right)$ b. $\left(4, \dfrac{5}{2}\right)$
11. a. (2, 7)   b. (−3, −3)
13. a. (3, 11)   b. (1, 3)
15. a. (0, 0)   b. (−3, −9)
17. a. (4, 13)   b. (1, 7)
19. a. (5, −14)   b. (3, −10)
21. a. (−2, −10)   b. (0, −2)
23. a. $\left(-2, \dfrac{7}{2}\right)$ b. $\left(5, \dfrac{-7}{4}\right)$
25. a. $\left(10, \dfrac{14}{3}\right)$ b. $\left(5, \dfrac{4}{3}\right)$
27. a. $\left(7, \dfrac{-21}{2}\right)$ b. $\left(5, \dfrac{-13}{2}\right)$
29. a. (−3, −3)   b. $\left(0, \dfrac{-3}{5}\right)$
31. $f(-2) = 3$, $f(2) = 7$
33. $f(-5) = -17$, $f(-1) = -9$
35. $f(-1) = 7$, $f(0) = 4$
37. $f(2) = 1$, $f(6) = 7$
39. $f(-5) = -4$, $f(0) = -1$
41. $f(-1) = -1$, $f\left(\dfrac{5}{2}\right) = \dfrac{2}{5}$

565

ANSWERS AND SOLUTIONS

EXERCISES 12.2

1.

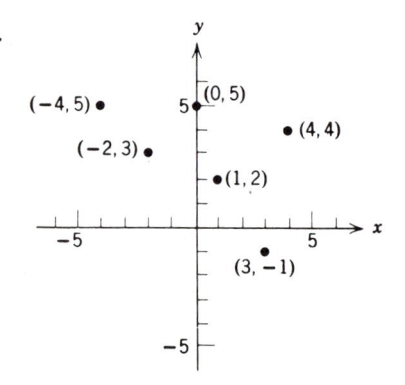

3.

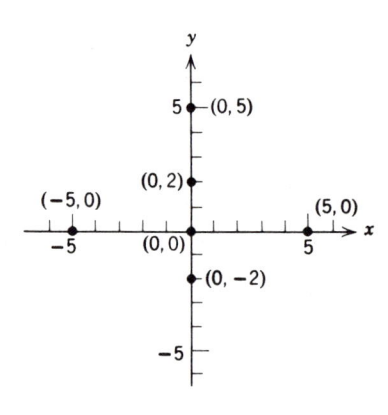

5.

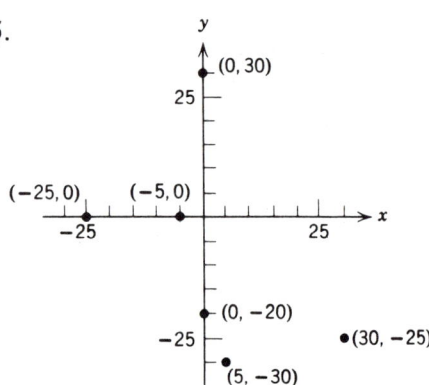

7.

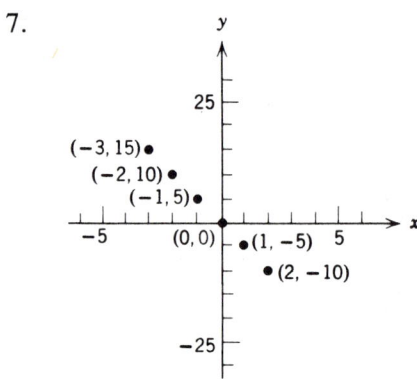

9. Result is a straight line
13. Origin

11. a. Ordinate   b. Abscissa
15. On a straight line bisecting the angles at the origin in the first and third quadrants.

17. 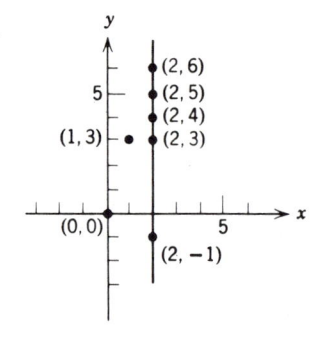
a. Yes
b. No
c. Yes
d. No

19. 2

EXERCISES 12.3

1. 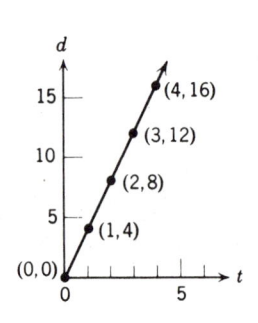
a. (0, 0)
b. (2, 8)
c. (4, 16)
d. On the line through these points
e. (1, 4)(3, 12)
f. Two

566

PART A: ANSWERS

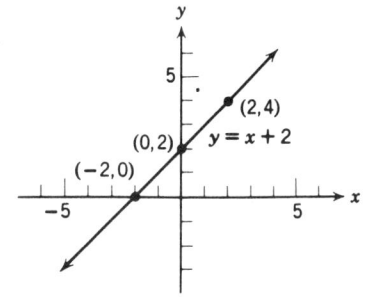

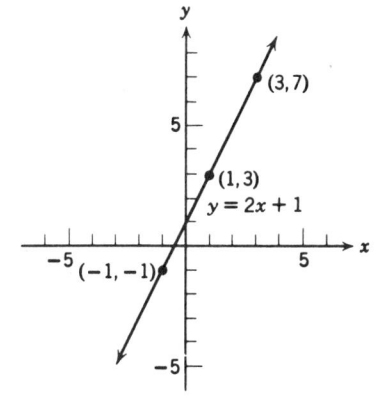

3. [graph: y = x + 2, points (−2,0), (0,2), (2,4)]

5. [graph: y = 2x + 1, points (−1,−1), (1,3), (3,7)]

7. [graph: y = 2x − 1, points (−1,−3), (0,−1), (3,5)]

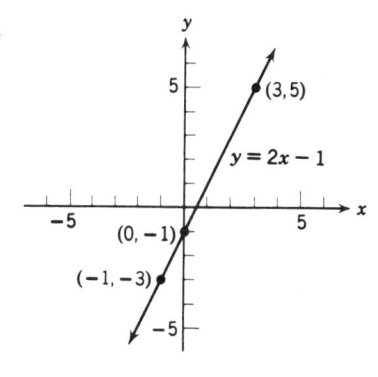

9. [graph: y − 3x = 0, points (−1,−3), (1,3), (2,6)]

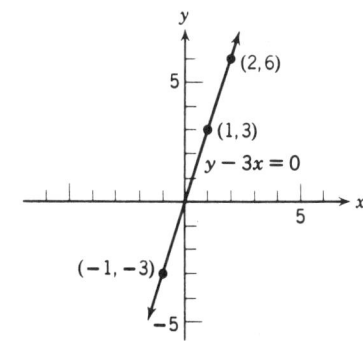

11. [graph: 3y = 4 − x, points (−5,3), (−2,2), (1,1)]

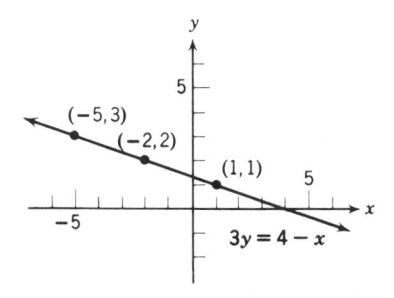

13. [graph: 2y + x − 6 = 0, points (−2,4), (0,3), (6,0)]

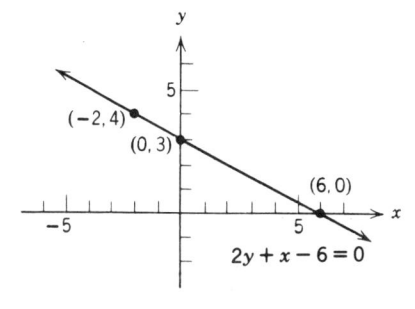

15. a. [graph: y = 3, points (2,3), (5,3)]
    b. Yes
    c. Yes
    d. No
    e. Yes
    f. Yes

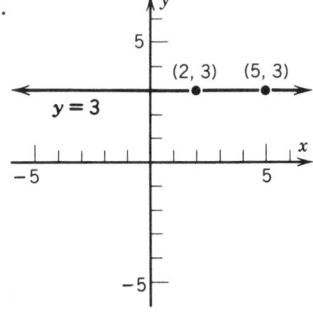

17. [graph: x = 4]

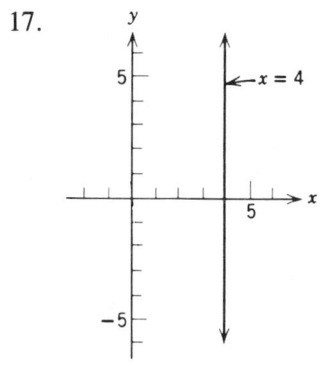

19. [graph: −2y = 8]

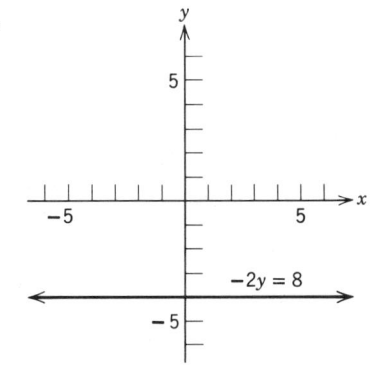

21. [graph: x = −2]

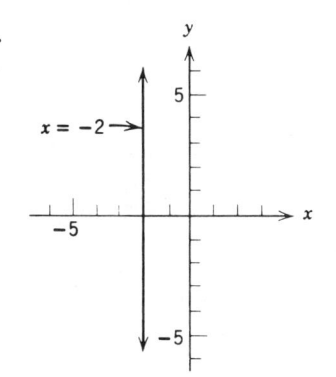

23. [graph: x = 0]

567

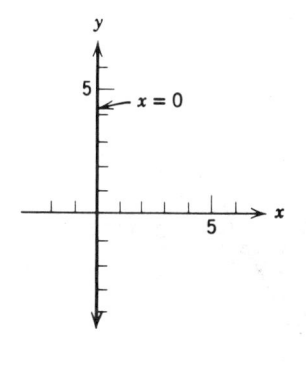

ANSWERS AND SOLUTIONS

25. [Graph of horizontal line $3y = 12$]

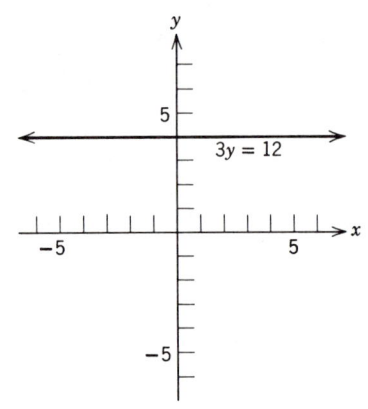

27. [Graph of vertical line $4x = -20$]

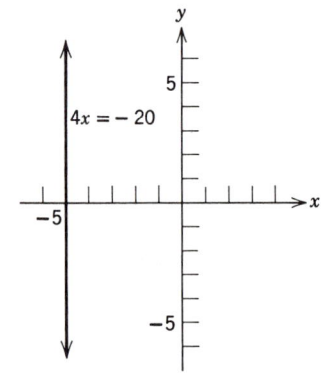

EXERCISES 12.4

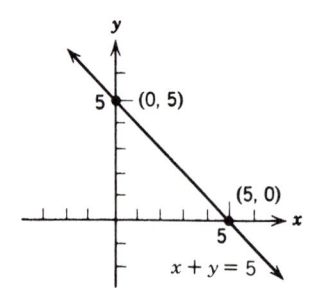

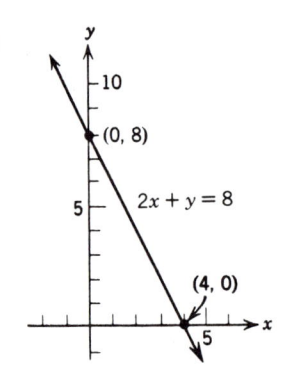

1. [Graph of $x + y = 5$ through (0, 5) and (5, 0)]

3. [Graph of $2x + y = 8$ through (0, 8) and (4, 0)]

5. [Graph of $3x - y = 6$ through (2, 0) and (0, -6)]

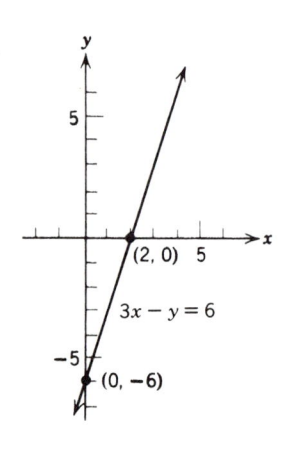

7. [Graph of $2x + 3y = 12$ through (0, 4) and (6, 0)]

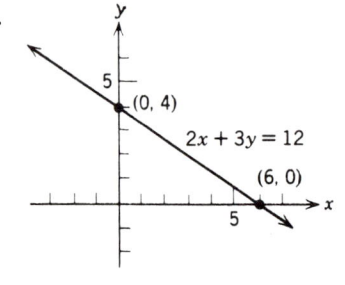

9. [Graph of $3x - 4y = 12$ through (4, 0) and (0, -3)]

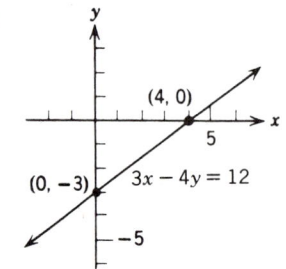

11. [Graph of $y = x + 6$ through (-6, 0) and (0, 6)]

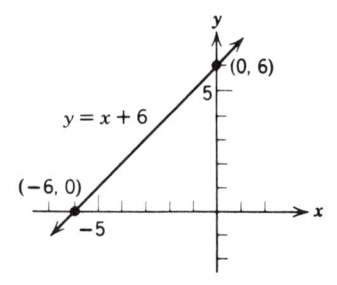

13. [Graph of $y = 2x - 4$ through (2, 0) and (0, -4)]

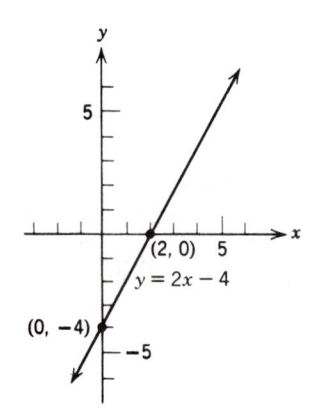

15. [Graph of $y = 2x + 5$ through $(-\frac{5}{2}, 0)$ and (0, 5)]

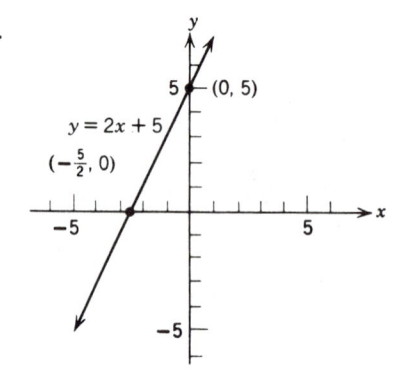

17. [Graph of $x = 4 + y$ through (4, 0) and (0, -4)]

568

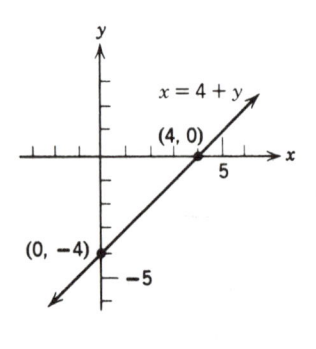

19.
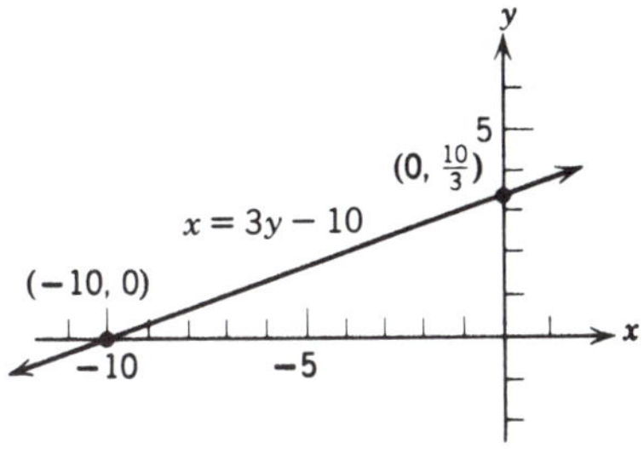

y
5
(0, 10/3)
x = 3y − 10
(−10, 0)
−10
−5
x

21.
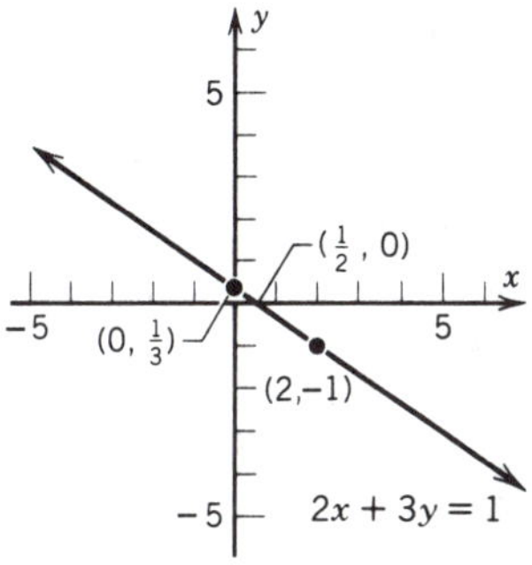

y
5
(1/2, 0)
x
−5
(0, 1/3)
5
(2,−1)
−5
2x + 3y = 1

23.
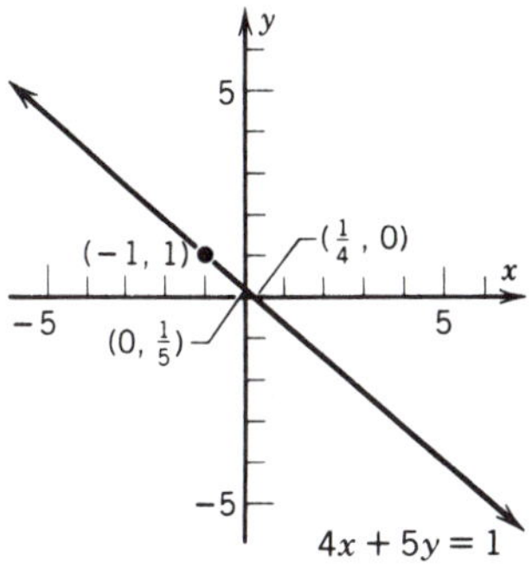

y
5
(−1, 1)
(1/4, 0)
x
−5
(0, 1/5)
5
−5
4x + 5y = 1

25.
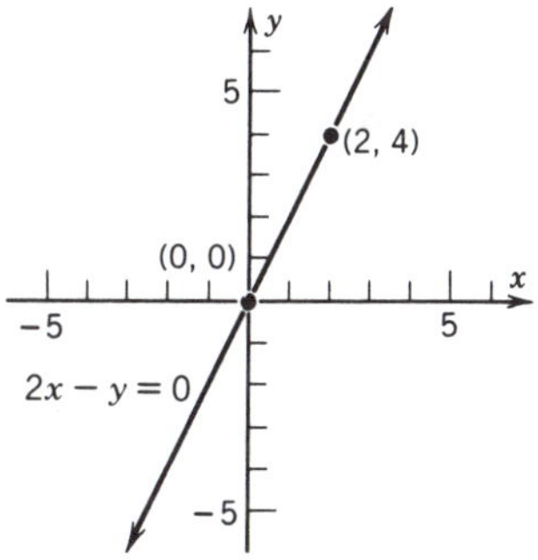

y
5
(2, 4)
(0, 0)
x
−5
5
2x − y = 0
−5

27.
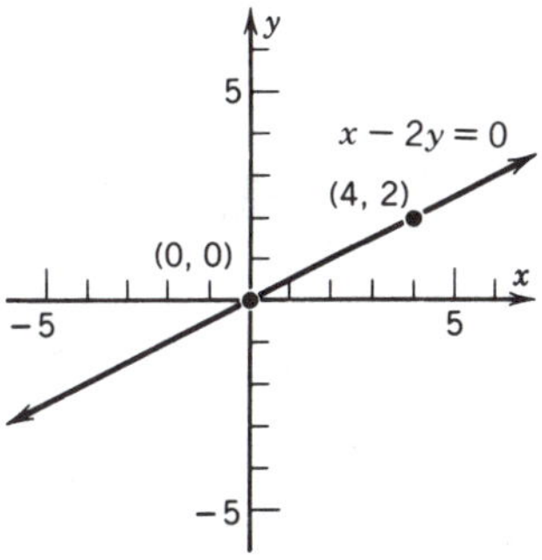

y
5
x − 2y = 0
(4, 2)
(0, 0)
x
−5
5
−5

29.
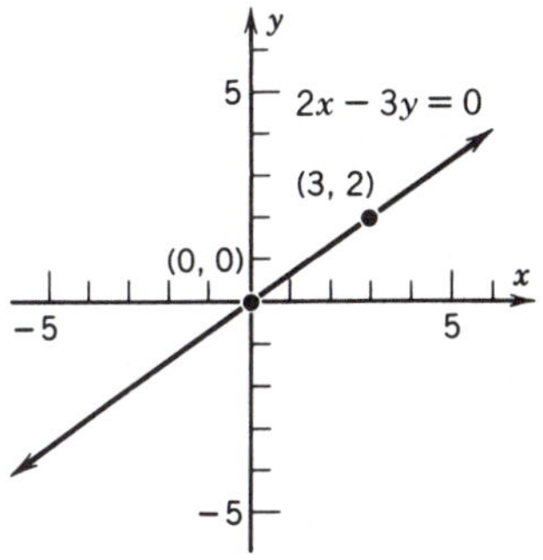

y
5
2x − 3y = 0
(3, 2)
(0, 0)
x
−5
5
−5

EXERCISES 12.5

1. $\frac{5}{3}$ 3. $\frac{-3}{4}$ 5. -1 7. $\frac{-5}{4}$ 9. Undefined 11. 0

13. a.
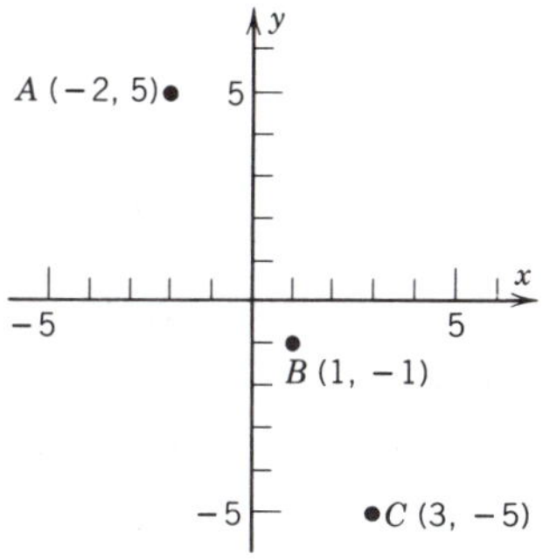

y
A (−2, 5)
5
x
−5
5
B (1, −1)
−5
C (3, −5)

b. -2
c. -2
d. Yes

15. a.
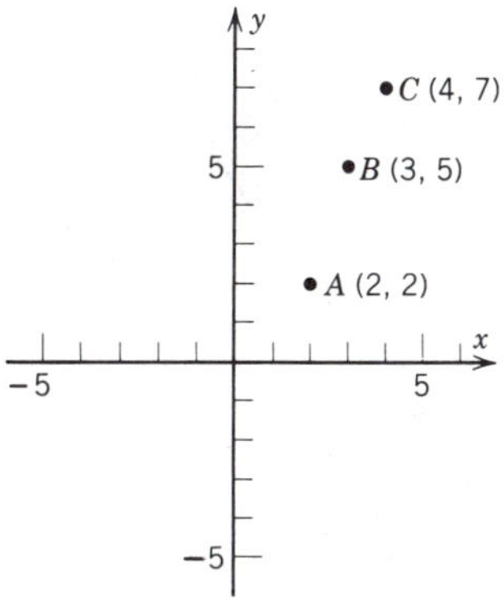

y
C (4, 7)
5
B (3, 5)
A (2, 2)
x
−5
5
−5

b. 3
c. 2
d. No

17. a. 2 b. 2 c.
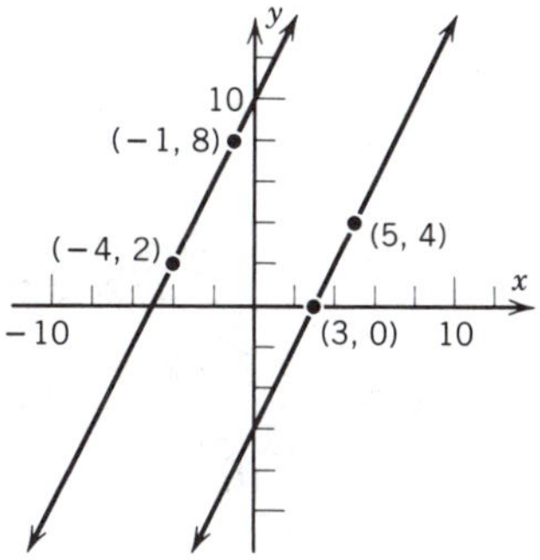

y
10
(−1, 8)
(−4, 2)
(5, 4)
x
−10
(3, 0)
10

d. Yes

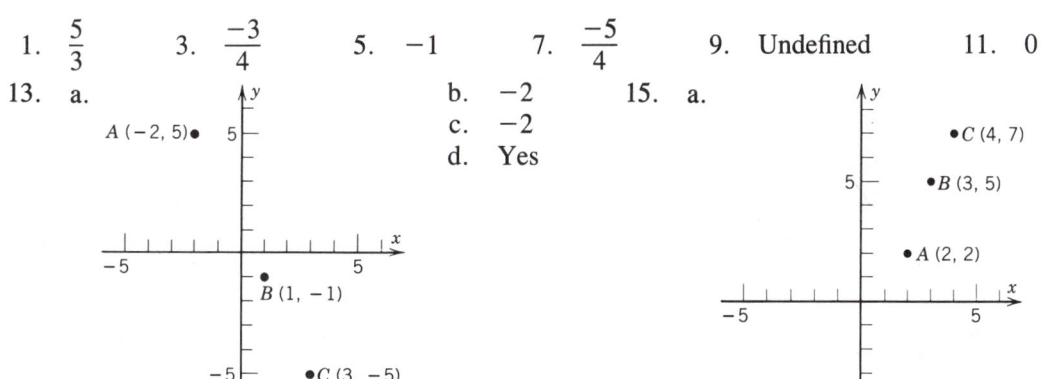

ANSWERS AND SOLUTIONS

19. a. $\frac{1}{2}$ b. $\frac{5}{8}$ c. d. No

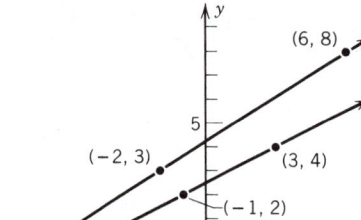

21. a. $\frac{1}{4}$ b. -4 c. d. Yes

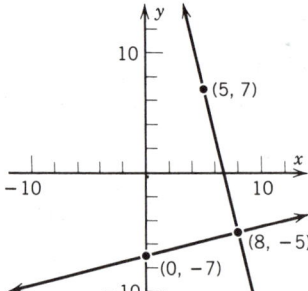

23. a. $\frac{5}{4}$ b. 2 c. d. No

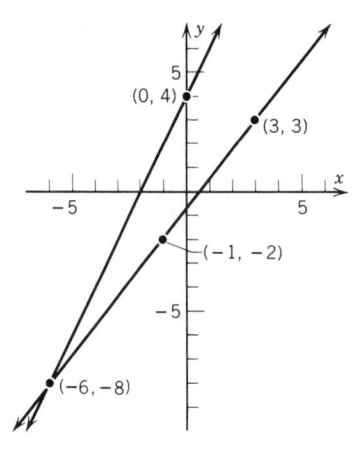

25. $\frac{1}{2}$ 27. $\frac{-3}{2}$ 29. 0

EXERCISES 12.6

1. $y = 3x + 4$
3. $y = -2x - 2$
5. $y = 1$
7. $y = \frac{1}{2}x - 5$
9. $y = -4x - 5$
11. $y = \frac{2}{3}x + 3$
13. $y = x + 6$
15. $y = 4x + 7$
17. $y = \frac{5}{3}x$
19. $y = 4$
21. $y = \frac{-1}{8}x + \frac{19}{4}$
23. Slope: 5; y intercept: -2
25. Slope: -4; y intercept: 3

570

PART A: ANSWERS

27. Slope: -3; y intercept: 4
29. Slope: -2; y intercept: $\frac{5}{3}$
31. Slope: $\frac{-5}{4}$; y intercept: $\frac{-3}{4}$
33. Slope: $\frac{2}{3}$; y intercept: -2
35. a. -3 b. -3 c. $y = -3x + 6$
37. a. -2 b. $\frac{1}{2}$ c. $y = \frac{1}{2}x$

EXERCISES 12.7

1. $y = kx$
3. $R = kL$
5. $c = kw$
7. $T = ks$
9. 21
11. 63
13. 168 miles
15. 94.29 volts
17. 30 lb per square foot
19. Circumference is doubled

EXERCISES 12.8

1. $(0, 1)$ is not a solution.
3. $(-1, 2)$ is not a solution.
5. $(-2, -1)$ is a solution.
7. $(-1, -1)$ is a solution.
9. $(0, -3)$ is not a solution.
11. $(-1, 1)$ is a solution.

13. $y - 3x \leq 5$

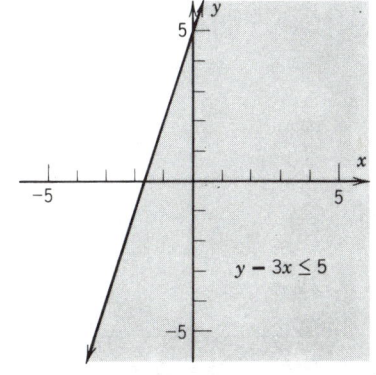

15. $3x - 4y \geq 12$

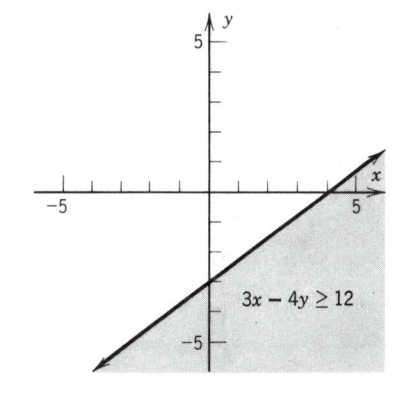

17. $y + 3x < 6$

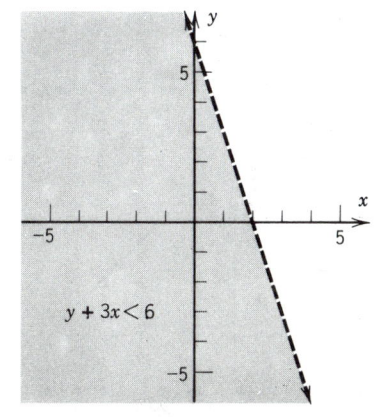

19. $y > 3x + 2$

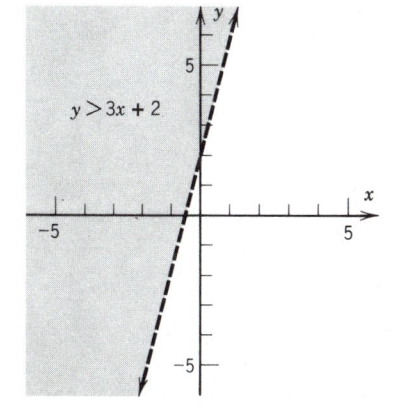

21. $y < x - 3$

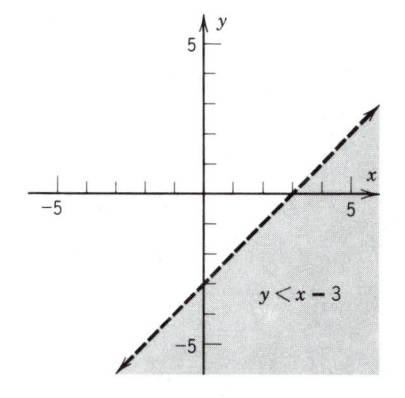

23. $6x + 4y \leq 12$

571

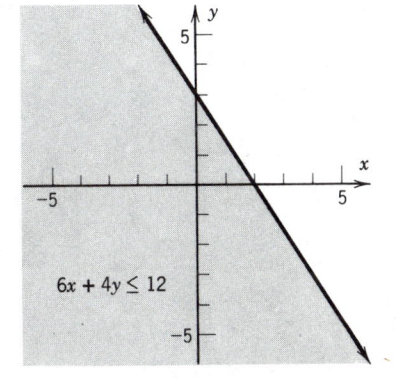

ANSWERS AND SOLUTIONS

25. $y \leq 2$

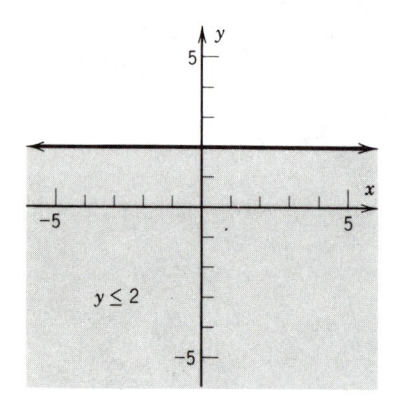

27. $x > -3$

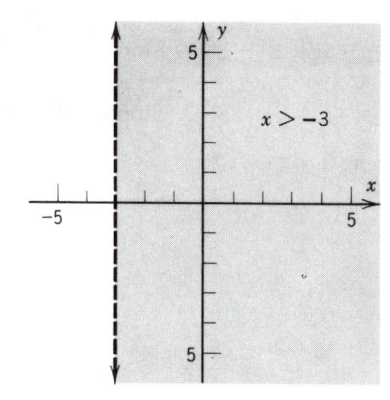

29. $y \leq 2x$

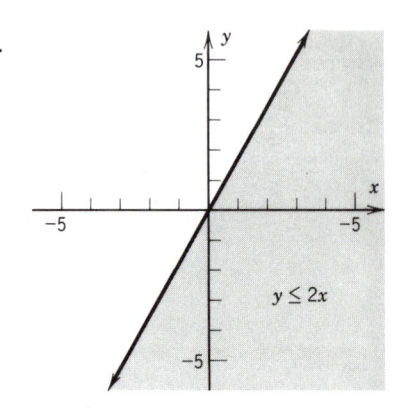

31. $y < 3x$

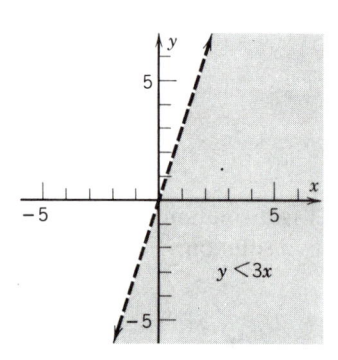

33. $y > 4x$

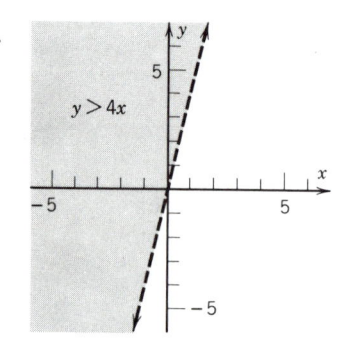

EXERCISES 13.1

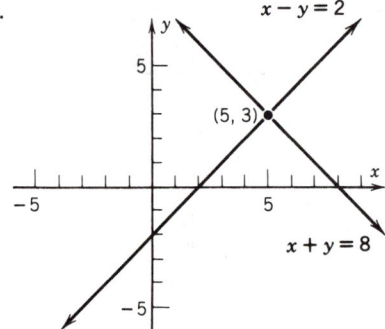

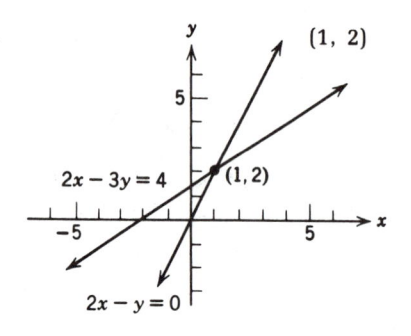

1. (5, 3); $x - y = 2$, $x + y = 8$

3. (1, 2); $2x - 3y = 4$, $2x - y = 0$

5. (−3, −1); $2x - y = -5$, $x - 3y = 0$

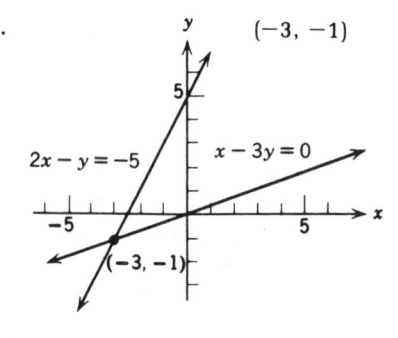

7. (4, 2); $x + y = 6$, $x - y = 2$

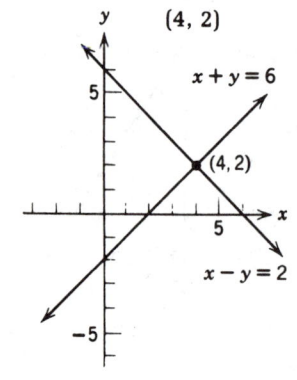

9. (2, 2); $y = 4 - x$, $y = x$

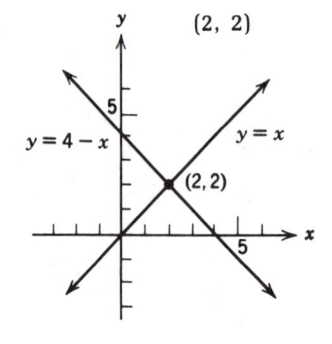

11. (−3, −2); $y - x = 1$, $y + x = -5$

572

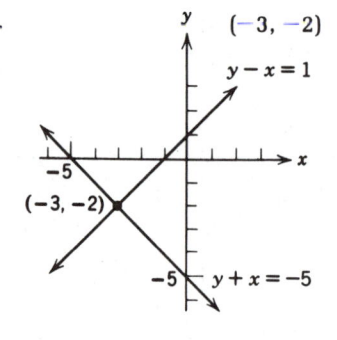

PART A: ANSWERS

13.

Graph showing lines $x - 3y = -20$ and $2x - y + 5 = 0$ intersecting at $(1, 7)$

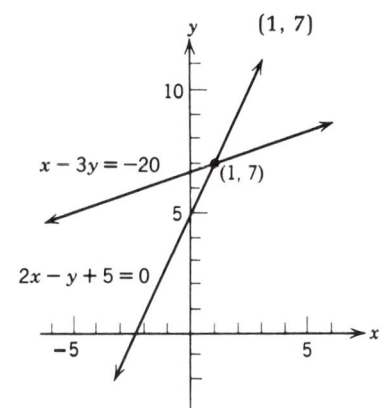

15.

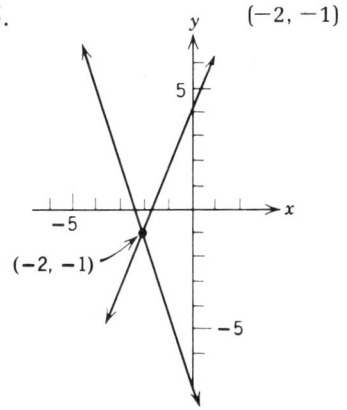

Graph showing lines intersecting at $(-2, -1)$

17. Dependent

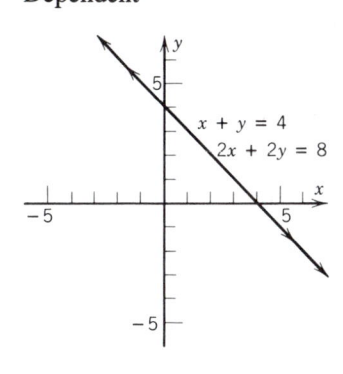

Graph showing $x + y = 4$ and $2x + 2y = 8$ (same line)

19. Inconsistent

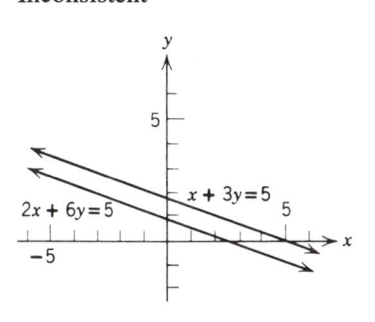

Graph showing parallel lines $x + 3y = 5$ and $2x + 6y = 5$

21. Dependent

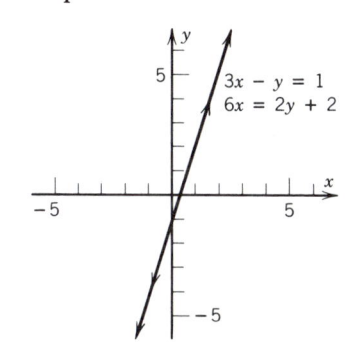

Graph showing $3x - y = 1$ and $6x = 2y + 2$ (same line)

23. Inconsistent

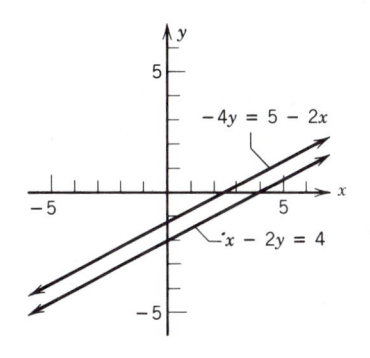

Graph showing parallel lines $-4y = 5 - 2x$ and $x - 2y = 4$

EXERCISES 13.2

1. (3, 2)   3. (−1, 4)   5. (6, 2)   7. (1, 1)   9. (7, −2)
11. (−1, 1)   13. (1, 1)   15. (5, 2)   17. (1, −1)   19. (−4, 10)
21. (5, −1)   23. (4, 3)   25. (−4, 3)   27. (−2, −1)   29. (0, 2)
31. (−4, 2)   33. (3, 1)   35. (4, 3)

EXERCISES 13.3

1. (1, 2)   3. (2, 2)   5. (−2, −1)   7. (−3, −2)   9. (5, −2)
11. (8, 5)   13. (1, 2)   15. (10, −2)   17. (1, 1)   19. (−1, −3)
21. (−1, 1)   23. (1, 0)   25. (1, −1)   27. (4, 3)   29. (4, 3)
31. (−3, 2)   33. (2, 3)   35. (−4, 10)   37. (6, 2)

EXERCISES 13.4

1. (2, 4)   3. (3, 1)   5. (−1, 2)   7. Dependent   9. (2, −2)
11. (1, −4)   13. (5, 3)   15. (5, 2)   17. (−3, 1)   19. (2, 2)

EXERCISES 13.5

1. 8, 17   3. 9 meters, 11 meters   5. 16 kilograms, 12 kilograms
7. Car: $8500, trailer: $3500   9. Walls: 9 hours, trim: 15 hours   11. 13 kilometers

573

ANSWERS AND SOLUTIONS

13. 60 flatcars
15. 6, 18
17. 3842, 3830
19. 20 dimes, 14 quarters
21. $2160 at 8%, $1440 at 12%

REVIEW OF FACTORING

1. $5(x + 2y)$
2. $3(x^2 + 2x - 1)$
3. $-2(x^2 + 2)$
4. $4x(x - 2)$
5. $4(y^2 + 2y - 2)$
6. $x(3x - 6 + x^2)$
7. $(x + 3)(x + 2)$
8. $(y - 3)^2$
9. $(y - 8)(y + 1)$
10. $(y + 7)(y - 5)$
11. $2(x + 5)(x - 2)$
12. $3(y + 4)(y - 1)$
13. $(2y - 3)(y + 1)$
14. $(3y - 1)(2y + 1)$
15. $(3x - 2)(2x - 3)$
16. $(3x + 7)(x - 5)$
17. $(3x + 4)(2x - 1)$
18. $(4y + 1)(2y - 1)$
19. $2(2y + 1)(y + 1)$
20. $3(2x + 1)(x + 3)$
21. $2(3x + 1)(3x - 8)$
22. $5(8x - 5)(2x + 1)$
23. $2(2x + 1)(x - 3)$
24. $3(3y - 2)(2y + 1)$
25. $(x + a)^2$
26. $(x - a)^2$
27. $(y + 3b)^2$
28. $(y - 2b)^2$
29. $(x + 4a)^2$
30. $(y - 5b)^2$
31. $(x + 4)(x - 4)$
32. $(y + 6)(y - 6)$
33. $(y + 2)(y - 2)$
34. $(x + 7)(x - 7)$
35. $(2x + 5)(2x - 5)$
36. $9(y + 3)(y - 3)$
37. $12(y + 2)(y - 2)$
38. $2(5x + 4)(5x - 4)$

EXERCISES 14.1

1. 7
3. 2
5. $\frac{1}{3}$
7. -3
9. 3, 5
11. 0, -4
13. 3, -3
15. $-\frac{1}{3}, \frac{3}{2}$
17. 0, $-\frac{4}{5}, \frac{7}{2}$
19. 2, 3
21. 0, 4
23. 0, -3
25. 2, -3
27. 1, -8
29. $-2, 3$
31. 0, 4
33. $-4, 3$
35. $\frac{3}{2}, -\frac{3}{4}$
37. $\frac{2}{3}, -\frac{2}{3}$
39. 0, $-\frac{3}{2}$
41. 3, 2, 1
43. 0, -2, 1
45. 0, -4, 3
47. 0, $-\frac{1}{2}, \frac{1}{2}$

EXERCISES 14.2

1. 0, -3
3. 0, $\frac{5}{2}$
5. 0, $\frac{9}{2}$
7. 0, 4
9. 1, -1
11. 2, -2
13. 4, -4
15. 8, -8
17. 3, -3
19. 3, -3
21. 1, 2
23. $-2, -2$
25. $-1, 4$
27. 3, -7
29. 2, -7
31. $-6, -6$
33. $-1, 6$
35. $-1, 4$
37. 5, -6
39. $-1, \frac{4}{3}$
41. $\frac{1}{3}, \frac{3}{2}$
43. $\frac{1}{2}, \frac{1}{2}$
45. $\frac{3}{2}, \frac{3}{2}$
47. $\frac{1}{2}, 1$
49. $-\frac{1}{2}, 2$
51. $-1, 2$
53. $-2, 3$
55. 1, 1
57. $-\frac{1}{2}, 2$
59. $\frac{1}{3}, -\frac{3}{2}$

EXERCISES 14.3

1. $\frac{1}{3}, -\frac{1}{3}$
3. $\frac{3}{2}, -\frac{3}{2}$
5. 4, -4
7. 0, -2
9. 0, -2
11. 0, 5
13. $\frac{3}{2}, -2$
15. $-\frac{3}{2}, -\frac{3}{2}$

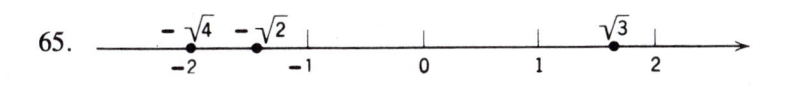

PART A: ANSWERS

17. $-\frac{3}{4}, -\frac{5}{2}$	19. $-5, 3$	21. $-1, -2$	23. $1, -3$
25. $1, 1$	27. $-3, 5$	29. $2, 2$	31. $10, 20$
33. $2, 13$	35. $5, -6$		

EXERCISES 14.4

1. $0, 5$
3. $8, 9$
5. 4
7. 5 seconds
9. 6 hours
11. 4 of $\frac{1}{4}$
13. $3, 5$
15. $3, 4$
17. 10 mph
19. Rate going: 45 mph, rate returning: 60 mph

EXERCISES 15.1

1. $9, -9$
3. $11, -11$
5. $\frac{4}{5}, \frac{-4}{5}$
7. 4
9. -9
11. ± 12
13. 3
15. $\frac{1}{6}$
17. $\pm \frac{2}{3}$
19. $x \geq 3$
21. $x \geq -7$
23. $x > 6$
25. x
27. $2x$
29. $-ac^2$
31. $\pm 6a^3$
33. $11ab$
35. $-(a+b)$
37. $\frac{a}{b}$
39. $\pm \frac{3}{xy}$

EXERCISES 15.2

1. $\sqrt{10}, -\sqrt{10}$
3. $\sqrt{17}, -\sqrt{17}$
5. $\sqrt{22}, -\sqrt{22}$
7. Rational; 6
9. Irrational
11. Irrational
13. Rational; 5
15. Rational; 12
17. Irrational
19. Rational; $\frac{2}{3}$
21. Irrational
23. Rational; 3
25. Irrational
27. 7.55
29. 1.73
31. 2.24
33. -5.10
35. 3.46
37. -8.66
39. 1.41
41. -1.73
43. 2.73
45. 1.59
47. 12.94
49. -22.88
51. 2.91
53. 0.51
55. 0.32
57. 0.72

59. number line with 7, $\sqrt{55}$, 8 marked between 7 and 8

61. number line with $-\sqrt{7}$ near -3, $\sqrt{3}$ and $\sqrt{5}$ between 0 and 3

63. number line with $-\sqrt{2}$ between -2 and -1, $\sqrt{1}$ at 1, $\sqrt{3}$ between 1 and 2

65. number line with $-\sqrt{4}$ at -2, $-\sqrt{2}$ between -2 and -1, $\sqrt{3}$ between 1 and 2

575

ANSWERS AND SOLUTIONS

67. [number line with points at $-\sqrt{40}$, $-\sqrt{20}$, and $\sqrt{30}$]

69. [number line with points at $-\sqrt{37}$, $-\sqrt{16}$, and $\sqrt{35}$]

71. [number line showing interval]

73. [number line showing interval]

75. [number line showing interval]

77. [number line showing interval]

EXERCISES 15.3

1. $2\sqrt{2}$
3. $3\sqrt{2}$
5. $-2\sqrt{5}$
7. $-6\sqrt{2}$
9. 8
11. $12\sqrt{2}$
13. $5\sqrt{5}$
15. $-6\sqrt{30}$
17. $\pm 12\sqrt{5}$
19. $\pm 18\sqrt{6}$
21. $x\sqrt{x}$
23. $y^3\sqrt{y}$
25. $-x^5\sqrt{x}$
27. x^3
29. $\pm x^4$
31. x^6
33. $2x$
35. $3x\sqrt{x}$
37. $-2y^2\sqrt{6y}$
39. $8y^2$
41. $7x^3\sqrt{x}$
43. $\pm 4x\sqrt{2x}$
45. $4x\sqrt{5}$
47. $-8\sqrt{x}$
49. $\pm y\sqrt{5y}$
51. $4x\sqrt{3y}$
53. $5xy\sqrt{x}$
55. $-3a^2\sqrt{5b}$
57. $\frac{3}{4}xy$
59. $\pm\frac{5}{6}y$
61. $2x$
63. $6\sqrt{x}$
65. $49y\sqrt{y}$
67. $2x^2\sqrt{y}$
69. $-a\sqrt{a}$
71. $\pm 2x^3$
73. 10.39
75. 16.59
77. 14.18
79. -12.32
81. -31.18
83. 36.59

EXERCISES 15.4

1. $3\sqrt{3}$
3. $\sqrt{5}$
5. 0
7. $5\sqrt{3}$
9. $-3\sqrt{2}$
11. $-4\sqrt{3}$
13. $4\sqrt{2} - 4\sqrt{3}$
15. $5\sqrt{3} + 3\sqrt{2}$
17. $8 + 6\sqrt{6}$
19. $5\sqrt{a}$
21. $12\sqrt{x}$
23. $2b\sqrt{b}$
25. $4\sqrt{3} + 4$
27. $-30 - 5\sqrt{7}$
29. $4\sqrt{2} - 4\sqrt{3}$
31. $3 + 9\sqrt{a}$
33. $x\sqrt{x} + 2x\sqrt{y}$
35. $xy^2\sqrt{x} + 2xy$
37. $2\sqrt{2} - 6\sqrt{3} - 10$
39. $-\sqrt{a} - \sqrt{b} + \sqrt{c}$
41. $2(1 + \sqrt{3})$
43. $4(\sqrt{2} - 3)$
45. $4(\sqrt{5} + 2)$
47. $8(1 - 4\sqrt{5})$
49. $6(1 + 4\sqrt{2})$
51. $3(1 + \sqrt{2})$
53. $4(1 - \sqrt{2})$
55. $3(7 + \sqrt{2})$
57. $4(1 + \sqrt{y})$
59. $3x(\sqrt{x} - 2\sqrt{y})$
61. $3x(\sqrt{y} - 3\sqrt{x})$
63. $2y(\sqrt{y} - 3y)$

EXERCISES 15.5

1. $2 + 3\sqrt{3}$
3. $3 - \sqrt{5}$
5. $-1 + \sqrt{2}$
7. $2 + 3\sqrt{3}$
9. $\frac{1 + \sqrt{2}}{2}$
11. $\frac{1 - \sqrt{3}}{2}$
13. $\frac{x - \sqrt{2}}{x}$
15. $1 - \sqrt{x}$

576

PART A: ANSWERS

17. $2 - \sqrt{y}$
19. $\dfrac{2 + \sqrt{2}}{3}$
21. $\dfrac{\sqrt{3} - 1}{5}$
23. $\dfrac{2\sqrt{10} - \sqrt{3}}{3}$
25. $\dfrac{\sqrt{11} + 1}{a}$
27. $\dfrac{5 + 6\sqrt{2}}{4}$
29. $\dfrac{3 + \sqrt{3}}{6a}$
31. $\dfrac{6 + 5\sqrt{3}}{15}$
33. $\dfrac{3\sqrt{3} - 4}{12}$
35. $\dfrac{4\sqrt{3} - 3\sqrt{2}}{6}$
37. $\dfrac{8 + 3\sqrt{2}}{2}$
39. $\dfrac{\sqrt{2} + 5}{5}$
41. $\dfrac{3\sqrt{3} + 6}{2}$
43. $\dfrac{1 + 2\sqrt{2}}{2x}$
45. $\dfrac{3 - 8\sqrt{y}}{4}$
47. $\dfrac{3\sqrt{x} + 2\sqrt{y}}{6}$

EXERCISES 15.6

1. $\sqrt{15}$
3. $\sqrt{30}$
5. $3\sqrt{2}$
7. 4
9. $x\sqrt{6}$
11. $2xy\sqrt{3y}$
13. $12a$
15. $5x\sqrt{6y}$
17. $\sqrt{30}$
19. 10
21. $6\sqrt{6}$
23. $6x\sqrt{x}$
25. $abc\sqrt{abc}$
27. x^4
29. $3\sqrt{2} + \sqrt{6}$
31. $3\sqrt{2} + 2\sqrt{3}$
33. $4\sqrt{5} + 5\sqrt{2}$
35. $\sqrt{6} + 3\sqrt{2}$
37. $3 + \sqrt{6}$
39. $2\sqrt{5} - 2$
41. $1 - 2\sqrt{2}$
43. $2 + 2\sqrt{5}$
45. 1
47. -11
49. $-4 - 3\sqrt{15}$
51. $12 + 5\sqrt{35}$

EXERCISES 15.7

1. 3
3. 5
5. 3
7. $2\sqrt{a}$
9. $\sqrt{3b}$
11. $\sqrt{b}$
13. $2\sqrt{7b}$
15. 1
17. a
19. $b\sqrt{3}$
21. $\dfrac{5\sqrt{2}}{2}$
23. $\dfrac{2\sqrt{x}}{x}$
25. $\dfrac{a\sqrt{b}}{b}$
27. $\dfrac{\sqrt{3}}{3}$
29. $\dfrac{\sqrt{2a}}{a}$
31. $\dfrac{\sqrt{3ab}}{b}$
33. $\dfrac{3\sqrt{x}}{x}$
35. $\dfrac{5\sqrt{y}}{y}$
37. $\sqrt{2a}$
39. $\sqrt{6x}$
41. $4\sqrt{a}$
43. $2\sqrt{3xy}$
45. $\dfrac{2\sqrt{6}}{3}$
47. $\dfrac{3\sqrt{2}}{2}$
49. $\dfrac{6\sqrt{10}}{5}$
51. $\dfrac{5\sqrt{x}}{x}$
53. $\dfrac{2\sqrt{2x}}{x}$
55. $\dfrac{x\sqrt{y}}{y}$
57. 0.38
59. 2.12
61. 1.34
63. 1.73

EXERCISES 16.1

1. $2, -2$
3. $4, -4$
5. $7, -7$
7. $\sqrt{3}, -\sqrt{3}$
9. $\sqrt{10}, -\sqrt{10}$
11. $\sqrt{6}, -\sqrt{6}$
13. $2\sqrt{3}, -2\sqrt{3}$
15. $3\sqrt{2}, -3\sqrt{2}$
17. $3\sqrt{2}, -3\sqrt{2}$
19. $2, -2$
21. $\sqrt{5}, -\sqrt{5}$
23. $0, 0$
25. $2\sqrt{5}, -2\sqrt{5}$
27. $3, -3$
29. $2\sqrt{2}, -2\sqrt{2}$
31. $3, -1$
33. $7, -3$
35. $6, 4$
37. $a + 5, a - 5$
39. $a + 3, -a + 3$
41. $a + b, a - b$

ANSWERS AND SOLUTIONS

43. $-3 + \sqrt{2}, -3 - \sqrt{2}$
45. $-5 + \sqrt{5}, -5 - \sqrt{5}$
47. $-10 + 2\sqrt{2}, -10 - 2\sqrt{2}$
49. $5 + \sqrt{a}, 5 - \sqrt{a}$
51. $-1 + \sqrt{b}, -1 - \sqrt{b}$
53. $b + \sqrt{a}, b - \sqrt{a}$
55. $\sqrt{a}, -\sqrt{a}$
57. $ab\sqrt{3a}, -ab\sqrt{3a}$
59. $\dfrac{\sqrt{30b}}{6}, -\dfrac{\sqrt{30b}}{6}$
61. $t = \dfrac{\sqrt{2gs}}{g}, t = \dfrac{\sqrt{2gs}}{g}$
63. $r = \dfrac{\sqrt{\pi A}}{2\pi}, r = \dfrac{\sqrt{\pi A}}{2\pi}$
65. $d = \dfrac{\sqrt{3Ik}}{I}, d = -\dfrac{\sqrt{3Ik}}{I}$

EXERCISES 16.2

1. $2, -6$
3. $1, 1$
5. $4, -5$
7. $-1, -2$
9. $-1, 4$
11. $-2, 5$
13. $1 + \sqrt{2}, 1 - \sqrt{2}$
15. $\dfrac{3 + \sqrt{21}}{2}, \dfrac{3 - \sqrt{21}}{2}$
17. $\dfrac{-1 + \sqrt{13}}{2}, \dfrac{-1 - \sqrt{13}}{2}$
19. $\dfrac{1}{2}, -\dfrac{3}{2}$
21. $\dfrac{1}{2}, -2$
23. $-\dfrac{5}{2}, 3$

EXERCISES 16.3

1. $a = 1, b = -3, c = 2$
3. $a = 1, b = -1, c = -30$
5. $a = 1, b = -2, c = 0$
7. $a = 4, b = 0, c = -3$
9. $a = 2, b = -7, c = 6$
11. $a = 6, b = -5, c = 1$
13. $a = 1, b = -8, c = 4$
15. $a = 2, b = -2, c = -1$
17. $a = 6, b = 7, c = -3$
19. $a = 9, b = 6, c = -8$
21. $1, 2$
23. $-2, 6$
25. $3, -5$
27. $\dfrac{-3 + \sqrt{13}}{2}, \dfrac{-3 - \sqrt{13}}{2}$
29. $\dfrac{3 + \sqrt{17}}{2}, \dfrac{3 - \sqrt{17}}{2}$
31. $0, 2$
33. $0, 5$
35. $0, 7$
37. $0, 3$
39. $\dfrac{\sqrt{3}}{2}, -\dfrac{\sqrt{3}}{2}$
41. $2, \dfrac{3}{2}$
43. $\dfrac{1}{3}, -\dfrac{1}{2}$
45. $\dfrac{5}{2}, -\dfrac{1}{3}$
47. $1 + \sqrt{2}, 1 - \sqrt{2}$
49. $2 + \sqrt{6}, 2 - \sqrt{6}$
51. $\dfrac{3 + \sqrt{17}}{4}, \dfrac{3 - \sqrt{17}}{4}$
53. $\dfrac{3}{2}, -\dfrac{5}{2}$
55. $\dfrac{3}{2}, \dfrac{2}{3}$
57. $-\dfrac{3}{2}, 3$

PART A: ANSWERS

EXERCISES 16.4

1. $(0, -3)$
3. $(-1, 0)$
5. $(-2, 5)$
7. $(0, -2)$
9. $(1, 0)$
11. $(-2, 0)$
13. $(0, 12)$
15. $(2, 2)$
17. $(4, 0)$
19. [graph of upward parabola]

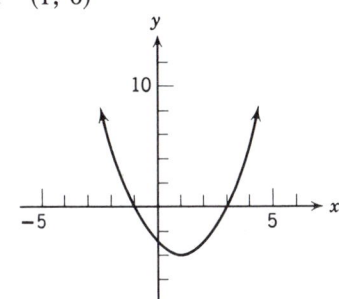

21. [graph of $y = x^2 - 2x$]

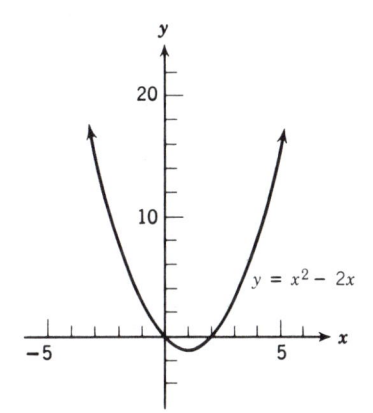

23. [graph of $y = x^2 + 2x$]

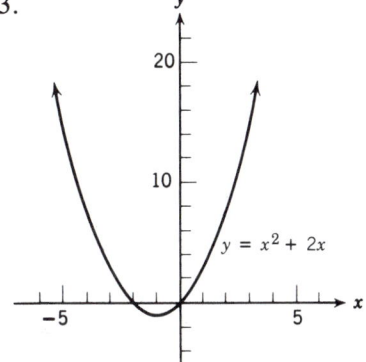

25. [graph of $9 - x^2 = y$]

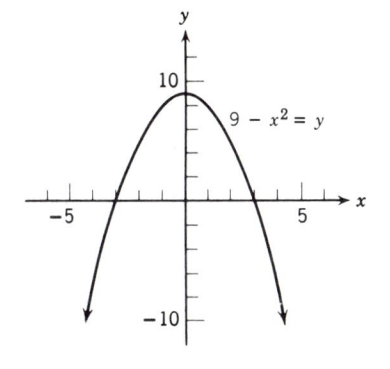

27. [graph of $y = x^2 - 5x + 4$]

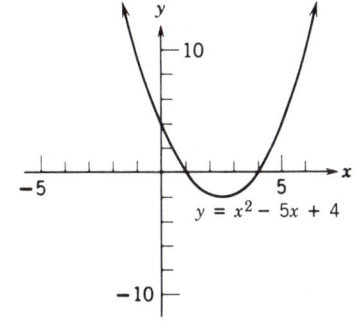

29. 0, 2; −2, 2; −2, 0; none

31. [graph of $y = x^2$]

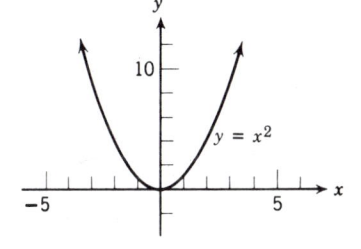

33. [graph of $y = 9 - x^2$]

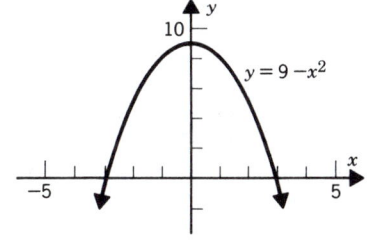

35. [graph of $x^2 - 2x - 3 = y$]

579

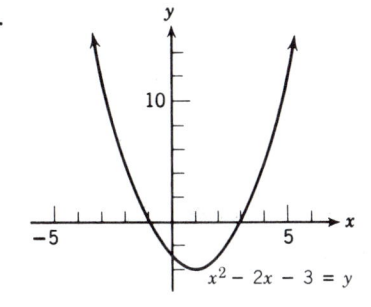

ANSWERS AND SOLUTIONS

CUMULATIVE REVIEW FOR PART II

1. $2 \cdot 2 \cdot 2 \cdot 3xxxyy$
2. 1
3. 1
4. $2a^2$
5. 0
6. (1, 2)
7. $\dfrac{3x + 4}{x(x + y)}$
9. $5x^2\sqrt{y}$
10. $a - x$
11. Two
12. Either $a = 0$, $b = 0$ or both a and $b = 0$
13. 27
14. 8, 32
15. 7, 9
16. -1
17. 0
18. $-3, \dfrac{-5}{2}, \dfrac{-5}{3}, 0, \dfrac{3}{7}, \dfrac{3}{8}, 3$
19. $x(x - 2)(x - 1)$
20. $2x^2 + 5x - 12$
21. $\dfrac{x + 2}{3}$
22. $2x - 6$
23. $\dfrac{3 - 2x}{x^2 + x}$
24. $\dfrac{3(x + 3)}{x^2 + 2x - 3}$
25. $\dfrac{3x + 2}{2x + 3}$
26. Two
27. -4
28. $-6ab^3c$
29. $\dfrac{5a - 9b}{6}$
30. $\dfrac{a - 1}{4x(2a + 5)}$
31. $x^2 + 3x - 1$
32. $(2x - 3)(2x + 3)$
33. -5
34. 6
35. $146 - n$
36. $25x$
37. Dependent
38. a. 3.27×10^6 b. 6.49×10^{-5}
39. $\dfrac{-1}{2}, 3$
40. 54,100
41. Slope: $\dfrac{-3}{2}$
43. 1
44. $\sqrt{6x}$
45. $\sqrt{6}$
46. $x = 3, y = 3$
47. 44
48. $y = 2x - 1$
49. 0, 1
50. 36.9
51. No
52. No
53. No
54. No
55. No
56. Yes
57. No
58. Yes
59. No
60. No
61. Yes
62. No
63. No
64. No
65. No
66. Yes
67. No
68. No
69. Yes
70. No
71. No
72. No
73. No
74. No
75. Yes
76. No
77. No
78. No
79. No
80. Yes

EXERCISES 17.1

1. $0.4°$
3. $6.6°$
5. $15.1°$
7. $107.15°$
9. $0°24'$
11. $5°54'$
13. $15°21'$
15. $154°39'$
17. a. $62°$ b. $152°$
19. a. $3°50'$ b. $93°50'$
21. a. $62.6°$ b. $152.6°$
23. a. $70.58°$ b. $160.58°$
25. a. $90°$ b. right angle
27. a. $75°$ b. acute angle
29. a. $135°$ b. obtuse angle
31. a. $150°$ b. obtuse angle
33. $\angle BOC$
35. $\angle DOB$
37. $75°$ and $15°$
39. $110°$ and $70°$

EXERCISES 17.2

1. 1 and 2, 3 and 8, 4 and 6, 5 and 7
3. 1 and 4, 2 and 6, 3 and 5, 7 and 8
5. $40°$
7. $40°$
9. $40°$
11. $90°$
13. $90°$
15. $145°$
17. $x°$
19. $x°$
21. $(180 - x)°$
23. $x°$
25. $x°$
27. $132°$
29. $156°$

EXERCISES 17.3

1. Equilateral; acute
3. Scalene; right
5. Isosceles; acute
7. $75°$
9. $45°$
11. $135°$
13. BF
15. FC
17. AE
19. $60°$

580

PART A: ANSWERS

21. 45°
23. $A = 24$, $P = 24$
25. $A = 4\sqrt{5}$, $P = 10 + 2\sqrt{5}$
27. $\sqrt{29}$
29. $2\sqrt{6}$
31. 6, 6, and 4 inches
33. 20° and 70°
35. 50 meters
37. $30\sqrt{2}$ meters

EXERCISES 17.4

1. ∠A and ∠B, ∠ACD and ∠BCD, ∠ADC and ∠BDC, AD and BD, AC and BC, DC and DC.
3. ∠CAE and ∠DBC, ∠ACE and ∠BCD, ∠AEC and ∠BDC, AC and BC, DC and EC, AE and BD.
5. ∠CAF and ∠CBD, ∠ACF and ∠BCD, ∠AFC and ∠BDC, AC and BC, DC and FC, BD and AF.
7. $x = 5$, $y = 6$
9. $x = 15$, $y = 14$
11. $x = 24\frac{1}{2}$
13. $y = 66\frac{2}{3}$
15. $x = 16$

EXERCISES 17.5

1. a. Square
   b. Rectangle, quadrilateral
   c. ∠B = 90°, ∠C = 90°, ∠D = 90°
3. a. Trapezoid
   b. Quadrilateral
   c. ∠C = 140°
5. 13.4
7. 10.12
9. 28°
11. 118°
13. 5.04
15. 5 centimeters
17. 127 feet
19. 12 inches
21. Width: 9 meters, length: 12 meters
23. Width: 2 millimeters, length: 5 millimeters
25. Width: 2 meters, length: 6 meters
27. Width: 6 centimeters, length: 7 centimeters
29. Width: 8 meters, length: 12 meters

EXERCISES 17.6

1. Tangent
3. Chord
5. Semicircle
7. 60°
9. 60°
11. $2\sqrt{3}$ centimeters
13. 37.7 square centimeters
15. 643.7 square centimeters
17. 42.1 square centimeters
19. 71.8 centimeters
21. 72.0 centimeters
23. 2400.0 square centimeters
25. 439.6 centimeters
27. 16 inches

EXERCISES 17.7

1. $V = 1620.0$ cubic feet, $S = 936.0$ square feet
3. $V = 28{,}938.2$ cubic centimeters, $S = 6028.8$ square centimeters
5. $V = 904.3$ cubic centimeters, $S = 936.0$ square centimeters
7. 946.1 cubic centimeters
9. 6.3 cubic inches
11. 27.6 cubic inches
13. 6.0 gallons
15. 179.5 cubic meters
17. 134.6 pounds

EXERCISES 17.8

1. $\sin A = \frac{3}{5}$, $\cos A = \frac{4}{5}$, $\tan A = \frac{3}{4}$
3. $\sin A = \frac{9}{\sqrt{130}}$, $\cos A = \frac{7}{\sqrt{130}}$, $\tan A = \frac{9}{7}$
5. $\sin A = \frac{14}{15}$, $\cos A = \frac{\sqrt{29}}{15}$, $\tan A = \frac{14}{\sqrt{29}}$
7. $\sin A = \frac{5}{\sqrt{74}}$, $\sin B = \frac{7}{\sqrt{74}}$
9. $\sin B = \frac{3\sqrt{5}}{7}$, $\tan B = \frac{3\sqrt{5}}{2}$
11. $\cos B = \frac{\sqrt{33}}{7}$, $\tan B = \frac{4}{\sqrt{33}}$

581

ANSWERS AND SOLUTIONS

| 13. 0.6157 | 15. 0.0872 | 17. 0.6820 | 19. 0.6947 |
| 21. 24° | 23. 70° | 25. 34° | 27. 82° |

EXERCISES 17.9

1. 5.6 inches 3. 5.0 centimeters 5. 11.2 centimeters
7. $AC = 5.6$ centimeters, $BC = 2.2$ centimeters, $\angle B = 68°$
9. $AC = 6.8$ centimeters, $CB = 6.5$ centimeters, $\angle B = 46°$
11. $AB = 9.1$ inches, $CB = 7.7$ inches, $\angle B = 58°$
13. 1122.3 feet 15. 11.5 meters 17. 31.4 feet 19. 7.5 centimeters 21. 24°

PART B: SOLUTIONS TO ALL CHAPTER REVIEW EXERCISES

CHAPTER 1 REVIEW

1. Two thousand five hundred ninety-seven
2. Seven hundred ninety-one thousand, two hundred nine
3. 7432   4. 21,206   5. 34.5
6. 0.0252   7. 0.015462
               ⇕⇕⇕⇕:
               0.015452

   0.015462 is greater

8. a. $5.8965 \longrightarrow 5.9$
   b. $5.8965 \longrightarrow 5.90$
   c. $5.8965 \longrightarrow 5.896$

9. $\$18.006 \longrightarrow \$18.01$
10. $\$157.49 \longrightarrow \$157.00$

11. 437
     581
     109

    1127

12. 54.98
     66.57
     78.62

    200.17

13.   942
     7667
     8984

    17,593

14. 1.4
    12.2
    17.4

    31.0

15. 19.7911
    46.217
     9.77612

    75.78422

    a. $75.78422 \longrightarrow 75.8$
    b. $75.78422 \longrightarrow 75.78$
    c. $75.78422 \longrightarrow 75.784$

16. 2.9507
    68.57353
     3.4142

    74.93843

    a. $74.93843 \longrightarrow 74.9$
    b. $74.93843 \longrightarrow 74.94$
    c. $74.93843 \longrightarrow 74.938$

17. 5439
    -1103

     4336

18. 9.409
    -2.358

     7.051

582

PART B: SOLUTIONS

19. $\quad 5479 - 87 - 3324$
$\quad\quad\quad 5392 - 3324$
$\quad\quad\quad\quad\quad = 2068$

20. $\quad \underbrace{916.07 - 72.892} - 5.117 + 4.39$
$\quad\quad\quad \underbrace{843.178 - 5.117} + 4.39$
$\quad\quad\quad\quad\quad 838.061 + 4.39$
$\quad\quad\quad\quad\quad\quad\quad = 842.451$

21. $D = 7$
22. $3.49 = D$

23. $\quad\quad S - 15.5 = 40.7$
$\quad 15.5 + S - 15.5 = 40.7 + 15.5$
$\quad\quad\quad\quad S = 56.2$

24. $\quad\quad\quad 1.82 = S - 9.05$
$\quad 9.05 + 1.82 = S - 9.05 + 9.05$
$\quad\quad\quad 10.87 = S$

25. $\quad\quad 518 + N - 9.7 = 672$
$\quad\quad\quad 508.3 + N = 672$
$\quad 508.3 + N - 508.3 = 672 - 508.3$
$\quad\quad\quad\quad N = 163.7$

26. The sum of a number and 701 is 811
$\quad\quad\quad\quad N \quad + 701 = 811$

   Ans. $N + 701 = 811$

27. The difference of 2.8 subtracted from a number is 94.2
$\quad\quad\quad\quad N \quad - \quad 2.8 \quad = 94.2$

   Ans. $N - 2.8 = 94.2$

28. a. Number: N
$\quad\quad N + 34 = 261$

   b. $N + 34 - 34 = 261 - 34$
$\quad\quad\quad N = 227$

29. a. Number: N
$\quad\quad 1524 - N = 976$

   b. $1524 - N + N = 976 + N$
$\quad\quad\quad 1524 = 976 + N$
$\quad 1524 - 976 = 976 + N - 976$
$\quad\quad\quad 548 = N$

30. a. Total cost: C
$\quad C = \$347.69 + \$84.50 + \$129.50 + \$15.75$

   b. $C = \$577.44$

CHAPTER 2   REVIEW

1. $\quad\quad 469$
$\quad\quad \times 38$
$\quad\quad \overline{3\ 752}$
$\quad\quad 14\ 07$
$\quad\quad \overline{17,822}$

2. $\quad\quad 4\ 6.7\ 3$
$\quad\quad \times 3.7\ 1$
$\quad\quad \overline{4\ 6\ 7\ 3}$
$\quad\quad 32\ 7\ 1\ 1$
$\quad\quad 140\ 1\ 9$
$\quad\quad \overline{173.3\ 6\ 8\ 3}$

a. $173.3\cancel{6}\cancel{8}\cancel{3} \longrightarrow 173.37$
b. $173.368\cancel{3} \longrightarrow 173.368$

3. a. $9.85 \times 0.1 \longrightarrow 9.85$
   Ans. 0.985

   b. $40.7 \times 10 \longrightarrow 40.7$
   Ans. 407

   c. $0.987 \times 0.01 \longrightarrow 0\ 0\ .987$
   Ans. 0.00987

4. a. $403 \times 58 \quad$ estimate
$\quad\quad 400 \times 60 = 24,000$

   b. $39.2 \times 29 \quad$ estimate
$\quad\quad 40 \times 30 = 1200$

5. a. $9^2 = 9 \times 9 = 81$
   b. $(2.4)^2 = 2.4 \times 2.4 = 5.76$
   c. $(4.1)^3 = 4.1 \times 4.1 \times 4.1$
$\quad\quad\quad = 68.921$

583

ANSWERS AND SOLUTIONS

6.a. $\begin{array}{r}73\\14\overline{)1022}\\\underline{98}\\42\\\underline{42}\\0\end{array}$
b. $\begin{array}{r}2.81\\5.7\overline{)16.0\,17}\\\underline{11\,4}\\4\,6\,1\\\underline{4\,5\,6}\\57\\\underline{57}\\0\end{array}$

7. $\begin{array}{r}1\,32.870\\7.35\overline{)976.60\,000}\\\underline{735}\\241\,6\\\underline{220\,5}\\21\,10\\\underline{14\,70}\\6\,40\,0\\\underline{5\,88\,0}\\52\,00\\\underline{51\,45}\\550\end{array}$

9.
a. $132.8\cancel{7}\cancel{0} \longrightarrow 132.9$
b. $132.87\cancel{0} \longrightarrow 132.87$

8.a. $68.3 \div 100 \longrightarrow 6\,8.3$
Ans. 0.683

b. $\dfrac{3.1}{1000} \longrightarrow 0\,0\,0\,3.1$
Ans. 0.0031

9.a. $1.5 \times \dfrac{62}{1.5} = 62$

b. $\dfrac{4.9 \times 3.1}{3.1} = 4.9$

10.a. $609 \div 22$ estimate
$\downarrow \quad \downarrow \quad\quad \downarrow$
$600 \div 20 = \quad 30$

b. $598.2 \div 19$ estimate
$\downarrow \quad\quad \downarrow \quad\quad \downarrow$
$600 \;\div 20 = \quad 30$

11. Miles traveled on 16 gallons: N
Miles $= \begin{pmatrix}\text{Number of}\\\text{gallons}\end{pmatrix} \times \begin{pmatrix}\text{Miles per}\\\text{gallon}\end{pmatrix}$
$N = 16 \times 27 = 432$
The car can travel 432 miles on 16 gallons of gasoline.

12. Amount: N
(Amount) $= \begin{pmatrix}\text{Price per}\\\text{month}\end{pmatrix} \times \begin{pmatrix}\text{Number of}\\\text{months}\end{pmatrix}$
$N = 108 \times 36$
$N = 3888$
The man pays $3888.

13. Cost of each set: N
(Cost) $= \begin{pmatrix}\text{Total}\\\text{price}\end{pmatrix} \div \begin{pmatrix}\text{Number}\\\text{of items}\end{pmatrix}$
$N = 6072 \div 24$
$N = 253$
Each set costs $253.

14. Items produced per minute: N
Items $= \begin{pmatrix}\text{Total items}\\\text{produced}\end{pmatrix} \div \begin{pmatrix}\text{Total time}\\\text{in minutes}\end{pmatrix}$
$N = 540 \div 15 = 36$
The machine can produce 36 items per minute.

15. $\dfrac{N}{32.1} = 26.7$
$32.1 \times \dfrac{N}{32.1} = 26.7 \times 32.1$
$N = 26.7 \times 32.1$
$N = 857.07$

16. $285 \times D = 118{,}845$
$\dfrac{285 \times D}{285} = \dfrac{118{,}845}{285}$
$D = \dfrac{118{,}845}{285}$
$D = 417$

17. $8 \times R \times 6 = 28.32$
$8 \times 6 \times R = 28.32$
$48 \times R = 28.32$
$\dfrac{48 \times R}{48} = \dfrac{28.32}{48}$
$R = \dfrac{28.32}{48}$
$R = 0.59$

18. $\dfrac{28 \times P}{27} = 420$
$27 \times \dfrac{28 \times P}{27} = 420 \times 27$
$28 \times P = 420 \times 27$
$\dfrac{28 \times P}{28} = \dfrac{420 \times 27}{28}$
$P = \dfrac{420 \times 27}{28}$
$P = 405$

19. **Cost** **Pears**
a. $\dfrac{c}{72} - \dfrac{8}{12}$
b. $c \times 12 = 8 \times 72$
$\dfrac{c \times 12}{12} = \dfrac{8 \times 72}{12}$
$c = 48$
8 pears will cost 48¢.

PART B: SOLUTIONS

20. Cost   Plants
  a. $\dfrac{c}{5.22} = \dfrac{10}{6}$
  b. $c \times 6 = 10 \times 5.22$
  $\dfrac{c \times 6}{6} = \dfrac{10 \times 5.22}{6}$
  $c = 8.7$
  Ten plants cost $8.70.

21. Cost   Batteries
  a. $\dfrac{c}{3.16} = \dfrac{14}{4}$
  b. $c \times 4 = 14 \times 3.16$
  $\dfrac{c \times 4}{4} = \dfrac{14 \times 3.16}{4}$
  $c = 11.06$
  14 batteries will cost $11.06.

22. After   Before
  a. $\dfrac{N}{10} = \dfrac{6}{4}$
  b. $N \times 4 = 10 \times 6$
  $\dfrac{N \times 4}{4} = \dfrac{10 \times 6}{4}$
  $N = 15$
  The 6-inch side will become 15 inches long.

CHAPTER 3   REVIEW

1. $936 \div 26 = 36$;   26 is a factor of 936.
2. $884 \div 48 = 18.41\overline{6}$;   48 is not a factor of 884.
3. $42 = 2 \times 21$
   $= 2 \times 3 \times 7$
4. $66 = 2 \times 33$
   $= 2 \times 3 \times 11$
5. $72 = 2 \times 36$
   $= 2 \times 4 \times 9$
   $= 2 \times 2 \times 2 \times 3 \times 3$
6. $108 = 2 \times 54$
   $= 2 \times 6 \times 9$
   $= 2 \times 2 \times 3 \times 3 \times 3$
7. $\dfrac{22}{72} = \dfrac{\cancel{2} \times 11}{\cancel{2} \times 2 \times 2 \times 3 \times 3} = \dfrac{11}{36}$
8. $\dfrac{24}{36} = \dfrac{2 \times \cancel{2} \times \cancel{2} \times \cancel{3}}{\cancel{2} \times \cancel{2} \times 3 \times \cancel{3}} = \dfrac{2}{3}$
9. $\dfrac{75}{105} = \dfrac{\cancel{3} \times \cancel{5} \times 5}{\cancel{5} \times \cancel{3} \times 7} = \dfrac{5}{7}$
10. $\dfrac{72}{126} = \dfrac{\cancel{3} \times \cancel{3} \times \cancel{2} \times 2 \times 2}{\cancel{2} \times 7 \times \cancel{3} \times \cancel{3}}$
    $= \dfrac{4}{7}$
11. $\dfrac{3}{8} = \dfrac{3 \times 4}{8 \times 4} = \dfrac{12}{32}$
    (building factor: $32 \div 8 = 4$)
12. $\dfrac{5}{12} = \dfrac{5 \times 15}{12 \times 15} = \dfrac{75}{180}$
    (building factor: $180 \div 12 = 15$)
13. $\dfrac{2}{5}, \dfrac{3}{7}$ LCD: $5 \times 7 = 35$
    $\dfrac{2 \times 7}{5 \times 7} = \dfrac{14}{35}; \dfrac{3 \times 5}{7 \times 5} = \dfrac{15}{35}$
14. $\dfrac{3}{4}, \dfrac{5}{12}$ LCD: $2 \times 2 \times 3 = 12$
    $\dfrac{3 \times 3}{4 \times 3} = \dfrac{9}{12}; \dfrac{5}{12}$
15. $\dfrac{2}{3}, \dfrac{5}{6}, \dfrac{2}{15}$
    LCD: $3 \times 2 \times 5 = 30$
    $\dfrac{2 \times 10}{3 \times 10} = \dfrac{20}{30}; \dfrac{5 \times 5}{6 \times 5} = \dfrac{25}{30};$
    $\dfrac{2 \times 2}{15 \times 2} = \dfrac{4}{30}$
16. $\dfrac{4}{1}, \dfrac{5}{18}, \dfrac{7}{24}$
    LCD: $3 \times 3 \times 2 \times 2 \times 2 = 72$
    $\dfrac{4 \times 72}{1 \times 72} = \dfrac{288}{72}; \dfrac{5 \times 4}{18 \times 4} = \dfrac{20}{72};$
    $\dfrac{7 \times 3}{24 \times 3} = \dfrac{21}{72}$
17. $\dfrac{6}{13}, \dfrac{5}{13}, \dfrac{8}{13}$
    5 is less than 6 and 6 is less than 8.
    Ans. $\dfrac{5}{13}, \dfrac{6}{13}, \dfrac{8}{13}$

585

ANSWERS AND SOLUTIONS

18. $\frac{2}{5}, \frac{1}{7}, \frac{9}{25}$

    LCD: $5 \times 7 \times 5 = 175$

    $\frac{2 \times 35}{5 \times 35} = \frac{70}{175}; \frac{1 \times 25}{7 \times 25} = \frac{25}{175};$

    $\frac{9 \times 7}{25 \times 7} = \frac{63}{175}$

    25 is less than 63 and 63 is less than 70.

    Ans. $\frac{25}{175}, \frac{63}{175}, \frac{70}{175}$ or $\frac{1}{7}, \frac{9}{25}, \frac{2}{5}$

22. $\frac{3}{20} + \frac{5}{12} + \frac{1}{6}$

    LCD: $2 \times 2 \times 5 \times 3 = 60$

    $\frac{3 \times 3}{20 \times 3} = \frac{9}{60}; \frac{5 \times 5}{12 \times 5} = \frac{25}{60};$

    $\frac{1 \times 10}{6 \times 10} = \frac{10}{60};$

    $\frac{9}{60} + \frac{25}{60} + \frac{10}{60} = \frac{9 + 25 + 10}{60}$

    $= \frac{\cancel{44}^{11}}{\cancel{60}_{15}} = \frac{11}{15}$

25. $\frac{25}{36} - \frac{11}{36} - \frac{7}{36} = \frac{25 - 11 - 7}{36}$

    $= \frac{7}{36}$

27. $9\frac{25}{32} - 2\frac{5}{32} = (9 - 2) + \left(\frac{25}{32} - \frac{5}{32}\right)$

    $= 7 + \frac{\cancel{20}^{5}}{\cancel{32}_{8}}$

    $= 7\frac{5}{8}$

29. $18 - 6\frac{5}{6}$ Rewrite 18 as $17 + 1 = 17 + \frac{6}{6}$

    $(17 - 6) + \left(\frac{6}{6} - \frac{5}{6}\right) = 11 + \frac{1}{6} = 11\frac{1}{6}$

19. $6\frac{3}{8} = \frac{(6 \times 8) + 3}{8} = \frac{51}{8}$

20. $4\frac{5}{6} = \frac{(4 \times 6) + 5}{6} = \frac{29}{6}$

21. $\frac{5}{24} + \frac{7}{24} = \frac{5 + 7}{24}$

    $= \frac{\cancel{12}^{1}}{\cancel{24}_{2}} = \frac{1}{2}$

23. $3\frac{2}{7} + 6\frac{3}{7} = (3 + 6) + \left(\frac{2}{7} + \frac{3}{7}\right)$

    $= 9 + \frac{5}{7} = 9\frac{5}{7}$

24. $15\frac{5}{16} + 2\frac{3}{8} + 4\frac{3}{4} = (15 + 2 + 4) + \left(\frac{5}{16} + \frac{3}{8} + \frac{3}{4}\right)$

    $= 21 + \left(\frac{5}{16} + \frac{3 \times 2}{8 \times 2} + \frac{3 \times 4}{4 \times 4}\right)$

    $= 21 + \left(\frac{5}{16} + \frac{6}{16} + \frac{12}{16}\right)$

    $= 21 + \frac{23}{16}$

    $= 21 + 1 + \frac{7}{16} = 22\frac{7}{16}$

26. $\frac{27}{32} - \frac{3}{8} - \frac{1}{4}$ LCD: $2 \times 2 \times 2 \times 2 \times 2 = 32$

    $= \frac{27}{32} - \frac{3 \times 4}{8 \times 4} - \frac{1 \times 8}{4 \times 8}$

    $= \frac{27 - 12 - 8}{32} = \frac{7}{32}$

28. $15\frac{5}{6} - 3\frac{1}{4}$ LCD: $2 \times 2 \times 3 = 12$

    $= (15 - 3) + \left(\frac{5}{6} - \frac{1}{4}\right)$

    $= 12 + \left(\frac{5 \times 2}{6 \times 2} - \frac{1 \times 3}{4 \times 3}\right)$

    $= 12 + \left(\frac{10}{12} - \frac{3}{12}\right)$

    $= 12 + \frac{7}{12} = 12\frac{7}{12}$

30. $5\frac{2}{3} + 3\frac{1}{5} - 4\frac{1}{6}$ LCD: $3 \times 5 \times 2 = 30$

    $= (5 + 3 - 4) + \left(\frac{2}{3} + \frac{1}{5} - \frac{1}{6}\right)$

    $= 4 + \left(\frac{2 \times 10}{3 \times 10} + \frac{1 \times 6}{5 \times 6} - \frac{1 \times 5}{6 \times 5}\right)$

    $= 4 + \left(\frac{20}{30} + \frac{6}{30} - \frac{5}{30}\right)$

    $= 4 + \frac{\cancel{21}^{7}}{\cancel{30}_{10}} = 4\frac{7}{10}$

PART B: SOLUTIONS

31. $\dfrac{3}{\cancel{4}_2} \times \dfrac{\cancel{2}^1}{5} = \dfrac{3}{10}$

32. $\dfrac{\cancel{6}^1}{\cancel{25}_5} \times \dfrac{\cancel{5}^1}{\cancel{12}_2} = \dfrac{1}{10}$

33. $\dfrac{\cancel{3}^1}{\cancel{7}_1} \times \dfrac{\cancel{14}^{\cancel{2}^1}}{\cancel{15}_{\cancel{3}_1}} \times \dfrac{\cancel{5}^1}{\cancel{8}_4} = \dfrac{1}{4}$

34. $16 \times 3\dfrac{1}{4} = \dfrac{\cancel{16}^4}{1} \times \dfrac{13}{\cancel{4}_1}$
$= \dfrac{52}{1} = 52$

35. $\dfrac{3}{8}$ of $512 = \dfrac{3}{\cancel{8}_1} \times \dfrac{\cancel{512}^{64}}{1}$
$= \dfrac{192}{1} = 192$

36. $2\dfrac{3}{5} = \dfrac{(2 \times 5) + 3}{5} = \dfrac{13}{5}$;
reciprocal: $\dfrac{5}{13}$

37. $\dfrac{15}{28} \div \dfrac{3}{7} = \dfrac{\cancel{15}^5}{\cancel{28}_4} \times \dfrac{\cancel{7}^1}{\cancel{3}_1}$
$= \dfrac{5}{4} = 1\dfrac{1}{4}$

38. $\dfrac{7}{8} \div 14 = \dfrac{\cancel{7}^1}{8} \times \dfrac{1}{\cancel{14}_2}$
$= \dfrac{1}{16}$

39. $3\dfrac{1}{2} \div 1\dfrac{3}{4} = \dfrac{7}{2} \div \dfrac{7}{4}$
$= \dfrac{\cancel{7}^1}{\cancel{2}_1} \times \dfrac{\cancel{4}^2}{\cancel{7}_1}$
$= \dfrac{2}{1} = 2$

40. $2\dfrac{3}{16} \div 2\dfrac{1}{2} = \dfrac{35}{16} \div \dfrac{5}{2}$
$= \dfrac{\cancel{35}^7}{\cancel{16}_8} \times \dfrac{\cancel{2}^1}{\cancel{5}_1}$
$= \dfrac{7}{8}$

41. $11\dfrac{2}{3} \times 3\dfrac{3}{5} \div 1\dfrac{1}{5} = \dfrac{35}{3} \times \dfrac{18}{5} \div \dfrac{6}{5}$
$= \dfrac{35}{\cancel{3}_1} \times \dfrac{\cancel{18}^{\cancel{6}^1}}{\cancel{5}_1} \times \dfrac{\cancel{5}^1}{\cancel{6}_1}$
$= \dfrac{35}{1} = 35$

42. $\dfrac{21}{35} \div \dfrac{7}{15} = \dfrac{\cancel{21}^3}{\cancel{35}_7} \times \dfrac{\cancel{15}^3}{\cancel{7}_1}$
$= \dfrac{9}{7} = 1\dfrac{2}{7}$

43. $\dfrac{3}{7} = 0.\overline{428571}$
a. 0.4 b. 0.43 c. 0.429

44. $\dfrac{2}{9} = 0.22\overline{2}$
a. 0.2 b. 0.22 c. 0.222

45. $\dfrac{7}{3} = 2.3\overline{3}$
a. 2.3 b. 2.33 c. 2.333

46. $\dfrac{13}{11} = 1.18\overline{18}$
a. 1.2 b. 1.18 c. 1.182

47. $2 + \dfrac{1}{2} = 2 + 0.5$
$= 2.5$

48. $5 + \dfrac{3}{4} = 5 + 0.75$
$= 5.75$

49. $12 + \dfrac{1}{5} = 12 + 0.2$
$= 12.2$

50. $9 + \dfrac{4}{5} = 9 + 0.8$
$= 9.8$

51. $6 + \dfrac{3}{8} = 6 + 0.375$
$= 6.375$

52. $12 + \dfrac{5}{8} = 12 + 0.625$
$= 12.625$

53. $2.25 = 2\dfrac{1}{4}$

54. $5.1 = 5\dfrac{1}{10}$

55. $6\dfrac{1}{2} \times 4.7 \times 3\dfrac{3}{4}$
$= 6.5 \times 4.7 \times 3.75$
$= 114.56 \left(\begin{array}{c}\text{nearest}\\\text{hundredth}\end{array}\right)$

56. $12\dfrac{2}{5} \times 6.31 \times 4\dfrac{1}{8}$
$= 12.4 \times 6.31 \times 4.125$
$= 322.76 \left(\begin{array}{c}\text{nearest}\\\text{hundredth}\end{array}\right)$

ANSWERS AND SOLUTIONS

57. $\dfrac{15}{24} = \dfrac{5}{8}$

58. $\dfrac{1}{2} = \dfrac{4}{8} = \dfrac{8}{16} = \dfrac{16}{32}$

    $\dfrac{3}{8}$ and $\dfrac{15}{32}$ are too small

59. Gallons: G

    $G = 3\dfrac{1}{2} + 2\dfrac{1}{4} + 4\dfrac{3}{8}$

    LCD: $2 \times 2 \times 2 = 8$

    $G = (3 + 2 + 4) + \left(\dfrac{1}{2} + \dfrac{1}{4} + \dfrac{3}{8}\right)$

    $= 9 + \left(\dfrac{1 \times 4}{2 \times 4} + \dfrac{1 \times 2}{4 \times 2} + \dfrac{3}{8}\right)$

    $= 9 + \left(\dfrac{4}{8} + \dfrac{2}{8} + \dfrac{3}{8}\right)$

    $= 9 + \dfrac{9}{8}$

    $= 9 + 1 + \dfrac{1}{8} = 10\dfrac{1}{8}$

60. Pounds remaining: P

    $P = 310 - 187\dfrac{1}{2} - 96\dfrac{3}{4}$

    LCD: $2 \times 2 = 4$

    $P = 309\dfrac{4}{4} - 187\dfrac{2}{4} - 96\dfrac{3}{4}$

    $= 122\dfrac{2}{4} - 96\dfrac{3}{4}$

    $= 121\dfrac{6}{4} - 96\dfrac{3}{4}$

    $= (121 - 96) + \left(\dfrac{6}{4} - \dfrac{3}{4}\right)$

    $= 25 + \dfrac{3}{4} = 25\dfrac{3}{4}$

61. Weight: W

    $W = 5\dfrac{1}{3} \times 6\dfrac{3}{4}$

    $= \dfrac{(5 \times 3) + 1}{3} \times \dfrac{(6 \times 4) + 3}{4}$

    $= \dfrac{\cancel{16}^{4}}{\cancel{3}_{1}} \times \dfrac{\cancel{27}^{9}}{\cancel{4}_{1}} = \dfrac{36}{1} = 36$

62. Cost per acre: C

    $C = 4200 \div 3\dfrac{1}{2}$

    $= 4200 \div \dfrac{7}{2}$

    $= \dfrac{\cancel{4200}^{600}}{1} \times \dfrac{2}{\cancel{7}_{1}} = 1200$

CHAPTER 4 REVIEW

1. $0.05\,6 = 5.6\%$

2. $18\% = 0.18$

3. $6\dfrac{3}{4}\% = 06.75\%$

   $= 0.0675$

4. $1.20 = 120\%$

5. 25% of 80
   $= 0.25 \times 80 = 20$

6. 64% of 75.6
   $= 0.64 \times 75.6$
   $= 48.384$

7. 4.5% of 600
   $= 0.045 \times 600$
   $= 27$

8. 120% of 45
   $= 1.2 \times 45$
   $= 54$

9. Percent: P

   $6 = P \times 8$

   $\dfrac{6}{8} = \dfrac{P \times 8}{8}$

   $0.75 = P$

   The required percent is 75%

PART B: SOLUTIONS

10. Percent: P
$P \times 125 = 40$
$\dfrac{P \times 125}{125} = \dfrac{40}{125}$
$P = 0.32$
The required percent is 32%

11. Number: N
$80 = 40\text{ of }N$
$80 = 0.40 \times N$
$\dfrac{80}{0.40} = \dfrac{0.40 \times N}{0.40}$
$200 = N$
The number is 200.

12. Number: N
$0.086 \times N = 7.74$
$\dfrac{0.086 \times N}{0.086} = \dfrac{7.74}{0.086}$
$N = 90$
The number is 90.

13. Transportation: T
$T = 9\%\text{ of }15{,}600$
$T = 0.09 \times 15{,}600$
$= 1404$
The transporation expense is $1404.

14. Percent: P
$P \times 15 = 12$
$\dfrac{P \times 15}{15} = \dfrac{12}{15}$
$P = 0.80$
80% of the questions are answered correctly.

15. Rate: R
$R \times 350 = 22.75$
$\dfrac{R \times 350}{350} = \dfrac{22.75}{350}$
$R = 0.065 = 6.5\%$
The tax rate is 6.5%.

16. Cost: C
$0.06 \times C = 300$
$\dfrac{0.06 \times C}{0.06} = \dfrac{300}{0.06}$
$C = 5000$
The cost was $5000.

17. Cost price: C
Discount: D
$D = 0.084 \times 60 = 5.04$
$C = 60 - 5.04 = 54.96$
The cost price is $54.96.

18. Selling price: S
Discount: D
$D = 0.15 \times 140 = 21$
$S = 140 - 21 = 119$
The selling price is $119.

19. Cost price: C
Discount: D
$D = 0.2 \times 185 = 37$
$C = 185 - 37 = 148$
The cost price is $148.

20. Cost price: C
Discount: D
a. $D = 0.24 \times 220 = 52.8$
$C = 220 - 52.8 = 167.2$
The cost is $167.20.

b. $D = 0.21 \times 220 = 46.2$
$C = 220 - 46.2 = 173.8$
The cost is $173.80.

21. $C = A \times P$
$C = 6400 \times 0.02$
$= 128$
The commission is $128.

22. $C = A \times P$
$507.50 = 14{,}500 \times P$
$\dfrac{507.5}{14{,}500} = \dfrac{14{,}500 \times P}{14{,}500}$
$0.035 = P$ or $3.5\% = P$
The commission rate is 3.5%

23. Percent: P
Amount of increase: A
$A = 210 - 140 = 70$
$P = \dfrac{70}{140} = 0.5 = 50\%$
The percent increase is 50%.

24. Percent: P
Amount of decrease: A
$A = 640 - 560 = 80$
$P = \dfrac{80}{640} = 0.125$
$= 12.5\%$
The percent decrease is 12.5%.

25. Percent: P
Amount of increase: A
$A = 1620 - 1500 = 120$
$P = \dfrac{120}{1500} = 0.08 = 8\%$
The percent of increase is 8%.

26. Amount: A
$I = P \times r \times t$
$I = 1000 \times 0.086 \times 3 = 258$
$A = 1000 + 258 = 1258$
The amount is $1258.

27. Amount: A
$I = P \times r \times t$
$I = 2400 \times 0.049 \times 1.5 = 176.4$
$A = 2400 + 176.4 = 2576.4$
The amount is $2576.40.

28. Amount: A
$I = P \times r \times t$
$I = 8600 \times 0.096 \times 0.5 = 412.8$
$A = 8600 + 412.8 = 9012.8$
The amount is $9012.80.

29. $I = P \times r \times t$
$62 = 1400 \times r \times 0.75$
$62 = 1050 \times r$
$\dfrac{62}{1050} = \dfrac{1050 \times r}{1050}$
$0.0590 = r$
$r = 5.90\%$
The rate of interest is 5.9%, to the nearest tenth of a percent.

589

ANSWERS AND SOLUTIONS

30. $I = P \times r \times t$
    $1360 = P \times 0.085 \times 2$
    $1360 = P \times 0.17$
    $\dfrac{1360}{0.17} = \dfrac{P \times 0.17}{0.17}$
    $8000 = P$
    $8000 must be invested.

CHAPTER 5 REVIEW

1. $\dfrac{156}{12} = 13$
   156 in. = 13 ft

2. $\dfrac{216}{4} = 54$
   216 qt = 54 gal

3. 3 mi = 3 × 5280 = 15,840 ft
   15,840 ft = 15,840 × 12
   = 190,080 in.
   Thus, 3 mi = 190,080 in.

4. 5 da = 5 × 24 = 120 hr
   120 hr = 120 × 60 = 7200 min
   7200 min = 7200 × 60
   = 432,000 sec
   Thus, 5 da = 432,000 sec.

5. 525 oz = $\dfrac{525}{16}$ = 32.8 lb

6. 5 ft 8 in. = (5 × 12 + 8) = 68 in.
   68 in. = $\dfrac{68}{4}$ = 17 hands
   Thus, 5 ft 8 in. = 17 hands.

7. 5 ft 6 in.
   8 ft 7 in.
   +4 ft 9 in.
   ----
   17 ft 22 in.
   = (17 + 1) ft + 10 in.
   = 18 ft 10 in.

8. 28 lb 6 oz − 10 lb 12 oz
   Write 28 lb 6 oz as
   27 lb + (16 + 6) oz
   = 27 lb + 22 oz;
   27 lb 22 oz
   −10 lb 12 oz
   ----
   17 lb 10 oz

9. 7 × (4 gal 3 qt)
   = (7 × 4) gal + (7 × 3) qt
   = 28 gal + 21 qt
   = (28 + 5) gal + 1 qt
   = 33 gal 1 qt

10. (16 yd 2 ft) ÷ 5
    16 yd 2 ft = (16 × 3 + 2) ft
    = 50 ft
    50 ft ÷ 5 = 10 ft

11. 16 ft − 4 ft 8 in. − 3 ft 10 in.
    16 ft − 4 ft 8 in.
    = 15 ft 12 in. − 4 ft 8 in.
    = 11 ft 4 in.;
    11 ft 4 in. − 3 ft 10 in.
    = 10 ft 16 in. − 3 ft 10 in.
    = 7 ft 6 in.

12. 87.0 cm = 870 mm
13. 30.7 hm = 30.7 km
14. 8.82 g = 8.82 day
15. 2031.0 kg = 20,310 hg
16. 94.30 hm = 9430 m
17. 8.370 kg = 8370 g

18. kg   lb
    $\dfrac{K}{0.454} = \dfrac{7.4}{1}$
    $K \times 1 = 0.454 \times 7.4 = 3.3596$
    To the nearest tenth, 7.4 lb
    equals 3.4 kg.

19. km   mi
    $\dfrac{K}{1.609} = \dfrac{16.3}{1}$
    $K \times 1 = 1.609 \times 16.3 = 26.2267$
    To the nearest tenth, 16.3 mi
    equals 26.2 km.

20. in.   cm
    $\dfrac{I}{1} = \dfrac{12.4}{2.54}$
    $I = 4.88$
    To the nearest tenth, 12.4 cm
    equals 4.9 in.

21. oz   g
    $\dfrac{Z}{1} = \dfrac{864.3}{28.35}$
    $Z = 30.48$
    To the nearest tenth, 864.3 g
    equals 30.5 oz.

PART B: SOLUTIONS

22. $\dfrac{\text{km}}{1.609} \dfrac{K}{1} = \dfrac{\text{mi}}{1}\dfrac{1049}{1}$
$K \times 1 = 1.609 \times 1049 = 1687.841$
To the nearest km, the distance was 1688 km.

23. $10 \times 3500 = 35{,}000$ calories associated with a 10-lb loss.
Walking for 60 minutes per day will use up 270 calories.
$\dfrac{\text{Days}}{1}\dfrac{D}{1} = \dfrac{C}{270}\dfrac{35{,}000}{270}$
$D = 129.6$
It will take 130 days to lose 10 pounds.

24. $8 \times 3500 = 28{,}000$ calories are associated with an 8-lb loss.
$\dfrac{C}{420} = \dfrac{\text{min}}{60}\dfrac{45}{60}$
$C \times 60 = 420 \times 45$
$\dfrac{C \times 60}{60} = \dfrac{420 \times 45}{60}$
$C = 315$

Bicycling for 45 minutes uses 315 calories.
$\dfrac{\text{days}}{1}\dfrac{D}{1} = \dfrac{C}{315}\dfrac{28{,}000}{315}$
$D = 88.8\overline{8}$
It will take 89 days to lose 8 pounds.

25. $\dfrac{90}{18} = 5$;
5¢ per ounce

26. $\dfrac{42.35}{5} = 8.47$
$8.47 per gallon

27. $\dfrac{25.98}{20{,}000} = 0.00130$ $\dfrac{36.84}{30{,}000} = 0.00123$
or 0.13¢ per mile.   or 0.123¢ per mile.
The 30,000 mile tire is a better buy.

28. $\dfrac{23.55}{36} = 0.65$; $\dfrac{38.00}{60} = 0.63$
or 65¢ per month.   or 63¢ per month.
The 5 yr (60 month) battery is the better buy.

29. Average: A
$A = \dfrac{20 + 18 + 15 + 17 + 24}{5} = 18.8$
Average is 18.8 points per game.

30. Grade on seventh test: G
$90 = \dfrac{87 + 89 + 88 + 90 + 84 + 92 + G}{7}$
$90 = \dfrac{530 + G}{7}$
$90 \times 7 = \dfrac{(530 + G)}{7} \times 7$
$630 = 530 + G$
$630 - 530 = 530 + G - 530$
$100 = G$
The student must receive 100% to earn a grade of A.

CHAPTER 6 REVIEW

1. [number line with points at −6, −2, −1, 2, 5, 7]

2. [number line with points at −5, −3, −1, 1, 3]

3. Arranged as on a line graph: $-4, -3, 0, 2, 5$

4. a. $-5 < 2$ b. $|-3| = 3$ c. $|-2| > 0$

5. a. -2, because $|-5| - |+3| = 2$, and $|-5|$ is greater than $|+3|$.
    b. -6, because $|-2| + |-4| = 6$, and both are negative.
    c. 1, because $|7| - |-6| = 1$, and $|7|$ is greater than $|-6|$.

6. a. -3 b. -3 c. -3

591

ANSWERS AND SOLUTIONS

7. a. $4 + 3 - 2$
        $= 7 - 2 = 5$
    b. $6 - 5 - 7$
        $= 1 - 7 = -6$
    c. $-4 - 3 - 8$
        $= -7 - 8 = -15$

8. a. Change to $\overset{6}{-8}$ and add.
        Ans. -2
    b. Change to $\overset{-7}{+3}$ and add.
        Ans. -4
    c. Change to $\overset{0}{+2}$ and add.
        Ans. 2

9. a. 6
    b. -12
    c. -10

10. a. $4(0)(-2)$
        $= 0(-2) = 0$
    b. $5(-1)(-3)$
        $= -5(-3) = 15$
    c. $-2(4)(-2)$
        $= -8(-2) = 16$

11. a. 5
    b. -4
    c. -3

12. a. -8
    b. -1
    c. 0

13. a. 4^2
    b. 5^3
    c. $(-2)^4$

14. a. $20 = 2 \cdot 10$
        $= 2 \cdot 2 \cdot 5$
    b. $-36 = -1 \cdot 2 \cdot 18$
        $= -1 \cdot 2 \cdot 2 \cdot 9$
        $= -1 \cdot 2 \cdot 2 \cdot 3 \cdot 3$
    c. $-56 = -1 \cdot 2 \cdot 28$
        $= -1 \cdot 2 \cdot 2 \cdot 14$
        $= -1 \cdot 2 \cdot 2 \cdot 2 \cdot 7$

15. a. $-8^2 = -1 \cdot 8^2$
        $= -1 \cdot 64 = -64$
    b. $(-8)^2 = (-8)(-8)$
        $= 64$
    c. $-5 \cdot 2^2 = -5 \cdot 4$
        $= -20$

16. a. $(-2)^2 \cdot 3^2 = (-2)(-2) \cdot 3 \cdot 3$
        $= 4 \cdot 9 = 36$
    b. $(-2)^3 \cdot (-3)^2 = (-2)(-2)(-2)(-3)(-3)$
        $= (-8)(9) = -72$
    c. $-2^3(-3)^2 = -1 \cdot 2 \cdot 2 \cdot 2 (-3)(-3)$
        $= -1 \cdot 8 \cdot 9 = -72$

17. a. $2 + (3)(4) = 2 + 12$
        $= 14$
    b. $4 - 2(3) = 4 - 6$
        $= -2$
    c. $4 \cdot 3 - 2 = 12 - 2$
        $= 10$

18. a. $\dfrac{12 + 3}{2 + 3} - \dfrac{10}{2} = \dfrac{15}{5} - \dfrac{10}{2}$
        $= 3 - 5$
        $= -2$
    b. $\dfrac{5 \cdot 4}{10} + \dfrac{2 + 8}{7 - 2} = \dfrac{20}{10} + \dfrac{10}{5}$
        $= 2 + 2$
        $= 4$
    c. $\dfrac{6 - 3}{3 - 2} - \dfrac{6 \cdot 3}{9} = \dfrac{3}{1} - \dfrac{18}{2}$
        $= 3 - 9$
        $= -6$

19. a. $\dfrac{(-2)^2 - 1}{3} - \dfrac{1 + 2^2}{5} = \dfrac{4 - 1}{3} - \dfrac{1 + 4}{5}$
        $= \dfrac{3}{3} - \dfrac{5}{5}$
        $= 1 - 1 = 0$
    b. $\dfrac{4 - (-2)^2}{3} + \dfrac{2^2 + 1}{5} = \dfrac{4 - 4}{3} + \dfrac{4 + 1}{5}$
        $= \dfrac{0}{3} + \dfrac{5}{5}$
        $= 0 + 1 = 1$
    c. $\dfrac{3^2 - 2^2}{5} - \dfrac{5 - 3^2}{2} = \dfrac{9 - 4}{5} - \dfrac{5 - 9}{2}$
        $= \dfrac{5}{5} - \dfrac{-4}{2}$
        $= 1 - (-2) = 3$

PART B: SOLUTIONS

20. a. $3(-2)^2 - 4 \cdot 3^2 = 3 \cdot 4 - 4 \cdot 9$
$= 12 - 36$
$= -24$

b. $-2(3)^2 + 4(-2)^2 = -2(9) + 4(4)$
$= -18 + 16 = -2$

c. $\dfrac{(-2)^2 + 2^2}{2} + (-3)^2 = \dfrac{4+4}{2} + 9$
$= \dfrac{8}{2} + 9$
$= 4 + 9 = 13$

CHAPTER 7 REVIEW

1. a. $4a^2b^3$ b. $-xy^2z^3$ c. 3^2c^2d

2. a. $2 \cdot 3xyyy$ b. $aaabb$ c. $3 \cdot 3 \cdot 3cdd$

3. a. $a^2 + c = (1)^2 + (-2)$
$= 1 + (-2) = -1$

b. $4a + 3b + c^2 = 4(1) + 3(0) + (-2)^2$
$= 4 + 0 + 4 = 8$

c. $\dfrac{c^2 - b}{2a} = \dfrac{(-2)^2 - (0)}{2(1)}$
$= \dfrac{4-0}{2} = 2$

4. a. $2(a+b)^2 = 2[2 + (-3)]^2$
$= 2(-1)^2$
$= 2(1) = 2$

b. $2a^2 + 2b^2 = 2(2)^2 + 2(-3)^2$
$= 2(4) + 2(9)$
$= 8 + 18 = 26$

c. $(2a)^2 + (2b)^2 = (2 \cdot 2)^2 + [2 \cdot (-3)]^2$
$= 4^2 + (-6)^2$
$= 16 + 36 = 52$

5. $P(x) = 2x^2 - x + 3$

a. $P(-2) = 2(-2)^2 - (-2) + 3$
$= 2(4) - (-2) + 3$
$= 8 + 2 + 3 = 13$

b. $P(0) = 2(0)^2 - (0) + 3$
$= 2(0) - 0 + 3$
$= 0 - 0 + 3 = 3$

c. $P(3) = 2(3)^2 - (3) + 3$
$= 2(9) - 3 + 3$
$= 18 - 3 + 3 = 18$

6. a. $\dfrac{-(-1)^2(-2)}{-(1)} = \dfrac{2}{-1} = -2$

b. $\dfrac{(-1)^2 - (1)^2}{-2(-2)} = \dfrac{1-1}{4} = 0$

c. $\dfrac{(-1)^2 - (-2)}{1} = \dfrac{1+2}{1} = 3$

7. a. $3xy + 2y + 3xy$
$= 6xy + 2y$

b. $6a^2 - a^2 - 3a$
$= 5a^2 - 3a$

c. $3r + 5s - r - s$
$= 2r + 4s$

8. a. $-2x^2 + 3x + x^2$
$= -x^2 + 3x$

b. $6x^2 + 2xy + 2x^2 - 3xy$
$= 8x^2 - xy$

c. $2xy^2 + 3xy - xy^2 + xy$
$= xy^2 + 4xy$

9. a. $(3x^2 - 2x) + (x^2 - x)$
$= 3x^2 - 2x + x^2 - x$
$= 4x^2 - 3x$

b. $(3x^2 - 1) - (2x^2 + 2)$
$= 3x^2 - 1 - 2x^2 - 2$
$= x^2 - 3$

c. $(4y^2 - 2y) - (y - 1)$
$= 4y^2 - 2y - y + 1$
$= 4y^2 - 3y + 1$

10. a. $(x + y + 2z) - (3x^2 + z)$
$= x + y + 2z - 3x^2 - z$
$= -3x^2 + x + y + z$

b. $(2a + 3b - 4c) - (a + b + c)$
$= 2a + 3b - 4c - a - b - c$
$= a + 2b - 5c$

c. $(x + y - 2z) - (2x + y - z)$
$= x + y - 2z - 2x - y + z$
$= -x - z$

593

ANSWERS AND SOLUTIONS

11. a. x^3y^3
    b. $12ab^4$
    c. r^4s^3

12. a. $3x^2 - 2x^3 + x^2$
       $= 4x^2 - 2x^3$
    b. $ab^3 - b^2$
    c. $2r^2s^2 - r^2s^2$
       $= r^2s^2$

13. a. $\dfrac{\cancel{4}a^{\cancel{2}\,2}b}{\cancel{2}\,\cancel{a}} = 2ab$
    b. $\dfrac{\cancel{3a^3b}}{\cancel{3a^3b}} = 1$
    c. $\dfrac{\cancel{12}^{3}xy^{\cancel{3}\,y}}{\cancel{4}y^{\cancel{2}}} = 3xy$

14. a. $\dfrac{\cancel{x}y}{-\cancel{x}} = -y$
    b. $\dfrac{-\cancel{x}\cancel{y}\cancel{z}}{\cancel{x}\cancel{y}\cancel{z}} = -1$
    c. $\dfrac{-\cancel{4}x}{-\cancel{4}} = x$

15. a. $\dfrac{3x-x}{2x} + 4 = \dfrac{2x}{2x} + 4$
       $= 1 + 4 = 5$
    b. $\dfrac{7x-4x}{3x} - 1 = \dfrac{3x}{3x} - 1$
       $= 1 - 1 = 0$
    c. $\dfrac{x^2}{x} - \dfrac{2x^3 + 4x^3}{2x^2} = \dfrac{x^2}{x} - \dfrac{6x^3}{2x^2}$
       $= x - 3x$
       $= -2x$

16. a. $\dfrac{3x^2 - 5x^2}{x} + 6x = \dfrac{-2x^2}{x} + 6x$
       $= -2x + 6x$
       $= 4x$
    b. $\dfrac{-x^3}{x^2} - \dfrac{12x}{3} = -x - 4x$
       $= -5x$
    c. $\dfrac{3x^2 - 4x^2 + x^2}{7} + 1 = \dfrac{0}{7} + 1$
       $= 1$

17. Binomial 18. Numerical coefficient 19. 1 (one) 20. 4 (four)

CHAPTER 8 REVIEW

1. a. $3 + x = 2x - 2$
   b. $12 - x = 2x$
   c. $\dfrac{3x}{4} = 21 - 6$

2. a. $2 + x = 8$
      $2 + x - 2 = 8 - 2$
      $x = 6$
   b. $5y = 2 + 4y$
      $5y - 4y = 2$
      $y = 2$
   c. $\dfrac{2a + 4a}{3} = 5a + 3$
      $\dfrac{6a}{3} = 5a + 3$
      $2a = 5a + 3$
      $2a - 5a = 5a + 3 - 5a$
      $-3a = 3$
      $a = -1$

3. a. $4x + 3x = 35$
      $7x = 35$
      $x = 5$
   b. $4x - 4 = 2x - 4$
      $4x - 4 - 2x + 4 = 2x - 4 - 2x + 4$
      $2x = 0$
      $x = 0$
   c. $8z + 6z = 2z - 12$
      $14z = 2z - 12$
      $12z = -12$
      $z = -1$

4. a. $\dfrac{2a}{3} = -12$
      $2a = -36$
      $a = -18$
   b. $\dfrac{b + 4b}{3} = 15$
      $\dfrac{5b}{3} = 15$
      $5b = 45$
      $b = 9$
   c. $\dfrac{6x - 2x}{3} = -4$
      $\dfrac{4x}{3} = -4$
      $4x = -12$
      $x = -3$

5. a. $\dfrac{-9a - a}{2} = 10$
      $\dfrac{-10a}{2} = 10$
      $-5a = 10$
      $a = -2$
   b. $\dfrac{3x + 5x}{2} = 6 + 3x$
      $\dfrac{8x}{2} = 6 + 3x$
      $4x = 6 + 3x$
      $4x - 3x = 6 + 3x - 3x$
      $x = 6$
   c. $\dfrac{8x - 4x}{2} = \dfrac{8 + 10}{3}$
      $\dfrac{4x}{2} = \dfrac{18}{3}$
      $2x = 6$
      $x = 3$

594

PART B: SOLUTIONS

6. a. $-5y + 1 < 26$
$-5y < 25$
$y > -5$

b. $3x + 2 > x - 10$
$2x > -12$
$x > -6$
$x \le -8$

c. $\dfrac{4x - 5x}{4} \ge 2$
$\dfrac{-x}{4} \ge 2$
$-x \ge 8$

7. a. $2 + x = 8;$
$2 + 6 \stackrel{?}{=} 8$
$8 = 8$

b. $5y = 2 + 4y$
$5(2) \stackrel{?}{=} 2 + 4(2)$
$10 = 10$

c. $\dfrac{2a + 4a}{3} = 5a + 3$
$\dfrac{2(-1) + 4(-1)}{3} \stackrel{?}{=} 5(-1) + 3$
$\dfrac{-6}{3} \stackrel{?}{=} -5 + 3$
$-2 = -2$

8. a. $\dfrac{f}{m} = \dfrac{ma}{m}$
$\dfrac{f}{m} = a$
or $a = \dfrac{f}{m}$

b. $v - k = k + gt - k$
$v - k = gt$
$\dfrac{v - k}{t} = \dfrac{gt}{t}$
$g = \dfrac{v - k}{t}$

c. $2M = 2 \cdot \dfrac{a + b}{2}$
$2M = a + b$
$2M - a = a + b - a$
$b = 2M - a$

9. $x + 2$ 10. $x + 2$ 11. $x + 1, x + 2, x + 3,$ and $x + 4$ 12. $3x$

13. taller: $x + 7$
shorter: $x - 7$

14. more: $x + 18$
less: $x - 12$

15. *Steps 1–2* First integer: x
$\left.\begin{array}{l}x + 1\\x + 2\\x + 3\end{array}\right\}$ the next three integers
Step 3 Not applicable.
Step 4 $x + x + 1 + x + 2 + x + 3 = 54$
Step 5 $4x + 6 = 54$
$4x = 48$
$x = 12$
Step 6 The integers are 12, 13, 14, and 15.

16. *Steps 1–2* Smaller integer: x
Second integer: $x + 2$
Third integer: $x + 4$
Step 3 Not applicable.
Step 4 $x + (x + 2) + (x + 4) = 5x$
Step 5 $3x + 6 = 5x$
$6 = 2x$
$3 = x$
Step 6 The integers are 3, 5, and 7.

17. *Steps 1–2* Number of papers for one girl: x
Number for the other: $x + 27$
Step 3 Not applicable.
Step 4 $x + (x + 27) = 431$
Step 5 $2x + 27 = 431$
$2x = 404$
$x = 202$
Step 6 202 papers for one girl, 229 for the other.

18. *Steps 1–2* Length of the shorter (second) piece: x
Length of the first piece: $x + 3$
Length of the third piece: $x + 5$
Step 3 Not applicable.
Step 4 $x + x + 3 + x + 5 = 32$
Step 5 $3x + 8 = 32$
$3x = 24$
$x = 8$
Step 6 The lengths are 11 ft, 8 ft, and 13 ft.

ANSWERS AND SOLUTIONS

19. *Steps 1–2* Miles driven: x
    Step 3 Not applicable.
    Step 4 $25 + .05x \leq 50$
    Step 5 $.05x \leq 25$
    $x \leq 500$
    Step 6 He can drive 500 miles or less.

20. *Steps 1–2* Cost of first meal: x
    Cost of second meal: $2x$
    Cost of third meal: $2(2x) = 4x$
    Step 3 Not applicable.
    Step 4 $x + 2x + 4x \leq 91$
    Step 5 $7x \leq 91$
    $x \leq 13$
    Step 6 The first meal must cost less than or equal to $13.

CHAPTER 9 REVIEW

1. a. $3x^3 + 3x^2$
   b. $2xy^2 - 2x^2y$
   c. $-x^2 + y - 1$

2. a. $2a - a^2$
   b. $-ab + b^2$
   c. $3ab + 3b^2 + 3bc$

3. a. $3a^2(1 - 2b)$
   b. $2x(x^2 + 2x + 3)$
   c. $-y^2(1 + y)$

4. a. $a^2(1 + b)$
   b. $4(b - 1)$
   c. $b(1 - b - b^2)$

5. a. $x^2 + x - 6$
   b. $6a^2 - 17a + 12$
   c. $4a^2 - 12a + 9$

6. a. $x^2 - ax - 2a^2$
   b. $2x^2 + bx - b^2$
   c. $4b^2 + 4b + 1$

7. a. $(x - 7)(x + 3)$
   b. $(2a + 3)(5a + 1)$
   c. $(2x + 3)(2x - 3)$

8. a. $(a - 7)(a - 3)$
   b. $(b + 1)(3b + 1)$
   c. $(2b - 1)(b + 2)$

9. a. $2(x^2 + 7x + 12) = 2(x + 3)(x + 4)$
   b. $3(y^2 + 8y - 20) = 3(y + 10)(y - 2)$
   c. $4x(x^2 - 1) = 4x(x + 1)(x - 1)$

10. a. $(x - a)(x - 2a)$
    b. $(x + a)(x - a)$
    c. $2(2b^2 + 3bc - 2c^2) = 2(2b - c)(b + 2c)$

11. a. $3x - 15 = 45$
    $3x = 60$
    $x = 20$
    b. $32 - 6b = 12 + 4b$
    $-10b = -20$
    $b = 2$
    c. $-b + 2 = 26 + 3b$
    $-4b = 24$
    $b = -6$

12. $A = 2kr(h + r)$
    $= 2(3.14)(7)(10 + 7)$
    $= 2(3.14)(7)(17)$
    $= 747.32$

13. $24 - x$

14. $10x$

15. $25(x + 3)$

16. $185(x + 4)$

17. *Steps 1–2* Amount invested at 10%: x
    Amount invested at 12%: $x + 500$

Step 3

Amount	Rate of interest	Amount of interest
x	0.10	0.10x
$x + 500$	0.12	0.12(x + 500)

Step 4 $0.10x + 0.12(x + 500) = 324$
Step 5 $10x + 12(x + 500) = 32{,}400$
$10x + 12x + 6000 = 32{,}400$
$22x = 26{,}400$
$x = 1{,}200$
Step 6 $1200 is invested at 10% and $1200 + 500 = $1700 is invested at 12%.

596

PART B: SOLUTIONS

18. *Steps 1–2* Smaller number: x
    Larger number: $x + 6$
    Step 3 Not applicable.
    Step 4 $10x - 4(x + 6) = 6$
    Step 5 $10x - 4x - 24 = 6$
    $6x = 30$
    $x = 5$
    Step 6 The numbers are 5 and $5 + 6 = 11$.

19. *Steps 1–2* Smaller number: x
    Larger number: $x + 10$
    Step 3 Not applicable.
    Step 4 $8x + 3(x + 10) = 129$
    Step 5 $8x + 3x + 30 = 129$
    $11x + 30 = 129$
    $11x = 99$
    $x = 9$
    Step 6 The numbers are 9 and $9 + 10 = 19$.

20. *Steps 1–2* Number of dimes: x
    Number of nickels: $x + 8$
    Step 3

Type of coin	Value of 1 coin in cents	Number of coins	Value of coins in cents
Dimes	10	x	$10x$
Nickels	5	$x + 8$	$5(x + 8)$

    Step 4 $10x + 5(x + 8) = 265$
    Step 5 $10x + 5x + 40 = 265$
    $15x = 225$
    $x = 15$
    Step 6 She has 15 dimes and $15 + 8 = 23$ nickels.

CHAPTER 10 REVIEW

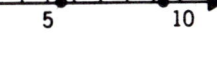

1. Number line with points at $-\frac{27}{4}$, $-\frac{5}{2}$, 2, $\frac{11}{2}$, $\frac{37}{4}$

2. a. $4 \cdot \dfrac{1}{9}$ b. $(x + 6) \cdot \dfrac{1}{3}$ c. $2y \cdot \dfrac{1}{x + y^2}$

3. a. $\dfrac{2(x - 3)}{3}$ b. $\dfrac{-(x^2 + 1)}{3}$ c. $\dfrac{-3(2x + y)}{4}$

4. a. $\dfrac{-3}{x + y}$ b. $\dfrac{a}{x}$ c. $\dfrac{-(b - 2)}{4}$ or $\dfrac{-b + 2}{4}$

5. a. $\dfrac{\cancel{x^2}\,y^{\cancel{2}}z^2}{\cancel{x}\,\cancel{y^3}} = \dfrac{xz^2}{y}$ b. $\dfrac{\cancel{(b-3)}}{(b+1)\cancel{(b-3)}} = \dfrac{1}{b+1}$ c. $\dfrac{a(a+1)}{a(a^2-1)} = \dfrac{\cancel{a}\cancel{(a+1)}}{\cancel{a}\cancel{(a+1)}(a-1)}$
$= \dfrac{1}{a-1}$

6. a. $\dfrac{\cancel{3x^3}^{x^2}}{\cancel{3x}} + \dfrac{\cancel{6x^2}^{2x}}{\cancel{3x}} - \dfrac{\cancel{9x}^3}{\cancel{3x}}$
$= x^2 + 2x - 3$
  b. $\dfrac{\cancel{8x^4}^{4x^2}}{\cancel{2x^2}} - \dfrac{\cancel{4x^2}^{2}}{\cancel{2x^2}} + \dfrac{3}{2x^2}$
$= 4x^2 - 2 + \dfrac{3}{2x^2}$
  c. $\dfrac{\cancel{6x^3}^{3x^2}}{\cancel{2x}} - \dfrac{\cancel{4x}^{2}}{\cancel{2x}} - \dfrac{1}{2x}$
$= 3x^2 - 2 - \dfrac{1}{2x}$

597

ANSWERS AND SOLUTIONS

7. a.
$$\begin{array}{r} 2x - 3 \\ x - 1 \overline{\smash{\big)}\, 2x^2 - 5x + 1} \\ \underline{2x^2 - 2x} \\ -3x + 1 \\ \underline{-3x + 3} \\ -2 \end{array}$$

Ans. $2x - 3 + \dfrac{-2}{x - 1}$

b.
$$\begin{array}{r} 2x^2 - 4x + 9 \\ x + 2 \overline{\smash{\big)}\, 2x^3 + 0x^2 + x - 3} \\ \underline{2x^3 + 4x^2} \\ -4x^2 + x \\ \underline{-4x^2 - 8x} \\ 9x - 3 \\ \underline{9x + 18} \\ -21 \end{array}$$

Ans. $2x^2 - 4x + 9 + \dfrac{-21}{x + 2}$

c.
$$\begin{array}{r} 2x + 3 \\ x + 1 \overline{\smash{\big)}\, 2x^2 + 5x + 3} \\ \underline{2x^2 + 2x} \\ 3x + 3 \\ \underline{3x + 3} \\ 0 \end{array}$$

Ans. $2x + 3$

8. a. $\dfrac{3}{x - y} = \dfrac{(2)3}{2(x - y)} = \dfrac{6}{2(x - y)}$

b. $x^2 - 3x + 2 = (x - 2)(x - 1)$

$\dfrac{x}{x - 2} = \dfrac{x(x - 1)}{(x - 2)(x - 1)} = \dfrac{x(x - 1)}{x^2 - 3x + 2}$

c. $x^2 - 9 = (x + 3)(x - 3)$

$\dfrac{2x}{x + 3} = \dfrac{2x(x - 3)}{(x + 3)(x - 3)} = \dfrac{2x(x - 3)}{x^2 - 9}$

9. a. $2 \cdot 5^{-2} = 2 \cdot \dfrac{1}{5^2} = \dfrac{2}{25}$

b. $3^0 \cdot x^{-2} = 1 \cdot \dfrac{1}{x^2} = \dfrac{1}{x^2}$

c. $4y^{-3} = 4 \cdot \dfrac{1}{y^3} = \dfrac{4}{y^3}$

10. a. 3.47×10^7 b. 8.73×10^{-4} c. 4.0×10^{-6}
    d. $48{,}300$ e. $0.000\,381$ f. $0.000\,000\,403$

CHAPTER 11 REVIEW

1. a. $\dfrac{\cancel{2}x\cancel{y^2}}{3} \cdot \dfrac{x}{\cancel{4}\cancel{y^2}} = \dfrac{x^2}{6}$

   (with 2 under 4)

   b. $\dfrac{x^2 - 2x}{5} \cdot \dfrac{25}{x^2} = \dfrac{\cancel{x}(x - 2)}{\cancel{5}} \cdot \dfrac{\overset{5}{\cancel{25}}}{\cancel{x^2}}$ (x under x^2)

   $= \dfrac{5(x - 2)}{x} = \dfrac{5x - 10}{x}$

   c. $\dfrac{x^2 - 7x + 6}{x^2 - 1} \cdot \dfrac{x + 1}{x - 6} = \dfrac{\cancel{(x - 1)}\cancel{(x - 6)}}{\cancel{(x + 1)}\cancel{(x - 1)}} \cdot \dfrac{\cancel{(x + 1)}}{\cancel{(x - 6)}} = 1$

2. a. $\dfrac{\cancel{2}\,y}{\cancel{8}\,\cancel{s}} \cdot \dfrac{\overset{7s}{\cancel{21s^2}}}{\cancel{2}\,r^2} = \dfrac{7s}{r}$ (with r under r)

   b. $\dfrac{(a + b)(a - b)}{\cancel{a}} \cdot \dfrac{\cancel{a}(a - 1)}{a(a + b)} = \dfrac{(a - b)(a - 1)}{a} = \dfrac{a^2 - a - ab + b}{a}$

   c. $\dfrac{\cancel{(2x + 1)}(x - 3)}{\cancel{x}\cancel{(2x + 1)}} \cdot \dfrac{\overset{x^3}{\cancel{x^4}}}{\cancel{(x - 3)}} = x^3$

3. a. $\dfrac{x}{5} - \dfrac{y}{5} + \dfrac{3}{5} = \dfrac{x - y + 3}{5}$

   b. $\dfrac{x + 3}{y} + \dfrac{-3}{y} = \dfrac{x + 3 + (-3)}{y} = \dfrac{x}{y}$

   c. $\dfrac{a - 2}{3} + \dfrac{-(a + 3)}{3} = \dfrac{a - 2 - a - 3}{3} = \dfrac{-5}{3}$

598

PART B: SOLUTIONS

4. a. LCD: 6
$$\frac{x}{3} + \frac{y}{6}$$
$$= \frac{2x}{6} + \frac{y}{6}$$
$$= \frac{2x + y}{6}$$

   b. LCD: $3x$
$$\frac{7}{x} - \frac{1}{3x}$$
$$= \frac{21}{3x} - \frac{1}{3x}$$
$$= \frac{21 - 1}{3x} = \frac{20}{3x}$$

   c. LCD: $2y$
$$\frac{5}{2y} - \frac{1}{y}$$
$$= \frac{5}{2y} - \frac{2}{2y}$$
$$= \frac{3}{2y}$$

5. a. LCD: $3x$
$$\frac{3}{x} - \frac{2}{3x}$$
$$= \frac{9}{3x} - \frac{2}{3x}$$
$$= \frac{7}{3x}$$

   b. LCD: $2rs$
$$\frac{3}{r} + \frac{5}{2s}$$
$$= \frac{6s}{2rs} + \frac{5r}{2rs}$$
$$= \frac{6s + 5r}{2rs}$$

   c. LCD: a^2b^2
$$\frac{2}{ab^2} - \frac{3}{a^2b}$$
$$= \frac{2a}{a^2b^2} - \frac{3b}{a^2b^2}$$
$$= \frac{2a - 3b}{a^2b^2}$$

6. a. LCD: $(a - b)(a + b)$
$$\frac{3(a + b)}{(a - b)(a + b)} + \frac{1(a + b)}{(a - b)(a + b)} = \frac{4a + 2b}{(a - b)(a + b)}$$

   b. $a^2 - 1 = (a + 1)(a - 1)$, $a^2 + a = a(a + 1)$
   LCD: $a(a + 1)(a - 1)$
$$\frac{a}{(a + 1)(a - 1)} + \frac{-1}{a(a + 1)} = \frac{a^2}{a(a + 1)(a - 1)} + \frac{-1(a - 1)}{a(a + 1)(a - 1)}$$
$$= \frac{a^2 - a + 1}{a(a + 1)(a - 1)}$$

   c. $x^2 - 25 = (x + 5)(x - 5)$, $x^2 - 4x - 5 = (x - 5)(x + 1)$
   LCD: $(x + 5)(x - 5)(x + 1)$
$$\frac{1}{(x + 5)(x - 5)} + \frac{5}{(x - 5)(x + 1)} = \frac{1(x + 1)}{(x + 5)(x - 5)(x + 1)} + \frac{5(x + 5)}{(x + 5)(x - 5)(x + 1)}$$
$$= \frac{6x + 26}{(x + 5)(x - 5)(x + 1)}$$

7. a. LCD: 18
$$\frac{\frac{3}{6}(18)}{\frac{2}{9}(18)} = \frac{9}{4}$$

   b. LCD: 6
$$\frac{(6)\frac{2}{3} + (6)\frac{1}{6}}{(6)\frac{1}{3} + (6)\frac{5}{6}}$$
$$= \frac{4 + 1}{2 + 5} = \frac{5}{7}$$

   c. LCD: 4
$$\frac{(4)1 + (4)\frac{1}{2}}{(4)3 - (4)\frac{1}{4}}$$
$$= \frac{4 + 2}{12 - 1} = \frac{6}{11}$$

8. a. LCD: b
$$\frac{(b)1 - (b)\frac{a}{b}}{(b)1 + (b)\frac{2}{b}}$$
$$= \frac{b - a}{b + 2}$$

   b. LCD: xy
$$\frac{(xy)x - (xy)\frac{x}{y}}{(xy)y - (xy)\frac{y}{x}}$$
$$= \frac{x^2y - x^2}{xy^2 - y^2}$$

   c. LCD: y
$$\frac{(y)\frac{1}{y} + 3(y)}{2(y) - (y)\frac{3}{y}}$$
$$= \frac{1 + 3y}{2y - 3}$$

599

ANSWERS AND SOLUTIONS

9. a. LCD: 6
    $$(6)\frac{x}{2} = (6)(-1) + (6)\frac{2x}{3}$$
    $$3x = -6 + 4x$$
    $$-x = -6$$
    $$x = 6$$

    b. LCD: 9
    $$(9)\frac{x}{3} + (9)\frac{7}{9} = (9)\frac{1}{3}$$
    $$3x + 7 = 3$$
    $$3x = -4$$
    $$x = \frac{-4}{3}$$

    c. LCD: 10
    $$(10)\frac{x+1}{2} = (10)\frac{2x-9}{5} + (10)3$$
    $$5(x+1) = 2(2x-9) + 30$$
    $$5x + 5 = 4x - 18 + 30$$
    $$x = 12 - 5 = 7$$

10. a. LCD: $x(x+5)$
    $$x(x+5)\frac{6}{x} = x(x+5)\frac{16}{x+5}$$
    $$(x+5)6 = 16x$$
    $$6x + 30 = 16x$$
    $$-10x = -30$$
    $$x = 3$$

    b. LCD: $2y$
    $$(2y)\frac{2+y}{y} = (2y)\frac{3}{2}$$
    $$2(2+y) = 3y$$
    $$4 + 2y = 3y$$
    $$4 = y \text{ or } y = 4$$

    c. LCD: $2y$
    $$(2y)\frac{y-2}{2y} = (2y)\frac{5}{2}$$
    $$y - 2 = 5y$$
    $$-4y = 2$$
    $$y = \frac{-1}{2}$$

11. a. LCD: $x(x+4)$
    $$x(x+4)\frac{10}{x+4} - x(x+4)\frac{6}{x} = x(x+4)\frac{-4}{x}$$
    $$10x - 6(x+4) = -4(x+4)$$
    $$10x - 6x - 24 = -4x - 16$$

    b. LCD: $x(x-1)$
    $$x(x-1)\frac{14}{x-1} + x(x-1)\frac{1}{x} = x(x-1)\frac{8}{x}$$
    $$14x + (x-1) = 8(x-1)$$
    $$14x + x - 1 = 8x - 8$$

12. a. LCD: 12
    $$(12)\frac{b}{3} = (12)\frac{2ax}{4}$$
    $$4b = 6ax$$
    $$\frac{4b}{6a} = x$$
    $$x = \frac{2b}{3a}$$

    b. LCD: 12
    $$(12)\frac{b-x}{4} - (12)\frac{b}{3} = (12)\frac{x}{2}$$
    $$3(b-x) - 4b = 6x$$
    $$3b - 3x - 4b = 6x$$
    $$-b = 9x$$
    $$\frac{-b}{9} = x \text{ or } x = \frac{-b}{9}$$

    c. LCD: $x(x-1)$
    $$x(x-1)\frac{a}{x-1} = x(x-1)\frac{2a}{x}$$
    $$ax = (x-1)2a$$
    $$ax = 2ax - 2a$$
    $$-ax = -2a$$
    $$x = 2$$

13. *Steps 1–2* The number: x
    Step 3 Not applicable.
    Step 4 $\dfrac{3x}{x+10} = \dfrac{1}{2}$
    Step 5 $2(x+10)\dfrac{3x}{x+10} = 2(x+10)\dfrac{1}{2}$
    $$6x = x + 10$$
    $$5x = 10$$
    $$x = 2$$
    Step 6 The number is 2.

14. *Steps 1–2* Number of defective parts: x
    Step 3

Defective	Total
x	276
3	92

    Step 4 $\dfrac{x}{3} = \dfrac{276}{92}$
    Step 5 $92x = 3(276)$
    $$x = 9$$
    Step 6 You would expect 9 defective parts in 276.

600

PART A: ANSWERS

15. *Steps 1–2* Rate of faster car: r
    Rate of slower car: $r - 10$

Step 3

	d	r	$t = d/r$
Faster car	90	r	$\dfrac{90}{r}$
Slower car	60	$r - 10$	$\dfrac{60}{r - 10}$

Step 4 The times are equal.

Step 5 $\dfrac{90}{r} = \dfrac{60}{r - 10}$

$90(r - 10) = 60r$
$90r - 900 = 60r$
$30r = 900$
$r = 30$

Step 6 The rates are 30 mph and $30 - 10 = 20$ mph.

CHAPTER 12 REVIEW

1. a. 0.05
   b. I and P
   c. It increases

2. $2x - 4 = y$
   or $y = 2x - 4$

3. $2y = 3x + 6$
   $y = \dfrac{3x + 6}{2}$

4. a. $y = 2(3) + 1 = 7$
      $(3, 7)$
   b. $y = 2(-2) + 1 = -3$
      $(-2, -3)$
   c. $y = 2(0) + 1 = 1$
      $(0, 1)$
   d. $y = 2\left(\dfrac{-1}{2}\right) + 1 = 0$
      $\left(\dfrac{-1}{2}, 0\right)$

5. a. $f(4) = 4 - 5 = -1$
   b. $f(-2) = -2 - 5 = -7$
   c. $f(0) = 0 - 5 = -5$
   d. $f(-6) = -6 - 5 = -11$

6. [Graph showing points $(0, 4)$, $(3, 4)$, $(-2, 3)$, $(3, -2)$ on coordinate axes]

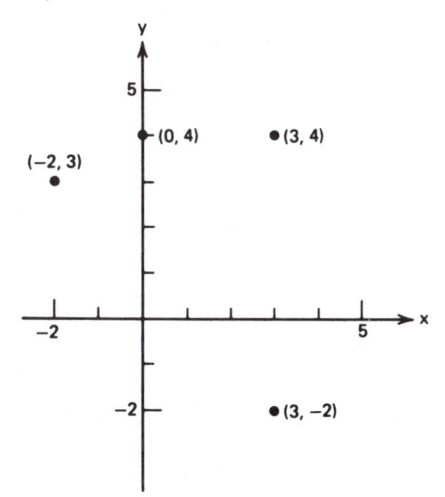

7. If $x = 0$, $y = 3$.
   If $y = 0$, $x = 3$.
   [Graph of line $x + y = 3$]

601

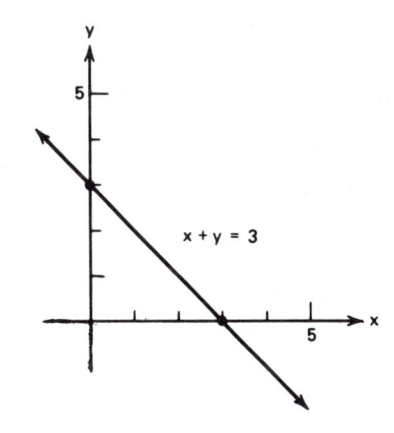

ANSWERS AND SOLUTIONS

8. Solve for x:
$2y - 4 = x$ or $x = 2y - 4$
If $y = 0$, $x = -4$.
If $y = 3$, $x = 2$.

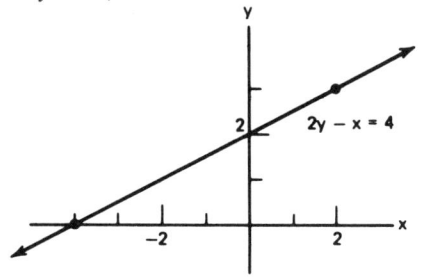

9.

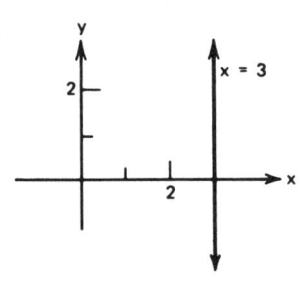

10. $3x + 2y = 6$
If $x = 0$, $y = 3$.
If $y = 0$, $x = 2$.

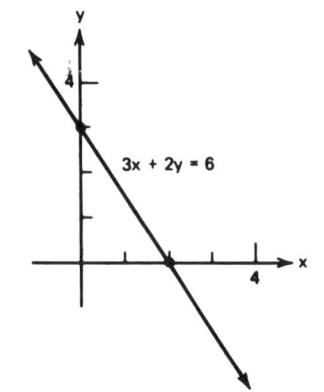

11. The graph crosses the x axis at $(8, 0)$ and the y axis at $(0, -8)$.

12. $m = \dfrac{5 - 3}{-4 - 2} = \dfrac{2}{-6} = \dfrac{-1}{3}$

Since the two slopes are different, the lines cannot be parallel.

13. $m_1 = \dfrac{4 - 3}{2 - (-1)} = \dfrac{1}{3}$

$m_2 = \dfrac{2 - (-1)}{-3 - 5} = \dfrac{3}{-8}$

14. $m = -2$; $(-2, -5)$
$y - (-5) = -2[x - (-2)]$
$y + 5 = -2(x + 2)$
$y + 5 = -2x - 4$
Ans. $y = -2x - 9$

15. $m = \dfrac{5 - 3}{1 - (-2)} = \dfrac{2}{3}$

$y - 5 = \dfrac{2}{3}(x - 1)$

$y - 5 = \dfrac{2}{3}x - \dfrac{2}{3}$

$y = \dfrac{2}{3}x - \dfrac{2}{3} + \dfrac{(5)3}{3}$

Ans. $y = \dfrac{2}{3}x + \dfrac{13}{3}$

16. $2y = 5x$
$y = \dfrac{5}{2}x + 0$
Slope: $m = \dfrac{5}{2}$
y intercept: 0

17. $y = kx$
$\dfrac{y}{x} = k$
$\dfrac{20}{6} = \dfrac{44}{x}$
$20x = 6(44)$
$x = \dfrac{6(44)}{20} = \dfrac{66}{5}$

602

PART B: SOLUTIONS

18. $2x + 3y \geq 6$
    Graph $2x + 3y = 6$.
    If $x = 0$, $y = 2$; $(0, 2)$
    If $y = 0$, $x = 3$; $(3, 0)$.
    Test $(0, 0)$.
    $2(0) + 3(0) \geq 6$ is false.
    Shade the half-plane not containing $(0, 0)$.

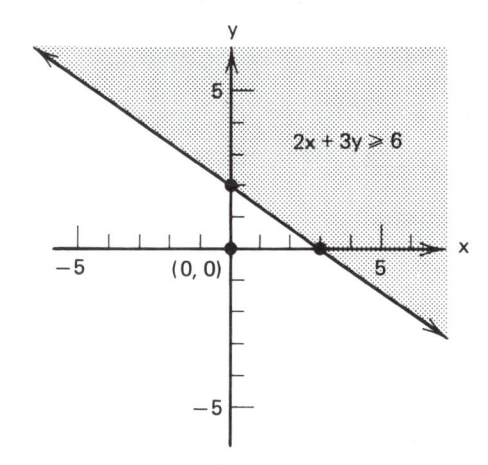

19. $x \leq -2$
    Graph $x = -2$.
    Test $(0, 0)$.
    $0 < -2$ is false.
    Shade the half-plane not containing $(0, 0)$.

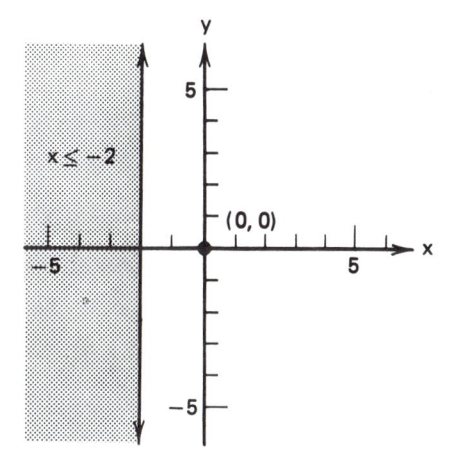

20. $y > -3x$
    Graph $y = -3x$ (use a dashed line).
    If $x = 0$, $y = 0$; $(0, 0)$.
    If $x = 1$, $y = -3$; $(1, -3)$.
    Test $(1, 1)$.
    $1 > -3(1)$ is true.
    Shade the half-plane containing $(1, 1)$.

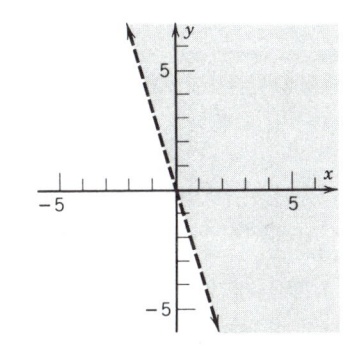

CHAPTER 13 REVIEW

1.

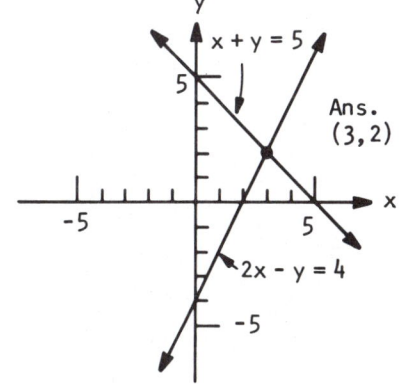

2.

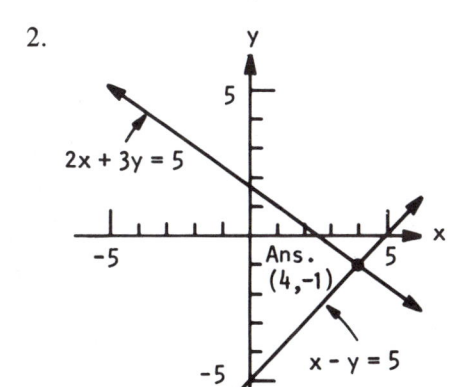

603

ANSWERS AND SOLUTIONS

3.

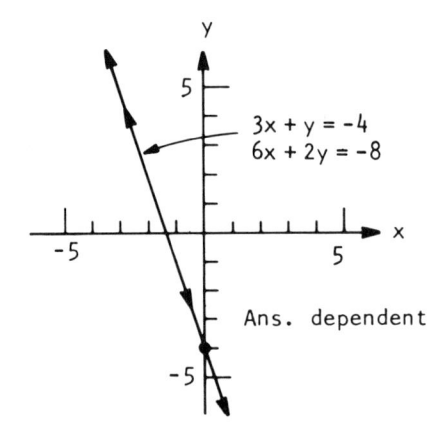

[Graph showing lines $3x + y = -4$ and $6x + 2y = -8$ coinciding. Ans. dependent]

4.

[Graph showing lines $2x = 3y - 1$ and $4x = 24 + 6y$ as parallel. Ans. inconsistent]

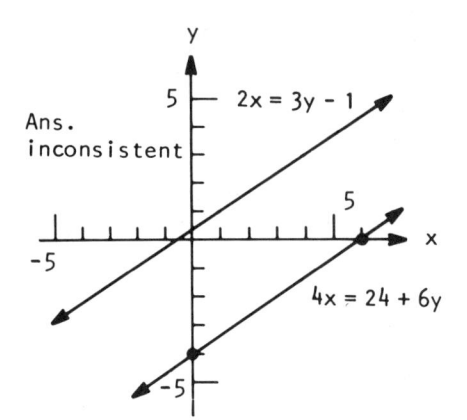

5. (1) $\quad x - y = 3$
   (2) $\quad x + y = 5$
   Add: $\overline{2x \quad\quad = 8}$
   $\quad\quad\quad x = 4$
   Substitute in (2):
   $\quad 4 + y = 5$
   $\quad\quad\quad y = 1$
   Ans. $(4, 1)$

6. (1) $\quad 2x + y = -6$
   (2) $\quad x - y = -3$
   Add: $\overline{3x \quad\quad = -9}$
   $\quad\quad\quad x = -3$
   Substitute in (2):
   $\quad -3 - y = -3$
   $\quad\quad\quad -y = 0$
   $\quad\quad\quad y = 0$
   Ans. $(-3, 0)$

7. (1) $\quad x - 3y = 1$
   (2) $\quad x + 2y = 1$
   Multiply Equation (1) by -1: $\quad -x + 3y = -1$
   $\quad\quad\quad\quad\quad\quad\quad\quad\quad\quad x + 2y = 1$
   $\quad\quad\quad\quad\quad\quad\quad\text{Add:} \quad \overline{5y = 0}$
   $\quad\quad\quad\quad\quad\quad\quad\quad\quad\quad\quad y = 0$
   Substitute in (2):
   $\quad x + 2(0) = 1$
   $\quad\quad\quad\quad x = 1$
   Ans. $(1, 0)$

8. (1) $\quad 4 = 2x - y$
   (2) $\quad -y = x + 7$
   Standard form: (1') $\quad 2x - y = 4$
   $\quad\quad\quad\quad\quad\quad$ (2') $\quad -x - y = 7$
   Multiply Equation (2') by -1: $\quad 2x - y = 4$
   $\quad\quad\quad\quad\quad\quad\quad\quad\quad\quad\quad x + y = -7$
   $\quad\quad\quad\quad\quad\quad\quad\text{Add:} \quad \overline{3x \quad\quad = -3}$
   $\quad\quad\quad\quad\quad\quad\quad\quad\quad\quad\quad x = -1$
   Substitute in (2):
   $\quad -y = -1 + 7 = 6$
   $\quad\quad y = -6$
   Ans. $(-1, -6)$

9. (1) $\quad 4y - 3x = -5$
   (2) $\quad x - 7 = -20y$
   Standard form: (1') $\quad -3x + 4y = -5$
   $\quad\quad\quad\quad\quad\quad$ (2') $\quad x + 20y = 7$
   Multiply Equation (2') by 3: $\quad -3x + 4y = -5$
   $\quad\quad\quad\quad\quad\quad\quad\quad\quad\quad\quad\quad 3x + 60y = 21$
   $\quad\quad\quad\quad\quad\quad\quad\text{Add:} \quad \overline{64y = 16}$
   $\quad\quad\quad\quad\quad\quad\quad\quad\quad\quad y = \frac{16}{64} = \frac{1}{4}$
   Substitute in (2):
   $\quad x - 7 = -\overset{5}{\cancel{20}}\left(\frac{1}{\cancel{4}}\right)$
   $\quad x - 7 = -5$
   $\quad\quad\quad x = 2$
   Ans. $\left(2, \frac{1}{4}\right)$

10. (1) $\quad 8y = -3x - 1$
    (2) $\quad 6 + 2x = -8y$
    Standard form: (1') $\quad 3x + 8y = -1$
    $\quad\quad\quad\quad\quad\quad$ (2') $\quad 2x + 8y = -6$
    Multiply Equation (2') by -1: $\quad 3x + 8y = -1$
    $\quad\quad\quad\quad\quad\quad\quad\quad\quad\quad\quad\quad -2x - 8y = 6$
    $\quad\quad\quad\quad\quad\quad\quad\text{Add:} \quad \overline{x \quad\quad = 5}$
    Substitute in (1):
    $\quad 8y = -3(5) - 1 = -16$
    $\quad\quad y = -2$
    Ans. $(5, -2)$

604

PART B: SOLUTIONS

11. $$\frac{y}{6} = \frac{-x}{2}$$
$7y + 2x = 19$
Multiply by LCD:
$(\cancel{6})\frac{y}{\cancel{6}} = (\cancel{6})^3\frac{-x}{\cancel{2}}$ or (1) $\quad y = -3x$
(2) $\quad 7y + 2x = 19$
Standard form: (1') $\quad 3x + y = 0$
(2') $\quad 2x + 7y = 19$
Multiply Equation (1') by -2 and Equation (2') by 3:
$-6x - 2y = 0$
$6x + 21y = 57$
Add: $\overline{\quad 19y = 57}$
$y = 3$
Substitute in (1'):
$3x + 3 = 0$
$3x = -3$
$x = -1$
Ans. $(-1, 3)$

12. $$\frac{2x}{3} + \frac{y}{2} = \frac{-19}{6}$$
$$\frac{x+3}{4} = \frac{-y}{10}$$
Multiply by LCD's:
$(\cancel{6})^2\frac{2x}{\cancel{3}} + (\cancel{6})^3\frac{y}{\cancel{2}} = (\cancel{6})\frac{-19}{\cancel{6}}$ or $\quad 4x + 3y = -19$
$(\cancel{20})^5\frac{x+3}{\cancel{4}} = (\cancel{20})^2\frac{-y}{\cancel{10}}$ or $\quad 5(x+3) = -2y$
and finally: (1) $\quad 4x + 3y = -19$
(2) $\quad 5x + 2y = -15$
Multiply Equation (1) by -2 and Equation (2) by 3:
$-8x - 6y = 38$
$15x + 6y = -45$
Add: $\overline{\quad 7x \quad = -7}$
$x = -1$
Substitute in (1):
$4(-1) + 3y = -19$
$3y = -15$
$y = -5$
Ans. $(-1, -5)$

13. (1) $\quad y = 3x - 5$
(2) $\quad y - x = -1$
Substitute $3x - 5$ for y in Equation (2):
$(3x - 5) - x = -1$
$2x = 4$
$x = 2$
Substitute 2 for x in Equation (1):
$y = 3(2) - 5 = 1$
Ans. $(2, 1)$

14. (1) $\quad y = 2x + 1$
(2) $\quad 2x + 3y = -21$
Substitute $2x + 1$ for y in Equation (2):
$2x + 3(2x + 1) = -21$
$2x + 6x + 3 = -21$
$8x = -24$
$x = -3$
Substitute -3 for x in Equation (1):
$y = 2(-3) + 1 = -5$
Ans. $(-3, -5)$

15. (1) $\quad x - y = 6$
(2) $\quad 3x - 4y = 16$
From (1):
(1') $\quad x = y + 6$
Substitute $y + 6$ for x in Equation (2):
$3(y + 6) - 4y = 16$
$3y + 18 - 4y = 16$
$-y = -2$
$y = 2$
From (1'):
$x = 2 + 6 = 8$
Ans. $(8, 2)$

16. (1) $\quad 4x + y = 5$
(2) $\quad 8x - 2y = -2$
From (1):
(1') $\quad y = -4x + 5$
Substitute $-4x + 5$ for y in Equation (2):
$8x - 2(-4x + 5) = -2$
$8x + 8x - 10 = -2$
$16x = 8$
$x = \frac{1}{2}$
From (1'):
$y = -\cancel{4}^2\left(\frac{1}{\cancel{2}}\right) + 5 = 3$
Ans. $\left(\frac{1}{2}, 3\right)$

605

ANSWERS AND SOLUTIONS

17. *Steps 1–2* Weight of larger package: x
    Weight of smaller package: y
    Step 3 Not applicable.
    Step 4 (1) $x + y = 84$
    (2) $x - y = 20$
    Step 5 Add: $\overline{2x = 104}$
    $x = 52$
    From (1):
    $y = 32$
    Step 6 The packages weigh 52 kg and 32 kg.

18. *Steps 1–2* Number of dimes: d
    Number of quarters: q
    Step 3 Not applicable.
    Step 4 (1) $d + q = 25$
    (2) $10d + 25q = 355$
    (1′) $-10d - 10q = -250$
    Step 5 Add: $\overline{15q = 105}$
    $q = 7$
    From (1): $d = 18$
    Step 6 There are 18 dimes and 7 quarters.

19. *Step 1–2* One number: x
    Other number: y
    Step 3 Not applicable.
    Step 4 (1) $x + y = -40$
    (2) $x - y = -8$
    Step 5 Add: $\overline{2x = -48}$
    $x = -24$
    From (1):
    $y = -16$
    Step 6 The numbers are -24 and -16.

20. *Steps 1–2* One number: x
    Other number: y
    Step 3 Not applicable.
    Step 4 (1) $x = 4y + 8$
    (2) $x + y = -2$
    Step 5 Substitute in (2):
    $(4y + 8) + y = -2$
    $5y + 8 = -2$
    $5y = -10$
    $y = -2$
    From (2): $x = 0$
    Step 6 The numbers are 0 and -2.

CHAPTER 14 REVIEW

1. a. $x - 2 = 0$; $x + 5 = 0$
   $x = 2$ $x = -5$

   b. $y = 0$; $y + 3 = 0$
   $y = -3$

2. a. $x(x - 2) = 0$
   $x = 0$; $x - 2 = 0$
   $x = 2$

   b. $3x^2 - 6x = 0$
   $3x(x - 2) = 0$
   $x = 0$; $x - 2 = 0$
   $x = 2$

3. a. $(x + 7)(x - 7) = 0$
   $x + 7 = 0$; $x - 7 = 0$
   $x = -7$ $x = 7$

   b. $3y^2 - 27 = 0$
   $3(y^2 - 9) = 0$
   $3(y + 3)(y - 3) = 0$
   $y + 3 = 0$; $y - 3 = 0$
   $y = -3$ $y = 3$

4. a. $(y - 5)(y + 1) = 0$
   $y - 5 = 0$; $y + 1 = 0$
   $y = 5$ $y = -1$

   b. $(y - 9)(y + 2) = 0$
   $y - 9 = 0$; $y + 2 = 0$
   $y = 9$ $y = -2$

5. a. $x^2 - 2x - 3 = 0$
   $(x - 3)(x + 1) = 0$
   $x - 3 = 0$; $x + 1 = 0$
   $x = 3$ $x = -1$

   b. $x^2 - 4x + 4 = 0$
   $(x - 2)(x - 2) = 0$
   $x - 2 = 0$
   $x = 2$

6. a. $2x^2 + 5x - 12 = 0$
   $(2x - 3)(x + 4) = 0$
   $2x - 3 = 0$; $x + 4 = 0$
   $x = \dfrac{3}{2}$ $x = -4$

   b. $2x^2 + 5x - 3 = 0$
   $(2x - 1)(x + 3) = 0$
   $2x - 1 = 0$; $x + 3 = 0$
   $x = \dfrac{1}{2}$ $x = -3$

PART B: SOLUTIONS

7. a. $3x^2 + 5x - 2 = 0$
$(3x - 1)(x + 2) = 0$
$3x - 1 = 0; \quad x + 2 = 0$
$x = \dfrac{1}{3} \quad\quad x = -2$

   b. $3x^2 - 14x + 8 = 0$
$(3x - 2)(x - 4) = 0$
$3x - 2 = 0; \quad x - 4 = 0$
$x = \dfrac{2}{3} \quad\quad x = 4$

8. a. $y^2 - 6y = 16$
$y^2 - 6y - 16 = 0$
$(y - 8)(y + 2) = 0$
$y - 8 = 0; \quad y + 2 = 0$
$y = 8; \quad\quad y = -2$

   b. $x^2 + 2x = 8$
$x^2 + 2x - 8 = 0$
$(x + 4)(x - 2) = 0$
$x + 4 = 0; \quad x - 2 = 0$
$x = -4 \quad\quad x = 2$

9. a. $x^2 - 3x - 40 = -36$
$x^2 - 3x - 4 = 0$
$(x - 4)(x + 1) = 0$
$x - 4 = 0; \quad x + 1 = 0$
$x = 4 \quad\quad x = -1$

   b. $x^2 - x - 2 = 4$
$x^2 - x - 6 = 0$
$(x - 3)(x + 2) = 0$
$x - 3 = 0; \quad x + 2 = 0$
$x = 3 \quad\quad x = -2$

10. a. $(4)\dfrac{x^2}{4} - (4)9 = (4)0$
$x^2 - 36 = 0$
$(x + 6)(x - 6) = 0$
$x + 6 = 0; \quad x - 6 = 0$
$x = -6 \quad\quad x = 6$

    b. $(25)3x^2 - (25)\dfrac{27}{25} = 0$
$75x^2 - 27 = 0$
$3(25x^2 - 9) = 0$
$3(5x + 3)(5x - 3) = 0$
$5x + 3 = 0; \quad 5x - 3 = 0$
$x = \dfrac{-3}{5} \quad\quad x = \dfrac{3}{5}$

11. a. $(6)\dfrac{y^2}{6} + (6)\dfrac{y}{6} - (6)2 = 0$
$y^2 + y - 12 = 0$
$(y + 4)(y - 3) = 0$
$y + 4 = 0; \quad y - 3 = 0$
$y = -4 \quad\quad y = 3$

    b. $(16)\dfrac{y}{2} - 16 = (16)\dfrac{y^2}{16}$
$8y - 16 = y^2$
$0 = y^2 - 8y + 16$
$0 = (y - 4)(y - 4)$
$y - 4 = 0; \quad y = 4$

12. a. $(y^2)\dfrac{15}{y^2} + (y^2)\dfrac{2}{y} = y^2$
$15 + 2y = y^2$
$-y^2 + 2y + 15 = 0$
$y^2 - 2y - 15 = 0$
$(y - 5)(y + 3) = 0$
$y - 5 = 0; \quad y + 3 = 0$
$y = 5 \quad\quad y = -3$

    b. $6x(x + 2)\left(\dfrac{2}{x}\right) - 6x(x + 2)\left(\dfrac{1}{6}\right) = 6x(x + 2)\dfrac{2}{x + 2}$
$12(x + 2) - x(x + 2) = 12x$
$12x + 24 - x^2 - 2x = 12x$
$-x^2 - 2x + 24 = 0$
$x^2 + 2x - 24 = 0$
$(x + 6)(x - 4) = 0$
$x + 6 = 0; \quad x - 4 = 0$
$x = -6 \quad\quad x = 4$

13. *Steps 1–2* Larger number: x
Smaller number: $x - 7$
Step 3 Not applicable.
Step 4 $x(x - 7) = 60$
Step 5 $x^2 - 7x - 60 = 0$
$(x + 5)(x - 12) = 0$
$x = -5$ (not positive), $x = 12$
Step 6 The numbers are 12 and $12 - 7 = 5$.

14. *Steps 1–2* One number: n
Other number: $11 - n$
Step 3 Not applicable.
Step 4 $n(11 - n) = 30$
Step 5 $-n^2 + 11n - 30 = 0$
$n^2 - 11n + 30 = 0$
$(n - 5)(n - 6) = 0$
$n = 5 \quad\quad n = 6$
Step 6 If $n = 5$, $11 - 5 = 6$;
if $n = 6$, $11 - 6 = 5$.
In either case, the numbers are 5 and 6.

607

ANSWERS AND SOLUTIONS

15. *Steps 1–2*   First integer: x
    Next consecutive integer: $x + 1$
    Step 3   Not applicable.
    Step 4
    $$\frac{1}{x} + \frac{1}{x+1} = \frac{11}{30}$$
    Step 5
    $$30\cancel{x}(x+1)\frac{1}{\cancel{x}} + 30x\cancel{(x+1)}\frac{1}{\cancel{x+1}} = \cancel{30}x(x+1)\frac{11}{\cancel{30}}$$
    $$30(x+1) + 30x = x(x+1)(11)$$
    $$60x + 30 = 11x^2 + 11x$$
    $$0 = 11x^2 - 49x - 30$$
    $$0 = (11x + 6)(x - 5)$$
    $$x = \frac{-6}{11} \text{ (not an integer)}, \quad x = 5$$
    Step 6   The integers are 5 and 5 + 1 = 6.

CHAPTER 15 REVIEW

1. $\sqrt{6}$ and $-\sqrt{\frac{2}{3}}$, while $-\sqrt{9} = -3$, $\sqrt{4} = 2$, and $\sqrt{\frac{9}{25}} = \frac{3}{5}$.

2. a. 9.644
   b. $2(6.856) = 13.712$
   c. $\frac{1}{4}(5.657) = 1.414$

3. a.
   b.

4. a. $\sqrt{3 \cdot 3 \cdot 2 \cdot 2 \cdot 2}$
      $= 3 \cdot 2\sqrt{2}$
      $= 6\sqrt{2}$
   b. $-\sqrt{3 \cdot 3 \cdot 2 \cdot 5} = -3\sqrt{10}$
   c. $\sqrt{5 \cdot 5 \cdot 7} = 5\sqrt{7}$

5. a. $\sqrt{y^4} = y^{4 \div 2}$
      $= y^2$
   b. $\sqrt{x^{15}} = \sqrt{x^{14}}\sqrt{x}$
      $= x^7\sqrt{x}$
   c. $\sqrt{x^3 y^7} = \sqrt{x^2 y^6}\sqrt{xy}$
      $= xy^3\sqrt{xy}$

6. a. $x\sqrt{27} = x\sqrt{3 \cdot 3 \cdot 3}$
      $= 3x\sqrt{3}$
   b. $3\sqrt{3x^2} = 3\sqrt{3xx}$
      $= 3x\sqrt{3}$
   c. $\frac{1}{3}\sqrt{27x^2} = \frac{1}{3}\sqrt{3 \cdot 3 \cdot 3xx}$
      $= \frac{1}{\cancel{3}} \cdot \cancel{3}x\sqrt{3}$
      $= x\sqrt{3}$

7. a. $\sqrt{7} + 3\sqrt{7} = 4\sqrt{7}$
   b. $4\sqrt{6} - 3\sqrt{6} + \sqrt{6} = 2\sqrt{6}$
   c. $7\sqrt{2} - 3\sqrt{2} + \sqrt{2} = 5\sqrt{2}$

8. a. $2\sqrt{12} - 4\sqrt{27} = 2\sqrt{2 \cdot 2 \cdot 3} - 4\sqrt{3 \cdot 3 \cdot 3}$
      $= 4\sqrt{3} - 12\sqrt{3}$
      $= -8\sqrt{3}$
   b. $\sqrt{9x} - \sqrt{4x} = \sqrt{3 \cdot 3x} - \sqrt{2 \cdot 2x}$
      $= 3\sqrt{x} - 2\sqrt{x}$
      $= \sqrt{x}$
   c. $\sqrt{x^2 y} - 3\sqrt{4x^2 y} + 7x\sqrt{y} = \sqrt{xxy} - 3\sqrt{2 \cdot 2xxy} + 7x\sqrt{y}$
      $= x\sqrt{y} - 6x\sqrt{y} + 7x\sqrt{y}$
      $= 2x\sqrt{y}$

9. a. $3\sqrt{y} - 18$
   b. $y\sqrt{xy} - 2y^2$
   c. $\sqrt{6} - \sqrt{18} = \sqrt{6} - \sqrt{3 \cdot 3 \cdot 2}$
      $= \sqrt{6} - 3\sqrt{2}$

10. a. $2 - 2\sqrt{2} = 2(1 - \sqrt{2})$
    b. $3\sqrt{3} - 3\sqrt{5} = 3(\sqrt{3} - \sqrt{5})$
    c. $3x^2 - x^2\sqrt{5} = x^2(3 - \sqrt{5})$

PART B: SOLUTIONS

11. a. $\dfrac{3-6\sqrt{3}}{3} = \dfrac{\cancel{3}(1-2\sqrt{3})}{\cancel{3}}$
$= 1 - 2\sqrt{3}$

b. $\dfrac{6-6\sqrt{3}}{3} = \dfrac{\overset{2}{\cancel{6}}(1-3)}{\cancel{3}}$
$= 2(1-\sqrt{3})$

c. $\dfrac{4-4\sqrt{2}}{4} = \dfrac{\cancel{4}(1-\sqrt{2})}{\cancel{4}}$
$= 1 - \sqrt{2}$

12. a. $\dfrac{3-\sqrt{3}}{2}$

b. $\dfrac{\sqrt{15}+2\sqrt{15}}{5} = \dfrac{3\sqrt{15}}{5}$

c. $\dfrac{4}{6} - \dfrac{\sqrt{3}}{6} = \dfrac{4-\sqrt{3}}{6}$

13. a. $\dfrac{2\sqrt{5}}{6} - \dfrac{9}{6} = \dfrac{2\sqrt{5}-9}{6}$

b. $\dfrac{2\sqrt{3}}{5} - \dfrac{5}{5} = \dfrac{2\sqrt{3}-5}{5}$

c. $\dfrac{3\sqrt{6}}{4} + \dfrac{8}{4} = \dfrac{3\sqrt{6}+8}{4}$

14. a. $\sqrt{15}$

b. $(4)(4\sqrt{2}) = 16\sqrt{2}$

c. $(x\sqrt{3})\sqrt{6x} = x\sqrt{18x}$
$= x\sqrt{3 \cdot 3 \cdot 2x}$
$= 3x\sqrt{2x}$

15. a. $\sqrt{5 \cdot 2 \cdot 15}$
$= \sqrt{150}$
$= \sqrt{5 \cdot 5 \cdot 2 \cdot 3}$
$= 5\sqrt{6}$

b. $2\sqrt{6 \cdot 3 \cdot 2} = 2\sqrt{36}$
$= 2(6)$
$= 12$

c. $(x\sqrt{3})(2\sqrt{x}) = 2x\sqrt{3x}$

16. a. $2\sqrt{3} - \sqrt{6}$

b. $\sqrt{14} - \sqrt{98} = \sqrt{14} - 7\sqrt{2}$

c. $\sqrt{16} - \sqrt{24} = 4 - 2\sqrt{6}$

17. a. $4 + 2\sqrt{3} - 2\sqrt{3} - 3 = 1$

b. $5 + \sqrt{35} - \sqrt{35} - 7 = -2$

c. $6 - 4\sqrt{3} - 3\sqrt{3} + 6 = 12 - 7\sqrt{3}$

18. a. $\sqrt{\dfrac{27}{3}} = \sqrt{9}$

b. $\sqrt{\dfrac{12a}{2}} = \sqrt{6a}$

c. $\sqrt{\dfrac{3a \cdot ab}{b}} = \sqrt{3aa}$
$= a\sqrt{3}$

19. a. $\sqrt{\dfrac{3}{5}} = \sqrt{\dfrac{3 \cdot 5}{5 \cdot 5}}$
$= \dfrac{\sqrt{15}}{5}$

b. $\sqrt{\dfrac{12}{5}} = \sqrt{\dfrac{12 \cdot 5}{5 \cdot 5}}$
$= \dfrac{\sqrt{60}}{5}$
$= \dfrac{2\sqrt{15}}{5}$

c. $\sqrt{\dfrac{3}{2x}} = \sqrt{\dfrac{3 \cdot 2x}{2x \cdot 2x}}$
$= \dfrac{\sqrt{6x}}{2x}$

20. a. $\dfrac{\sqrt{85}}{\sqrt{5x}} = \sqrt{\dfrac{85}{5x}}$
$= \sqrt{\dfrac{17}{x}}$
$= \dfrac{\sqrt{17 \cdot x}}{\sqrt{x \cdot x}}$
$= \dfrac{\sqrt{17x}}{x}$

b. $\dfrac{\sqrt{3y}}{\sqrt{2xx}} = \dfrac{\sqrt{3y}}{x\sqrt{2}}$
$= \dfrac{\sqrt{3y}}{x\sqrt{2}} \dfrac{\sqrt{2}}{\sqrt{2}}$
$= \dfrac{\sqrt{6y}}{2x}$

c. $\dfrac{4\sqrt{7x}}{\sqrt{2x}} \dfrac{\sqrt{2x}}{\sqrt{2x}} = \dfrac{4\sqrt{14xx}}{2x}$
$= \dfrac{\overset{2}{\cancel{4}}\,\cancel{x}\sqrt{14}}{\cancel{2}\,\cancel{x}}$
$= 2\sqrt{14}$

609

ANSWERS AND SOLUTIONS

CHAPTER 16 REVIEW

1. a. $x^2 = 25$
    $x = \pm\sqrt{25}$
    $x = 5; \quad x = -5$

   b. $3x^2 = 27$
    $x^2 = 9$
    $x = \pm\sqrt{9}$
    $x = 3; \quad x = -3$

2. a. $6x^2 = 42$
    $x^2 = 7$
    $x = -\sqrt{7}; \quad x = \sqrt{7}$

   b. $2y^2 = 12$
    $y^2 = 6$
    $y = -\sqrt{6}; \quad y = \sqrt{6}$

3. a. $(z - 2)^2 = 9$
    $z - 2 = \pm 3$
    $z = 2 \pm 3$
    $z = 5; \quad z = -1$

   b. $(z + 3)^2 = 1$
    $z + 3 = \pm 1$
    $z = -3 \pm 1$
    $z = -2; \quad z = -4$

4. a. $(x - 7)^2 = 16$
    $x - 7 = \pm 4$
    $x = 7 \pm 4$
    $x = 11; \quad x = 3$

   b. $(x - a)^2 = c^2$
    $x - a = \pm c$
    $x = a \pm c$
    $x = a + c; \quad x = a - c$

5. a. $(x + 3)^2 = a$
    $x + 3 = \pm\sqrt{a}$
    $x = -3 \pm \sqrt{a}$
    $x = -3 + \sqrt{a}; \quad x = -3 - \sqrt{a}$

   b. $(y + a)^2 = 4$
    $y + a = \pm 2$
    $y = -a \pm 2$
    $y = -a + 2; \quad y = -a - 2$

6. a. $x^2 + 3x - 4 = 0$
    $(x + 4)(x - 1) = 0$
    $x = -4; \quad x = 1$

   b. $y^2 - 3y - 3 = 0$
    $a = 1, \quad b = -3, \quad c = -3$
    $y = \dfrac{-(-3) \pm \sqrt{(-3)^2 - 4(1)(-3)}}{2(1)} = \dfrac{3 \pm \sqrt{21}}{2}$
    $y = \dfrac{3 + \sqrt{21}}{2}; \quad y = \dfrac{3 - \sqrt{21}}{2}$

7. a. $2x^2 + 4x + 2 = 0$
    $2(x^2 + 2x + 1) = 0$
    $2(x + 1)(x + 1) = 0$
    $x = -1$

   b. $y^2 - y - 2 = 0$
    $(y - 2)(y + 1) = 0$
    $y = 2; \quad y = -1$

8. $y^2 - 4y + 4 = 5 + 4$
   $(y - 2)^2 = 9$
   $y - 2 = \pm 3$
   $y = 2 \pm 3$
   $y = 5; \quad y = -1$

9. $(\cancel{x})\dfrac{x^2}{\cancel{x}} + (\cancel{x})\dfrac{2x}{\cancel{x}} - (\cancel{x})\dfrac{15}{\cancel{x}} = 0$
   $x^2 + 2x - 15 = 0$
   $a = 1, \quad b = 2, \quad c = -15$
   $x = \dfrac{-2 \pm \sqrt{2^2 - 4(1)(-15)}}{2(1)} = \dfrac{-2 \pm \sqrt{64}}{2}$
   $= \dfrac{-2 \pm 8}{2}$
   $x = 3; \quad x = -5$

10. $x^2 + 3x + 1 = 0$
    $a = 1, \quad b = 3, \quad c = 1$
    $x = \dfrac{-3 \pm \sqrt{3^2 - 4(1)(1)}}{2(1)}$
    $= \dfrac{-3 \pm \sqrt{5}}{2}$
    $x = \dfrac{-3 + \sqrt{5}}{2}; \quad x = \dfrac{-3 - \sqrt{5}}{2}$

PART B: SOLUTIONS

11. $(\cancel{12})^3 \dfrac{x^2}{\cancel{4}} - (\cancel{12})\dfrac{13}{\cancel{12}}x + (12)1 = 0$

    $3x^2 - 13x + 12 = 0$

    $a = 3, \quad b = -13, \quad c = 12$

    $x = \dfrac{-(-13) \pm \sqrt{(-13)^2 - 4(3)(12)}}{2(3)}$

    $= \dfrac{13 \pm \sqrt{25}}{6} = \dfrac{13 \pm 5}{6}$

    $x = 3; \quad x = \dfrac{4}{3}$

12. $y = x^2 + 3x$

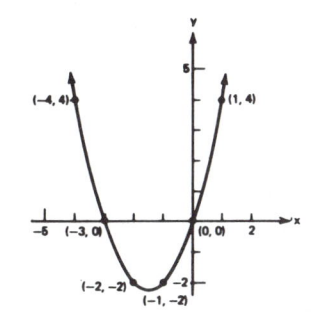

    Points: $(-4, 4)$, $(1, 4)$, $(-3, 0)$, $(0, 0)$, $(-2, -2)$, $(-1, -2)$

13. $y = x^2 + 3x - 4$

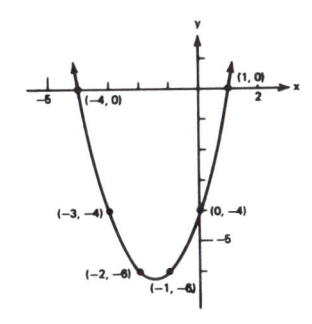

    Points: $(-4, 0)$, $(1, 0)$, $(-3, -4)$, $(0, -4)$, $(-2, -6)$, $(-1, -6)$

14. $y = x^2 - x - 6$

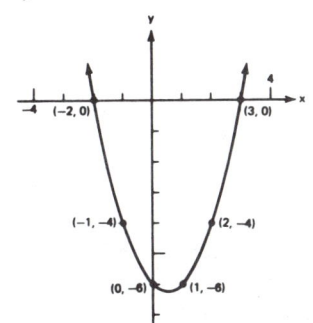

    Points: $(-2, 0)$, $(3, 0)$, $(-1, -4)$, $(2, -4)$, $(0, -6)$, $(1, -6)$

15. y is 0 when $x = 3$ or $x = -2$.

CHAPTER 17 REVIEW

1. a. $30' = \dfrac{30}{60}(1)°$
       $= 0.5°$
       Ans. $32.5°$

    b. $42' = \dfrac{42}{60}(1)°$
       $= 0.7°$
       Ans. 0.7

2. a. $0.75° = (.75)(60')$
       $= 45'$
       Ans. $4°45'$

    b. $0.2° = (.2)(60')$
       $= 12'$
       Ans. $= 83°12'$

3. a. Compliment is $90° - 47°21' = 42°39'$.

    b. Supplement is $180° - 47°21' = 132°39'$.

4. $\angle 2 = 90 - 35 = 55°$

5. $\angle 3 = \angle 1 = 35°$

6. $\angle 4 = 180 - (\angle 3)$
    $= 180 - 35 = 145°$

7. $\angle A = \angle C = 50;$
    $\angle A + \angle C + \angle ABC = 180;$
    $50 + 50 + \angle ABC = 180$
    $\angle ABC = 80°$

8. $DC^2 + BD^2 = BC^2$
    $DC^2 + 5^2 = 6^2$
    $DC^2 = 36 - 25 = 11$
    $DC = \sqrt{11}$ or $DC = -\sqrt{11}$
    Only meaningful answer $\sqrt{11}$

611

ANSWERS AND SOLUTIONS

9. a. Area = $\frac{1}{2}(AC)(BD)$, AC is twice DC; thus $AC = 2\sqrt{11}$.
Area = $\frac{1}{2}(2\sqrt{11})(5) = 5\sqrt{11}$

   b. Perimeter = $AB + BC + AC$;
$AB = BC = 6$
Perimeter = $6 + 6 + 2\sqrt{11}$
$= 12 + 2\sqrt{11}$

10.

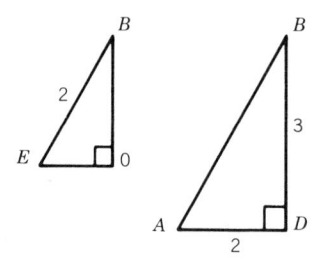

   a. By the Pythagorean theorem
$AB^2 = 2^2 + 3^2 = 13$
$AB = \sqrt{13}$

   b. $\frac{EO}{AD} = \frac{BE}{AB}$; $\frac{EO}{2} = \frac{2}{\sqrt{13}}$
$EO = \frac{4}{\sqrt{13}} \frac{\sqrt{13}}{\sqrt{13}} = \frac{4\sqrt{13}}{13}$

   c. $\frac{BO}{BD} = \frac{BE}{BA}$; $\frac{BO}{3} = \frac{2}{\sqrt{13}}$
$BO = \frac{6}{\sqrt{13}} \frac{\sqrt{13}}{\sqrt{13}} = \frac{6\sqrt{13}}{13}$

11. a. $\sin \angle DBC = \frac{DC}{BC} = \frac{4}{6} = \frac{2}{3}$

   b. $\cos \angle C = \frac{DC}{BC} = \frac{4}{6} = \frac{2}{3}$

   c. $\tan \angle DBC = \frac{DC}{BD}$; $BD^2 + DC^2 = BC^2$; $BD^2 = 36 - 16 = 20$
$BD^2 + 4^2 = 6^2$ $BD = \sqrt{20} = 2\sqrt{5}$
$\tan \angle DBC = \frac{4}{2\sqrt{5}} = \frac{2}{\sqrt{5}} \frac{\sqrt{5}}{\sqrt{5}} = \frac{2\sqrt{5}}{5}$

12. a. $\cos 81° = 0.1564$

   b. $\tan 29° = 0.5543$

13. a. $\angle A = 63°$

   b. $\angle B = 11°$

14.

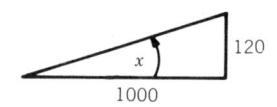

Angle: x
$\tan x = \frac{120}{1000} = 0.12$; $x = 7°$

15.

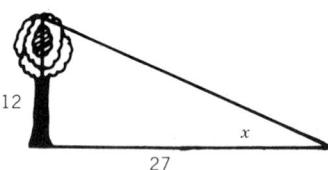

Angle: x
$\tan x = \frac{12}{27} = 0.4444$; $x = 24°$

16.

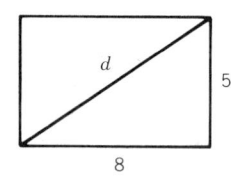

Diagonal: d
$d^2 = 8^2 + 5^2 = 64 + 25$
$= 89$
$d = -\sqrt{89}$; $d = \sqrt{89}$ (only meaningful answer)
Ans. The diagonal is $\sqrt{89}$ meters.

17.

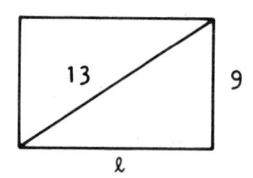

Length: ℓ
$13^2 = \ell^2 + 9^2$
$169 - 81 = \ell^2$
$88 = \ell^2$
$\ell = -\sqrt{88}$, $\ell = \sqrt{88}$ (only meaningful answer)

PART B: SOLUTIONS

18. Diameter: d

    $C = \pi d$

    $37.68 = 3.14d$; $\dfrac{37.68}{3.14} = d$; $d = 12$

    Ans. The diameter is 12 centimeters.

20. $V = \dfrac{4}{3}\pi r^3$; $r = \dfrac{1}{2}(5) = 2.5$

    $V = \dfrac{4}{3}(3.14)(2.5)^3 = 65.4$

    Ans. It will hold 65.4 cubic centimeters.

19. $V = \pi r^2 h$

    $400 = (3.14)(4^2)h$; $h = \dfrac{400}{(3.14)(4^2)}$

    $ = 7.96$

    Ans. Height is 7.96 meters.

613

INDEX

Abscissa, 391
Absolute value, 198
Abstract numbers, 155
Addition, associative law of, 203
  commutative law of, 12, 203
  of decimal numbers, 13
  of denominate numbers, 155
  of fractions, 101
  of irrational numbers, 464
  of mixed numbers, 104
  of signed numbers, 201
Addition, subtraction rule, 26, 248
Algebraic expressions, 219, 220
  equivalent, 222
  evaluation of, 219
Angles, 507
  acute, 509
  adjacent, 508
  alternate exterior, 513
  alternate interior, 513
  bisector of, 508
  central, 532
  complementary, 509
  corresponding, 513
  degree of, 508
  inscribed, 532
  obtuse, 509
  right, 509
  straight, 509
  supplementary, 509
  vertical, 512
Area, of a circle, 532
  of a parallelogram, 528
  of a rectangle, 528
  of a square, 528
  of a trapezoid, 528
  of a triangle, 519
Associative law, of addition, 203
  of multiplication, 209
Average, 176
Axes, coordinate, 390

Base of a power, 48, 219
Binomials, 220
  products of, 285, 295
Bisector, 514
Building factor, 93
Building fractions, 93, 335

Cartesian coordinates, 391
Circle, 532
  arc of, 532
  area of, 532
  center of, 532
  central angle of, 532
  chord of, 532
  circumference of, 532

INDEX

Circle (*Continued*)
   diameter of, 532
   inscribed angle of, 532
   radius of, 532
   secant of, 532
   semicircle, 532
   tangent of, 532
Coefficient, 221
   numerical, 221
Combining like terms, 229
Combining radicals, 464
Commutative law, of addition, 203
   of multiplication, 40, 209
Comparison shopping, 174
Completely factored form, 88, 291
Completing the square, 486
Complex fraction, 366
Components of an ordered pair, 384
Composite numbers, 88
Constant, 220
   of variation, 78
Conversions, in metric system, 157
   in United States system, 157
Coordinates, Cartesian or rectangular, 391
Cross, multiplication rule, 78

Decimal equivalents, 127
Decimal numbers, 4
   rounding-off, 8
Decimal-percent equivalents, 133
Degree, 508
   of polynomial, 221
Denominate numbers, 155
   addition and subtraction, 161
   multiplication and division, 162
Denominator, 87
   lowest common, 97
   rationalization of a, 494
Dependent equations, 384
Dependent variable, 384
Diagonal of a quadrilateral, 528
Difference, of integers, 204
   of fractions, 110
   of two squares, 303
Digit, round off, 8
Direct variations, 410
Distributive law, 229, 279
Dividend, 51
Division, 50
   of fractions, 121, 349
   long, 331
   of monomials, 241
   of polynomials, 330
   of radicals, 473
   of signed numbers, 211
   by zero, 51
Divisor, 51, 330
   common, 326

Equality, 25
   symmetric property of, 249
Equations, 25, 71
   dependent, 418
   equivalent, 26, 248, 385
   first-degree, *see* First-degree equations
   fractional, 370
   graphing, 417
   inconsistent, 418
   involving fractions, 370
   involving parentheses, 305
   linear, 393
   members of, 25
   quadratic, *see* Quadratic equations
   solutions of, 25, 26, 71, 258, 371
   of straight lines, 405
   systems of, 418
   in two variables, 383
Equivalent equations, 26, 248, 385
Equivalent expressions, 222
Equivalent fractions, 91, 126, 325
Equivalent systems, 420
Estimating:
   differences, 20
   products, 47
   quotients, 61
   sums, 14
Evaluation, numerical, 225
Even integers, 196
Exponential notation, 48, 210, 219
Exponents, 48
   integer, 339
   laws of, 239
Expression, algebraic, 220
Expression, radical, 451
Extraction of roots, 481
Extremes of a proportion, 78

Factor, 39, 88, 208, 279
   binomial, 93
   common, 326
   monomial, 283
   prime, 88
Factoring, 87
   difference of two squares, 303
   monomials from polynomials, 283
   trinomials, 289, 298

INDEX

First-degree equations, 247
    graphing, 393
    solution of, 247, 258, 383
        by addition, 258
        by division, 258
        by multiplication, 258
First law of exponents, 239
Form, scientific, 340
Formula(s), 261
    for Pythagorean relationships, 519
    for solving a quadratic equation, 491
Fraction-percent equivalents, 134
Fractions, 87, 98, 319, 345
    addition of, 102, 352, 356
    building, 92, 335
    building factor, for, 93
    changing signs of, 320
    comparing, 98
    complex, 366
    division of, 121, 349
    equations containing fractions, 370
    equivalent, 91, 126, 325
    fundamental principle of, 92, 241, 324
    graphical representation of, 319
    improper, 87
    multiplication of, 119, 345
    proper, 87
    reducing, 92, 324
    signs of, 320
    standard forms for, 335
    subtraction of, 110, 361
Function notation, 386

Gram, 165
Graphing, intercept method, 399
Graphs, of first-degree equations, 393
    of fractions, 319, 323
    of inequalities, 273
    of integers, 195
    of linear inequalities, 413
    of ordered pairs, 390
    of quadratic equations, 495
Greater than, 197
Grouping numbers, 5

Horizontal axis, 390
Hypotenuse of a right triangle, 517

Inconsistent equations, 418
Independent variable, 384
Inequalities:
    in one variable, 271
    in two variables, 383, 412
Integers, 195, 265
    differences of, 204
    graphical representation of, 195
    products of, 208
    quotients of, 208
    sums of, 201
Intercept method of graphing, 399
Interest, 148
Irrational numbers, 455

Least common denominator, 96
Less than, 196
Like terms, 228
    addition of, 229
    subtraction of, 234
Line, 507
Linear equation in two variables, 393
Linear inequalities in two variables, 412
Lines, 403
    intersecting, 512
    parallel, 403, 513
    perpendicular, 403
Line segment, 507
Liter, 165
Lowest common denominator, 96
Lowest term of a fraction, 92, 324

Means of a proportion, 78
Measurement, 155
    of angles, 507
Meter, 165
Metric system, 155
    conversions in, 166
Mixed numbers, 103
    addition of, 104
    division of, 121
    multiplication of, 120
    subtraction of, 111
Monomials, 220
    addition of, 229
    division of, 241
    multiplication of, 238
    subtraction of, 234
Multiplication, 39
    associative law of, 209
    of binomials, 285
    commutative law of, 40, 209
    of decimals, 42
    of denominate numbers, 162
    division rule, 72, 250
    of fractions, 119
    of mixed numbers, 120

INDEX

Natural numbers, 195
Negative numbers, 196
Notation, scientific, 339
Number line, 196
Numbers, absolute value of, 198
  abstract, 155
  composite, 88
  decimal, 4
  denominate, 155
  irrational, 455
  mixed, 103
  natural, 195
  negative, 196
  positive, 196
  prime, 88
  rational, 454
  real, 456
  signed, 196
  whole, 3, 195
Numerator, 87
Numerical coefficient, 221
Numerical evaluation, 225

Odd integers, 196
Ordered pairs, 383
  graphs of, 390
Order of operations, 214
Ordinate, 391
Origin, 390

Parabola, 496
Parallel lines, 403, 513
Parallelogram, 529
  area of, 528
  perimeter of, 528
Parentheses, in equations, 305
  removing, 280
Percent, Case I, 136
  Case II, 137
  Case III, 138
Perimeter, of a parallelogram, 528
  of a rectangle, 528
  of a square, 528
  of a trapezoid, 528
  of a triangle, 518
Place valve, 3
Point, 507
Polynomials, 220, 283
  addition of, 229
  division of, 330
  subtraction of, 234
Positive number, 196

Powers, descending, 220
  of a number, 48
  products of, 238
Prime factors, 88
Prime number, 88
Principal, 148
Principal square root, 452
Product, 39, 279, 295
  of fractions, 119, 345
  of integers, 208
  law of exponents, 239
  of radicals, 470
  of variables, 103
Proportion, 78, 372
  extremes, 78
  means, 78
  terms of a, 78
Pythagorean theorem, 519

Quadrant, 390
Quadratic equations, 435
  graph of, 496
  solution of, by completing the square, 486
  by extraction of roots, 481
  by factoring, 438, 442
  by formula, 491
  by graphing, 495
  standard form for, 435
Quadratic formula, 491
Quadrilaterals, 528
  diagonal of, 528
  parallelogram, 528
  rectangle, 528
  square, 528
  trapezoid, 528
Quotient, 50
  of fractions, 121
  of integers, 208
  of polynomials, 330
  of powers, 238
  of radical expressions, 473

Radicals, 451, 467
  adding or combining, 464
  division of, 473
  multiplication of, 465, 470
  simplification of, 459, 464
Radicand, 452
Ratio, 77
Rationalizing denominators, 474
Rational number, 454

INDEX

Real numbers, 456
Reciprocal, 120
Rectangle, area of a, 528
    perimeter of a, 528
Rectangular coordinate system, 391
Relationship, 270
Right circular cylinder, 535
Right rectangular prism, 535
Right triangle, 517
    hypotenuse of, 517
    legs of, 517
Root, square, 451
Rounding off, decimals, 8
    whole numbers, 8

Scientific form, 339
Scientific notation, 340
Signed numbers, 196
    addition of, 201
    division of, 211
    graphical representation of, 196
    multiplication of, 209
    subtraction of, 204
Signs of fractions, 320
Slope of a line, 401
    Slope-intercept form, 407
Solids, 535
    right circular cylinder, 535
    right rectangular prism, 535
    sphere, 535
Solution of, equations, 23, 247, 383
    addition-subtraction rule, 26, 248
    multiplication-division rule, 71, 250
    quadratic equations, 491
    systems of linear equations, 420
Sphere, 535
Square, area of a, 528
    of a number, 48
    perimeter of a, 528
Square roots, 451
    principal, 452
    product of, 465
    quotient of, 473
Standard form, for fractions, 321
    for quadratic equations, 435
    for systems of equations, 424
Subtraction, 17
    of decimals, 18
    of denominate numbers, 161, 162
    of fractions, 361
    of like terms, 234
    of mixed numbers, 103
    of signed numbers, 204
Sums, 11
    of fractions, 102, 357
    of integers, 201
Symbols, of grouping, 5
Symmetric property of equality, 249
Systems of equations, 418
    equivalent, 420
    solution of, 418
        by addition or subtraction, 420
        by graphing, 418
        by substitution, 427
    standard form, 424

Terms, algebraic, 220
    coefficients of, 229
    combining like, 229
    degree of, 221
    like, 229
    of a proportion, 77
Test digit, 8
Trapezoid, area of, 528
    perimeter of, 528
Triangle, 516
    acute, 517
    altitude of, 518
    area of, 519
    base of, 518
    centroid of, 518
    congruent, 523
    corresponding parts of, 523
    equilateral, 517
    isosceles, 517
    median of, 518
    obtuse, 517
    perimeter of, 518
    right, 517
    scalene, 517
    similar, 523
    solving, 542
Trigonometric ratios, 538
Trinomial, 220
    factoring a, 289, 298, 300

United States, metric conversions, 155
United States, conversions in, 155
  system, 155

INDEX

Unit-product rule, 65
Unit-quotient rule, 66
Unlike terms, 228

Variable, 25, 234
  dependent, 384
  independent, 384
Variation, direct, 410
Vertical angles, 512
Vertical axis, 390

Whole numbers, 3, 195
  rounding off, 8

x-intercept, 399

y-intercept, 399

Zero, division by, 211

APPENDIX: TABLES OF MEASUREMENTS

U.S. MEASUREMENTS

1. Weight
   - 16 ounces (oz) = 1 pound (lb)
   - 2000 pounds = 1 ton (T)

2. Time
   - 60 seconds (sec) = 1 minute (min)
   - 60 minutes = 1 hour (hr)
   - 24 hours = 1 day (da)
   - 7 days = 1 week (wk)
   - 30 days = 1 month (mo)
   - 52 weeks = 1 year (yr)
   - 12 months = 1 year
   - 365 days = 1 calendar year
   - 366 days = 1 leap year
   - 360 days = 1 business year

3. Counting
   - 20 units = 1 score
   - 12 units = 1 dozen (doz)
   - 12 dozen = 1 gross (gro)
   - 12 gross = 1 great gross

4. Length
   - 12 inches (in.) = 1 foot (ft)
   - 3 feet = 36 inches = 1 yard (yd)
   - 5280 feet = 1760 yards = 1 mile (mi)

5. Area
   - 144 square inches (sq in.) = 1 square foot (sq ft)
   - 9 square feet = 1 square yard (sq yd)
   - 43,560 square feet = 1 acre
   - 640 acres = 1 square mile (sq mi)

6. Volume
   - 1728 cubic inches (cu in.) = 1 cubic foot (cu ft)
   - 27 cubic feet = 1 cubic yard (cu yd)
   - 231 cubic inches = 1 gallon (gal)

7. Dry Capacity
   - 2 pints (pt) = 1 quart (qt)
   - 8 quarts = 1 peck (pk)
   - 4 pecks = 1 bushel (bu)

8. Liquid Capacity
   - 16 fluid ounces (fl oz) = 1 pint (pt)
   - 2 pints = 1 quart (qt)
   - 4 quarts = 1 gallon (gal)
   - $31\frac{1}{2}$ gallons = 1 barrel (bbl)

METRIC MEASUREMENTS

9. Length
   - 10 millimeters (mm) = 1 centimeter (cm)
   - 10 centimeters = 1 decimeter (dm)
   - 10 decimeters = 1 meter (m)
   - 10 meters = 1 dekameter (dam)
   - 10 dekameters = 1 hectometer (hm)
   - 10 hectometers = 1 kilometer (km)

10. Weight
    - 10 milligrams (mg) = 1 centigram (cg)
    - 10 centigrams = 1 decigram (dg)
    - 10 decigrams = 1 gram (g)
    - 10 grams = 1 dekagram (dag)
    - 10 dekagrams = 1 hectogram (hm)
    - 10 hectograms = 1 kilogram (kg)

11. Liquid Capacity
    - 10 milliliters (ml) = 1 centiliter (cl)
    - 10 centiliters = 1 deciliter (dl)
    - 10 deciliters = 1 liter (l)
    - 10 liters = 1 dekaliter (dal)
    - 10 dekaliters = 1 hectoliter (hl)
    - 10 hectoliters = 1 kiloliter (kl)

12. Area
    - 100 square millimeters = 1 square centimeter
    - 10,000 square centimeters = 1 square meter
    - 1,000,000 square meters = 1 square kilometer

13. Volume
    - 1000 cubic millimeters = 1 cubic centimeter
    - 1,000,000 cubic centimeters = 1 cubic meter

APPROXIMATE METRIC EQUIVALENTS OF U.S. UNITS

14. Length
    - 1 inch = 2.540 centimeters
    - 1 foot = 30.48 centimeters = 0.305 meter
    - 1 yard = 0.914 meter
    - 1 mile = 1.609 kilometers

15. Weight
    - 1 ounce = 28.350 grams
    - 1 pound = 453.592 grams = 0.454 kilogram

16. Volume
    - 1 pint = 0.473 liter
    - 1 quart = 0.946 liter
    - 1 gallon = 3.785 liters